Lignocellulosic Biomass II

Lignocellulosic Biomass II

Alejandro Rodríguez Pascual
Eduardo Espinosa Víctor
Carlos Martín

MDPI • Basel • Beijing • Wuhan • Barcelona • Belgrade • Novi Sad • Cluj • Manchester

Editors
Alejandro Rodríguez Pascual
Inorganic Chemistry and Chemical Engineering
Instituto Químico para la Energía y el Medioambiente
University of Córdoba
Córdoba
Spain

Eduardo Espinosa Víctor
Inorganic Chemistry and Chemical Engineering
Instituto Químico para la Energía y el Medioambiente
University of Córdoba
Córdoba
Spain

Carlos Martín
Department of Biotechnology
Inland Norway University of Applied Sciences
Hamar
Norway

Editorial Office
MDPI AG
Grosspeteranlage 5
4052 Basel, Switzerland

This is a reprint of articles from the Special Issue published online in the open access journal *Molecules* (ISSN 1420-3049) (available at: www.mdpi.com/journal/molecules/special_issues/lignocellulosic_II).

For citation purposes, cite each article independently as indicated on the article page online and as indicated below:

Lastname, A.A.; Lastname, B.B. Article Title. *Journal Name* **Year**, *Volume Number*, Page Range.

ISBN 978-3-7258-2256-0 (Hbk)
ISBN 978-3-7258-2255-3 (PDF)
doi.org/10.3390/books978-3-7258-2255-3

Contents

About the Editors

Alejandro Rodríguez Pascual

Pascual has evaluated four research periods and supervised eleven doctoral theses; he is currently supervising three theses. His Google Scholar h index is 45, and he has 6152 total cites; his Scopus h index is 40, and he has 4486 total cites. He has 137 JCR articles. One-hundred percent of his published works from 2008 were cited. More than 40 percent of his articles published in the last 8 years have been international collaborations. He obtained his doctorate from the University of Córdoba in 2002, commenced his "Ramón y Cajal" Contract in 2006, and has been a full Professor since 2021. In 2007, he began to integrate biorefinery processes in the research line of his thesis. The objective was the valorization of lignocellulosic waste from agri-food activities. In 2013, he began working on obtaining cellulose nanofibers from agricultural residue, being one of the first groups at the international level to do so. He also demonstrated with several publications the benefit of having hemicelluloses in the celluloses to be nanofibrillated. He also obtained nanocelluloses with lignin content by means of mechanical pretreatment, which was not studied before, concluding that the presence of lignin is indeed beneficial. In recent years, he has conducted research in various applications, focusing on the food packaging sector, where different packaging solutions have been developed (rigid, films, coatings) with excellent results in food preservation. Currently, he is collaborating with Andalusian SMEs in the search for strategies for the valorization of lignocellulosic wastes from agri-food activities. He has organized three editions of the International Congress on Biorefinery of Lignocellulosic Materials. He is also an expert evaluator for several international and national research agencies (REA, Romania, Latvia, Chile, Uruguay, Cyprus, etc.), as well as a founding partner of Spin off OrganInk SL and a lead of the RNM 940 group.

Eduardo Espinosa Víctor

Dr. Espinosa is an Assistant Professor in the Chemical Engineering Area of the Universidad de Córdoba (UCO) in the Bioproducts and Process Engineering group (BIOPREN). He received his PhD in Biosciences and Agri-Food Sciences from the UCO in 2019. His research focuses on the valorization of agri-food by-products through biorefinery processes to obtain high value-added bioproducts and their application. Among the research developed, the use of bioactive compounds, nanocellulose, and cellulose in applications in the paper industry, food packaging, and biomedicine stands out. This research activity has resulted in the publication of 50 scientific articles. He has participated in >50 international conferences through invited lectures, oral communications, and posters, and he is the co-author of three book chapters. He has carried out different stays in national centers, such as the University of Girona (2015), and international ones, such as the LGP2 group of INP-Pagora in Grenoble, France (2017), and the RISE PFI research center in Trondheim, Norway (2018, 2019). This broad international collaboration has resulted in 50% of his published scientific papers being the result of international collaborations, and the published articles show a high scientific impact, placing 78.9% of the papers among 25% of the most cited worldwide (according to the Scopus database). Among his research projects, his participation as a coordinator of a European project in call PRIMA 2023 and PI in a regional public-private transfer project 2021 stands out. He has participated as part of a research team in five projects (one European, three national, and one regional). He also carries out intense work of knowledge transfer to society. In the transfer to the private sector, CIRCULARIZA was created as a spin-off of Universidad de Córdoba dedicated to the development of technologies for the valorization of agroindustrial waste.

Carlos Martín

Carlos Martín Medina is a Professor at Inland Norway University of Applied Sciences and Umeå University (Sweden) and an Alexander von Humboldt Foundation fellow. He has an MSc degree from the Leningrad Forest Technical Academy (USSR) and a PhD degree from the University of Matanzas (Cuba) in a sandwich program with Lund University (Sweden). He carried out postdoctoral stays at Risø DTU National Laboratory for Sustainable Energy (Denmark) and Karlstad University (Sweden). He has also worked at Thünen Institute of Wood Research (Germany). Martín's research interests are biomass and bioresources' utilization, biorefineries, wood chemistry, lignocellulose pretreatment, and biofuels.

Preface

Continuing the success of its predecessor, "Lignocellulosic Biomass I", the second volume of this Special Issue embarks on a journey to explore the latest research and advancements in lignocellulosic biomass utilization. This collection of articles reflects on the collaborative efforts of leading researchers dedicated to addressing the challenges and opportunities of biomass conversion and sustainable resource management. Our Special Issue offers a comprehensive compilation that delves into emerging trends, novel approaches, and significant discoveries in the field of lignocellulosic biomass research.

Despite its long history, lignocellulosic biomass research demands persistent efforts from the scientific community. The emergence of transformative technologies, the integration of circular bioeconomy principles, and the pressing need to combat climate change highlight the necessity to spare no efforts or resources in this domain.

With the objective of inspiring new ideas, fostering interdisciplinary collaboration, and promoting transformative solutions for global environmental sustainability, "Lignocellulosic Biomass II" aspires to contribute significantly to this field.

As Guest Editors, we express our sincere gratitude to all authors for their exceptional research contributions to this Special Issue. Additionally, we extend our thanks to the reviewers for their rigorous evaluations, which have been instrumental in maintaining the scientific rigor and quality of the submitted articles.

With immense enthusiasm and anticipation, we invite you to explore the diverse contributions of "Lignocellulosic Biomass II" and join us in shaping a sustainable future for our planet.

Alejandro Rodríguez Pascual, Eduardo Espinosa Víctor, and Carlos Martín
Editors

molecules

Editorial

Special Issue "Lignocellulosic Biomass II"

Alejandro Rodríguez [1,*], Eduardo Espinosa [1] and Carlos Martín [2,3]

1 BioPrEn Group, Instituto Químico para la Energía y el Medioambiente (IQUEMA), Chemical Engineering Department, Universidad de Córdoba, 14014 Córdoba, Spain; a02esvie@uco.es
2 Department of Biotechnology, Inland Norway University of Applied Sciences, N-2317 Hamar, Norway; carlos.medina@inn.no
3 Department of Chemistry, Umeå University, SE-901 87 Umeå, Sweden
* Correspondence: a.rodriguez@uco.es; Tel.: +34-957-21-22-74

1. Introduction

As a result of human population growth, the availability of residual lignocellulosic materials from agriculture, forestry, food- and wood-processing industries, and other waste streams is continuously increasing. The composition and abundance of lignocellulosic biomass make it a realistic option as feedstock for biorefineries, producing the bio-based fuels, materials, and chemicals required for the sustainable development of society without depending on fossil-based resources [1]. Lignocellulosic biomass represents a practically infinite feedstock source for satisfying the demands of the industry, and could thus contribute to achieving a carbon-neutral future and alleviating the current threats related to climate change. In previous collections, we compiled relevant contributions on the chemistry and technology of lignocellulosic biomass [2] and on the bioconversion of crop residues to biofuels and other bio-based products [3]. With this Special Issue, we present a new collection of articles dealing with cutting-edge topics in the field of lignocellulosic biomass.

2. This Issue

In this Special Issue, twelve original research papers and three reviews, covering some of the latest advances in research on lignocellulosic materials, are presented. Included in the collection are research results related to the by-products from agriculture and crop processing, e.g., corn stover, barley straw, rice husks, agave bagasse, apricot seed husks, walnut shells, cacao pods waste, and spent coffee grounds. The collection also features research on woody biomass, e.g., softwood and hardwood chips of different origin, teak wood, *Paulownia* bark, eucalyptus kraft pulp, and hardwood-based spent mushroom substrate, as well as herbaceous biomass, e.g., hemp and tall fescue.

Three articles deal with novel lignocellulosic materials of interest in biorefining. Wawro et al. [4] investigated the suitability of four Polish varieties of industrial hemp as raw materials for the production of lignocellulosic ethanol. After the alkaline treatment of samples of the four materials, biomass from the Tygra and Rajan varieties was submitted to enzymatic saccharification following either separated hydrolysis and fermentation or simultaneous saccharification and fermentation schemes. The authors concluded that both Tygra and Rajan varieties are promising raw materials for bioethanol production. Halysh et al. [5] evaluated apricot seed husks and walnut shells as potential raw materials for biorefineries in Ukraine. Both materials were pretreated with either H_2SO_4, NaOH, or by steam explosion, and then subjected to enzymatic saccharification with commercial cellulases. The results showed the potential of apricot seed husks and walnut shells for producing value-added products using affordable and environmentally friendly chemical technologies. It was concluded that alkaline pretreatment is the most advantageous method not only because of its glucose yield, but also because it allows an easy lignin regeneration from the spent liquor and the simple recovery of sodium hydroxide, which can be returned to the technological process. Batog et al. [6] investigated the use of annually renewable

check for updates

Citation: Rodríguez, A.; Espinosa, E.; Martín, C. Special Issue "Lignocellulosic Biomass II". *Molecules* **2023**, *28*, 6230. https://doi.org/10.3390/molecules28176230

Received: 21 August 2023
Accepted: 22 August 2023
Published: 24 August 2023

plant biomass from saline soils, namely tall fescue and hemp, for biorefining processes leading to biofuels and novel materials. After alkaline pretreatment and simultaneous saccharification, both tall fescue and hemp resulted in yields of 14 g ethanol per 100 g of raw material. Both types of halophyte biomass, were shown to be suitable as fillers for green composites displaying good mechanical properties.

Two papers bring up important aspects of the biorefinery processing of lignocellulosic biomass. In one of these, Poveda-Giraldo et al. [7] used data from the literature and simulation tools for developing a multi-criterion weighting of the best-performing schemes for the isolation of the main lignocellulose constituents. It was concluded that dilute acid is the most effective method for cellulose isolation and hemicelluloses removal, while the kraft process is the best option for lignin removal and its future use in biorefineries. The results are of high value for pretreatment systematization in the design of biorefineries. In the second of these papers, Allegretti et al. [8] combined hydrothermal pretreatment with a treatment with deep eutectic solvents in an environmentally friendly process for separating the main components of rice husks. The produced lignin fractions were characterized and evaluated as water reducers in cement pastes. The proposed strategy is very promising for the valorization of husks of raw and parboiled rice from a circular economy perspective.

Two papers address the extraction of bioactive compounds from lignocellulosic biomass. Rodríguez-Seoane et al. [9] applied pressurized extraction for recovering phenolics from *Paulownia* bark. Subcritical water extraction (SWE) was carried out under a non-isothermal operation by heating up to temperatures ranging between 140 and 240 °C, while supercritical CO_2 ($scCO_2$) extraction was performed at different pressures, temperatures, and ethanol concentrations. SWE reached higher extraction yield and antiradical capacity than $scCO_2$ extraction. Extraction yields up to 30%, with extracts containing up to 7% phenolics, were achieved with SWE under heating to 160 °C while, at 240 °C, the yield decreased to 20%, but the phenolic content increased to 21% and the antiradical activity was considerably higher. Gallic acid, vanillic acid, vanillin, and apigenin were the major phenolics found in the extracts. Klausen et al. [10] evaluated the extraction of bioactive compounds and cellulose saccharification as valorization strategies for spent mushroom substrate (SMS). Four extraction methods, namely Soxhlet, reflux, SWE, and ultrasound-assisted extraction, were used. SWE at 150 °C resulted in the best extraction parameters among all the tested methods. Vanillic and chlorogenic acids were the primary phenolic acids identified in the extracts, and the concentration of caffeic acid correlated well with the antioxidant activity. The enzymatic saccharification of cellulose was enhanced after the extraction of bioactive compounds.

Three articles discussed the use of wood-processing products and agroindustrial residues for developing novel materials. Valente et al. [11] developed green composites by reinforcing biopolymeric matrices of poly(lactic acid) and poly(hydroxybutyrate) with micronized bleached eucalyptus kraft pulp fibers. The produced materials displayed superior mechanical performance and lower water uptake compared with the composites with non-micronized pulp fibers. The results showed the potential of micronization as a simple and sustainable alternative for the manufacturing of entirely bio-based composites. Naydenova et al. [12] used hydrolysis lignin, the residue of the industrial saccharification of woody materials, for producing biochars by hydrothermal liquefaction at 500–700 °C. The biochar produced at 600 and 700 °C displayed properties typical for microporous adsorbents, suitable for selective adsorption purposes. The authors also proposed to use the produced biochars as catalysts. Lavado-Meza et al. [13] present a study on producing biosorbents for the removal of heavy metals from aqueous solutions. The biosorbents were produced through the alkaline modification of coffee and cocoa agroindustrial waste. The adsorbtion capacity and other characteristics of the biosorbents were thoroughly assessed. Their efficiency on Pb (II) removal was higher than that of comparable biosorbents reported in the literature.

Two papers on the bioconversion of lignocellulosic materials are included in this Special Issue. Sierra-Ibarra et al. [14] reported on the fermentation of the hydrolysates of five different lignocellulosic materials with the ethanologenic strain *Escherichia coli* MS04. The hydrolysates contained different glucose and xylose concentrations. It was found that deleting the *xylR* regulator plays an important role in xylose consumption and that acetate has a positive effect on the co-consumption rates of glucose and xylose in hydrolysates. Madubuike and Ferry [15] reported on the bioprospecting of novel enzymes for designing cocktails to deconstruct lignocellulose. They characterized a novel acetyl xylan esterase from the gut microbiota of the common black slug. The enzyme showed high thermal stability and potential for hydrolysing acetylated xylan.

Three review papers on crucial topics related to lignocellulose processing complete this Special Issue. Broda et al. [16] summarize the state of the art in bioethanol production from lignocellulose. The paper highlights the most challenging steps of the process, presents recent advances in the area, and discusses future perspectives for second-generation biorefineries. In a paper on processing biomass for hydrogen fermentation, Zhila et al. [17] review aspects related to the efficiency of hydrogen production through the fermentation of lignocellulosic hydrolysates. The effect of process parameters, both prior to and after fermentation, are discussed. The paper examines the formation of inhibitory compounds as a result of lignocellulose degradation during pretreatment and the effects of those compounds on the microorganisms involved in dark fermentation and photo-fermentation and on the management of post-fermentative liquid streams. A review on the durability of cellulosic-fibers-reinforced geopolymer composites (CFGC) is presented by Liu and Lv [18]. The paper analyzes the recent literature on the influence of nanomaterials on the properties of geopolymer composites. The effect of the degradation of cellulosic fibers and other factors on CFGC durability is summarized.

3. Conclusions

The content of this Special Issue shows the relevance of lignocellulosic biomass as a sustainable feedstock niche for bio-based industries. The guest editors of this Special Issue acknowledge all the contributing authors. Their contributions show the recent advancements in the research area of lignocellulosic biomass and provide indications of the routes to follow towards a bio-based society.

The heterogeneity of lignocellulosic biomass sources remains challenging for science and innovation today. New contributions regarding novel methods and technologies that would allow the efficient processing of different lignocellulosic materials are expected to be explored in a near future.

Author Contributions: Conceptualization, A.R., E.E. and C.M.; methodology, A.R., E.E. and C.M.; writing—original draft preparation, C.M.; writing—review and editing, A.R., E.E. and C.M. All authors have read and agreed to the published version of the manuscript.

Funding: This research was funded by Sparebankstiftelsen Hedmark (through grant number 362001-10), Bio4Energy strategic research environment (through grant number 550080300), and the Department of Economic Transformation, Industry, Knowledge and Universities (Regional Government of Andalusia). This research is part of the project P18-RT-4064.

Acknowledgments: The Department of Biotechnology, Inland Norway University of Applied Sciences, and the BioPren Research Group, from the Instituto Químico para la Energía y el Medioambiente at the Chemical Engineering Department, Universidad de Córdoba, are thanked for their institutional support.

Conflicts of Interest: The authors declare no conflict of interest.

References

1. Casau, M.; Dias, M.F.; Matias, J.C.O.; Nunes, L.J.R. Residual Biomass: A Comprehensive Review on the Importance, Uses and Potential in a Circular Bioeconomy Approach. *Resources* **2022**, *11*, 35.
2. Rodríguez, A.; Espinosa, E. Special Issue "Lignocellulosic Biomass". *Molecules* **2021**, *26*, 1483. [CrossRef] [PubMed]
3. Martín, C. Pretreatment of Crop Residues for Bioconversion. *Agronomy* **2021**, *11*, 924. [CrossRef]

4. Wawro, A.; Batog, J.; Gieparda, W. Polish Varieties of Industrial Hemp and Their Utilisation in the Efficient Production of Lignocellulosic Ethanol. *Molecules* **2021**, *26*, 6467. [CrossRef] [PubMed]
5. Halysh, V.; Romero-García, J.M.; Vidal, A.M.; Kulik, T.; Palianytsia, B.; García, M.; Castro, E. Apricot Seed Shells and Walnut Shells as Unconventional Sugars and Lignin Sources. *Molecules* **2023**, *28*, 1455. [CrossRef] [PubMed]
6. Batog, J.; Bujnowicz, K.; Gieparda, W.; Wawro, A.; Rojewski, S. Effective Utilisation of Halophyte Biomass from Saline Soils for Biorefinering Processes. *Molecules* **2021**, *26*, 5393. [CrossRef] [PubMed]
7. Poveda-Giraldo, J.A.; Garcia-Vallejo, M.C.; Cardona Alzate, C.A. Analysis of Single-Step Pretreatments for Lignocellulosic Platform Isolation as the Basis of Biorefinery Design. *Molecules* **2023**, *28*, 1278. [CrossRef] [PubMed]
8. Allegretti, C.; Bellinetto, E.; D'Arrigo, P.; Ferro, M.; Griffini, G.; Rossato, L.A.M.; Ruffini, E.; Schiavi, L.; Serra, S.; Strini, A.; et al. Fractionation of Raw and Parboiled Rice Husks with Deep Eutectic Solvents and Characterization of the Extracted Lignins towards a Circular Economy Perspective. *Molecules* **2022**, *27*, 8879. [CrossRef]
9. Rodríguez-Seoane, P.; Díaz-Reinoso, B.; Domínguez, H. Pressurized Solvent Extraction of Paulownia Bark Phenolics. *Molecules* **2022**, *27*, 254. [CrossRef] [PubMed]
10. Klausen, S.J.; Falck-Ytter, A.B.; Strætkvern, K.O.; Martin, C. Evaluation of the Extraction of Bioactive Compounds and the Saccharification of Cellulose as a Route for the Valorization of Spent Mushroom Substrate. *Molecules* **2023**, *28*, 5140. [CrossRef]
11. Valente, B.F.A.; Silvestre, A.J.D.; Neto, C.P.; Vilela, C.; Freire, C.S.R. Effect of the Micronization of Pulp Fibers on the Properties of Green Composites. *Molecules* **2021**, *26*, 5594. [CrossRef] [PubMed]
12. Naydenova, I.; Radoykova, T.; Petrova, T.; Sandov, O.; Valchev, I. Utilization Perspectives of Lignin Biochar from Industrial Biomass Residue. *Molecules* **2023**, *28*, 4842. [CrossRef] [PubMed]
13. Lavado-Meza, C.; De la Cruz-Cerrón, L.; Asencios, Y.J.O.; Marcos, F.C.F.; Dávalos-Prado, J.Z. Alkaline Modification of Arabica-Coffee and Theobroma-Cocoa Agroindustrial Waste for Effective Removal of Pb(II) from Aqueous Solutions. *Molecules* **2023**, *28*, 683. [CrossRef]
14. Sierra-Ibarra, E.; Vargas-Tah, A.; Moss-Acosta, C.L.; Trujillo-Martínez, B.; Molina-Vázquez, E.R.; Rosas-Aburto, A.; Valdivia-López, Á.; Hernández-Luna, M.G.; Vivaldo-Lima, E.; Martínez, A. Co-Fermentation of Glucose–Xylose Mixtures from Agroindustrial Residues by Ethanologenic Escherichia Coli: A Study on the Lack of Carbon Catabolite Repression in Strain MS04. *Molecules* **2022**, *27*, 8941. [CrossRef]
15. Madubuike, H.; Ferry, N. Characterisation of a Novel Acetyl Xylan Esterase (BaAXE) Screened from the Gut Microbiota of the Common Black Slug (Arion Ater). *Molecules* **2022**, *27*, 2999. [CrossRef]
16. Broda, M.; Yelle, D.J.; Serwańska, K. Bioethanol Production from Lignocellulosic Biomass—Challenges and Solutions. *Molecules* **2022**, *27*, 8717. [CrossRef] [PubMed]
17. Honarmandrad, Z.; Kucharska, K.; Gębicki, J. Processing of Biomass Prior to Hydrogen Fermentation and Post-Fermentative Broth Management. *Molecules* **2022**, *27*, 7658. [CrossRef] [PubMed]
18. Liu, J.; Lv, C. Durability of Cellulosic-Fiber-Reinforced Geopolymers: A Review. *Molecules* **2022**, *27*, 796. [CrossRef] [PubMed]

molecules

MDPI

Article

Evaluation of the Extraction of Bioactive Compounds and the Saccharification of Cellulose as a Route for the Valorization of Spent Mushroom Substrate

Sarah J. Klausen [1], Anne Bergljot Falck-Ytter [1], Knut Olav Strætkvern [1] and Carlos Martin [1,2,*]

[1] Department of Biotechnology, Inland Norway University of Applied Sciences, N-2317 Hamar, Norway; sa.schnell@gmail.com (S.J.K.); anne.falckytter@inn.no (A.B.F.-Y.); knut.stratkvern@inn.no (K.O.S.)

[2] Department of Chemistry, Umeå University, SE-901 87 Umeå, Sweden

* Correspondence: carlos.medina@inn.no

Abstract: The extraction of bioactive compounds and cellulose saccharification are potential directions for the valorization of spent mushroom substrate (SMS). Therefore, investigating the suitability of different extraction methods for recovering bioactive compounds from SMS and how the extraction affects the enzymatic saccharification is of uppermost relevance. In this work, bioactive compounds were extracted from *Pleurotus* spp. SMS using four extraction methods. For Soxhlet extraction (SoE), a 40:60 ethanol/water mixture gave the highest extraction efficiency (EE) (69.9–71.1%) among the seven solvent systems assayed. Reflux extraction with 40:60 ethanol/water increased the extraction yield and EE compared to SoE. A shorter reflux time yielded a higher extraction of carbohydrates than SoE, while a longer time was more effective for extracting phenolics. The extracts from 240 min of reflux had comparable antioxidant activity (0.3–0.5 mM GAE) with that achieved for SoE. Ultrasound-assisted extraction (UAE) at 65 °C for 60 min allowed an EE (~82%) higher than that achieved by either reflux for up to 150 min or SoE. Subcritical water extraction (SWE) at 150 °C resulted in the best extraction parameters among all the tested methods. Vanillic acid and chlorogenic acid were the primary phenolic acids identified in the extracts. A good correlation between the concentration of caffeic acid and the antioxidant activity of the extracts was found. Saccharification tests revealed an enhancement of the enzymatic digestibility of SMS cellulose after the extraction of bioactive compounds. The findings of this initial study provide indications on new research directions for maximizing the recovery of bioactive compounds and fermentable sugars from SMS.

Keywords: spent mushroom substrate; *Pleurotus ostreatus*; bioactive compounds; ultrasound-assisted extraction; subcritical-water extraction; enzymatic saccharification; cellulose

Citation: Klausen, S.J.; Falck-Ytter, A.B.; Strætkvern, K.O.; Martin, C. Evaluation of the Extraction of Bioactive Compounds and the Saccharification of Cellulose as a Route for the Valorization of Spent Mushroom Substrate. *Molecules* **2023**, *28*, 5140. https://doi.org/10.3390/molecules28135140

Academic Editor: Xuebing Zhao

Received: 31 May 2023
Revised: 23 June 2023
Accepted: 25 June 2023
Published: 30 June 2023

1. Introduction

Edible mushrooms are climate-smart protein-rich food sources with the potential to partially substitute meat, whose production has a high climate impact [1]. Due to the high content of bioactive compounds in mushrooms, their consumption exerts benefits on human health. Mushroom production is a fast-expanding industrial activity involving the cultivation of more than fifty fungal species [2]. *Pleurotus* spp. mushrooms are among the most commercialized in the world market [3].

The generation of crop residues is continuously increasing due to the agricultural expansion driven by global population growth. Disposal by burning is common practice for managing the accumulation of plant residues [4]. However, this practice is against sustainability principles and results in a waste of bioresources that are valuable for bioconversion into products of high economic and social value [5]. Since mushrooms are cultivated on substrates based on plant biomass, using crop and forest residues and by-products from food and wood processing as substrates for mushroom cultivation is a rational bioconversion alternative contributing to sustainable agriculture and forestry.

Mushroom cultivation is directed to produce fruitbodies of edible fungi. After harvesting the fruitbodies, an exhausted residual substrate, i.e., spent mushroom substrate (SMS), is generated. SMS is the main by-product of the mushroom-producing business Depending on the substrate formulation, the fungal species, and the production system, between three and five kilograms of SMS are formed per kg of cultivated mushrooms [6] The amount of SMS generated globally by the mushroom industry is expected to reach around 100 million tons by 2026 [7].

SMS is currently regarded as a residue with little value, and its accumulation challenges mushroom producers. Transporting SMS, a bulky material with high moisture content, to disposal sites is expensive, while drying it at the mushroom farm is technically demanding and energy-intensive. Additionally, SMS accumulation is of high environmental concern due to foul odors, greenhouse gases emission from spontaneous anaerobic digestion, and leachate drainage to water sources [8].

SMS valorization following a circular-economy model is critical for the sustainability of the mushroom industry. A valorization option to be considered is based on the potential of SMS as a source of bioactive compounds and polysaccharides [9]. SMS contains bioactive molecules originating from different sources, e.g., (i) the fungal mycelium; (ii) substances secreted by fungal growth; (iii) lignocellulose phytochemicals; and (iv) products of partial degradation of polysaccharides and lignin. Possible bioactive compounds in SMS include polyphenols, polysaccharides, sterols, proteins, vitamins, and other substances.

Recovering bioactive compounds and using them as raw materials for developing novel high-added value products is a promising direction for valorizing SMS. However, while the extraction of bioactive molecules from fungal fruitbodies has been broadly investigated [10], the study of bioactive compounds available in SMS is an entirely new research field. Consequently, the literature has not yet covered essential knowledge on extracting bioactive compounds from SMS.

Bioactive compounds can be recovered from biomass materials using conventional techniques, such as Soxhlet or reflux extraction, performed with various solvents at their boiling temperatures. However, those techniques are time- and energy-consuming, can affect the properties of thermosensitive compounds, and the used organic solvents often present safety, environmental, and toxicological risks. To face that challenge, green-extraction intensification techniques, e.g., ultrasonic assistance, microwaves, or pressurized extractions, with lower energy consumption, shorter extraction times, and non-pollutant solvents, can be used [11].

Saccharification of the polysaccharides contained in the solid residue remaining after extraction of the bioactive compounds and using the resulting sugars in microbial fermentations is another direction for valorizing SMS. We have previously shown that cellulose contained in the SMS of shiitake mushroom (*Lentinula edodes*) is highly susceptible to enzymatic saccharification [12] and that the produced hydrolysates are readily fermentable by baker's yeast [13]. However, there are no systematic studies on the enzymatic saccharification of SMS of *Pleurotus* spp. mushrooms, and it is unknown how the removal of bioactive compounds by different extraction techniques can affect cellulose saccharification. The main novelty of this work is that bioactive compounds are extracted from *Pleurotus* spp. SMS and that the enzymatic saccharification of the extraction residues is evaluated.

In the current work, the recovery of bioactive compounds from SMS of *Pleurotus* spp. mushrooms by conventional methods, e.g., Soxhlet and reflux extraction, and by green-extraction techniques, e.g., ultrasound-assisted extraction and subcritical water extraction, were investigated. The composition and antioxidant activity of the produced extracts was evaluated using standard methods. Spectrophotometric methods were used to analyze phenolic compounds, carbohydrates, and sterols, and high-performance liquid chromatography (HPLC) was used to identify individual compounds. The enzymatic digestibility of the raw SMS and the extraction residues was assessed using an analytical enzymatic saccharification protocol and HPLC.

2. Results and Discussion

This study included the extraction of bioactive compounds from SMS and the enzymatic saccharification of extraction residues (Figure 1). The SMS investigated was generated from cultivating two strains of oyster mushrooms on a hardwood-based substrate. The used strains were a commercial *Pleurotus ostreatus* × *Pleurotus eryngii* hybrid strain known as Black Pearl King Oyster (BPKO) and a wild strain of *P. ostreatus* isolated in Norway, which is hereafter referred to as Norwegian Oyster (NO). The initial substrate contained around 36% (*w*/*w*) cellulose, 28% hemicelluloses, 19% lignin and 9% extractives (Table S1).

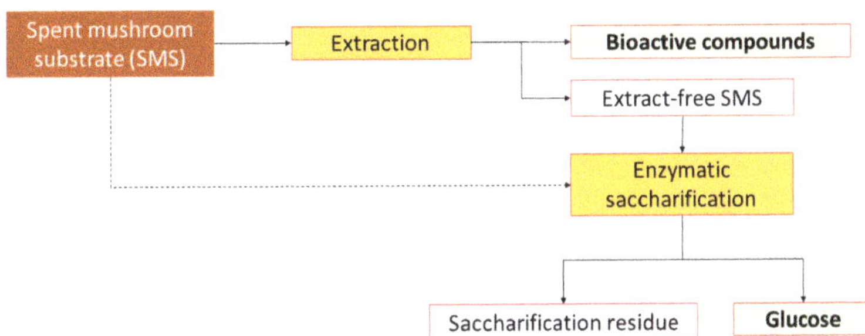

Figure 1. Schematic layout of the experimental procedure followed for the extraction of bioactive compounds from SMS and enzymatic saccharification of the cellulose contained in the extract-free SMS.

2.1. Characterization of the Spent Mushroom Substrate

Cellulose was the main constituent of SMS from both fungal strains (Table 1). Cellulose content accounted for 38.3% of the dry mass of BPKO SMS and 37.5% of that of NO SMS. That is slightly higher than the cellulose content in the initial substrate. BPKO SMS also had a higher xylan content (18.5%) than NO SMS (16.6%). The content of anhydroarabinose was comparable for both SMSs. The combined content of xylan and anhydroarabinose in both SMSs was lower than the hemicelluloses' content of the initial substrate. SMS from BPKO had a lower lignin content (16.0%) than NO SMS (18.4%). For both SMSs, lignin content was lower than in the initial substrate. Extractive content was higher for the SMS of NO (15.0%) than for BPKO SMS (13.5%) and the initial substrate. Water extractives (12.0–13.7% of the SMS dry mass) were predominant over the ethanol extractives (up to 1.5%). The content of water extractives was higher for NO than for BPKO, while that of ethanol extractives was comparable for the SMS from both strains. Carbohydrates dominated the water extractives, while phenolics represented a significant share of the ethanol extractives. Carbohydrates were 72 and 89% of the mass share of the water extractives for NO and BPKO SMS, respectively. Phenolic compounds represented around 20% of the mass share of ethanol extractives. However, it should be noted that a large part of the phenolics had already been solubilized by the water extraction, which was performed before the ethanol extraction. Ergosterol was identified after alkali-assisted ethanol extraction. Ash content was higher for the NO SMS (2.7%) than for the BPKO one (2.0%).

The lower content of lignin and hemicelluloses in the SMSs than in the initial substrate can be attributed to degradation during cultivation. *Pleurotus* spp., like other white-rot fungi, degrade, either partially or entirely, wood structural components and use the degradation products as nutrients [14]. The higher content of extractives in the SMSs than in the initial substrate is also a consequence of the fungal growth. Some extractive compounds originated from the fungal degradation of the lignocellulosic components of the substrate. For example, the phenolic compounds result from lignin degradation by oxidative enzymes, such as peroxidases and laccases, that are secreted by fungi during cultivation [3]. Other extractive compounds originated from fungal biomass. That includes carbohydrates, such as β-glucans [15], and proteins and sterols from the fungal mycelium, which remains in the SMS by the end of cultivation [16].

Table 1. Chemical composition of SMS from the cultivation of BPKO and NO strains, mass fractions in % (dry weight). Mean values from triplicate analyses are presented for all the components except extractives. Standard deviation is shown in parenthesis.

	BPKO	NO
Cellulose	38.3 (0.6)	37.5 (0.6)
Hemicelluloses	22.1	20.4
Xylan	18.5 (0.6)	16.6 (0.6)
Anhydroarabinose	3.6 (0.1)	3.8 (0.3)
Klason lignin	16.0 (0.6)	18.4 (0.9)
Water extractives	12.0	13.7
Total carbohydrates	10.7(1.0)	9.8 (0.3)
Total phenolics	1.0 (0.2)	1.0 (<0.1)
Ethanol extractives	1.5	1.3
Total phenolics	0.3 (<0.1)	0.2 (<0.1)
Total extractives	13.5	15.0
Ergosterol	<0.1	<0.1
Ash	2.0 (0.3)	2.7 (0.5)

2.2. *Extraction of Bioactive Compounds from SMS Using Conventional Methods*

The high content of extractive compounds in SMS makes it of interest as a source of bioactive compounds. The extraction of bioactive compounds, provided that appropriate methods are identified, might provide a suitable alternative for resource recovery from SMS [16]. In this work, bioactive compounds were extracted from SMS using both conventional and modern techniques, and the remaining solids were then subjected to enzymatic saccharification (Figure 1).

Conventional methods, including maceration, percolation, and reflux extraction, have been used to extract bioactive compounds from plant-based material for decades [17]. Reflux is more effective than percolation and maceration, but the three methods require a long extraction time using large volumes of organic solvents. Another conventional extraction method is the Soxhlet technique, which is a reflux system with an incorporated siphoning device. It is a more efficient approach since it allows continuous extraction combining reflux and percolation methods. However, it still requires high temperature and long extraction time, which increase the possibilities of thermal degradation.

In this study, Soxhlet extraction with seven solvent systems was performed for SMS from BPKO mushroom. After that, the solvent mixture giving the best results on yield, composition, and antioxidant activity was used in reflux for several extraction times.

2.2.1. Soxhlet Extraction

In Soxhlet extraction applied to SMS from the BPKO strain, the seven solvent systems resulted in different extraction yields (EYs). The highest EY (9.5% (*w*/*w*)) was achieved with the 40:60 ethanol/water mixture (E40 in Figure 2a), which corresponded to an extraction efficiency of 69.9%. The EY decreased proportionally with the increase in the share of any solvent component in the mixture.

As the content of either water or ethanol increased, the recovery of total phenolic compounds decreased. The highest recovery of total phenolic compounds (9.1 mg GAE/g biomass) in the Soxhlet extracts was detected in the sample from the extraction with a 40:60 ethanol/water mixture (E40), and they were only slightly lower for the E50 and E60 mixtures (Figure 2b). The recovery of phenolic compounds from SMS was higher than that previously reported for the Soxhlet extraction of dehydrated *P. ostreatus* fruitbodies (3.9–6.4 mg GAE/g), but it was lower than that achieved for fresh fruitbodies (14.3–21.9 mg GAE/g) [18]. The recovery of phenolics in the current study was comparable with reported results for other plant-biomass materials using a similar extraction approach, but some particularities were observed. For example, the range of recovery of total phenolics in this work (3.0–9.1 mg GAE/g biomass) was comparable with that achieved by Espinosa et al. [19] by Soxhlet extraction of orange-peel waste with different ethanol/water mixtures (2.43–8.55 mg GAE/g biomass). However, the extractability profile was different in both studies. In Espinosa et al.'s study, the highest recoveries were achieved with solvent systems containing at least 75% (v/v) ethanol, whereas in the current work, the highest values were reached with mixtures containing between 40 and 60% (v/v) ethanol. The different behavior can be attributed to the different compositions of the phenolic fraction in both materials. Apparently, more polar compounds are predominant in the phenolic fraction of SMS than in that of orange-peel waste. That is in line with the high share of phenolic compounds found in the water extractives' fraction during the SMS characterization of the SMSs (Table 1).

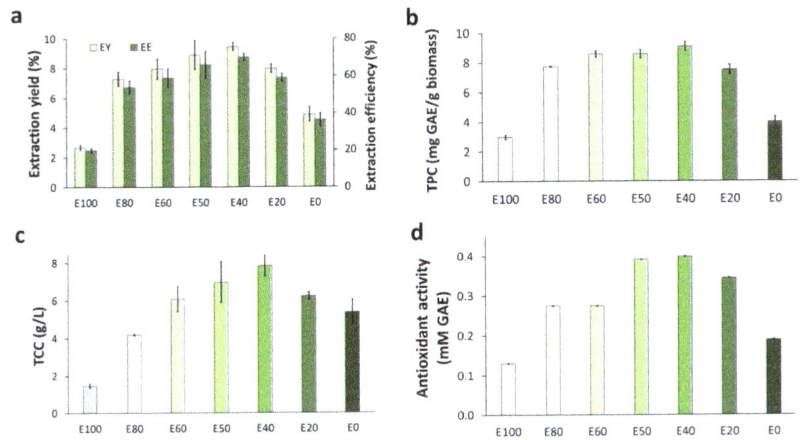

Figure 2. Soxhlet extraction applied to BPKO SMS. The extraction was held for four hours using different ethanol/water mixtures. Extraction yield and extraction efficiency (**a**), recovery of total phenolics, (**b**) concentration of total carbohydrates (**c**), and antioxidant activity (**d**). The numerals in the horizontal axis indicate ethanol (E) volumetric share in the solvent mixture. Mean values from triplicate experiments were used to build the graphs. The error bars show the standard deviations.

The concentration of total carbohydrates was the lowest when only ethanol was used as the solvent, and it increased as the volumetric share of ethanol in the solvent system decreased from 100 to 40% (Figure 2c). The maximum value was reached for the E40 mixture. Then, a decrease was observed for E20 and E0. The increased concentrations in extractions with decreasing ethanol shares in the solvent mixture agree with the existing knowledge on the solubility of carbohydrates in ethanol/water mixtures [20]. The observed decrease in the concentrations in extractions with solvent systems with lower ethanol shares (20 and 0%) might be related to the predominance of polysaccharides or relatively large oligosaccharides, which typically have lower ethanol solubility than that of monosaccharides [21]. The low solubility of SMS carbohydrates in ethanol can be exploited for their isolation from water extracts using ethanol precipitation.

The highest antioxidant activity (0.4 mM GAE) was found in the extract from the E40 solvent mixture (Figure 2d). The lowest value (0.1 mM GAE) was observed when ethanol was the only solvent (E100). Antioxidant activity correlated well with the concentration of total phenolics in the extracts (see Section 3.5).

2.2.2. Reflux Extraction

Reflux extraction applied to BPKO SMS lasted between 60 and 240 min, and it was performed with the E40 solvent system, which gave the best results in the Soxhlet experiment (see Section 2.2.1). The extraction yield and efficiency increased proportionally with the reflux time for the first three hours (Figure 3a). After that, a plateau was reached Reflux lasting 150 min or longer resulted in higher EY and EE than Soxhlet extractions. For facilitating the comparison, horizontal lines denoting the best results achieved by Soxhlet extraction are included in Figure 3.

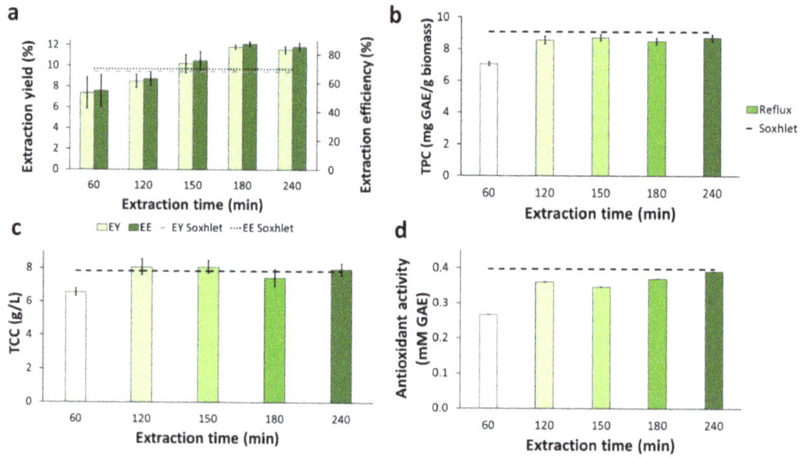

Figure 3. Reflux extraction applied to BPKO SMS. Extraction yield and extraction efficiency (**a**), recovery of total phenolics (**b**), concentration of total carbohydrates, and (**c**) antioxidant activity (**d**). Mean values from triplicate experiments were used to build the graphs. The error bars show the standard deviations.

The extract from the 60-min reflux had a significantly lower (p-value < 0.05) concentration of total phenolic compounds (Figure 3b) and total carbohydrates (Figure 3c) than those achieved with Soxhlet extraction using the same solvent mixture (Figure 2b). For the concentration of phenolic compounds from more extended reflux extractions, the difference was less remarkable, but it was still slightly lower than for Soxhlet extraction. Notably, no significant differences were observed between the recoveries of total phenolics in extracts from reflux extractions lasting 120 min or more, which were all in the range between 7.1 and 8.7 mg GAE/g biomass. For total carbohydrates, the concentrations in all the extracts from reflux lasting 120 min or more were comparable to the highest value achieved with Soxhlet extraction.

The antioxidant activity of all reflux extracts of the BPKO SMS, except for that from the 60-min lasting extraction, showed comparable results (Figure 3d). The antioxidant activity of the 60-min extract (0.3 mM GAE) was significantly lower (p-value < 0.05) than the highest value achieved in Soxhlet extraction (0.4 mM GAE).

2.3. *Extraction of Bioactive Compounds from SMS Using Green Techniques*

The conventional extraction methods are usually performed with organic solvents and require a high volume of solvent and a long time, which can affect the properties of the extracted molecules. Green, or non-conventional, extraction methods have features such as lower consumption of organic solvents, shorter extraction times, and higher efficiency and selectivity, which allow overcoming the disadvantages of conventional extraction methods [22]. Two green-extraction methods, namely ultrasound-assisted extraction and subcritical water extraction, were evaluated for SMS in this study.

2.3.1. Ultrasound-Assisted Extraction

UAE under optimized conditions is a valuable green technique for extracting bioactive compounds from plant biomass [23]. In this work, a full-factorial 3^2 experimental design was used to investigate the effect of temperature and time on the yield and efficiency of UAE of BPKO SMS (Table 2). Since UAE has not been reported before for SMS, the selected operational conditions were based on reported data for other plant-based residual materials [11,19,24,25].

Table 2. Experimental design used in the ultrasound-assisted extraction applied to BPKO SMS. Operational conditions of the design and extraction yields resulting from the experiment.

No.	Temperature, °C	Time, Min	Extraction Efficiency [1], g/100 g Biomass
1	35	30	10.9 (4.7)
2	35	45	20.1 (8.0)
3	35	60	12.4 (1.4)
4	50	30	61.6 (2.6)
5	50	45	60.7 (3.9)
6	50	60	67.5 (1.1)
7	65	30	69.4 (6.0)
8	65	45	80.8 (3.1)
9	65	60	81.7 (0.5)

[1] Mean values from triplicate measurements. The standard deviations are shown in parentheses.

Ultrasound-assisted extraction applied to BPKO SMS at 35 °C resulted in comparable responses independently of the extraction time. Increasing the temperature resulted in an increase in the EY from 1.5–2.7% at 35 °C to 9.4–11.1% at 65 °C (Figure 4a). The differences between results from different temperatures were statistically significant (p-value < 0.05). For the experiments at 35 and 50 °C, the extraction yield was lower than the highest value achieved with Soxhlet extraction. However, UAE at 65 °C resulted in comparable and higher yields than Soxhlet extraction. This result agrees well with a previous report on orange peel waste, where UAE at 80 °C resulted in higher extraction yield than Soxhlet extraction but compared negatively at either 30 or 60 °C [19]. For UAE at 50 and 65 °C, an increase in the EY with time was observed as a general trend. The highest EY (11.1%) was observed for the extraction at 65 °C for 60 min. The trend observed for the EE was similar to the one observed for the EY (Table 2). For EY and EE in UAE, less time and lower temperature were required to reach comparable results as those obtained by Soxhlet and reflux extractions.

The significance analysis revealed that the temperature was the independent factor exerting the most important estimated effect on the extraction efficiency, as shown by the Pareto chart of standardized effects (Figure 4b). The positive sign indicates that a temperature increase enhances the extraction efficiency. The quadratic term of the temperature also exerted a significant effect but had a negative sign. No significant effects were exerted by the extraction time, its quadratic term, and its interaction with the temperature. Previous reports on longer UAE performed for orange peel waste showed that for that material and under those conditions, time significantly affected extraction yield in the range of 10–60 min but had only a marginal effect in the range of 60–180 min [19].

An empirical model (Equation (1), R^2 = 97.5%) describing the effect of the operational conditions on extraction efficiency was proposed based on the experimental results. The factors exerting no significant effects were excluded from the model. The model is shown as Equation (1), in which EE is the extraction efficiency (in %), and T is the temperature (in °C).

$$EE = 63.27 + 62.83\,T - 34.77\,T^2 \quad (1)$$

The response contour plot provides a better description of the estimated effect of the temperature on the extraction efficiency (Figure 4c). The graph shows that the EE increased sharply with temperature increase from 35 °C to around 50 °C. In contrast, further temperature increases exerted a relatively moderate effect. A region with maximal EE can be reached at around 60–65 °C and extraction times above 50 min.

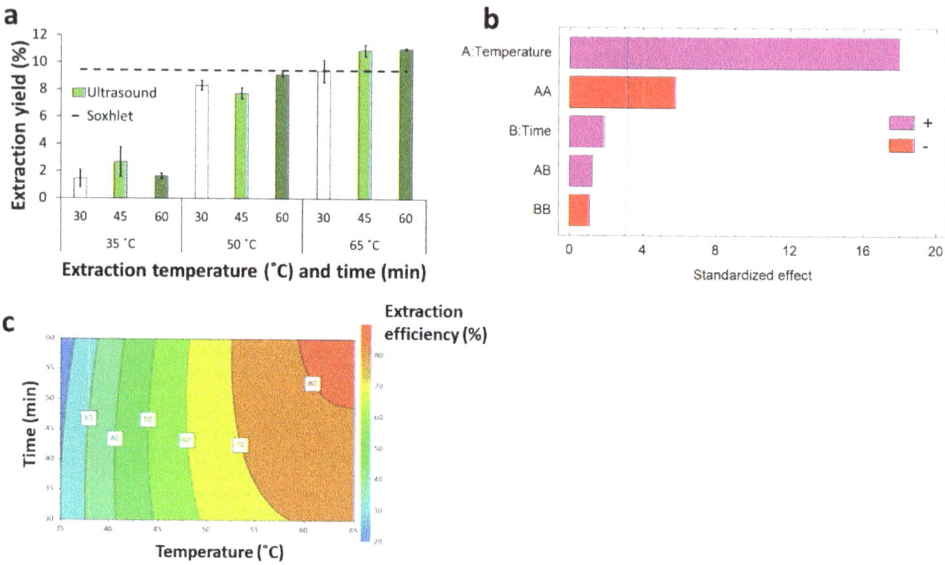

Figure 4. Extraction yield and efficiency (%) for ultrasound-assisted extraction applied to BPKO SMS. Extraction yield (**a**), Pareto chart of standardized effects (**b**), response contour plot (**c**).

The recovery of total phenolic compounds ranged from 2.5 mg GAE/g biomass in the extracts from UAE at 35 °C for 30 min to 5.6 mg GAE/g in those from extractions at 65 °C for 60 min (Figure 5a). Those concentrations were significantly lower (p-value < 0.05) than the highest values observed for Soxhlet (Figure 2b) and reflux (Figure 3b) extractions. Nonetheless, the concentrations of phenolics achieved in all UAE experiments at 65 °C were higher than those from Soxhlet extractions that used either pure ethanol (E100) or water (E0) as solvents.

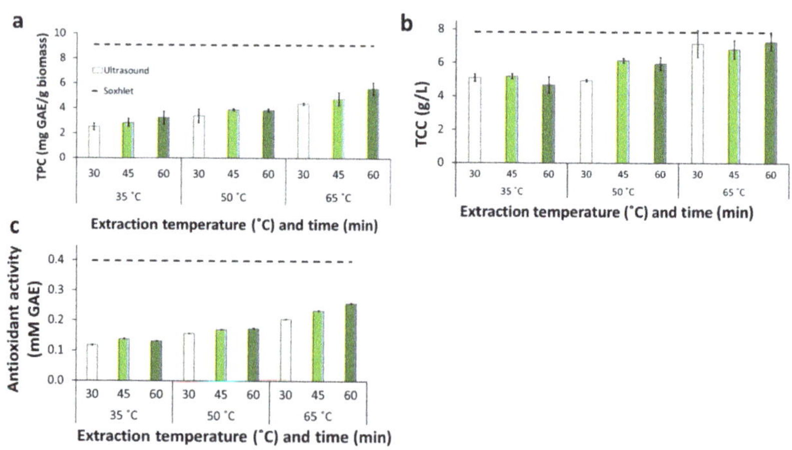

Figure 5. Ultrasound-assisted extraction applied to BPKO SMS. Recovery of total phenolics (**a**), concentration of total carbohydrates (**b**), and antioxidant activity (**c**). The extraction was performed using the E40 solvent mixture. Mean values from triplicate experiments were used to build the graphs. The error bars show the standard deviations.

For UAE at 35 and 50 °C, the concentrations of total carbohydrates in the extracts were between 4.7 and 6.0 g/L (Figure 5b). That concentration range is clearly lower than the highest values found for Soxhlet (Figure 2c) and reflux (Figure 3c) extractions. There was a general increase in concentration with the increase in temperature, as can be shown by the highest values for 35 °C (5.2 g/L), 50 °C (6.0 g/L), and 65 °C (7.3 g/L). As a rule, extracts from UAE at a given temperature displayed carbohydrate concentrations in a comparable concentration regardless of the extraction time. The highest total carbohydrate concentrations were observed in extracts from experiments at 65 °C (6.8–7.3 g/L). Those values were only slightly lower than the results in the best-performing Soxhlet and reflux extractions.

The antioxidant activity of UAE extracts increased with temperature (Figure 5c). For experiments at 35 and 50 °C, the results were somewhat comparable for different extraction times at a given temperature except for extractions at 65 °C, which showed a general increase with extraction time. In addition, even for the extract from UAE at 65 °C and for 60 min, the antioxidant activity was lower than for Soxhlet and reflux extractions.

2.3.2. Subcritical-Water Extraction

In a final experiment, SWE, e.g., extraction with water in a pressurized reactor, was performed for the BPKO SMS. SWE was performed either non-isothermally (SWE 0) (by heating an SMS suspension followed by cooling it immediately after reaching 150 °C) or with a holding time of 10 min at 150 °C (SWE 10). The temperature profiles of both experiments are shown in Figure 6a. The severity factor (SF) was 1.8 for non-isothermal SWE and 2.3 for the experiment with a holding time of 10 min.

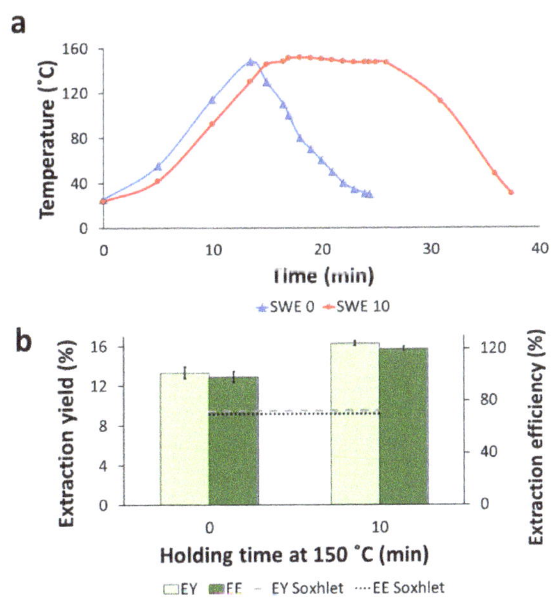

Figure 6. Subcritical-water extraction applied to BPKO SMS. Temperature profiles (**a**), extraction yield, and extraction efficiency (**b**). The extraction was held with either 0 or 10 min holding at 150 °C. Mean values from triplicate experiments were used to build the graph in panel (**b**). The error bars show the standard deviations.

The extraction yield was significantly higher (p-value < 0.05) for SWE with 10-min holding at 150 °C (16.3%) than for the non-isothermal experiment (13.3%) (Figure 6b). The same trend was observed for the extraction efficiency. The EY and EE were higher for SWE than the other extraction methods, and the EE of SWE 10 exceeded 100% (w/w). That phenomenon can be attributed to an increase in the apparent mass of extractives by the partial solubilization of cell wall constituents, mainly hemicelluloses. It is known that the partial solubilization of the hemicelluloses can occur in subcritical water at temperatures around 150 °C, especially if the temperature is held for a certain time [26].

The recovery of total phenolic compounds (Figure 7a) and the concentration of total carbohydrates (Figure 7b) in the SWE extracts were higher for SWE 10 than SWE 0. For both extracts, the content of phenolics was comparable with the values resulting from Soxhlet and reflux extraction, but it was higher than for UAE. The content of total carbohydrates was significantly higher (p-value < 0.05) than for Soxhlet extraction, as illustrated by the position of the horizontal dashed line in Figure 7b. It was also higher than for reflux and UAE. The extract from SWE 10 had the highest concentration of carbohydrates (16.8 g/L) among the extracts from all extraction methods. That high concentration might be linked with the presence of sugars resulting from the degradation of hemicelluloses.

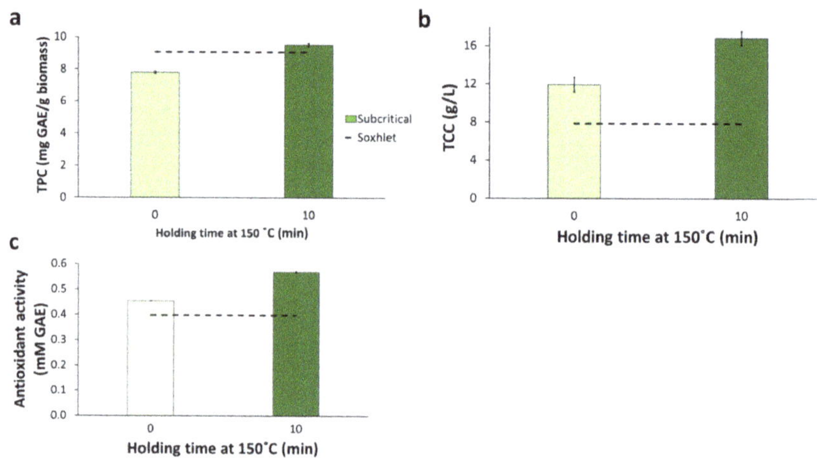

Figure 7. Subcritical-water extraction applied to BPKO SMS. Recovery of total phenolics (**a**), concentration of total carbohydrates (**b**), and antioxidant activity (**c**). The extraction was held with a holding time of either 0 or 10 min at 150 °C. Mean values from triplicate experiments were used to build the graphs. The error bars show the standard deviations.

As for other parameters, the antioxidant activity was higher in the extract from the SWE 10 experiment than in the SWE 0 one (Figure 7c), and both extractions displayed higher antioxidant activity than the extracts from Soxhlet, reflux, and ultrasound extractions.

The solvent used in SWE was distilled water, the so-called E0 solvent system in Soxhlet extractions (see Section 2.2.1). When comparing the E0 Soxhlet extraction with both SWE experiments, it is evident that SWE resulted in higher yield, efficiency, total phenolics and carbohydrates concentration, and antioxidant activity than Soxhlet extraction. Furthermore, those results were achieved in a much shorter extraction time. That confirms the potential of subcritical-water extraction as an efficient method for achieving suitable extraction parameters more time-effectively than conventional extraction under atmospheric conditions [27].

2.4. Extraction of Bioactive Compounds from SMS of the Wild Strain of *P. ostreatus*

Soxhlet-, reflux-, and ultrasound-assisted extraction were applied to the NO SMS using the conditions resulting in the best results for the BPKO SMS. A comparison of the results for the extraction of both SMSs is shown in Table 3. The E40 solvent mixture was used in all the experiments.

Table 3. Comparison of the results of different extraction methods applied to SMS of two *Pleurotus* spp. strains. Selected extraction conditions: Soxhlet with the E40 solvent mixture; E40 reflux for 120 min; E40 UAE at 65 °C for 60 min. Mean values from triplicate analyses. The standard deviations are shown in parentheses.

Parameter/Extraction Method	BPKO SMS	NO SMS
Extraction yield, % (*w*/*w*)		
Soxhlet extraction	9.5 (0.3)	10.7 (0.2)
Reflux extraction	8.5 (0.6)	12.6 (0.1)
UAE	11.1 (0.1)	12.3 (0.3)
Extraction efficiency, % (*w*/*w*)		
Soxhlet extraction	69.9 (1.9)	71.1 (1.5)
Reflux extraction	62.9 (4.6)	83.7 (0.5)
UAE	81.7 (0.5)	82.0 (2.2)
Total phenolics, mg GAE/g biomass		
Soxhlet extraction	9.1 (0.3)	8.3 (0.2)
Reflux extraction	8.6 (0.2)	9.5 (0.2)
UAE	5.6 (0.5)	7.4 (0.3)
Total carbohydrates, g/L		
Soxhlet extraction	7.8 (0.5)	6.1 (0.5)
Reflux extraction	8.1 (0.5)	7.9 (0.9)
UAE	7.3 (0.5)	7.4 (0.6)
Antioxidant activity, mM GAE		
Soxhlet extraction	0.4 (<0.1)	0.5 (<0.1)
Reflux extraction	0.4 (<0.1)	0.5 (<0.1)
UAE	0.3 (<0.1)	0.4 (<0.1)

The extraction yield and extraction efficiency were higher for NO SMS than for BPKO SMS (Table 3). The differences were statistically significant (p-value < 0.05) for the EY for all the extraction methods. The EE differences were significant for reflux extraction but not for Soxhlet extraction or UAE.

The recovery of total phenolics in Soxhlet extracts was significantly higher for BPKO SMS than for NO SMS, but in reflux extracts and UAE extracts, it was the other way around (Table 3). The highest recovery of phenolics (9.5 mg GAE/g biomass) was found in the reflux extracts of the NO SMS.

On the other hand, in Soxhlet extracts, the concentration of total carbohydrates was higher for BPKO SMS than for NO SMS. In contrast, the values were comparable for both SMSs in reflux extracts and UAE extracts. The highest concentration of total carbohydrates (8.1 g/L) was found in the reflux extract of the BPKO SMS. The antioxidant activity of the extracts from all the extraction methods was significantly higher (p-value < 0.05) for NO SMS than for BPKO SMS (Table 3).

The higher EY, EE, concentration of total phenolic compounds, and antioxidant activity observed in the extracts of the SMS of the wild strain are in accordance with the content of extractive compounds in the SMS (Table 1). Furthermore, these results might also be interpreted as an indication that NO SMS has a higher susceptibility to being extracted and to yielding extractive compounds with higher antioxidant activity than BPKO SMS.

2.5. Phenolic Acids in the Extracts and Their Correlation with the Antioxidant Activity

Six phenolic acids were identified in the extracts obtained by different methods (Figure 8). Vanillic, gallic, and chlorogenic acids were the compounds detected in the highest concentrations in most of the extracts (Table 4). The concentrations of 2,3-dihydrobenxoic acid and protocatechuic acid were below the detection limits. The highest concentrations of vanillic acid (16.4 mg/L) and gallic acid (11.1 mg/L) were found in the SWE extract, while the highest concentration of chlorogenic acid (7.3 g/L) was detected in the Soxhlet extract using the E40 solvent mixture.

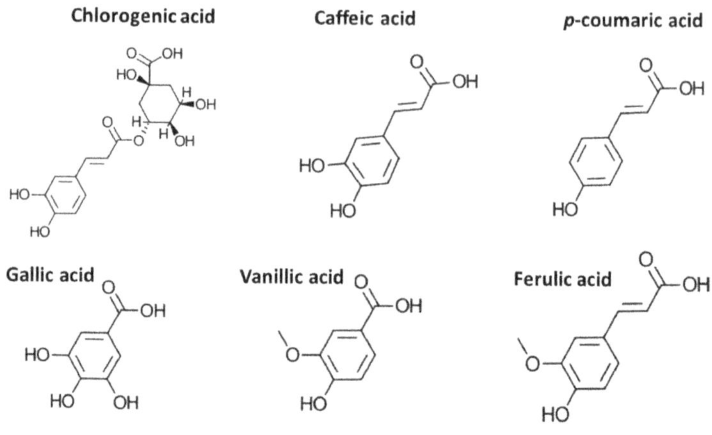

Figure 8. Phenolic acids identified in extracts of BPKO SMS.

Table 4. Concentration of phenolic acids [1] in selected extracts from BPKO SMS with different extractions methods, mg/L. Mean values from triplicate analyses. The standard deviations are shown in parentheses.

Extract Sample [2]	Chlorogenic Acid	Caffeic Acid	*p*-Coumaric Acid	Ferulic Acid	Gallic Acid	Vanillic Acid
SoE-80	6.0 (0.2)	0.7 (0.2)	2.3 (0.1)	0.9 (0.1)	5.5 (0.4)	12.4 (1.0)
SoE-40	7.3 (0.3)	1.1 (0.2)	2.3 (0.1)	1.1 (0.1)	3.6 (0.4)	8.2 (2.5)
SoE-20	4.6 (0.3)	0.6 (<0.1)	1.6 (<0,1)	0.8 (<0.1)	3.0 (0.6)	6.9 (1.3)
ReE-60	4.8 (0.3)	0.7 (0.1)	1.7 (0.1)	0.7 (<0.1)	2.8 (0.3)	3.8 (0.4)
ReE-120	4.2 (0.2)	0.7 (0.1)	1.1 (0.1)	0.6 (<0.1)	2.8 (0.4)	4.5 (0.4)
ReE-240	4.8 (<0.1)	0.7 (0.1)	1.5 (0.1)	0.7 (0.1)	2.5 (0.3)	5.8 (0.6)
UAE 35–30	3.1 (<0.1)	0.2 (0.1)	1.3 (0.1)	0.5 (<0.1)	2.6 (0.1)	2.9 (0.3)
UAE 50–45	3.6 (<0.1)	0.3 (<0.1)	1.3 (0.1)	0.6 (<0.1)	3.2 (0.2)	1.8 (0.4)
UAE 65–60	4.6 (0.1)	0.3 (<0.1)	1.7 (<0.1)	0.8 (0.1)	3.0 (0.3)	3.0 (0.2)
SWE 10	5.4 (<0.1)	1.7 (0.4)	1.2 (0.1)	0.7 (0.1)	11.1 (0.5)	16.4 (0.2)

[1] Determined by HPLC; [2] Sample codification: for Soxhlet extraction (SoE), the numeral indicates the volumetric share of ethanol in the solvent system; for reflux extraction (ReE), the numeral indicates the time (in min); for UAE, the numerals indicate, respectively, the temperature (in °C) and the time (in min); for SWE, the numeral indicate the holding time (in min).

The Soxhlet E40 extract also had the highest concentrations of *p*-coumaric acid (2.3 mg/L) and ferulic acid (1.1 mg/L) (Table 4). For Soxhlet extraction, the concentration of total identified phenolic acids and the concentrations of gallic and vanillic acids increased with the increase in ethanol share in the solvent mixture. Gallic acid concentration increased from 3.0 mg/L in the E20 extract to 5.5 g/L in the E80 extract, while vanillic acid increased from 6.9 to 12.4 mg/L. This is an expected result, as gallic and vanillic acids are more soluble in ethanol than water [28,29]. For other compounds, no clear trend was observed. For caffeic acid (0.6–0.7 mg/L), ferulic acid (0.8–1.1 mg/L), and *p*-coumaric acid (1.6–2.3 mg/L), the concentrations remained comparable for all the extracts.

For reflux extraction, regardless of the extraction time, comparable concentrations were observed for each identified phenolic acid (Table 4). The exception was vanillic acid, which showed increased concentration with extended extraction time. The phenolic acid displaying the highest concentration in any reflux extract was vanillic acid (5.8 mg/L in the 240-min extract). The concentrations found in reflux extracts were generally comparable to those from Soxhlet extractions with the E20 solvent mixture.

For UAE, a general increase in the concentration of most phenolic acids was observed with the increase in temperature and time (Table 4). The exceptions were gallic acid, whose highest concentration (3.2 mg/L) was observed for the extraction at 50 °C and 45 min, and vanillic acid, which did not show a clear trend. The highest concentration for any of the identified compounds in UAE was found for chlorogenic acid (4.6 mg/L) in the extract from the experiment at 65 °C and 60 min. UAE yielded lower concentrations for all the phenolic acids than those observed for the E40 Soxhlet extraction and reflux experiments. The most remarkable case was caffeic acid, whose concentration in UAE extracts corresponded to only around half the value in the reflux extracts and one-fourth in Soxhlet extracts. That phenomenon can be attributed to the degradation of caffeic acid under ultrasound treatment, as shown previously in a study of sonication of phenolic acids in a model system [30]. The highest combined concentration of all identified phenolic acids (36.5 mg/L) was found for SWE (Table 4). The sum of the phenolic acids identified in the Soxhlet, reflux, and UAE extracts was within the range of, respectively, 17–27.8, 14.0–16.1, and 10.6–13.5 mg/L. SWE also resulted in the highest concentrations of several of the individual compounds, e.g., caffeic acid (1.7 mg/L), gallic acid (11.1 mg/L), and vanillic acid (16.4 mg/L).

Some phenolic acids have been reported before in fruitbodies of edible mushrooms [31]. The content of each phenolic acid identified in this work was lower than those previously reported for extracts *P. ostreatus* fruitbodies [32]. The partial identification of phenolic acids in SMS has been reported for shiitake mushrooms [33]. However, to our knowledge, there are no previous studies on identifying phenolic acids extracted from the SMS of oyster mushrooms.

The concentrations of total phenolic compounds, individual phenolic acids, and total carbohydrates in all the extracts were correlated to the corresponding antioxidant activity, which was expressed as ferric-ion reducing antioxidant power (FRAP). A strong, significant, and positive correlation between the concentration of total phenolic compounds (mg/L) and the ferric-reducing power (mM GAE) ($R^2 = 0.80$, Pearson correlation coefficient (PCC): 0.89, p-value < 0.05) was revealed (Figure 9a). That is an indication of a strong contribution of the total phenolic compounds to the observed antioxidant activity in the samples. This result agrees with previous studies reporting correlations between antioxidant activity, using various criteria, and total phenolic content (TPC) in extracts of fungal biomass and other materials. For example, Martínez-Flores et al. [18] found a strong positive correlation between TPC and antioxidant activity, determined by both the radical scavenging activity (DPPH) and the 2,2′-azino-bis(3-ethylbenzothiazoline-6-sulfonic acid) (ABTS) assay, in aqueous ethanol extracts of *P. ostreatus* and *Pleurotus djamor* fruitbodies. Yim et al. [34] reported a strong significant correlation in aqueous extracts of *P. ostreatus* fruitbodies using the FRAP assay. For other biomass materials, Showkat et al. [35] reported a significant correlation between the TPC of extracts of Jerusalem artichoke and their antioxidant activity using the DPPH assay, and a similar result was reported by Espinosa et al. [19] for extracts of orange peels using the ABTS assay. No correlations between antioxidant activity and concentration of phenolics or other compounds have been reported for SMS extracts.

Our results also revealed a positive correlation between the antioxidant activity and the concentration of caffeic acid ($R^2 = 0.81$, PCC = 0.90, p-value < 0.05) (Figure 9b), indicating that caffeic acid is an essential contributor to the antioxidant activity observed in the SMS extracts. This agrees well with the existing knowledge on the antioxidant activity exerted by caffeic acid [36]. The low caffeic acid concentration of the UAE extracts (Table 4) might be behind their lower antioxidant activity compared to the other extracts (Figure 5c). Similarly, the higher antioxidant activity of SWE extracts (Figure 7c) might be linked to their higher caffeic acid concentration compared with that of the extracts obtained by other methods. On the other hand, the correlations found between the antioxidant activity and other phenolic acids had relatively low R^2 values (Tables S7 and S8). Some correlation ($R^2 = 0.68$, PCC = 0.82, p-value < 0.05) was observed between the antioxidant activity and the concentration of total carbohydrates (Table S9). That might be related to specific carbohydrates, such as β-glucans [37], which are typically contained in fungal fruitbodies and mycelium, and are known to show antioxidant activity [32,38].

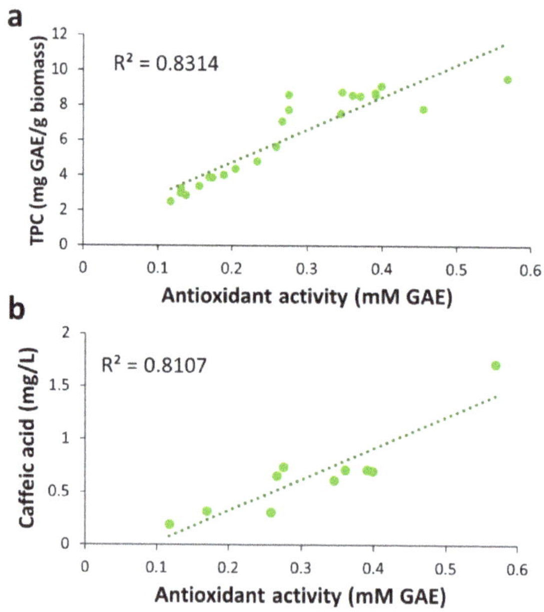

Figure 9. Scatter plots of the antioxidant activity in extracts from BPKO SMS versus concentrations of total phenolics (**a**) and caffeic acid (**b**). Samples are from different extraction methods.

2.6. Enzymatic Saccharification of SMS

The enzymatic saccharification was evaluated by comparing the cellulose digestibility, i.e., the mass percentage of cellulose being saccharified, of different SMS samples upon treating them with cellulolytic enzymes. An analytical enzymatic saccharification assay [39] was applied. The initial substrate used for mushroom cultivation was also included in the experiment. The enzymatic digestibility of cellulose was 9.4% (*w*/*w*) for the initial substrate. In comparison, it was 12.5% and 15.8% for NO SMS and BPKO SMS, respectively (Figure 10). That corresponds to an enhancement of the enzymatic digestibility by 33% for NO SMS and 68% for BPKO SMS compared to the initial substrate. The enhanced enzymatic digestibility of cellulose in SMS compared with the initial lignocellulosic substrate is a consequence of fungal cultivation acting as a biological pretreatment [40]. While growing on lignocellulose, fungi remove part of lignin and hemicelluloses, and that improves the accessibility of cellulose to the enzymes. The digestibility pattern for both SMSs is in good agreement with their lignin content: BPKO SMS, which contained 16% lignin (Table 1), had better digestibility than NO SMS, which contained 18.4% lignin.

The observed enhancement of the enzymatic digestibility is higher than the previously reported by Zadrazil [41] for SMS of *P. ostreatus* grown on a wheat-straw substrate, but our absolute digestibility values were lower. However, the enzymatic digestibility achieved in the current work was below the expectations that one could have for a biologically pretreated material. It is expected that the delignification caused by fungal cultivation, i.e., a biological pretreatment, leads to an enhancement of the enzyme access to the substrate, which results in an improvement of the enzymatic saccharification [13]. The poor enzymatic digestibility observed in the current work might be attributed to pretreatment-related issues, such as the fungal strains or the cultivation conditions. It might be so that the cultivation time was not long enough to ensure an effective delignification. The lignin content in the SMS (16.0–18.4% (Table 1) was only slightly lower than in the initial substrate (19%,

Table S1). The enzymatic digestibility values were lower than our previously reported results for *Lentinula edodes* SMS [12]. The lower enzymatic digestibility of the *Pleurotus* spp. SMS than that of *L. edodes* SMS is in line with the results of a previous study with the SMS of another oyster mushroom [42]. In that study, the enzymatic saccharification of SMS from the cultivation of *Pleurotus pulmonarius* on a birch-based substrate resulted in lower glucose yields than typical values for an *L. edodes* birch-based SMS. The result was attributed to the fact that the longer time required for shiitake cultivation provides a better pretreated SMS than that of oyster mushrooms, which requires less time for cultivation and results in a lower delignification degree.

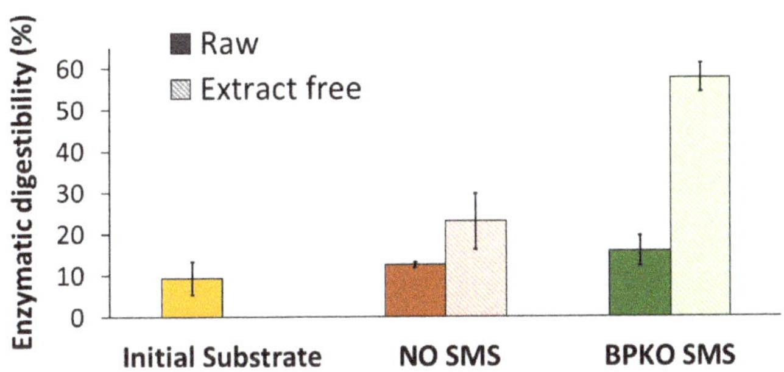

Figure 10. Enzymatic digestibility of cellulose contained in the initial substrate, raw and extract-free SMS of the BPKO and NO strains of *Pleurotus* spp. Mean values from triplicate experiments were used to build the graphs. The error bars show the standard deviations.

The enzymatic saccharification was also assayed for the SMS after exhaustive extraction. For both SMSs, the enzymatic digestibility of cellulose was improved after the removal of the extractive compounds (Figure 10). For the NO SMS, the enzymatic digestibility increased approximately two-fold (from 12.5 to 23.0%), while for the BPKO SMS, an almost four-fold increase was observed (from 15.8 to 57.6%). The enhancement of the enzymatic digestibility after extraction can be attributed to the partial removal of extractive compounds that are inhibitory to the enzymes. It is known that certain phenolic compounds formed by lignin degradation or by solubilization of the extractive fraction of lignocellulose inhibit cellulolytic enzymes [43]. An investigation of the effect of SMS phenolic compounds on the inhibition of enzymatic saccharification is underway by this group.

Even if clear improvements were observed due to the extraction, the achieved digestibility values are still relatively low. The current study is a pioneering investigation on upgrading SMS by recovering bioactive compounds and producing sugars to be converted into high-added value products following a biorefinery approach. Developing an SMS-based biorefinery requires further improvement for producing enough sugars to back microbial fermentation processes of industrial relevance. A strategy to follow in that direction is to investigate how different extraction methods and operational conditions can improve enzymatic saccharification. Another strategy to consider is applying a post-treatment, e.g., with an acid or an alkali, to the extract-free SMS to improve its enzymatic digestibility further. It has been shown that treating the SMS of *Flammulina velutipes* with dilute sulfuric acid results in a high yield of reducing sugars [44]. It has also been reported that alkaline treatment of *Pleurotus florida* SMS facilitated the enzymatic saccharification to produce enough sugar for producing ethanol by yeast fermentation [45]. Furthermore, since the 5% (*w*/*w*) biomass load used in the enzymatic saccharification assay in this study is intended only for analytical purposes, larger loads are required for producing fermentable sugar amounts of industrial relevance. Therefore, preparative enzymatic saccharification protocols with biomass loads in the 10–20% range need to be developed and optimized for SMS from different sources.

3. Materials and Methods

3.1. Materials

SMS from the cultivation of two fungal strains was provided by Husbonden AS (Disenå, Norway). The cultivated fungi were a commercial strain, which is a *Pleurotus ostreatus* × *Pleurotus eryngii* hybrid, referred to as the Black Pearl King Oyster (BPKO) mushroom, and a wild *P. ostreatus* strain isolated in Norway, which is hereafter referred to as the Norwegian Oyster (NO) mushroom. The initial substrate (IS) used for mushroom cultivation was composed of 32% oak (*Quercur robur*) sawdust, 8% wheat (*Triticum aestivum*) bran, ~60% water, and 0.08% $CaCO_3$. SMS and IS samples were dried at room temperature in a ventilation hood until a dry-matter (DM) content above 90% (w/w) was reached. The dry samples were shredded using a Robot Coupe shredder (Robot-Coupe SNC, Montceau-en-Bourgogne, France) and sieved to a particle size below 0.5 mm. The resulting powdered material was stored in sealable plastic bags until further use.

The chemicals were acquired from Sigma-Aldrich Chemie GmbH (Steinheim, Germany) if not stated otherwise. The enzyme blend Cellic CTec2, procured from Sigma-Aldrich, was used in the enzymatic saccharification experiments. It contains cellulases, β-glucosidases, other enzymes involved in cellulose degradation, and hemicellulolytic enzymes. The CMCase activity of the used enzyme preparation was 330 U/mL.

3.2. Characterization of Biomass Materials

The content of dry matter, mineral components (ash), extractive compounds, structural polysaccharides, and lignin in raw- and extract-free SMS and the initial substrate was determined using standard protocols. All the analyses were performed in triplicates, and mean values and standard deviations were reported.

Dry matter was determined gravimetrically by drying overnight at 105 °C as previously described [46]. A Termaks TS4057 oven (Termaks, Bergen, Norway) and a Sartorius analytical balance (Sartorius Lab Instruments GmbH, Göttingen, Germany) were used. For determining the content of mineral components, 2 g (dry weight (DW)) samples were incinerated at 575 °C based on the National Renewable Energy Laboratory (NREL) standard [47]. A Carbolite CWF 1100 muffle furnace (Carbolite Gero, Sheffield, UK) was used.

Extractive compounds were determined by sequential Soxhlet extraction using water and absolute ethanol according to NREL standard protocol [48]. Extraction was performed using 7 g of biomass and 250 mL of solvent at a 35:1 (v/w) solvent-to-biomass ratio (SBR). Water extraction was run for eight hours, and ethanol extraction was run for six hours. After extraction, the total volume of the liquid fraction was brought back to 250 mL, and vacuum filtration was performed using ashless filter paper (S & S, Dassel, Germany). Then, 10 mL samples were withdrawn and stored frozen for further analyses. After that, the solvent was recovered by vacuum evaporation using a Büchi R-114 rotary evaporator (BÜCHI Labortechnik AG, Flawil, Switzerland). The liquid remaining at the end of the operation was left to evaporate in pre-weighed beakers in a ventilation hood. Finally, the extracts were dried in an oven (Termaks) overnight at 35 °C for multiple days until constant weight.

The determination of structural carbohydrates and lignin was performed by analytical acid hydrolysis based on an NREL protocol [49]. Aliquots of 300 mg (DW) were mixed with 3 mL 72% sulfuric acid and hydrolyzed at 30 °C for 1 h in a water bath. Then, the samples were transferred to previously weighed 100 mL screw-top flasks, 84 mL distilled water was added to bring the concentration of H_2SO_4 down to 4%, and hydrolysis was run at 121 °C for 1 h in a CertoClav CV-EL 18L autoclave (CertoClav Sterilizer GmbH, Leonding, Austria). After that, the flasks with the hydrolysates were cooled to room temperature in an ice bath, and the weight was readjusted to the initial value by distilled water addition. The hydrolysates were then separated from the solid residue (Klason lignin) by filtering through pre-dried and pre-weighed filters. Klason lignin was determined gravimetrically after drying overnight at 105 °C in a Termaks oven. Then, 5 mL hydrolysate samples were stored frozen in plastic tubes for later analysis.

3.3. Extractions

3.3.1. Soxhlet Extraction

A Soxhlet apparatus with an extractor chamber capacity of 200 mL was used with a heating mantle. Extractions with solvent systems containing water and absolute ethanol in 100:0, 80:20, 60:40, 50:50, 40:60, 20:80, and 0:100 (*v*/*v*) ratios were held for 4 h for the SMS of the BPKO strain. The SMS of the NO strain was extracted only with the 40:60 ethanol/water mixture. Eight grams (DW) of biomass samples were used, and the solvent was added at a volume ensuring an SBR of 25:1 (*v*/*w*). The extraction with 50:50 solvent mixtures was performed in triplicate, and the standard error of the experiment was calculated.

After all extractions, the liquid fractions containing the solvent and the extract were brought back to their starting volume by adding the corresponding solvent and then vacuum filtered as described above. Samples were taken to determine the extraction yield and extraction efficiency, and around 10 mL was stored frozen in Falcon tubes until further analyses. The remaining extract was concentrated by rotary evaporation (Büchi). For determining the EY, 2 mL aliquots of the liquid sample were dried overnight at 105 °C (Termaks), and the extract was weighed using an analytical balance (Sartorius). The EY was calculated as the percentage of extract from the mass of the SMS sample submitted to extraction. The EE was calculated as the mass percentage of the extract out of the total content of extractive compounds determined in the characterization. The concentrated extracts were frozen in 50 mL falcon tubes and used later to identify individual phenolic acids by HPLC. The extract-free solids were air-dried at room temperature. After that, their dry matter content was determined, and the rest was stored in plastic bags.

3.3.2. Reflux Extraction

Reflux extraction was performed with a 250 mL round flask and a 250 mm reflux condenser together with a heating mantle, using a 40:60 (*v*/*v*) absolute ethanol/water mixture as a solvent system at the same solvent-to-biomass ratio as in the Soxhlet extraction. The extraction time was 60, 120, 150, 180, or 240 min. The extraction lasting 150 min was performed in triplicate. The extracts were processed following the same procedures used for Soxhlet extraction.

3.3.3. Ultrasound-Assisted Extraction

UAE was performed with a 40:60 (*v*/*v*) ethanol/water mixture as a solvent system. A 25:1 (*w*/*v*) solvent-to-biomass ratio was used. A 2^3 experimental design was applied, with the temperature (35, 50, and 65 °C) and extraction times (30, 45, and 60 min) as independent factors. Extraction was performed by immersing a sonication probe (Sonifier SFX250, Brandson, Mexico) in a 250 mL round wide-neck flask containing a suspension of the SMS in the solvent system. The ultrasonic frequency was 20 kHz, the maximum power was 250 W, and the amplitude was 50%. A cold-water bath was used to avoid overheating of the suspension. The extracts were separated and processed as described in previous sections.

3.3.4. Subcritical Water Extraction

Subcritical water extraction was performed by heating a mixture containing 12 g (DW) of dry SMS suspended in 300 mL of distilled water in a pressurized Parr 4520 reactor (Parr Instrument Company, Moline, IL, USA). Two SWE regimes were applied. In one run, the SMS suspension was treated non-isothermally by heating it to 150 °C and cooling it to room temperature immediately afterward. In the second run, the temperature was held at 150 °C for 10 min before cooling to room temperature. The severity factor (log R_0) was 1.8 in the first run and 2.3 in the second. The reactor was cooled by immersion in icy water.

3.4. Analytical Enzymatic Saccharification (AES)

Around 50 mg (DM) of raw and extract-free SMS was suspended in 900 μL of 50 mM sodium citrate buffer (pH 5.2) at a 5% (*w*/*w*) solids content in Eppendorf tubes. The tubes with the suspensions were mixed at 100 rpm for one hour in a Thermo Scientific compact digital waving rotator (Thermo Fischer Scientific, Waltham, MA, USA) placed in a Termaks B4115 incubator set at 45 °C. After that, 50 μL of a previously prepared stock solution of the cellulase blend Cellic CTec2 (Sigma-Aldrich, Steinheim, Germany) was added at a load of 80 CMCase units/g biomass. The reaction mixtures were then incubated under the same conditions for 72 h. Enzyme-free substrate controls and substrate-free enzyme blanks were run in parallel. After elapsing the reaction time, the resulting slurry was separated using a Heraeus Pico 21 centrifuge (Thermo Fisher Scientific, Osterode, Germany) at 20,200× g for 5 min at 4 °C. The supernatant, hereafter referred to as hydrolysate, was transferred to new Eppendorf tubes and stored frozen until further analysis, and the solid material was discarded. Glucose in the hydrolysates was determined by high-pressure liquid chromatography (HPLC). Glucose concentration was used for calculating the enzymatic digestibility. The enzymatic digestibility was calculated as the mass percentage of the cellulose contained in the AES assay saccharified to glucose.

3.5. *Analytical Methods*

3.5.1. Total Carbohydrates

Determination of the concentration of total carbohydrates was performed based on a modification of the phenol–sulfuric acid protocol [50]. The method is based on reading the absorbance of a 1:3 (v/v) mixture of extract and concentrated sulfuric acid at 315 nm in a quartz cuvette using a UV 3100P spectrophotometer (VWR, Leuven, Belgium). The analysis was performed in triplicates.

3.5.2. Total Phenolics

To determine the concentration of total phenolics, the Folin–Ciocalteau method [51] was used. The calibration standard was gallic acid dissolved in 10% ethanol. The absorbance was measured at 765 nm with a UV 3100P spectrophotometer (VWR). The analysis was performed in triplicates, expressing the results as gallic acid equivalents (GAE). The results were used for calculating the recovery of total phenolics in extractions from SMS samples.

3.5.3. Antioxidant Activity

Antioxidant activity was determined essentially according to Guo et al. [52] using the ferric-ion reducing antioxidant power (FRAP) assay. The FRAP solution was freshly prepared by mixing 40 mM tripyridyltriazine (TPTZ) with 20 mM $FeCl_3$ and acetate buffer (pH 3.6) in a ratio of 1:1:10. Distilled water (30 μL) was pipetted into each well of a Nunclon 96-well plate (Nalge Nunc International, Roskilde, Denmark) before adding 6 μL of sample, standard or blank (distilled water). The FRAP solution was preheated in a water bath at 37 °C for 10 min before adding 270 μL to each well. The microplate was incubated at 37 °C for 10 min before reading absorbance using a FluroStar Optima spectrophotometer (BMG Labtech, Offenburg, Germany) at 595 nm over 30 min. Sample measurements were performed in triplicates. TPTZ, FRAP solution, and gallic acid standards (0.01, 0.1, 0.25, 0.5, and 1.0 mM) were prepared fresh for every run.

3.5.4. Ergosterol Determination

Ergosterol was extracted by mixing 0.5 g (DM) biomass with 15 mL ethanolic solution of KOH in a 30:1 (v/w) SBR. The mixture was held in a water bath at 85 °C for 1 h for saponification and then cooled to room temperature and vacuum filtered. In a separation funnel, ergosterol was recovered from the saponification mixture by two-fold liquid–liquid extraction with cyclohexane (Merck, Darmstadt, Germany). The resulting organic fractions from both extraction steps were pooled together. Quantification was performed by reading the absorbance at 285 nm with a UV 3100P spectrophotometer (VWR) and using analytical-grade ergosterol as the calibration standard.

3.5.5. HPLC Determination of Sugars and Phenolic Acids

Glucose, xylose, and arabinose in the analytical acid hydrolysis, glucose in the enzymatic saccharification, and phenolic acids in the extract were determined by HPLC using an Ultimate 3000 (Dionex Softron, Germering, Germany) system. The separation of sugars in the analytical acid hydrolysates samples was completed with an Aminex HPX-87H column (Bio-Rad Laboratories, Hercules, CA, USA) held at 60 °C, and detection was performed with a Series 200 refractive index (RI) detector (Showa Denko, Tokyo, Japan). The mobile phase was a 5 mM sulfuric acid solution (VWR Chemicals, France), eluted at 0.6 mL/min flow for 40 min per sample. Glucose in the enzymatic saccharification was determined using a Rezex-RPM Monosaccharide Pb^{+2} column (Phenomenex, Torrance, CA, USA) heated to 85 °C and an RI detector (Showa Denko). The eluent was distilled water, filtered through a 0.45 µm membrane filter of regenerated cellulose (GE Healthcare, Buckinghamshire, United Kingdom), and eluted at 0.6 mL/min for 40 min per sample. The phenolic acids were separated on a Kinetex 5u C18 100A column (Phenomenex). A UV detector (Dionex Softron) at either 330 nm or 280 nm was used. HPLC was run at 25 °C with a flow of 1.5 mL/min for 18 min per sample. The mobile phase contained 95% of 0.1% trifluoroacetic acid (FLUKA Chemie AG, Buchs, Switzerland) and 5% acetonitrile (VWR Chemicals, France). HPLC-grade chemicals were used as standards. Before HPLC analysis, all the samples were appropriately diluted and filtered using 0.45 µm Nylon syringe filters (VWR, Radnor, PA, USA).

3.6. Statistical Analysis

The statistical significance was calculated using one-way and two-way ANOVA and Student's *t*-test. Correlations between antioxidant activity and either concentration of total phenolic compounds or individual phenolic acids were determined using Microsoft Excel (version 2102), and the R^2, Pearson's correlation coefficient and p-value were calculated. Statistical analysis tables are provided as Supplementary Material (Tables S2–S9).

Microsoft Excel, Statgraphics Plus 5.0 for Windows (Manugistics Inc., Rockville, MD, USA), and MODDE 11.0 (Umetrics AB, Umeå, Sweden) were used for processing the results.

3.7. Formatting of Chemical Structures

Chemical structures were drawn using the molecular editing software ChemDoodle 10.3.0 (https://www.ichemlabs.com/, accessed on 10 April 2023).

4. Conclusions

This study provided basic knowledge on the suitability of different extraction methods for recovering bioactive compounds from spent mushroom substrate. The potential of ultrasound-assisted extraction and subcritical-water extraction for extracting bioactive compounds from the spent mushroom substrate of two *Pleurotus* spp. strains was shown. UAE allowed a comparable extraction yield under relatively milder conditions than reflux or Soxhlet extraction at higher temperatures and longer times. Among the four tested methods, SWE resulted in the highest extraction yield and efficiency, higher concentrations of phenolic compounds and total carbohydrates, and higher antioxidant activity of the extracts. The results of this study reveal clear directions about how the research ought to proceed for maximizing the recovery of bioactive compounds from SMS. Further experiments based on these findings are currently underway in our group.

A comparison of the extraction parameters for SMS of two fungal strains revealed a higher susceptibility of a wild *P. ostreatus* strain to yield extracts with higher antioxidant activity than that of a commercial *Pleurotus* spp. strain.

Several phenolic acids were identified in the extracts of SMS. The antioxidant activity of SMS extracts was found to be strongly correlated with their concentration of caffeic acid and total phenolic compounds.

Lastly, it was demonstrated that the enzymatic digestibility of cellulose contained in SMS is enhanced after the extraction of bioactive compounds.

Supplementary Materials: The following supporting information can be downloaded at: https://www.mdpi.com/article/10.3390/molecules28135140/s1, Table S1. Chemical composition of the initial substrate used for the cultivation of the BPKO and NO strains. Mass fractions in % (dry weight). Mean values from triplicate analyses are presented for all the components, except extractives. Standard deviation is shown in parenthesis. Table S2. One-factor ANOVA analysis with replication of total phenolic content in Soxhlet extractions. Based on triplicate measurements from nine extractions (E0-E100). Table S3. Two-factor ANOVA analysis with replication of total carbohydrate content in ultrasound-assisted extractions. Based on triplicate measurements from nine extractions at three extraction times and three temperatures (30, 45 and 60 min) (35, 50 and 65 °C). Table S4. One-factor ANOVA analysis with replication, of antioxidant activity (measured by FRAP) in reflux extractions. Based on triplicate measurements from seven extractions at different extraction times (60–240 min). Table S5. Correlation between caffeic acid concentration and antioxidant activity (FRAP). Table S6. Correlation between total phenolic content and antioxidant activity (FRAP). Data points from all extractions were included. Table S7. Correlation of ferulic acid concentration with antioxidant activity (FRAP). Table S8. Correlation of chlorogenic acid concentration with antioxidant activity (FRAP). Data points from all extractions were included. Table S9. Correlation between total carbohydrate content and antioxidant activity (FRAP). Data points from all extractions were included.

Author Contributions: Conceptualization, C.M. and K.O.S.; methodology, C.M. and K.O.S.; software, S.J.K.; validation, S.J.K.; formal analysis, S.J.K. and A.B.F.-Y.; investigation, S.J.K.; resources, C.M. and K.O.S.; data curation, S.J.K.; writing—original draft preparation, S.J.K.; writing—review and editing, C.M. and K.O.S.; visualization, S.J.K.; supervision, C.M. and K.O.S.; project administration, C.M., K.O.S. and A.B.F.-Y.; funding acquisition, C.M. and K.O.S. All authors have read and agreed to the published version of the manuscript.

Funding: This research was funded by RFF Innlandet (grant number 341800), NordForsk (grant number 132066), Inland Norway University of Applied Sciences (INN) through *Grønn Forskning* project (INN project number 342001) and MSc thesis support program, and Sparebankstiftelsen Hedmark (https://sparebankstiftelsenhedmark.no, accessed on 10 April 2023).

Institutional Review Board Statement: Not applicable.

Informed Consent Statement: Not applicable.

Data Availability Statement: Data will be made available on request.

Acknowledgments: Husbonden A.S. (https://husbonden.no/, accessed on 10 May 2023) is acknowledged for supplying the SMS samples. The Department of Biotechnology, Inland Norway University of Applied Sciences, is thanked for its institutional support. In memory of the MSc program coordinator Svein Birger Wærvågen who passed away during the implementation of this work.

Conflicts of Interest: The authors declare no conflict of interest. The funding agencies had no role in the design of the study; in the interpretation of data; in the writing of the manuscript, or in the decision to publish the results.

References

1. El-Ramady, H.; Abdalla, N.; Badgar, K.; Llanaj, X.; Törős, G.; Hajdú, P.; Eid, Y.; Prokisch, J. Edible Mushrooms for Sustainable and Healthy Human Food: Nutritional and Medicinal Attributes. *Sustainability* **2022**, *14*, 4941. [CrossRef]
2. Royse, D.J.; Baars, J.; Tan, Q. Current Overview of Mushroom Production in the World. In *Edible and Medicinal Mushrooms*; John Wiley & Sons, Ltd: Hoboken, NJ, USA, 2017; pp. 5–13. ISBN 978-1-119-14944-6.
3. Kumla, J.; Suwannarach, N.; Sujarit, K.; Penkhrue, W.; Kakumyan, P.; Jatuwong, K.; Vadthanarat, S.; Lumyong, S. Cultivation of Mushrooms and Their Lignocellulolytic Enzyme Production through the Utilization of Agro-Industrial Waste. *Molecules* **2020**, *25*, 2811. [CrossRef] [PubMed]
4. Kaushal, L.A.; Prashar, A. Agricultural Crop Residue Burning and Its Environmental Impacts and Potential Causes—Case of Northwest India. *J. Environ. Plan. Manag.* **2021**, *64*, 464–484. [CrossRef]
5. Martín, C. Pretreatment of Crop Residues for Bioconversion. *Agronomy* **2021**, *11*, 924. [CrossRef]
6. Zisopoulos, F.K.; Ramírez, H.A.B.; van der Goot, A.J.; Boom, R.M. A Resource Efficiency Assessment of the Industrial Mushroom Production Chain: The Influence of Data Variability. *J. Clean. Prod.* **2016**, *126*, 394–408. [CrossRef]
7. Atallah, E.; Zeaiter, J.; Ahmad, M.N.; Leahy, J.J.; Kwapinski, W. Hydrothermal Carbonization of Spent Mushroom Compost Waste Compared against Torrefaction and Pyrolysis. *Fuel Process. Technol.* **2021**, *216*, 106795. [CrossRef]

8. Beyer, D.M. Impact of the Mushroom Industry on the Environment. Pennsylvania State University 2011. Available online: https://extension.psu.edu/impact-of-the-mushroom-industry-on-the-environment (accessed on 30 March 2023).
9. Martin, C.; Zervakis, G.I.; Xiong, S.; Koutrotsios, G.; Strætkvern, K.O. Spent Substrate from Mushroom Cultivation: Exploitation Potential towards Various Applications and Value-Added Products. *Bioengineered* **2023**, in press.
10. Kumar, K.; Mehra, R.; Guiné, R.P.F.; Lima, M.J.; Kumar, N.; Kaushik, R.; Ahmed, N.; Yadav, A.N.; Kumar, H. Edible Mushrooms: A Comprehensive Review on Bioactive Compounds with Health Benefits and Processing Aspects. *Foods* **2021**, *10*, 2996. [CrossRef]
11. Sanz, V.; Torres, M.D.; Lopez Vilarino, J.M.; Dominguez, H. Green Extraction of Phenolic Compounds from *Perle Hallertau* and *Nuggets* Hop Pellets. *Food Biosci.* **2022**, *50*, 102044. [CrossRef]
12. Xiong, S.; Martín, C.; Eilertsen, L.; Wei, M.; Myronycheva, O.; Larsson, S.H.; Lestander, T.A.; Atterhem, L.; Jönsson, L.J. Energy-Efficient Substrate Pasteurisation for Combined Production of Shiitake Mushroom (*Lentinula edodes*) and Bioethanol. *Bioresour. Technol.* **2019**, *274*, 65–72. [CrossRef]
13. Chen, F.; Xiong, S.; Gandla, M.L.; Stagge, S.; Martín, C. Spent Mushroom Substrates for Ethanol Production—Effect of Chemical and Structural Factors on Enzymatic Saccharification and Ethanolic Fermentation of *Lentinula edodes*-Pretreated Hardwood. *Bioresour. Technol.* **2022**, *347*, 126381. [CrossRef]
14. Li, T.; Cui, L.; Song, X.; Cui, X.; Wei, Y.; Tang, L.; Mu, Y.; Xu, Z. Wood Decay Fungi: An Analysis of Worldwide Research. *J. Soils Sediments* **2022**, *22*, 1688–1702. [CrossRef]
15. Bak, W.C.; Park, J.H.; Park, Y.A.; Ka, K.H. Determination of Glucan Contents in the Fruiting Bodies and Mycelia of *Lentinula edodes* Cultivars. *Mycobiology* **2014**, *42*, 301–304. [CrossRef]
16. He, P.; Li, F.; Huang, L.; Xue, D.; Liu, W.; Xu, C. Chemical Characterization and Antioxidant Activity of Polysaccharide Extract from Spent Mushroom Substrate of *Pleurotus eryngii*. *J. Taiwan. Inst. Chem. Eng.* **2016**, *69*, 48–53. [CrossRef]
17. Zhang, Q.-W.; Lin, L.-G.; Ye, W.-C. Techniques for Extraction and Isolation of Natural Products: A Comprehensive Review. *Chin. Med.* **2018**, *13*, 20. [CrossRef]
18. Martínez-Flores, H.E.; Contreras-Chávez, R.; Garnica-Romo, M.G. Effect of Extraction Processes on Bioactive Compounds from *Pleurotus ostreatus* and *Pleurotus djamor*: Their Applications in the Synthesis of Silver Nanoparticles. *J. Inorg. Organomet. Polym.* **2021**, *31*, 1406–1418. [CrossRef]
19. Espinosa, E.; Rincón, E.; Morcillo-Martín, R.; Rabasco-Vílchez, L.; Rodríguez, A. Orange Peel Waste Biorefinery in Multi-Component Cascade Approach: Polyphenolic Compounds and Nanocellulose for Food Packaging. *Ind. Crops Prod.* **2022**, *187*, 115413. [CrossRef]
20. Balto, A.S.; Lapis, T.J.; Silver, R.K.; Ferreira, A.J.; Beaudry, C.M.; Lim, J.; Penner, M.H. On the Use of Differential Solubility in Aqueous Ethanol Solutions to Narrow the DP Range of Food-Grade Starch Hydrolysis Products. *Food Chem.* **2016**, *197*, 872–880. [CrossRef]
21. Bouchard, A.; Hofland, G.W.; Witkamp, G.-J. Properties of Sugar, Polyol, and Polysaccharide Water−Ethanol Solutions. *J. Chem. Eng. Data* **2007**, *52*, 1838–1842. [CrossRef]
22. Rodríguez García, S.L.; Raghavan, V. Green Extraction Techniques from Fruit and Vegetable Waste to Obtain Bioactive Compounds—A Review. *Crit. Rev. Food Sci. Nutr.* **2022**, *62*, 6446–6466. [CrossRef]
23. Kumar, K.; Srivastav, S.; Sharanagat, V.S. Ultrasound Assisted Extraction (UAE) of Bioactive Compounds from Fruit and Vegetable Processing by-Products: A Review. *Ultrason. Sonochem.* **2021**, *70*, 105325. [CrossRef] [PubMed]
24. Esteban-Lustres, R.; Sanz, V.; Domínguez, H.; Torres, M.D. Ultrasound-Assisted Extraction of High-Value Fractions from Fruit Industrial Processing Waste. *Foods* **2022**, *11*, 2089. [CrossRef] [PubMed]
25. Martínez-Patiño, J.C.; Gómez-Cruz, I.; Romero, I.; Gullón, B.; Ruiz, E.; Brnčić, M.; Castro, E. Ultrasound-Assisted Extraction as a First Step in a Biorefinery Strategy for Valorisation of Extracted Olive Pomace. *Energies* **2019**, *12*, 2679. [CrossRef]
26. Cabeza, A.; Piqueras, C.M.; Sobrón, F.; García-Serna, J. Modeling of Biomass Fractionation in a Lab-Scale Biorefinery: Solubilization of Hemicellulose and Cellulose from Holm Oak Wood Using Subcritical Water. *Bioresour. Technol.* **2016**, *200*, 90–102. [CrossRef]
27. Rodríguez-Seoane, P.; Díaz-Reinoso, B.; Domínguez, H. Pressurized Solvent Extraction of *Paulownia* Bark Phenolics. *Molecules* **2022**, *27*, 254. [CrossRef]
28. Daneshfar, A.; Ghaziaskar, H.S.; Homayoun, N. Solubility of Gallic Acid in Methanol, Ethanol, Water, and Ethyl Acetate. *J. Chem. Eng. Data* **2008**, *53*, 776–778. [CrossRef]
29. Vilas-Boas, S.M.; Vieira, V.; Brandão, P.; Alves, R.S.; Coutinho, J.A.P.; Pinho, S.P.; Ferreira, O. Solvent and Temperature Effects on the Solubility of Syringic, Vanillic or Veratric Acids: Experimental, Modeling and Solid Phase Studies. *J. Mol. Liq.* **2019**, *289*, 111089. [CrossRef]
30. Qiao, L.; Ye, X.; Sun, Y.; Ying, J.; Shen, Y.; Chen, J. Sonochemical Effects on Free Phenolic Acids under Ultrasound Treatment in a Model System. *Ultrason. Sonochem.* **2013**, *20*, 1017–1025. [CrossRef]
31. Valanciene, E.; Jonuskiene, I.; Syrpas, M.; Augustiniene, E.; Matulis, P.; Simonavicius, A.; Malys, N. Advances and Prospects of Phenolic Acids Production, Biorefinery and Analysis. *Biomolecules* **2020**, *10*, 874. [CrossRef]
32. Koutrotsios, G.; Kalogeropoulos, N.; Stathopoulos, P.; Kaliora, A.C.; Zervakis, G.I. Bioactive Compounds and Antioxidant Activity Exhibit High Intraspecific Variability in *Pleurotus ostreatus* Mushrooms and Correlate Well with Cultivation Performance Parameters. *World J. Microbiol. Biotechnol.* **2017**, *33*, 98. [CrossRef]
33. Ishihara, A.; Goto, N.; Kikkawa, M.; Ube, N.; Ushijima, S.; Ueno, M.; Ueno, K.; Osaki-Oka, K. Identification of Antifungal Compounds in the Spent Mushroom Substrate of *Lentinula edodes*. *J. Pestic. Sci.* **2018**, *43*, 108–113. [CrossRef]

34. Yim, H.S.; Chye, F.Y.; Tan, C.T.; Ng, Y.C.; Ho, C.W. Antioxidant Activities and Total Phenolic Content of Aqueous Extract of *Pleurotus ostreatus* (Cultivated Oyster Mushroom). *Malays. J. Nutr.* **2010**, *16*, 281–291.
35. Showkat, M.M.; Falck-Ytter, A.B.; Strætkvern, K.O. Phenolic Acids in Jerusalem Artichoke (*Helianthus tuberosus* L.): Plant Organ Dependent Antioxidant Activity and Optimized Extraction from Leaves. *Molecules* **2019**, *24*, 3296. [CrossRef]
36. Genaro-Mattos, T.C.; Maurício, Â.Q.; Rettori, D.; Alonso, A.; Hermes-Lima, M. Antioxidant Activity of Caffeic Acid against Iron-Induced Free Radical Generation—A Chemical Approach. *PLoS ONE* **2015**, *10*, e0129963. [CrossRef]
37. Nasir, A.M.; Ruslan, N.R.N.; Zakaria, Z.; Hassan, S.A.M.; Ishak, N.; Rohaizad, N.M.; Gunny, A.A.N. β-Glucan Extraction from Mycelium in Spent Mushroom Substrate of *Pleurotus ostreatus* and *Schizophyllum commune*. *IOP Conf. Ser. Earth Environ. Sci.* **2021**, *765*, 012012. [CrossRef]
38. Synytsya, A.; Míčková, K.; Synytsya, A.; Jablonský, I.; Spěváček, J.; Erban, V.; Kovaříková, E.; Čopíková, J. Glucans from Fruit Bodies of Cultivated Mushrooms *Pleurotus ostreatus* and *Pleurotus eryngii*: Structure and Potential Prebiotic Activity. *Carbohydr. Polym.* **2009**, *76*, 548–556. [CrossRef]
39. Gandla, M.L.; Martín, C.; Jönsson, L.J. Analytical Enzymatic Saccharification of Lignocellulosic Biomass for Conversion to Biofuels and Bio-Based Chemicals. *Energies* **2018**, *11*, 2936. [CrossRef]
40. Chen, F.; Martín, C.; Lestander, T.A.; Grimm, A.; Xiong, S. Shiitake Cultivation as Biological Preprocessing of Lignocellulosic Feedstocks—Substrate Changes in Crystallinity, Syringyl/Guaiacyl Lignin and Degradation-Derived by-Products. *Bioresour. Technol.* **2022**, *344*, 126256. [CrossRef]
41. Zadrazil, F. Changes in In Vitro Digestibility of Wheat Straw During Fungal Growth and after Harvest of Oyster Mushrooms (*Pleurotus* spp.) on Laboratory and Industrial Scale. *J. Appl. Anim. Res.* **1997**, *11*, 37–48. [CrossRef]
42. Chen, F.; Xiong, S.; Sundelin, J.; Martín, C.; Hultberg, M. Potential for Combined Production of Food and Biofuel: Cultivation of *Pleurotus pulmonarius* on Soft- and Hardwood Sawdusts. *J. Clean. Prod.* **2020**, *266*, 122011. [CrossRef]
43. Jönsson, L.J.; Martín, C. Pretreatment of Lignocellulose: Formation of Inhibitory by-Products and Strategies for Minimizing Their Effects. *Bioresour. Technol.* **2016**, *199*, 103–112. [CrossRef]
44. Wu, S.; Lan, Y.; Wu, Z.; Peng, Y.; Chen, S.; Huang, Z.; Xu, L.; Gelbič, I.; Guan, X.; Zhang, L.; et al. Pretreatment of Spent Mushroom Substrate for Enhancing the Conversion of Fermentable Sugar. *Bioresour. Technol.* **2013**, *148*, 596–600. [CrossRef] [PubMed]
45. Devi, R.; Kapoor, S.; Thakur, R.; Sharma, E.; Tiwari, R.K.; Joshi, S.J. Lignocellulolytic Enzymes and Bioethanol Production from Spent Biomass of Edible Mushrooms Using *Saccharomyces cerevisiae* and *Pachysolen tannophilus*. *Biomass Convers. Biorefin.* **2022**. [CrossRef]
46. Martin, C.; Lopez, Y.; Plasencia, Y.; Hernandez, E. Characterisation of Agricultural and Agro-Industrial Residues as Raw Materials for Ethanol Production. *Chem. Biochem. Eng. Q.* **2006**, *20*, 443–447.
47. Sluiter, A.; Hames, B.; Ruiz, R.; Scarlata, C.; Sluiter, J.; Templeton, D. *Determination of Ash in Biomass. Laboratory Analytical Procedure (LAP)*; Technical Report; National Renewable Energy Laboratory: Golden, CO, USA, 2008.
48. Sluiter, A.; Ruiz, R.; Scarlata, C.; Sluiter, J.; Templeton, D. *Determination of Extractives in Biomass. Laboratory Analytical Procedure (LAP)*; Technical Report; National Renewable Energy Laboratory: Golden, CO, USA, 2005.
49. Sluiter, A. *Determination of Structural Carbohydrates and Lignin in Biomass. Laboratory Analytical Procedure (LAP)*; Version 7 August 2011; Technical Report; National Renewable Energy Laboratory: Golden, CO, USA, 2008; p. 18.
50. Albalasmeh, A.A.; Berhe, A.A.; Ghezzehei, T.A. A New Method for Rapid Determination of Carbohydrate and Total Carbon Concentrations Using UV Spectrophotometry. *Carbohydr. Polym.* **2013**, *97*, 253–261. [CrossRef]
51. Sánchez-Rangel, J.C.; Benavides, J.; Heredia, J.B.; Cisneros-Zevallos, L.; Jacobo-Velázquez, D.A. The Folin–Ciocalteu Assay Revisited: Improvement of Its Specificity for Total Phenolic Content Determination. *Anal. Methods* **2013**, *5*, 5990–5999. [CrossRef]
52. Guo, C.; Yang, J.; Wei, J.; Li, Y.; Xu, J.; Jiang, Y. Antioxidant Activities of Peel, Pulp and Seed Fractions of Common Fruits as Determined by FRAP Assay. *Nutr. Res.* **2003**, *23*, 1719–1726. [CrossRef]

molecules

MDPI

Article

Utilization Perspectives of Lignin Biochar from Industrial Biomass Residue

Iliyana Naydenova [1], Temenuzhka Radoykova [2], Tsvetelina Petrova [1,*], Ognyan Sandov [1] and Ivo Valchev [2]

1 Department of Energy and Mechanical Engineering, Technical College-Sofia, Technical University of Sofia, 1000 Sofia, Bulgaria
2 Department of Pulp, Paper and Printing Arts, Faculty of Chemical Technologies, University of Chemical Technology and Metallurgy, 1000 Sofia, Bulgaria
* Correspondence: tzvetelina.petrova@tu-sofia.bg

Abstract: The present study aimed at utilizing technically hydrolyzed lignin (THL), industrial biomass residue, derived in high-temperature diluted sulfuric acid hydrolysis of softwood and hardwood chips to sugars. The THL was carbonized in a horizontal tube furnace at atmospheric pressure, in inert atmosphere and at three different temperatures (500, 600, and 700 °C). Biochar chemical composition was investigated along with its HHV, thermal stability (thermogravimetric analysis), and textural properties. Surface area and pore volume were measured with nitrogen physisorption analysis often named upon Brunauer–Emmett–Teller (BET). Increasing the carbonization temperature reduced volatile organic compounds (40 ÷ 96 wt. %), increased fixed carbon (2.11 to 3.68 times the wt. % of fixed carbon in THL), ash, and C-content. Moreover, H and O were reduced, while N- and S-content were below the detection limit. This suggested biochar application as solid biofuel. The biochar Fourier-transform infrared (FTIR) spectra revealed that the functional groups were gradually lost, thus forming materials having merely polycyclic aromatic structures and high condensation rate. The biochar obtained at 600 and 700 °C proved having properties typical for microporous adsorbents, suitable for selective adsorption purposes. Based on the latest observations, another biochar application was proposed—as a catalyst.

Keywords: technically hydrolyzed lignin; carbonization; biochar characterization

check for updates

Citation: Naydenova, I.; Radoykova, T.; Petrova, T.; Sandov, O.; Valchev, I. Utilization Perspectives of Lignin Biochar from Industrial Biomass Residue. *Molecules* **2023**, *28*, 4842. https://doi.org/10.3390/molecules28124842

Academic Editors: Alejandro Rodriguez Pascual, Eduardo Espinosa Víctor and Carlos Martín

Received: 20 April 2023
Revised: 8 June 2023
Accepted: 15 June 2023
Published: 18 June 2023

1. Introduction

Biomass residues are generated annually in huge amounts as a result of different human activities. The plants' structure contains three main components that are in different proportions—cellulose, hemicellulose, and lignin. Generally, the cellulose content is predominant, followed by lignin [1]. Actually, lignin is the main by-product obtained in a plentiful amount from numerous industrial processes [2], such as the food and paper industries, lignocellulose-based biorefinery, etc. [3–5]. For example, only the pulping industry generates around 40 million tons of lignin annually [2]. Due to its calorific value, lignin is often used as solid biofuel in industrial boilers. However, lignin has the potential as feedstock that substitutes the petroleum-based products utilized to manufacture industrial coatings, gels, emulsifiers, etc. [5,6]. In fact, lignocellulose biomass is considered a major source of value-added products and a bioenergy carrier worldwide [7].

In recent decades, the utilization of lignocellulosic matter as a source of renewable fuel, chemicals, or porous biochar derivatives is gaining considerable attention due to its neutral carbon cycle [8]. Comprehensive utilization of lignocellulosic biomass is possible after solving the issue related to its decomposition. The bio-refinery might be more effective if, along with ethanol production from cellulose and hemicellulose, the factory succeeds to obtain value-added products also from the rested hemicellulose and lignin [5]. On the other hand, lignin has been widely studied and typically processed for producing bio-based fuels and chemicals [9]. The significant interest in ethanol production from vegetal raw materials places the question of adequate utilization of the resulting biomass residue. The large internal surface of the lignocellulosic material and the availability of different functional groups could suggest the possible usage of such materials as adsorbents of metal ions, e.g., for water purification purposes. Renewable agricultural residues are produced in bulk as waste, and their storage and management create an environmental problem. The application of agricultural wastes as biosorbents is possible directly or after activation [10–12]. Previous investigations [13] proved that some of the biosorbents' advantages are biodegradability and good adsorption properties due to their morphology and surface functional groups distribution.

Biochar is often generated from lignocellulosic biomass residue by applying a thermal conversion technique, such as gasification, pyrolysis, torrefaction, carbonization, or hydrothermal liquefaction [14,15].

Pyrolysis is a process of thermal degradation at a limited amount of oxygen and the initial lignocellulosic material can be transformed into solid, liquid, and gaseous products [16]. The pyrolysis-based technologies show great promise for converting lignin and other wood components into biochemicals, biomaterials, and biofuels [17]. The process can be divided into fast and slow pyrolysis, which give different yields of the desirable products [18]. Arni [18] examines the yield of gaseous products during both fast and slow pyrolysis of lignocellulosic feedstock (sugarcane bagasse). The findings are that the low temperature is a better condition for producing methane other than hydrogen for both processes, while high temperature aids in obtaining hydrogen.

Carbonization is a slow pyrolysis technology in which the processed biomass is heated and turned into biochar after thermal decomposition under inert conditions [19].

The valorization of lignin can be performed via its chemical modification to obtain bioactive derivatives, i.e., sulfated lignin, which has anticoagulant and antiplatelet activity and can be used in the treatment of thrombotic disorders [20]. The authors optimize experimentally and numerically the process of sulfation of ethanol lignin birch wood with a mixture of sulfamic acid and urea in a 1,4-dioxane medium. The aim is to characterize the structure and thermochemical properties of the sulfated ethanol lignin. The findings are that obtained sulfated birch ethanol lignin has properties for use in the production of new sorbents, biocomposites, and nanomaterials, as well as in the development of new anticoagulant and antiviral medicines. Kazachenko et al. [21] examined the effect of a type of solid acid catalyst on the sulfation of wheat straw soda lignin with sulfamic acid in a 1,4-dioxane medium, to elucidate the possibility of recycling and to examine the composition and structure of the obtained products. The conclusions are that the solid catalysts used in the sulfation process cause hydrolysis reactions and reduce the molecular weight and polydispersity index.

Hydrogenation is an efficient and reliable technology for lignin valorization, aiming to diminish the difficulties, related to the extraction of its functional phenolic compound [22]. Abdullah et al. [23] present lignin hydrogenation as a depolymerization method, which uses hydrogen as a reductant under mild conditions. The authors aim at obtaining aromatic products with low oxygen content and increased products' stability. The most critical decision for such processes is the selection of suitable catalysts [24,25]. The phenolic compounds can be increased using the catalytic liquefaction reactions, and such an example is well described in [26], testing various bimetal selective catalysts and alcoholic solvents.

Currently, a great research effort is imposed to establish barely studied and effective materials for the disposal of harmful/pathogenic elements or organisms in the air and water environment. In this respect, activated carbon has emerged as promising material. Activated carbon has been used for such purposes as a stand-alone material or as a carrier of active ingredients. However, the requirements and regulations for its production are continuously increasing, especially in terms of its porous texture. It is necessary to create the structure with pores of a specific size, which would increase the material's selectivity in relation to certain components that need to be removed. Various types of feedstocks are used to produce activated carbon. When waste matter is utilized, it reduces the feedstock's cost and solves related environmental problems. In view of this, technically hydrolyzed lignin (THL) is of particular interest because it is typically generated in large quantities as a hardly utilized by-product of certain industrial processes. According to Shiraki et al. [27], the concentrated sulfuric acid can completely swell and hydrolyze cellulose. The authors consider the concentrated sulfuric acid hydrolysis as the most effective process capable of recovering the maximum yield of monomeric sugars from woody biomass. Further, they discuss the difficulties of utilizing the solid by-product (lignin) because of the self-condensation between the lignin molecules under acidic conditions and propose a method for lignin valorization using a unique additive, t-butyl alcohol. The results show unchanged sugar yields along with a lignin yield higher than 40%. Thermoplastic lignin with good solubility is successfully recovered in acetone, and the method is foreseen as a new candidate for implementation in sugar platform biorefineries.

The present work aimed at investigating a utilization path for industrial biomass residue, namely technically hydrolyzed lignin (THL). In the area of Razlog, Republic of Bulgaria, a 140-acre landfill of hydrolysis lignin residue is located. The THL had been deposed outdoor for many years, and its total amount is evaluated to be about 350,000–400,000 tons. This THL was formed as by-product of a high-temperature diluted sulfuric acid hydrolysis of softwood and hardwood chips to sugars, which were further subjected to yeast fodder production. The accumulated huge amount of THL releases different gaseous air pollutants including greenhouse emissions. During the summer period, when the outdoor temperature significantly increases, this matter is also self-igniting. Therefore, the present case study aims to propose a THL utilization method. This type of biomass residue is of potential harm to the surrounding environment and population. For that purpose, the THL was carbonized in a horizontal tube furnace (HTF) at a temperature range between 500 and 700 °C. The experimental set up is described elsewhere [19,28]. The obtained biochar was chemically characterized through a set of chemical and physical analyses [13,29–31]. The possible biochar applications were discussed in line with the present European effort for circular economy and climate change preservation (e.g., Regulation (EU) 2018/1999 and Directive (EU) 2018/2001). The suggested methodology was based on a well-established technique for THL thermal conversion and methods for the product's characterization. It provides the basis for detailed investigations nationwide on both optimized biomass conversion and products utilization, thus reducing the negative footprint of the local biorefineries.

2. Results and Discussion

2.1. Effect of Carbonization Temperature on Biochar Yield and Its Chemical Composition

Several analytical methods were used (proximate, ultimate, ash, calorimetric, and lignocellulosic analyses) to characterize both THL and biochar. The obtained results were summarized in Table 1. Increasing the carbonization temperature led to significantly reduced content of volatile organic compounds (from 40 to 96 wt. %) and increased fixed carbon (from 2.11 to 3.68 times the FC wt. % in THL) and ash content. Farrokh et al. [32] report similar effects. The authors examine lignin biochar produced at three different temperatures (300, 500, and 650 °C). In addition, H- and O-content was considerably reduced along with the N- and S-content, which for some samples was measured below the detection limit. The results from the ultimate analysis are in accordance with [32,33]. However, the higher heating value (HHV) and the biochar yield slightly decreased with increasing the carbonization temperature, due to the structural transformations in the carbonization process, relevant to the chosen experimental conditions. The effect was observed also in [34], concerning biochar samples, obtained at temperatures above 500 °C. The results were in line with the investigations of [35]. The authors proved that increasing the carbonization temperature and/or residence time often leads to lower biochar mass yield and HHV.

Table 1. Chemical characteristics of the samples.

Parameter Studied	THL [2]	Biochar 500 °C	Biochar 600 °C	Biochar 700 °C
Proximate analysis, wt. %				
Volatiles, db	65.27	38.36	6.42	2.16
Fixed carbon, db [1]	23.34	49.32	81.41	85.99
Moisture	7.78	4.98	3.74	3.48
Ash, db	3.61	7.35	8.00	8.37
Ultimate analysis, wt. %, db				
C	55.54	76.70	83.72	85.39
H	7.10	3.5	2.65	1.64
N	0.26	-	-	-
S	0.74	0.05	0.05	-
O [1]	24.97	7.42	1.84	1.12
HHV, db, MJ/kg	23.27	31.36	29.17	29.20
Lignocellulosic analysis, wt. %, db				
Cellulose	25.5	-	-	-
Lignin	72.6	-	-	-
Mineral substances	2.8	-	-	-
Cellulose	25.5	-	-	-
Biochar mass yield, wt. %	-	42.95	40.10	37.99

[1] By difference; [2] Reported in [36].

The lignocellulosic analysis confirmed that during the diluted sulfuric acid hydrolysis of the initial biomass, the hemicellulose was hydrolyzed and the THL became rich in lignin and resistant to hydrolysis cellulose fraction.

The ICP-OES spectroscopy allowed determining the ash composition of THL and its carbonized products. Table 2 summarizes the mean values from three independent repetitions of each experiment. Except for Pb, Si, and Na, the rest of the elements were concentrated in the biochar, showing significant temperature dependence. According to [34], increasing the carbonization temperature might lead to the volatilization of some metals. Expectedly, increasing the carbonization temperature led to a higher concentration of most of the measured elements in the biochar, generated at 700 °C.

Table 2. Ash composition of THL and biochar.

Chemical Elements, g/kg	THL	Biochar 500 °C	Biochar 700 °C
Al	1.329	4.045	4.860
Ba	0.069	0.135	0.132
Ca	1.790	3.937	4.267
Cu	0.028	0.047	0.049
Fe	0.362	0.729	0.807
Pb	0.002	<0.01	<0.01
Mg	0.142	0.324	0.354
Mn	0.010	0.020	0.025
K	0.378	1.330	1.458
Na	0.093	0.128	0.058
Sr	0.016	0.047	0.049
Zn	0.006	0.014	<0.01
Si	0.138	0.041	0.074
C	<0.01	<0.01	<0.01
Ti	<0.01	<0.01	<0.01
S	0.587	1.776	1.680

2.2. Thermal Analysis

Thermal stability analyses, such as Thermogravimetric (TG), Differential Thermal Analysis (DTA), and Differential Scanning Calorimetry (DSC) are typically used to estimate the processes of thermal degradation of biomass and its derivatives.

Herein, simultaneous TG-DTA/DSC study was carried out, and the thermal conversion of THL and biochar (derived at 500, 600, and 700 °C) was investigated along with the effects of weight loss and thermal stability. The graphic interpretation of the TG-DTG-DSC temperature dependence is illustrated in Figure 1. The following three global stages were identified:

- Stage 1—Water vaporization was determined in the temperature range between room temperature (RT) and 246 °C. Typical for this stage, an endothermic peak was observed, which normally corresponds to the elimination of humidity, followed by broad exothermal peaks.
- Stage 2—Devolatilization and dehydrogenation (of some hydroxides in the mineral composition) took place in the following temperature range: 175 ÷ 900 °C.
- Stage 3—Fixed carbon combustion was observed at temperatures between 520 and 950 °C. The TGA curves of biochar showed that this stage overlapped with stage 2.

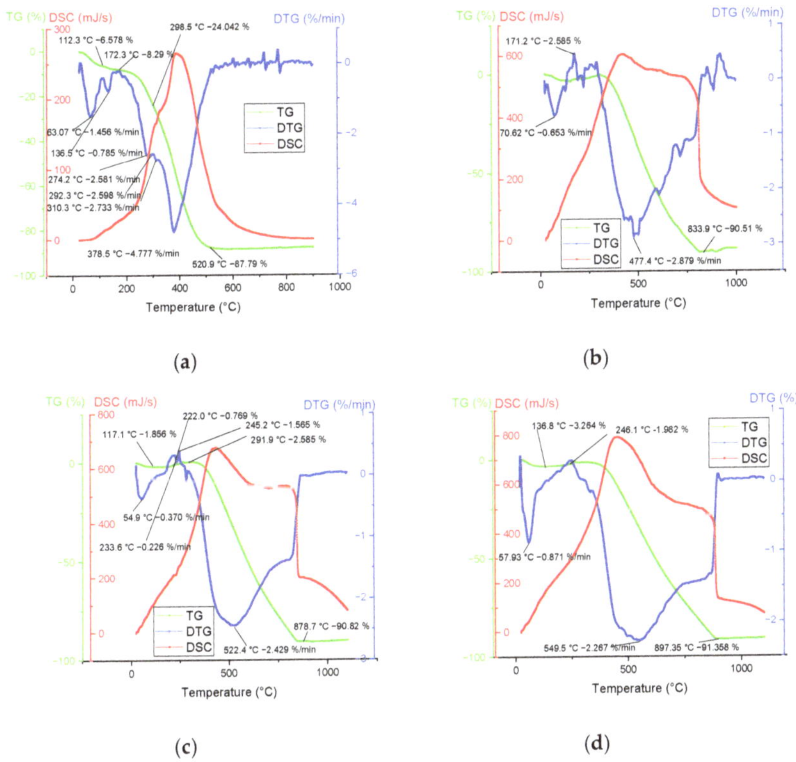

Figure 1. Thermal analysis of THL (**a**) and biochar, obtained at 500 °C (**b**), 600 °C (**c**), and 700 °C (**d**) in dynamic heating mode (20 ÷ 1000 °C), and constant heating (10 K/min) and air flow (100 mL/min) rates.

The lignin decomposes slower and over a broader temperature range [37] in comparison to cellulose and hemicellulose [38]. The effect is attributed to the specific thermal stability of some oxygen-containing functional groups with scission occurring at lower temperatures [39].

As expected, the present thermal analyses showed the occurrence of mostly exothermic reactions. The DSC peaks coincided well with the appearance of the maximum mass loss rates (Table 3). The peaks at higher temperatures were associated with the thermal decomposition of both lignin and difficult to hydrolyze polysaccharides [38].

Table 3. Thermal characteristics of the studied materials.

Stage	Mass Loss	Temperature at Max Loss Rate	Max Mass Loss Rate	Total Mass Loss	Heat Effect
No.	wt. %	°C	%/min	wt. %	MJ/kg
THL					
1	6.58	63.7	1.456		
2	1.71	136.5	0.785	87.79	26.48
3	79.5	378.5	4.777		
Biochar 500 °C					
1	2.59	70.62	0.653	90.51	73.28
2	87.9	477.4	2.879		
Biochar 600 °C					
1	1.6	54.9	0.370	90.82	74.01
2	89.22	522.4	2.429		
Biochar 700 °C					
1	3.26	57.93	0.871	91.36	87.59
2	88.1	549.5	2.267		

The complex decomposition of THL (see, e.g., its DTG curve in Figure 1 and Table 3) resulted in at least five overlapped steps with maximum mass loss rate at 378.5 °C and a long tail beyond 500 °C. Instead of one simple peak at 300 °C, the THL showed a complex destruction process between 270 and 310 °C, which was related to cellulose degradation [40].

The maximum mass loss rate of THL (4.78 %/min) was observed at 378.5 °C. This behavior was related to the low cellulose content in the examined material (Table 1). The THL thermal decomposition finishes at about 530 °C. The biochar degradation showed that increasing the carbonization temperature broadened the interval to the total decomposition from 835 °C (biochar, obtained at 500 °C) to 895 °C (biochar—at 700 °C) as well as increased the peak temperature in the same order.

2.3. Fourier-Transform Infrared (FTIR) Spectroscopy

The effect of temperature on the functional groups of THL and its carbonized products was studied also with FTIR spectroscopy (Figure 2). The FTIR spectrum of THL was influenced by the higher content of lignin and polysaccharides. The wide band at 3400 cm^{-1}, belonging to the zone 3500–3100 cm^{-1}, was due to the valence vibrations of alcoholic (phenolic) and hydroxyl groups included in hydrogen bonds [41]. The intensive bands at 2931 and 2800 cm^{-1} referred to different types of valence vibrations of CH bonds in the methyl and methylene groups. The band at 1710 cm^{-1}, falling in the range 1600–1760 cm^{-1}, was characteristic of the vibrational oscillation of the group C=O in alkyl-aromatic ketones. In particular, a ketocarbonyl group is typically supported by β-carbon atom of a propane chain [42]. The bands at 1600 and 1509 cm^{-1} were associated with vibrations of aromatic nuclei [43]. The bands at 1459 cm^{-1} and 1382 cm^{-1} denote deformation vibrations of CH in the methyl and methylene groups, while the band at 1155 cm^{-1} referred to C-O-C asymmetric vibrational oscillation in ether groups. Further investigation of the THL–FTIR spectrum attributed the 862 cm^{-1} band to deformation vibrations of the CH bonds in a three-substituted aromatic nucleus, and the one at 771 cm^{-1} was due to deformation vibrations of the CH bonds in a mono-substituted aromatic nucleus [43].

The FTIR spectra of biochar proved that their functional groups were gradually lost with increasing the carbonization temperature, where the role of the polycyclic aromatic structures was significant [34]. This conclusion is in line with the observations, reported by [44]. The authors discussed that such results are helping to explore the applicability of different types of biochar for the immobilization of specific environmental pollutants, carbon sequestration, etc.

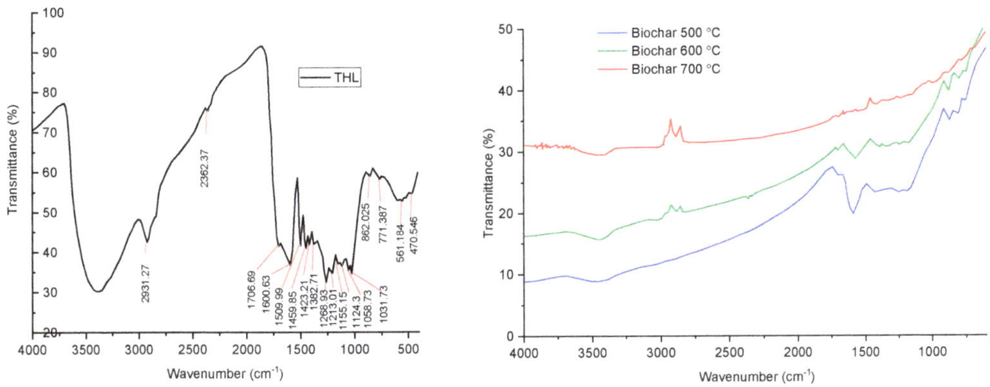

(a) (b)

Figure 2. FTIR spectrum of THL (**a**) and the biochar, derived at 500 °C, 600 °C, and 700 °C (**b**).

The band at 3443 cm^{-1} referred to O–H stretching of H-bonded hydroxyl groups, while the one at 2870 cm^{-1} was ascribed to symmetric C–H stretching of aliphatic hydrocarbon (e.g., from the propane chain of the monomer units in lignin). The band at 1695 cm^{-1} referred to C=O stretching vibrations of alkyl-aromatic ketones, whereas the bands at 1600 cm^{-1} and 1430 cm^{-1} were connected with C=C stretching vibrations of aromatic components. The results correspond well with Li et al. [45]. A similar observation is reported by Wang et al. [46] reporting the split of the phenolic groups at temperatures above 500 °C. The bands at 870 cm^{-1}, 811 cm^{-1}, and 757 cm^{-1} were due to C–H bending vibrations from three-substituted, di-substituted, and mono-substituted aromatic nuclei, respectively [43]. The band at 1191 cm^{-1} was attributed to C–O–C symmetric stretching vibrations in ester groups, while the bands, detected between 870 cm^{-1} and 675 cm^{-1} were associated with C–H bending vibrations [45].

2.4. Surface Area and Pore Volume

The evaluation of the specific surface area of the THL and biochar was carried out by adsorption of nitrogen at −196 °C. Nitrogen adsorption–desorption isotherms were used to calculate the specific surface area using the BET equation [13]. The results are summarized in Table 4 and Figure 3.

Table 4. Basic adsorption-textural parameters of THL and biochar.

Sample	S_{BET}, m^2/g	V_t, cm^3/g	V_{MI}, cm^3/g	D_{AV}, nm	S_{MI}, m^2/g	S_{EXT}, m^2/g
THL	4	0.03	-	28	-	-
Biochar 600 °C	378	0.19	0.10	2.0	267	111
Biochar 700 °C	430	0.17	0.13	1.6	383	47

The isotherm of THL is of type II, according to IUPAC classification, evidencing the material is nonporous or microporous (Figure 3). The hysteresis loop is of type H3, which could be attributed to aggregates of plate-like particles giving rise to slit-shaped pores.

The isotherms of the samples of biochar are of type I, indicating that the micropores were dominating the textural properties of the biochar derived at 600 and 700 °C. The hysteresis loop of H4 type herein started at relatively higher pressure, due to which two types of pores were considered: mesoporous and microporous. The H4 loop is often attributed to narrow slit-like pores. The hysteresis loops do not close for the biochar samples, derived at 600 and 700 °C. This could be due to hindered evaporation of the trapped nitrogen because of the heterogeneous coal surface or ink-bottled pores [47,48]. These results were confirmed by the specific surface area estimations. The BET equation, used for determining the surface area, was applied in the interval of relative pressure (P/Po) between 0.05 and 0.35, considering partial surface occupation. The BET surface area obtained in the present work is in line with the data reported in earlier investigations of Wang et al. [46] on the characterized biochar, produced from (bamboo and elm) woody residue, pyrolyzed at 500 or 700 °C. They confirm that increasing the carbonization temperature results in an increased BET surface area. Similar temperature dependence was reported also by Shaaban et al. [49] in their characterization of biochar, derived during slow pyrolysis of rubber wood sawdust (300–700 °C). The pore size distribution was also plotted for all examined biochar samples and generally confirmed the results for the porous texture, which were deduced from the adsorption isotherms. The adsorption isotherms of the samples obtained at 500 °C denote that the formation of micropores began at this temperature but the mesopores dominated over the micropores. The biochar, obtained at 600 and 700 °C showed a narrower interval of pore-diameter variations (1 ÷ 2 nm). Thus, an opportunity is foreseen for selective adsorption of molecules, having particular size and/or chemical structure, typical for some microporous adsorbents.

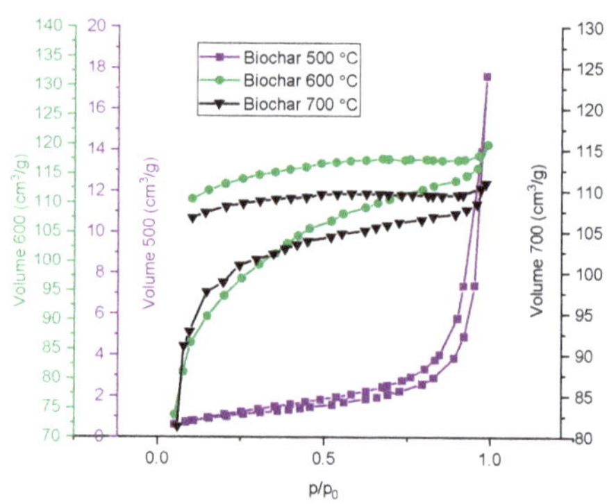

(a)

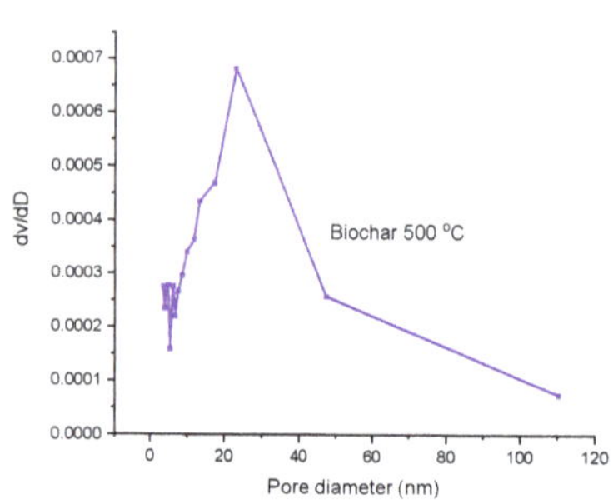

(b)

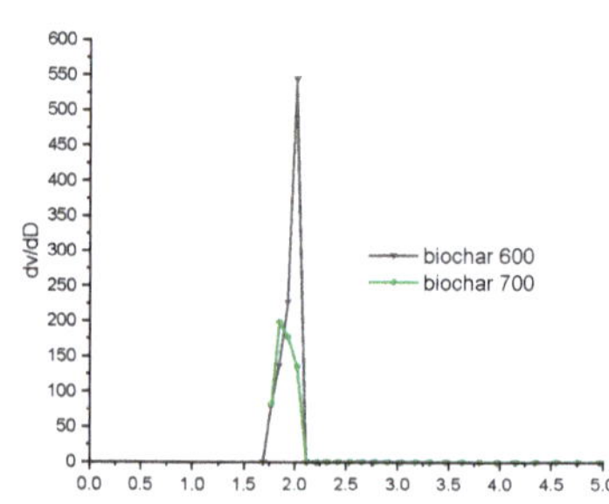

(c)

Figure 3. (**a**) Nitrogen adsorption–desorption isotherms and pore size distribution of (**b**) biochar—500 °C and (**c**) biochar—600 and 700 °C.

3. Materials and Methods

3.1. Feedstock Origin

The investigated THL was a typical example of industrial biomass residue, derived during high-temperature diluted sulfuric acid hydrolysis of softwood and hardwood chips to sugars. In order to develop an efficient biomass processing technology, it is crucial to understand its characteristics and decomposition behavior. Herein, the THL was carbonized at well-controlled conditions.

3.2. Experimental Equipment

The THL was carbonized in HTF (Figure 4) at atmospheric pressure and at three different temperatures, 500, 600, and 700 °C. The residence time of a single sample within the reaction zone was one hour [19]. The carbonization process was carried out in inert atmosphere (nitrogen), with nitrogen flow rate of 1 L/min, and heating rate of 24 °C/min. The HTF was thoroughly described elsewhere [28].

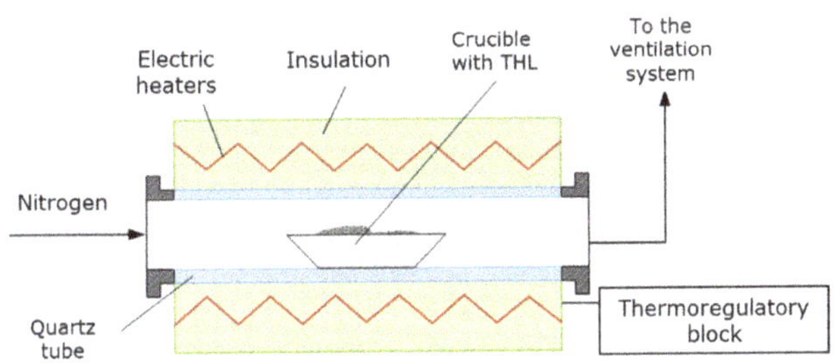

Figure 4. Schematic interpretation of the experimental setup—HTF.

At the end of the process the crucibles, containing biochar, were covered and tempered in a desiccator for at least an hour. Then, the samples were weighed with an analytical balance. Thus, the *biochar mass yield* was obtained according to the following equation:

$$Biocharmassyield = \frac{mass\ of\ biochar\ (g)}{mass\ of\ THL\ (g)} \cdot 100,\ wt.\%, \tag{1}$$

3.3. Feedstock and Biochar Characterization

The feedstock (THL) and the obtained biochar were chemically characterized through proximate, ultimate, ash, lignocellulosic, and calorimetric analyses. The lignocellulosic composition of the THL was determined according to the following methods: cellulose [29] and lignin [30].

The ultimate analysis (C, N, S, and H) of all types of samples was performed with an Elemental Analyzer Eurovector EA 3000.

Inductively Coupled Plasma Optical Emission Spectroscopy (ICP-OES) was applied for the ash analysis of both THL and its carbonized products. The analysis was performed by pre-acid decomposition, and the elemental content was evaluated by Prodigy High Dispersion ICP-OES, Telledyne Leeman Labs using US and BDS EN ISO 11885:2009 Standard [31].

In the present work, simultaneous thermal analyses were carried out with STA PT 1600 TG-DTA/DSC analyzer (LINSEIS Messgeräte GmbH, Germany) in dynamic heating mode from room T (RT = 20 °C) to 1000 °C, with constant heating (10 °C/min) and air flow rates (100 mL/min) and in static oxidizing conditions (still air).

The biochar was examined through Fourier-transform infrared (FTIR) spectroscopy, as well as thermal and nitrogen physisorption analysis. The FTIR spectroscopy was carried out using Varian 660 IR spectrometer. The infrared spectra were collected in the mid-infrared region (4000–400 cm^{-1}). The samples were prepared by the standard KBr pellets method. The specific surface area of the biochar was determined by low-temperature

(77.4 K) nitrogen adsorption in a Quantachrome Instruments NOVA 1200e (USA) apparatus. Before the analyses, the samples were outgassed (argon) at 120 °C for 16 h in a vacuum. The nitrogen adsorption–desorption isotherms were used to evaluate the following parameters: the specific surface area (SBET) was determined through the Brunauer, Emmett, and Teller (BET) equation [13]; the total pore volume (Vt) was estimated in accordance with the Gurvich rule at a relative pressure close to 0.99; the volume of the micropores (V_{MI}) and the specific surface area connected to micropores (S_{MI}), as well as the external specific surface area (S_{EXT}), were evaluated according to V–t-method; additionally, the pore size distributions (PSD) were calculated by equilibrium nonlocal density functional theory (NLDFT) method using slit shape kernel for carbons.

4. Conclusions

In the present study, technically hydrolyzed lignin was utilized, and the effect of carbonization temperature on the physicochemical properties of biochar was experimentally measured. Increasing the carbonization temperature (500 ÷ 700 °C) led to the following general conclusions, in view of which the possible use of biochar was proposed, as follows:

- The biochar mass yield slightly decreased with increasing the carbonization temperature. The chemical characterization showed biochar with gradually reduced content of volatile matter, between 40 and 96 wt. % in contrast to THL. The fixed carbon content was increased from 2.11 to 3.68 times the wt. % of fixed carbon in the THL, along with slightly increased ash content. Besides Pb, Si, and Na, most of the elements showed increased concentration in the biochar ash, with increasing the carbonization temperatures. As expected, the ultimate analysis showed significant increase in the C-content, but considerably reduced H- and O-composition, whereas the reduction of the N- and S-content in the high-temperature biochar showed values below the detection limit. This suggested possible biochar application as solid biofuel as well as for soil amendment (e.g., as compensatory fertilizer for trace elements) as discussed in [50].
- The textural analysis (FTIR spectroscopy) showed that the functional groups were gradually lost thus, forming materials characterized merely by polycyclic aromatic structures and high condensation rate.
- The results from the nitrogen physisorption analysis along with those from the FTIR spectroscopy suggested that the proposed utilization technology of THL (specifically the carbonization at 600 and 700 °C) produced biochar, having the properties typical for the microporous adsorbents, which allows for selective adsorption of specific molecules. Based on the latest observations, another possible application was assumed—as a catalyst.

The perspective proposed herein for THL covers a narrow line of opportunities generally based on the applied conversion technique and characterization methods. According to Ramos et al. [51], at present, about 60% of the lignin (obtained often by kraft pulping) is utilized for heat and chemicals in large-scale industries. The authors summarized earlier investigations on lignin utilization where depending on the chosen conversion process [52] lignin can be converted to a great variety of valuable chemicals and materials, such as: (a) hydrocarbons, phenols and catechols, benzylic aldehydes, quinones, alkyl benzenes, bio-oil, carbon fibers [53]; (b) activated carbon and polymeric materials such as phenol-formaldehyde resins, which is precursor for carbon fibers production [54]; (c) highly functionalized molecules, such as phenolic aldehydes, phenolic ketones, phenolic acids, and many others [20,51].

In view of the extensive research carried out during the last decades and the current legislation framework, it is more than ever time to stimulate the Member States to consider deposited lignocellulosic residue as a valuable resource of goods, which least existence is the natural degradation and the related environmental concern.

Author Contributions: I.N.: conceptualization, methodology, experimental work, results analysis and visualization, writing and editing original draft, funding acquisition; T.R.: methodology, results analysis and visualization, writing—original draft preparation, funding acquisition; T.P.: experimental work, writing—review and editing O.S.: biomass carbonization, biochar characterization, results analysis and visualization, I.V.: biochar chemical characterization, results analysis and visualization. All authors have read and agreed to the published version of the manuscript.

Funding: This research was funded by Ministry of Education and Science (MES) of Bulgaria and the Bulgarian National Science Fund and the publication process was supported by the Research and Development Sector at the Technical University of Sofia.

Acknowledgments: The authors would like to acknowledge the financial support from from the National Science Program "Environmental Protection and Reduction of Risks of Adverse Events and Natural Disasters", approved by the Resolution of the Council of Ministers No. 577/17.08.2018 and supported by the Ministry of Education and Science (MES) of Bulgaria (Agreement No. 01-230/06.12.2018; Agreement No. 01-322/18.12.2019; Agreement No. 01-363/17.12.2020; Agreement No. 01-279/03.12.2021), supported by the Bulgarian National Science Fund (BNSF), MES for ensuring the technical maintenance of the laboratory equipment needed for biomass thermal conversion, as well as the Research and Development Sector at the Technical University of Sofia for the financial support for the present publication. Iliyana Naydenova expresses her gratitude also to WIRE—COST Action CA20127 "Waste biorefinery technologies for accelerating sustainable energy processes" and the National co-funding Contract КП-06-КОСТ/12/6.12.2021.

Conflicts of Interest: The authors declare no conflict of interest.

Sample Availability: Not available.

References

1. Tao, J.; Li, S.; Ye, F.; Zhou, Y.; Lei, L.; Zhao, G. Lignin—An underutilized, renewable and valuable material for food industry. *Crit. Rev. Food Sci. Nutr.* **2020**, *60*, 2011–2033. [CrossRef] [PubMed]
2. Gul, E.; Alrawashdeh, K.A.B.; Masek, O.; Skreiberg, Ø.; Corona, A.; Zampilli, M.; Wang, L.; Samaras, P.; Yang, Q.; Zhou, H.; et al. Production and use of biochar from lignin and lignin-rich residues (such as digestate and olive stones) for wastewater treatment. *J. Anal. Appl. Pyrolysis* **2021**, *158*, 105263. [CrossRef]
3. Ha, J.-M.; Hwang, K.-R.; Kim, Y.-M.; Jae, J.; Kim, K.H.; Lee, H.W.; Kim, J.-Y.; Park, Y.-K. Recent progress in the thermal and catalytic conversion of lignin. *Renew. Sustain. Energy Rev.* **2019**, *111*, 422–441. [CrossRef]
4. Cotana, F.; Cavalaglio, G.; Nicolini, A.; Gelosia, M.; Coccia, V.; Petrozzi, A.; Brinchi, L. Lignin as co-product of second generation bioethanol production from lignocellulosic biomass. *Energy Procedia* **2011**, *45*, 52–60. [CrossRef]
5. Devi, A.; Bajar, S.; Kour, H.; Kothari, R.; Pant, D.; Singh, A. Lignocellulosic Biomass Valorization for Bioethanol Production: A Circular Bioeconomy Approach. *Bioenergy Res.* **2022**, *15*, 1820–1841. [CrossRef] [PubMed]
6. De Corato, U.; de Bari, I.; Viola, E.; Pugliese, M. Assessing the main opportunities of integrated biorefining from agro-bioenergy co/by-products and agroindustrial residues into high-value added products associated to some emerging markets: A review. *Renew. Sust. Energy Rev.* **2018**, *88*, 326–346. [CrossRef]
7. Yadav, V.; Kumar, A.; Bilal, M.; Nguyen, T.A.; Iqbal, H.M.N. Chapter 12—Lignin removal from pulp and paper industry waste streams and its application. In *Nanotechnology in Paper and Wood Engineering*; Bhat, R., Kumar, A., Nguyen, T.A., Sharma, S., Eds.; Micro and Nano Technologies; Elsevier: Amsterdam, The Netherlands, 2022; pp. 265–283, ISBN 9780323858359. [CrossRef]
8. Kumar, A.; Anushree, K.J.; Bhaskar, T. Utilization of lignin: A sustainable and eco-friendly approach. *J. Energy Inst.* **2020**, *93*, 235–271. [CrossRef]
9. Cao, Y.; Chen, S.S.; Zhang, S.; Ok, Y.S.; Matsagar, B.M.; Wu, K.C.-W.; Tsang, D.C.W. Advances in lignin valorization towards bio-based chemicals and fuels: Lignin biorefinery. *Bioresour. Technol.* **2019**, *291*, 121878. [CrossRef]
10. Sumathi, K.M.S.; Mahimairaja, S.; Naidu, R. Use of low-cost biological wastes and vermiculite for removal of chromium from tannery effluent. *Bioresour. Technol.* **2005**, *96*, 309–316. [CrossRef]
11. Dias, J.; Alvim-Ferraz, M.; Almeida, M.; Rivera-Utrilla, J.; Sanchez-Polo, M. Waste materials for activated carbon preparation and its use in aqueous-phase treatment: A review. *J. Environ. Manag.* **2007**, *85*, 833–846. [CrossRef]
12. Ioannidou, O.; Zabaniotou, A. Agricultural residues as precursors for activated carbon production—A review. *Renew. Sustain. Energy Rev.* **2007**, *11*, 1996–2005. [CrossRef]
13. Vassileva, P.S.; Radoykova, T.H.; Detcheva, A.K.; Avramova, I.A.; Aleksieva, K.I.; Nenkova, S.K.; Valchev, I.V.; Mehandjiev, D.R. Adsorption of Ag+ ions on hydrolyzed lignocezllulosic materials based on willow, paulownia, wheat straw and maize stalks. *Int. J. Environ. Sci. Technol.* **2016**, *13*, 1319–1328. [CrossRef]
14. Low, Y.W.; Yee, K.F. A review on lignocellulosic biomass waste into biochar-derived catalyst: Current conversion techniques, sustainable applications and challenges. *Biomass Bioenergy* **2021**, *154*, 106245. [CrossRef]

15. Zhou, N.; Thilakarathna, W.P.D.W.; He, Q.S.; Rupasinghe, H.P.V. A Review: Depolymerization of Lignin to Generate High-Value Bio-Products: Opportunities, Challenges, and Prospects. *Front. Energy Res.* **2022**, *9*, 758744. [CrossRef]
16. Tan, H.; Lee, C.T.; Ong, P.Y.; Wong, K.Y.; Bong, C.P.C.; Li, C.; Gao, Y. A Review on the Comparison Between Slow Pyrolysis and Fast Pyrolysis on the Quality of Lignocellulosic and Lignin-Based Biochar. In *IOP Conference Series: Materials Science and Engineering*; IOP Publishing: Bristol, UK, 2021; Volume 1051, p. 012075. [CrossRef]
17. Kawamoto, H. Lignin pyrolysis reactions. *J. Wood Sci.* **2017**, *63*, 117–132. [CrossRef]
18. Arni, S.A. Comparison of slow and fast pyrolysis for converting biomass into fuel. *Renew. Energy* **2018**, *124*, 197–201. [CrossRef]
19. Ferreira, A.F.; Ribau, J.P.; Costa, M. A decision support method for biochars characterization from carbonization of grape pomace. *Biomass Bioenergy* **2021**, *145*, 105946. [CrossRef]
20. Levdansky, A.V.; Vasilyeva, N.Y.; Malyar, Y.M.; Kondrasenko, A.A.; Fetisova, O.Y.; Kazachenko, A.S.; Levdansky, V.A.; Kuznetsov, B.N. Method of Birch Ethanol Lignin Sulfation with a Sulfaic Acid-Urea Mixture. *Molecules* **2022**, *27*, 6356. [CrossRef]
21. Kazachenko, A.S.; Akman, F.; Vasilieva, N.Y.; Malyar, Y.N.; Fetisova, O.Y.; Lutoshkin, M.A.; Berezhnaya, Y.D.; Miroshnikova, A.V.; Issaoui, N.; Xiang, Z. Sulfation of Wheat Straw Soda Lignin with Sulfamic Acid over Solid Catalysts. *Polymers* **2022**, *14*, 3000. [CrossRef]
22. Romanenko, I.; Kurz, F.; Baumgarten, R.; Jevtovikj, I.; Lindner, J.-P.; Kundu, A.; Kindler, A.; Schunk, S.A. Lignin Depolymerization in the Presence of Base, Hydrogenation Catalysts, and Ethanol. *Catalysts* **2022**, *12*, 158. [CrossRef]
23. Abdullah, B.; Muhammad, S.A.F.; Ad, S.; Mahmood, N.A.N. Production of Biofuel via Hydrogenation of Lignin from Biomass. In *New Advances in Hydrogenation Processes*; Ravanchi, M.T., Ed.; IntechOpen: London, UK, 2017. [CrossRef]
24. Shu, R.; Zhang, Q.; Xu, Y.; Long, J.; Ma, L.; Wang, T.; Chen, P.; Wu, Q. Hydrogenation of lignin-derived phenolic compounds over step by step precipitated Ni/SiO_2. *RSC Adv.* **2016**, *6*, 5214–5222. [CrossRef]
25. Wang, Y.-Y.; Ling, L.-L.; Jiang, H. Selective hydrogenation of lignin to produce chemical commodities by using a biochar supported Ni–Mo 2 C catalyst obtained from biomass. *Green Chem.* **2016**, *18*, 4032–4041. [CrossRef]
26. Biswas, B.; Kumar, A.; Kaur, R.; Krishna, B.B.; Bhaskar, T. Catalytic hydrothermal liquefaction of alkali lignin over activated bio-char supported bimetallic catalyst. *Bioresour. Technol.* **2021**, *337*, 125439. [CrossRef] [PubMed]
27. Shiraki, Y.; Goto, T.; Nonaka, H. Concentrated sulfuric acid hydrolysis of softwood with t-butyl alcohol. *Biomass Convers. Biorefinery* **2021**, *11*, 937–941. [CrossRef]
28. Sandov, O. Construction of a flow reactor for the combustion/pyrolysis of biomass fuels. In Proceedings of the XXIV Scientific Conference with International Participation FPEPM 2019, Sozopol, Bulgaria, 17–20 September 2019; pp. 258–268.
29. Kürschner, K.; Hoffer, A. Cellulose and cellulose derivatives. *Fresenius' J. Anal. Chem.* **1933**, *92*, 145–154. [CrossRef]
30. Technical Association of the Pulp and Paper Industry. *TAPPI T 222 om-11: Acid-Insoluble Lignin in Wood and Pulp*; Technical Association of the Pulp and Paper Industry: Peachtree Corners, GA, USA, 2011.
31. *ISO 11885:2007*; Water Quality—Determination of Selected Elements by Inductively Coupled Plasma Optical Emission Spectrometry (ICP-OES). International Organization for Standardization: Geneva, Switzerland, 2007. Available online: https://bds-bg.org/bg/project/show/bds:proj:80192 (accessed on 14 June 2022).
32. Farrokh, N.T.; Suopajärvi, H.; Sulasalmi, P.; Fabritius, T. A thermogravimetric analysis of lignin char combustion. *Energy Procedia* **2019**, *158*, 1241–1248. Available online: https://www.sciencedirect.com/science/article/pii/S1876610219304357 (accessed on 14 June 2023). [CrossRef]
33. Wan, J.; Liu, L.; Ayub, K.S.; Zhang, W.; Shen, G.; Hu, S.; Qian, X. Characterization and adsorption performance of biochars derived from three key biomass constituents. *Fuel* **2020**, *269*, 117142. [CrossRef]
34. Zhao, S.-X.; Ta, N.; Wang, X.-D. Effect of Temperature on the Structural and Physicochemical Properties of Biochar with Apple Tree Branches as Feedstock Material. *Energies* **2017**, *10*, 1293. [CrossRef]
35. Sadaka, S.; Sharara, M.A.; Ashworth, A.; Keyser, P.; Allen, F.; Wright., A. Characterization of Biochar from Switchgrass Carbonization. *Energies* **2014**, *7*, 548–567. [CrossRef]
36. Valchev, I.; Yordanov, Y.; Savov, V.; Antov, P. Optimization of the Hot-Pressing Regime in the Production of Eco-Friendly Fibreboards Bonded with Hydrolysis Lignin. *Period. Polytech. Chem. Eng.* **2022**, *66*, 125–134. [CrossRef]
37. Álvarez, A.; Pizarro, C.; García, R.; Bueno, J.L.; Lavín, A.G. Determination of kinetic parameters for biomass combustion. *Bioresour. Technol.* **2016**, *216*, 36–43. [CrossRef] [PubMed]
38. Brebu, M.; Vasile, C. Thermal degradation of lignin—A review. *Cellul. Chem. Technol.* **2010**, *44*, 353–363.
39. Petrova, T.; Naydenova, I.; Ribau, J.; Ferreira, A.F. Biochar from Agro-Forest Residue: Application Perspective Based on Decision Support Analysis. *Appl. Sci.* **2023**, *13*, 3240. [CrossRef]
40. Cagnon, B.; Py, X.; Guillot, A.; Stoeckli, F.; Chambat, G. Contributions of hemicellulose, cellulose and lignin to the mass and the porous properties of chars and steam activated carbons from various lignocellulosic precursors. *Bioresour. Technol.* **2009**, *100*, 292–298. [CrossRef] [PubMed]
41. Adinata, D.; Daud, M.A.W.; Aroua, M.K. Preparation and characterization of activated carbon from palm shell by chemical activation with K2CO3. *Bioresour. Technol.* **2007**, *98*, 145–149. [CrossRef]
42. Draganova, R.; Nenkova, S. *Chemistry and Structure of Plant Tissues*; University of Chemical Technology and Metallurgy: Sofia, Bulgaria, 2002; p. 372, ISBN 954-8954-16-8. (In Bulgarian)
43. Chowdhury, Z.Z.; Karim, M.Z.; Ashraf, M.A.; Khalid, K. Influence of carbonization temperature on physicochemical properties of biochar derived from slow pyrolysis of durian wood (*Durio zibethinus*) sawdust. *BioResources* **2016**, *11*, 3356–3372. [CrossRef]

44. Janu, R.; Mrlik, V.; Ribitsch, D.; Hofman, J.; Sedlaček, P.; Bielska, L.; Soja, G. Biochar surface functional groups as affected by biomass feedstock, biochar composition and pyrolysis temperature. *Carbon Resour. Convers.* **2021**, *4*, 36–46. [CrossRef]
45. Li, J.; Li, Y.; Wu, Y.; Zheng, M. A comparison of biochars from lignin, cellulose and wood as the sorbent to an aromatic pollutant. *J. Hazard. Mater.* **2014**, *280*, 450–457. [CrossRef]
46. Wang, Y.; Hu, Y.; Zhao, X.; Wang, S.; Xing, G. Comparisons of biochar properties from wood material and crop residues at different temperatures and residence times. *Energy Fuel* **2013**, *27*, 5890–5899. [CrossRef]
47. Tang, X.; Wang, Z.; Ripepi, N.; Kang, B.; Yue, G. Adsorption affinity of different types of coal: Mean isosteric heat of adsorption. *Energy Fuels* **2015**, *29*, 3609–3615. [CrossRef]
48. Qi, L.; Tang, X.; Wang, Z.; Peng, X. Pore characterization of different types of coal from coal and gas outburst disaster sites using low temperature nitrogen adsorption approach. *Int. J. Min. Sci. Technol.* **2017**, *27*, 371–377. [CrossRef]
49. Shaaban, A.; Sea, S.-M.; Mitan, N.M.M.; Dimin, M.F. Characterization of biochar derived from rubber wood sawdust through slow pyrolysis on surface porosities and functional groups. *Procedia Eng.* **2013**, *68*, 365–371. [CrossRef]
50. Schmidt, H.P.; Wilson, K. The 55 uses of biochar. *Biochar J.* Arbaz, Switzerland, ISSN 2297-1114. 2014. Available online: www.biochar-journal.org/en/ct/2 (accessed on 7 June 2023).
51. Ramos, L.P.; Suota, M.J.; Pavaneli, G.; Corazza, M.L. The role of biomass pretreatment for sustainable biorefineries. *Bulg. Chem. Commun.* **2019**, *51*, 62–68.
52. Hermansson, F.; Janssen, M.; Svanström, M. Prospective study of lignin-based and recycled carbon fibers in composites through meta-analysis of life cycle assessments. *J. Clean. Prod.* **2019**, *223*, 946. [CrossRef]
53. Mahmood, S.F.; Batchelor, B.L.; Jung, M.; Park, K.; Voit, W.E.; Novak, B.M.; Yang, D. Study of a melt processable polymer precursor for carbon fiber. *Carbon Lett.* **2019**, *29*, 605–612. [CrossRef]
54. Ragauskas, A.J.; Beckham, G.T.; Biddy, M.J.; Chandra, R.; Chen, F.; Davis, M.F.; Davison, B.H.; Dixon, R.A.; Gilna, P.; Keller, M.; et al. Lignin valorization: Improving lignin processing in the biorefinery. *Science* **2014**, *344*, 1246843. [CrossRef] [PubMed]

molecules

MDPI

Article

Apricot Seed Shells and Walnut Shells as Unconventional Sugars and Lignin Sources

Vita Halysh [1,2], Juan Miguel Romero-García [3,4,*], Alfonso M. Vidal [3], Tetiana Kulik [2], Borys Palianytsia [2], Minerva García [5] and Eulogio Castro [3,4,*]

1 Department of Ecology and Technology of Plant Polymers, Faculty of Chemical Engineering, Igor Sikorsky Kyiv Polytechnic Institute, Peremogy Avenu 37/4, 03056 Kyiv, Ukraine
2 Laboratory of Kinetics and Mechanisms of Chemical Reactions on the Surface of Solids, Chuiko Institute of Surface Chemistry, National Academy of Sciences of Ukraine, General Naumov Str., 17, 03164 Kyiv, Ukraine
3 Department of Chemical, Environmental and Materials Engineering, Universidad de Jaén, Campus Las Lagunillas s/n, 23071 Jaén, Spain
4 Center for Advanced Studies in Earth Sciences, Energy and Environment (CEACTEMA), Universidad de Jaén, Campus Las Lagunillas s/n, 23071 Jaén, Spain
5 Tecnológico Nacional de México/Instituto Tecnológico de Zitácuaro, Av. Tecnológico No. 186 Manzanillos, Zitácuaro 61534, Michoacán, Mexico
* Correspondence: jrgarcia@ujaen.es (J.M.R.-G.); ecastro@ujaen.es (E.C.); Tel.: +34-9532182163 (E.C.)

Abstract: The present study focuses on using apricot seeds shells and walnut shells as a potential renewable material for biorefinery in Ukraine. The goal of the research work was to determine the relationship between the chemical composition of solid residues from biomass after acid pretreatment with H_2SO_4, alkaline pretreatment with NaOH, and a steam explosion pretreatment and the recovery of sugars and lignin after further enzymatic hydrolysis with the application of an industrial cellulase Cellic CTec2. Apricot seeds shells and walnut shells consist of lots of cellulose (35.01 and 24.19%, respectively), lignin (44.55% and 44.63%, respectively), hemicelluloses (10.77% and 26.68%, respectively), and extractives (9.97% and 11.41%, respectively), which affect the efficiency of the bioconversion of polysaccharides to sugars. The alkaline pretreatment was found to be more efficient in terms of glucose yield in comparison with that of acid and steam explosion, and the maximum enzymatic conversions of cellulose reached were 99.7% and 94.6% for the solids from the apricot seeds shells and the walnut shells, respectively. The maximum amount of lignin (82%) in the residual solid was obtained during the processing of apricot seed shells submitted to the acid pretreatment. The amount of lignin in the solids interferes with the efficiency of enzymatic hydrolysis. The results pave the way for the efficient and perspective utilization of shells through the use of inexpensive, simple and affordable chemical technologies, obtaining value-added products, and thus, reducing the amount of environmental pollution (compared to the usual disposal practice of direct burning) and energy and material external dependency (by taking advantage of these renewable, low-cost materials).

Keywords: biorefinery; enzymatic hydrolysis; lignin; pretreatment; sugars

Citation: Halysh, V.; Romero-García, J.M.; Vidal, A.M.; Kulik, T.; Palianytsia, B.; García, M.; Castro, E. Apricot Seed Shells and Walnut Shells as Unconventional Sugars and Lignin Sources. *Molecules* **2023**, *28*, 1455. https://doi.org/10.3390/molecules28031455

Academic Editors: Alejandro Rodriguez Pascual, Eduardo Espinosa Víctor and Carlos Martín

Received: 19 December 2022
Revised: 23 January 2023
Accepted: 30 January 2023
Published: 2 February 2023

1. Introduction

Global population growth is associated with an increasing demand for energy resources. In this regard, the use of oil is growing every year [1]. As a fossil energy source, the use of oil presents both environmental and availability concerns, making the search for unconventional alternative energy sources necessary. Renewable, lignocellulosic materials such as forest, agricultural, and agroindustrial residues constitute prominent examples of those potential energy sources. Contrary to fossil energy sources, renewable sources are spread all over the world, and their use does not increase the net greenhouse gases emissions.

In addition to energy, the chemical composition of these materials can be used, through different fractionation and conversion schemes, to obtain a wide range of bio-based products that can be substituted for their similar ones currently derived from fossil sources. Biofuels such as bio-oil, biochar, non-condensable gases, and bioethanol can be obtained from different types of biomasses via thermal, thermocatalytic, and biological conversion [2]. Ethanol is regarded as the most attractive and clean fuel, and it can be used in combustion engines [3]. First- and second-generation biofuels are obtained mainly from renewable plant materials. Food crops, e.g., wheat [4], corn [5], sugarcane [6], etc., are the main materials for first-generation biofuels, while lignocellulosic materials, e.g., inexpensive wastes of food and non-food industries, can be used for the biochemical processing into second-generation biofuels [7,8]. The growing interest in the development and application of new goods and materials, chemicals of an aromatic nature, and energy and fuels from renewable vegetable wastes can be observed worldwide [9–11] due to their widespread availability and low cost.

Different technologies, including hydrolysis, extraction, pyrolysis, and chemical modifications have been recently reviewed to summarize the progress on the production of value-added products such as polymers, bioactive compounds, and bioplastics, among others, starting with biomass and waste feedstocks [12].

Another way of converting lignocellulosic materials into useful products is the biochemical rout, which includes processing via a pretreatment, the enzymatic hydrolysis of the resulting solid, and the fermentation of hydrolysates. This procedure allows us to converse the polysaccharide component of the lignocellulose into biofuels and to obtain lignin as a residue [13]. In spite of the recent acquisition of knowledge about the bioconversion processes, it is still necessary to study the processes of biochemical conversion for individual plant materials because much of it is still not fully understood. For instance, the resistance of carbohydrates to degradation by enzymes (the so-called recalcitrance effect), which is closely related to the presence of extractives of a different nature and lignin [14,15], the amount of which in the biomass depends on its type and the growing and harvesting conditions, as well as many other factors. Pretreatment is an important stage before biological conversion, which allows it to partially overcome biomass recalcitrance. The main task of any pretreatment of biomass is the mitigation of the recalcitrance and increase in the accessibility of cellulose [16]. Even though many researchers have worked on developing different methods, including a mechanical one [17], extraction with organic solvents [18], a hydrothermal one [19], a dilute acid one [20], and an alkaline one [21], most of these papers are related to the study of the sugar yield without considering the structural changes of each biomass constituent, including lignin.

The most typical methods for raw materials pretreatment before enzymatic hydrolysis are an acid pretreatment with the application of diluted sulfuric acid at a concentration of 1%, an alkaline pretreatment with the application of 15% sodium hydroxide solution, and also, a steam explosion pretreatment [22–24].

The great advantage of lignocellulosic waste, including agricultural waste, is its wide distribution throughout the earth, rather than the specific locations of fossil energy sources such as oil. This means that the number of raw materials that can be used is very large. Among the different agricultural residues, those derived from the cultivation of apricots and walnuts, specifically the shells, have been studied only in a limited way up to now. Ukraine is an agro-industrial country, which annually produces a huge amount of plant wastes from the agricultural and forestry industries. According to FAO [25], during the period 2017–2021, the average production in Ukraine of apricot seed shells was 81,666 tons, and that of walnut shells was 111,088 tons, representing 8.32 and 32.74% of the European production total, as well as a disposal problem. That is why these materials were chosen as raw materials for the experimental work. Agricultural residues and forestry wastes are a lignocellulose complex [26,27] which does not have further industrial applications. Most of these residues are burned, which causes significant damage to the environment. However, their sugar composition indicates great potential for chemical processing. It is promising, from an economic point of view, to develop efficient methods to utilize them and obtain valuable materials, chemicals, and fuels [28,29].

As it is known, the lignocellulosic biomass is composed of cellulose, hemicelluloses, and lignin in different proportions. Lignin is composed of a heterogeneous polymer of phenyl-propane, and it has a recalcitrant and complex nature due to it having strong bonds. The effect of lignin on the release of sugars during the enzymatic hydrolysis of pretreated biomass has been widely studied especially because of the inhibition of enzyme action; in general, two strategies to separate the lignin fraction can be performed, depending on the pretreatment applied. On the one side, lignin can be obtained as the first fraction following an alkaline pretreatment [30] or it can constitute the solid residue after the fractionation of hemicellulose and conversion of cellulose into glucose. A delignification of biomass can improve the saccharification yield [31].

The aim of the of present research work was to assess the impact of the acid, alkaline, and steam explosion pretreatments of apricot seeds shells and walnut shells on the efficiency of enzymatic hydrolysis with an industrial cellulase Cellic CTec2 enzyme complex for the recovery of sugars and lignin, which are products used by a biorefinery in Ukraine.

2. Results and Discussion

2.1. Raw Material Composition

The chemical compositions of the initial apricot seed shells and walnut shells are given in Table 1. The analysis of the apricot seed shells composition showed that they contain less extractives compared to those in walnut shells. The amount of cellulose in the apricot seed shells is higher than it is in walnut shells, but the walnut shells are composed of a higher proportion of hemicelluloses. The amount of cellulose in the apricot seed shells agrees with data presented by other scientists, who gave a value of 34.31% [32]. Most of the glucoses are related to cellulose. The amount of xylose is the highest among all of the low-molecular-weight polysaccharides, and it is 9.41% and 24.13% for apricot seed shells and walnut shells, respectively. Both of the plant materials are composed of a high percentage of lignin (44.5 and 44.6% dry weight). Acid-insoluble lignin makes up 44.4 and 44.5% of the total lignin, with the rest being acid-soluble lignin. The amount of ash in the studied materials is lower than 0.5%. The amount of acetyl groups in walnut shells is higher than that of apricot seed shells.

Table 1. Chemical composition of apricot seed shells and walnut shells (% dry weight).

Components	Apricot Seed Shells	Walnut Shells
Extractives	9.97 ± 0.51	11.41 ± 2.23
Cellulose	35.01 ± 0.42	24.19 ± 0.68
Hemicellulose	10.77 ± 0.14	25.68 ± 1.70
Xylan	8.33 ± 0.09	21.35 ± 1.83
Galactan	1.28 ± 0.05	2.78 ± 0.07
Arabinan	1.16 ± 0.02	1.55 ± 0.01
Mannan	-	-
Lignin	44.55 ± 1.05	44.63 ± 1.01
Acid-Soluble Lignin	1.12 ± 0.01	1.14 ± 0.01
Acid-Insoluble Lignin	43.43 ± 1.04	43.49 ± 1.00
Ash	0.23 ± 0.01	0.37 ± 0.01
Acid-insoluble ash	-	0.12 ± 0.00
Acetyl groups	0.85 ± 0.02	3.87 ± 0.09
Acetic acid	1.21 ± 0.03	5.53 ± 0.12

Compared to the data provided by Queirós et al. [33], the amount of the sugars in the walnut shells are quite similar, while the other components are a little bit different. Compared to the chemical composition of walnut shells reported by other scientists [34], the difference in the amount of the main components is significant. Such differences can be related to the differences in the plant variety, growth conditions, climate, soils, and other factors.

2.2. Effect of Pretreatment on Solid Recovery

Hemicelluloses contribute to biomass recalcitrance as they cover and protect cellulose fibrils from enzymatic destruction [35]. An acid pretreatment of biomass with diluted sulfuric acid is an efficient means to solubilize low-molecular-weight polysaccharides. At the same time, an alkaline pretreatment is one of the most attractive techniques for the biomass' preparation for further enzymatic hydrolysis because of its high selectivity for lignin separation [36]. The steam explosion pretreatment is regarded as an alternative and environmentally friendly technique that can be used to reduce biomass recalcitrance, resulting in desirable cellulose accessibility [37]. It happens due to the fact that saturated steam heats the lignocellulosic biomass and releases pressure, causing the cleavage of fibers. Additionally, the removal of extractives and the partial solubilization of hemicellulosic constituents is observed.

The effect of the raw materials pretreatments with 1% H_2SO_4 at 180 °C in an agitated tank reactor and with 15% NaOH at 180 °C in agitated tank reactor and with the application of steam explosion for solid recovery is shown in Figures 1 and 2. Table 2 shows the results of the acid, alkali, and steam explosion (180 °C, 10 min) pretreatments in terms of solid characterization, e.g., sugars contents.

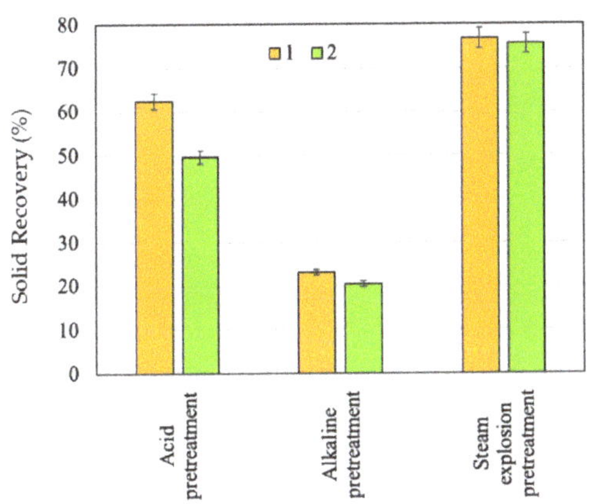

Figure 1. Effect of pretreatments on solid recovery: 1—apricot seed shells; 2—walnut shells.

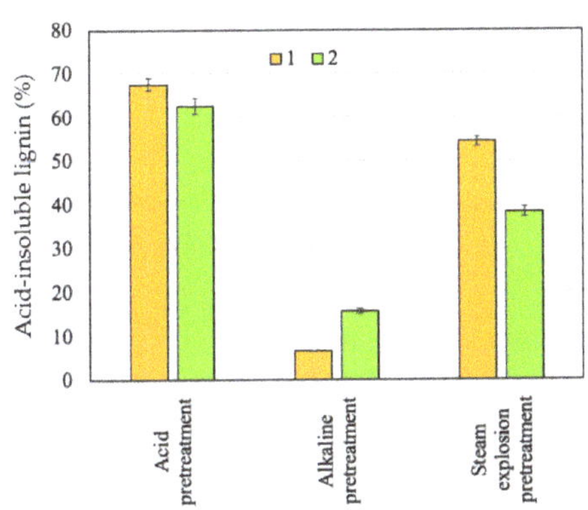

Figure 2. Effect of pretreatments on residual lignin content: 1—apricot seed shells; 2—walnut shells.

Table 2. Solid characterization after acid, alkaline, and steam explosion pretreatment of shells.

Material	Sugars Content (%)					Acid-Insoluble Ash (%)
	Glucan	Xylan	Galactan	Arabinan	Mannan	
Acid pretreatment						
Apricot seed shells	27.43 ± 0.29	-	-	-	-	-
Walnut shells	30.30 ± 0.64	0.90 ± 0.06	-	-	-	0.05 ± 0.00
Alkaline pretreatment						
Apricot seed shells	75.58 ± 1.14	3.59 ± 0.28	-	-	-	-
Walnut shells	73.38 ± 1.58	5.57 ± 0.09	-	-	-	-
Steam explosion pretreatment						
Apricot seed shells	32.32 ± 0.64	7.75 ± 0.29	0.57 ± 0.09	-	-	-
Walnut shells	24.35 ± 0.12	23.25 ± 0.39	2.58 ± 0.06	1.24 ± 0.05	-	-

The pretreatment of apricot seed shells and walnut shells with 1% H_2SO_4 at 180 °C in an agitated tank reactor caused the removal of 37.54% and 50.35% by weight from raw material, respectively, and resulted in the formation of solids with more lignin, while the amount of cellulose is still high (based on glucose content, it is greater than 30%), but it is slightly lower than it was in the initial material. Lignin and cellulose were less solubilized, thus both of the components are the major constituents in the solid. Similar results in terms of solid recovery were obtained by Benjamin et al. [38] during a bagasse sugarcane pretreatment with diluted acid: the solid recovery ranged from 50.1% to 76.5% depending on the parameters of the process. A high solid recovery value after the acid pretreatment was obtained for the apricot seed shells. This may be due to there being more cellulose in the initial biomass.

The loss of 76.79% and 79.5% of the mass during the alkali pretreatment of the apricot seed shells and walnut shells with 15% NaOH at 180 °C in an agitated tank reactor, respectively, is related to delignification. The amount of lignin in the solids was 6.7 and 2.8 times lower in comparison with those of the initial apricot seed shells and walnut shells, respectively. Under the same conditions, the apricot seed shells were more easily delignified than the walnut shells were. The corn stover alkaline pretreatment resulted in the dissolution of 50% of the hemicellulose and of 60–80% of the lignin [39]. The reduction in lignin content to 9.50% and to 10.88% is also observed during corncob and sweet sorghum bagasse alkaline pretreatment [40]. It is expected that lignin removal will have a positive effect on biomass enzymatic degradation as the presence of lignin in the raw materials limits the enzymatic hydrolysis by impeding the cellulose accessibility [41]. As it can be seen, the removal of the non-cellulosic components of a polysaccharide nature also took place, resulting in the full removal of galactose, arabinose, and mannose and the partial removal of xylose. The removal of ash was observed.

The analysis of the solids after the alkaline pretreatment indicated that they contained a high percentage of cellulose (based on the glucose content, it was greater than 80%). It is expected that the cellulose in the solids after the alkaline pretreatment of the raw materials will most effectively transform into glucose.

According to the obtained results, the steam explosion pretreatment produced a minimal effect on the amount of glucose of both the apricot seed shells and walnut shells. The dissolution of part of low-molecular-weight polysaccharides took place. As it can be seen, the amount of dissolved hexoses, pentoses, and oligosaccharides was higher in liquid after the steam explosion pretreatment of the walnut shells due to their higher initial biomass in comparison with that of the apricot seed shells. The effect of the steam explosion pretreatment is not uniform with the amount of lignin in the pretreated materials. That of apricot seed shells increased probably due to reactions of condensation between the lignin, extractives of a different nature, and products of the sugar degradation [42], which agrees with experimental results of other studies [43,44]. At the same conditions, the amount of lignin in the walnut shell solid decreased; the same fact was observed during the steam explosion of spruce bark [45] and corn stalks [46]. Such differences in the behavior of lignin can be associated with the fact that steam explosion can cause the melting of lignin, and its partial depolymerization occurs through the cleavage of mainly ß-O-4 ether, resulting in the formation of alcohol derivatives, as well as condensation byproducts, the amount and composition of which largely depends on the type of feedstock [47].

2.3. *Effect of Pretreatment on Hydrolysate Characterization*

The effect of raw materials pretreatment with 1% H_2SO_4 at 180 °C in an agitated tank reactor and with the application of steam explosion on the hydrolysate composition is shown in Table 3. Hydrolysate after the alkaline pretreatment could not be analyzed because of the large amount of soluble and suspended substances that could not be recovered by centrifugation and membrane filtration.

Table 3. Hydrolysate characterization after acid and steam explosion pretreatment of raw materials.

Composition	Acid Pretreatment		Steam Explosion Pretreatment	
	Apricot Seed Shells	Walnut Shells	Apricot Seed Shells	Walnut Shells
Hemicellulosic sugars recovery (%)	4.21	10.91	2.02	5.24
Glucose recovery (%)	37.16	35.67	0.02	3.52
Sugars content (g/L)				
Glucose	27.86 ± 0.08	18.98 ± 0.07	0.02 ± 0.00	0.68 ± 0.01
Xylose	1.35 ± 0.02	6.33 ± 0.02	0.08 ± 0.00	0.21 ± 0.01
Galactose	0.17 ± 0.00	1.55 ± 0.01	0.10 ± 0.00	0.46 ± 0.01
Arabinose	-	0.19 ± 0.00	0.02 ± 0.00	0.82 ± 0.01
Mannose	0.17 ± 0.00	-	-	-
Inhibitors (g/L)				
Formic acid	2.45 ± 0.01	1.96 ± 0.02	1.42 ± 0.01	1.87 ± 0.02
Acetic acid	2.91 ± 0.01	9.92 ± 0.03	2.16 ± 0.02	1.59 ± 0.01
Furfural	3.74 ± 0.02	9.74 ± 0.03	0.00 ± 0.00	0.02 ± 0.00
Hydroxymethylfurfural	1.47 ± 0.01	1.06 ± 0.01	0.06 ± 0.00	0.11 ± 0.00
Levulinic acid	3.31 ± 0.02	2.05 ± 0.02	0.31 ± 0.01	0.16 ± 0.01

As it can be seen, high glucose concentrations in the liquids after the acid pretreatment were obtained, which indicates a partial degradation of the cellulose during such a pretreatment. The amount of hemicellulosic sugars recovered in the liquids after the acid pretreatment (in form of monosaccharides) of both of the raw materials is much lower compared to those of other agricultural residues; for instance, the hemicellulosic sugars recovered after an olive tree pruning biomass pretreatment with diluted sulfuric acid was 78.5% [48]. The amount of recovered sugar in the liquid fractions obtained from the steam explosion pretreatment was very low, as can be seen in Table 3. A small amount of sugars recovered from the apricot seed shells and walnut shells is attributed to the complete depolymerization of hemicelluloses fraction into soluble monosaccharides and low-molecular-weight compounds during the acid pretreatment with 1% H_2SO_4 at 180 °C. Part of the acetyl groups in the hemicelluloses were cleaved, leading to the formation of acetic acid. Additionally, the formation of organic acids, such as formic and levulinic, as well as furfural and hydroxymethylfurfural, is detected as the result of monosaccharide degradation [49]. The amount of xylan, which is the largest constituent part of hemicelluloses, is the main indicator of the acid hydrolysis pretreatment. As it can be observed, xylan virtually disappeared in both of the raw materials when they were subjected to the acid pretreatment (Table 2).

Table 3 shows that liquid after the walnut shells acid pretreatment contained more inhibitors, which are the products of sugar degradation. The analysis shows that the liquid after the walnut shells pretreatment contained a high percentage of furfural and acetic acid (9.74 and 9.92% dry weight, respectively). Concerning the apricot seed shells, the liquid after the acid pretreatment contained far fewer inhibitor compounds due to the lower amount of hemicelluloses in the initial biomass.

The resulting liquid after the steam explosion pretreatment contained fewer inhibitors, indicating the preservation of sugars in the biomass.

2.4. Enzymatic Hydrolysis

The pretreated solids were submitted to enzymatic hydrolysis with the industrial cellulase complex called Cellic CTec2, and the effects of time on the efficiency of the process for the different solids were compared (Figure 3).

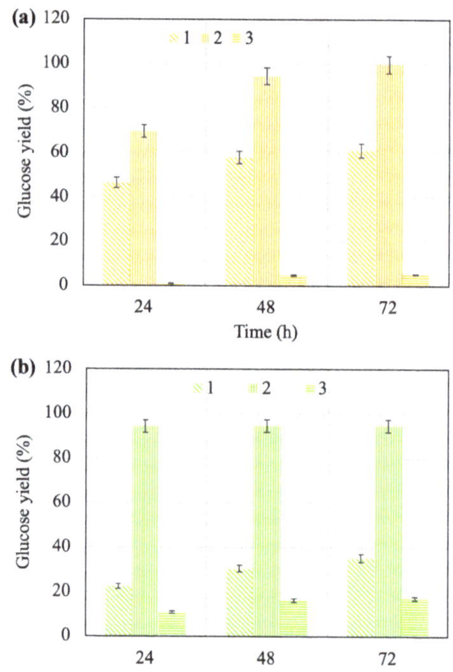

Figure 3. Effect of time on glucose yield during enzymatic hydrolysis of pretreated apricot seed shells (**a**) and walnut shells (**b**): 1—after acid pretreatment; 2—after alkaline pretreatment; 3—after steam explosion pretreatment.

As expected, the cellulose hydrolysis yields increased with an increasing duration of the process for all of the samples, except for the solid after the alkaline pretreatment of the walnut shells, for which the maximum efficiency of hydrolysis was achieved in the first 24 h of contact, and a further increase in time almost did not lead to its change. As it can be seen, the alkaline and acid pretreatments are more efficient at increasing the yield of glucose than the steam explosion is. For the both of the samples of solids after the steam explosion pretreatment of the apricot seed shells and walnut shells, the maximum yields of glucose did not exceed 10 and 20%, respectively, while the yields for the solids after the acid pretreatment did not exceed 60 and 35%, respectively.

Maximum glucose yields (99.7% and 94.6%) from the solids were achieved with industrial cellulase enzymatic hydrolysis after the alkaline pretreatment of the raw materials.

Hemicellulose removal and lignin removal from the raw materials during the acid and alkaline pretreatments probably created additional pores and sites in the biomass available for cellulolytic enzymes, which affected the cellulose. The effect is most pronounced with a low lignin content [50].

2.5. Solid Characterization after Enzymatic Hydrolysis

The maximum amount of lignin (82%) was achieved during the enzymatic hydrolysis of the solid after the acid pretreatment of the apricot seed shells due to the greater amount of lignin in the solid before the process (Figure 4a). In the case of walnut shells, the highest lignin values in the solids after enzymatic hydrolysis are around 70% for the cases of the acid and alkaline pretreated solids (Figure 4b). Samples subjected to 72 h of enzymatic hydrolysis were further analyzed.

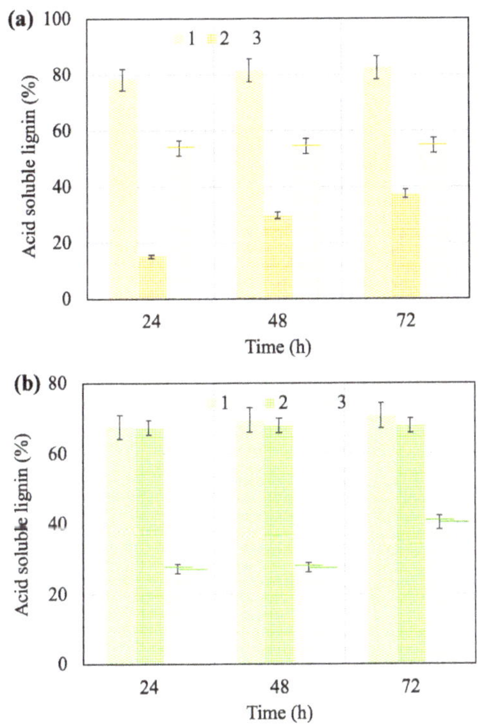

Figure 4. Effect of time of the process on amount of acid-soluble lignin in solids after hydrolysis of materials based on apricot seed shells (**a**) and walnut shells (**b**): 1—solid after acid pretreatment; 2—solid after alkaline pretreatment; 3—solid after pretreatment with steam explosion.

According to Figure 4, the effect of time on the amount of acid-soluble lignin is limited, except for the apricot seed shells submitted to the alkaline pretreatment; in this case, an increase over time is observed. On the contrary, the amount remains virtually the same for the other pretreatments with this raw material and also for walnut shells in all of the cases.

The FTIR spectra of all of the samples are characterized by the presence of different peaks for both the aromatic and polysaccharide components, but their intensities vary (Figure 5). The peaks between 3600–3000 cm^{-1} are presented in all of the spectra, and they are caused by the OH stretching vibration. The peaks between 2945–2855 cm^{-1} are attributed to C-H stretching vibration in all of the studied materials. The peaks at 1648, 1431, 1417, 1375, 1165, 1060, and 897 cm^{-1} are related to polysaccharides. The bands at 1605, 1509, and 1425 cm^{-1} correspond to the aromatic skeleton vibrations of lignin. Solids obtained thorough the alkaline pretreatment of the shells are characterized by an absence of bands at 1509 and 1425 cm^{-1} due to delignification. The bands at 1509 and 1425 cm^{-1} related to lignin become more intense in the sample obtained after the enzymatic hydrolysis of the solids after the steam explosion and acid pretreatment, and they also show that there is more lignin in these samples. The peak at 1249 cm^{-1} is related to C=C bonds in the lignin aromatic rings and C-O-C stretching in the hemicelluloses. This peak is less pronounced in the pretreated solids, which indicates the removal of hemicelluloses. The main difference between the FTIR spectra of the initial materials, the pretreated solids, and the solids after the hydrolysis is the increase in the intensity of the bands related to lignin (1509 and 1425 cm^{-1}), which confirms that the resulting solids contain a large amount of lignin.

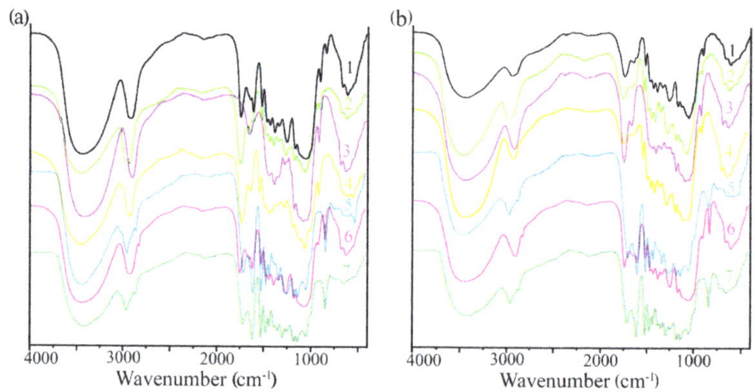

Figure 5. FTIR spectra for initial, pretreated and hydrolyzed apricot seed shells (**a**) and walnut shells (**b**): 1—initial materials; 2–4—solids, alkaline and steam explosion pretreatment, respectively; 5–7—respective solids after enzymatic hydrolysis with industrial cellulase Cellic CTec2 enzyme complex.

The samples were analyzed using temperature-programmed desorption mass spectrometry (TPD-MS). An analysis of the P/T curves of the pressure of the volatile pyrolysis products against the temperature of the sample shows significant differences from the behavior of the P/T curves for the studied samples depending on both the type of biomass and the type of pretreatment, as shown in Figure 6.

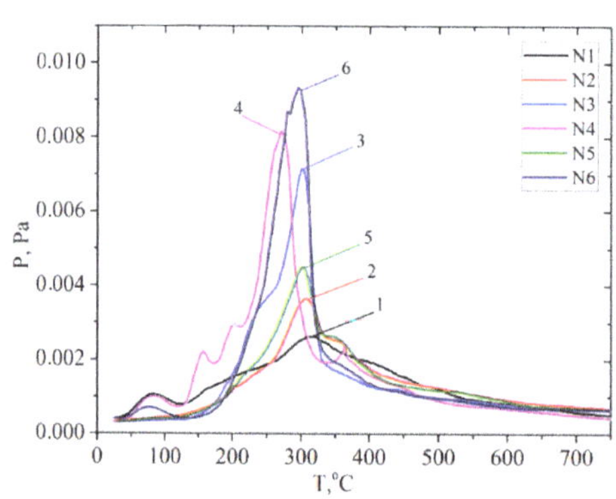

Figure 6. P/T curves of vapor pressure of gaseous products measured at temperature of pyrolysis: solids after hydrolysis of materials based on apricot seed shells (N1–3) and walnut shells (N4–6): 1 and 4—solid after alkaline pretreatment; 2 and 5—solid after acid pretreatment; 3 and 6—solid after pretreatment with steam explosion.

A comparison of P/T curves for the apricot samples shows that the highest intensity of the desorption of pyrolysis products is observed for the sample after the pretreatment with a steam explosion (Figure 7). In this case, the intensity of the peak of desorption of pyrolysis products increases most significantly at a temperature of Tmax~300 °C. The intensity of this TPD peak for the sample apricot seed shells after the pretreatment with steam explosion is 2–3 times higher than the intensity for the samples after the acid and alkaline pretreatments. The same pattern is observed for the materials based on the walnut shells (Figure 8). Probably, such a treatment leads to the efficient depolymerization of lignocellulose.

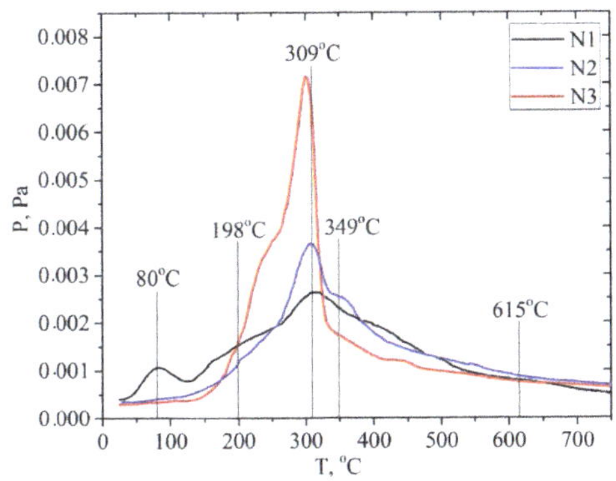

Figure 7. P/T curves of vapor pressure of gaseous products measured at temperature of pyrolysis: solids after hydrolysis of materials based on apricot seed shells: N1—solid after alkaline pretreatment; N2—solid after acid pretreatment; N3—solid after pretreatment with steam explosion.

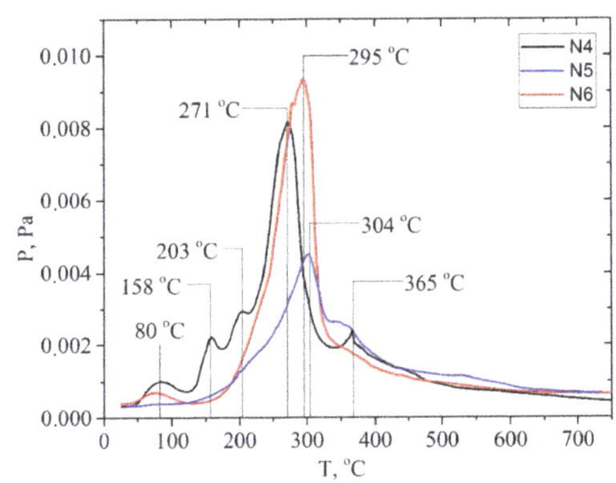

Figure 8. P/T curves of vapor pressure of gaseous products measured at temperature of pyrolysis: solids after hydrolysis of materials based on walnut shells: N4—solid after alkaline pretreatment; N5—solid after acid pretreatment; N6—solid after pretreatment with steam explosion.

The P/T curve has overlapping peaks that correspond to the release of gaseous products due to thermal transformations of individual components of lignocellulose—hemicellulose, cellulose, and lignin. Decomposing the P/T curve into separate Gaussians makes it possible to identify the main stages of pyrolysis at 81, 198, 309, 348, and 615 °C, etc., (Figure 9).

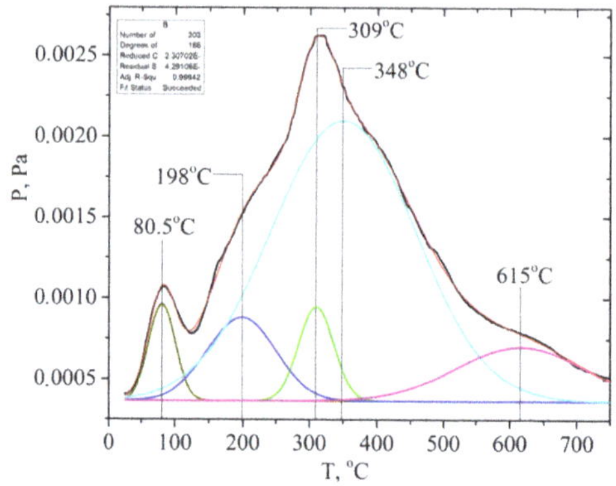

Figure 9. Deconvolution of the P/T curve for the solid based on apricot seed shells after alkaline pretreatment into separate Gaussians. Cellulose: apricot seed shell ——; walnut shell ——; hemicellulose: apricot seed shell ——; walnut shell ——. In addition, the ion with m/z 60 ($[H_2C{=}C(OH)_2]^+$) is the most intense one in the mass spectrum of acetic acid. Hemicellulose pyrolysis is accompanied by the intensive desorption of acetic acid due to the elimination of acetyl groups. Therefore, according to the intensity of peaks on the TPD curve for the ion with m/z 60 at ~200 °C (150–250 °C), one can identify the relative amount of hemicelluloses in the biomass samples, and according to the intensity of the peak at ~300 °C (250–350 °C), one can identify the relative amount of cellulose, as shown in Figures S1–S6 (see SM file). The TPD MS data are in good agreement with the data in Table 3. The highest peak intensity on the TPD curve for m/z 60 at around ~200 °C is observed for the samples of apricot seed shells and walnut shells after the pretreatment with steam explosion (Figures S3 and S6). The lowest intensities are observed for the samples of apricot seed shells and walnut shells after the acid pretreatment (Figures S2 and S5), since acid hydrolysis leads to the almost complete dissolution and removal of hemicellulose from the biomass. In addition, the acid treatment leads to the hydrolysis of acetyl groups with the formation of acetic acid. This process can be used in "green technologies" to produce renewable bio-based acetic acid.

It is known that the decomposition of the main components of plant biomass—hemicellulose, cellulose, and lignin—occurs at specific temperature ranges and is characterized by the temperatures of the maximum desorption rate, Tmax. Pyrolysis begins with the decomposition of the carbohydrate components of the biomass at a temperature of >150 °C, namely hemicellulose. The polymer chain of hemicellulose consists predominantly of monosaccharide five-membered xylose cycles, which are less stable than the six-membered glucose cycles are that make up cellulose. Therefore, the decomposition of cellulose is characterized by a higher Tmax~300 °C than that of hemicelluloses Tmax~200 °C. The most intense marker ion in the mass spectra of carbohydrate pyrolysis products is a fragment ion with m/z 60 ($HOCHCHOH^+$) [51].

The thermal degradation of the lignin polymer chain occurs by the formation of many aromatic compounds. Previously, we found that lignin pyrolysis proceeds in two main stages [52]. Stage I is due to thermal transformations of external phenol–propanoic blocks. It is characterized by the temperature of the maximum desorption rate at about Tmax~320 ± 25 °C. It proceeds in the range of 220–440 °C depending on the type of biomass and its pretreatment (Figures S1–S6). The alkali pretreatment shifts the Tmax of this pyrolysis stage towards lower temperatures. The main pyrolysis products formed during stage I are syringol (m/z 154), methylguaiacol (m/z 138), 4-vinylpyrocatechol (m/z 136), guaiacol (m/z 124), pyrocatechol (m/z 110), phenol (m/z 94), etc.

Pyrolysis stage II is due to the decomposition of the internal phenol–propanoic blocks of lignin, which are interconnected by stronger chemical bonds. In this case, products with a lower molecular weight are formed: o-cresol or p-cresol (m/z 107, 108), phenol (m/z 94), toluene (m/z 91), benzene (m/z 78), etc. This process can be used in "green technologies" to produce renewable bio-based aromatic compounds.

2.6. Comparison with Previous Studies

A few works have been found in which fermentable sugars are produced from walnut shells; Table 4 shows some of these.

In the first of these works, a more complex two-stage process was used: first, a p-TsOH pretreatment, and then an H_2O_2 treatment and enzymatic hydrolysis with the same enzyme as that which was used in our work, Cellic CTec 2, but at a higher dose (40 FPU/g cellulose), achieving a hydrolysis yield of 94.4% [53], which is similar to what we have obtained when the alkaline pretreatment was performed. In the second work, a pretreatment with HNO_3 and subsequent enzymatic hydrolysis with another enzyme ZSL cellulose (40 FPU/g solid) is carried out, reaching a hydrolysis yield of 80% [54], which is higher than that which was obtained when H_2SO_4 was used in our work, approx. 60%, but using a much lower dose of the enzyme than that which we used (15 FPU/g solid). In the last of the works, a pretreatment with a deep eutectic solvent was carried out, but the hydrolysis yield was low (16.94%) [55], as also happened in our case when we were using steam explosion. Table 4 shows more products that have been obtained from walnut shells, such as xylooligosaccharides, lignin, lignin nano-particles, and cellulose nanocrystals. In the case of the production of xylooligosaccharides, different strategies have been proposed. In one of the cases, the researchers proposed a first stage of delignification with $NaClO_2$-CH_3COOH and a second stage of alkaline extraction to recover the hemicelluloses that will later be enzymatically hydrolyzed [56], and on the other hand, the use of a single hydrothermal pretreatment stage [57]. In this latest work, lignin and cellulose nanocrystals are also produced after a second crganosolv delignification stage. To improve the recovery of lignin in another work, sequential organosolv delignification (3n) has been proposed, achieving close to 70% recovery of the initial lignin compared to less than 60% [58].

In the case of apricot seed shells, no previous reports have been found in the reviewed bibliography in which this raw material was used to produce sugars. This raw material has been used mainly to produce biomaterials such as nanocarbon, lignin-derived activated carbon, or cellulose basic-ion exchangers (Table 4). To produce these biomaterials, different processes have been used, for example, in the case of the production of the biomaterial which was used as a biosorbent, the apricot seed shells were subjected to a two-stage process, first, with NH_4OH. and then, they were phosphorylated in an aqueous solution [59]. In the case of the production of cellulose basic-ion exchangers, a two-stage process has also been used, first, a treatment with ammonium, and then, amination using pyridine [60]. For the production of nanocarbons, simpler processes have been used in a single stage, including some more traditional ones using KOH [61] or H_3PO_4/KOH [62] for carbonization, and other ones that are more innovative such as "sol-gel" technology, with the use of catalysts of the type $(CuO)x^*(CoO)y^*(NiO)z^*(Fe_2O_3)k^*(MoO_3)m$/HSZ [63].

2.7. Mass Balance

Mass balance was built for each material and pretreatment type, and it is shown in Figures 10–12. In all of the cases, the results are given for 100 g of raw material for ease of comparison. In the pretreatment stage, the type of pretreatment is the difference between the scheme (i.e., acid, alkaline, and steam explosion ones). As can be seen, in each scheme 62.5 g, 23.2 g, and 76.8 g of pretreated solids from the apricot seed shells and 49.7 g, 20.5 g, and 75.5 g of pretreated solids from then walnut shells were generated. The higher solid yield after steam explosion can be attributed to the fact that steam explosion is a hydrothermal pretreatment used to disrupt the bonding between components in a lignocellulosic complex through the use of high-pressure steam for some time and rapid explosive decompression. This is a type of physicochemical treatment, as high-pressure steam break the lignocellulose structure resulting in the partial degradation of hemicellulose and the removal of extractives. During the acid pretreatment, mostly the hydrolysis of polysaccharides (10.7% in initial apricot seed shells and 25.7% in walnut shells), as well as the removal of extractives, took place. During the alkali pretreatment, mostly the oxidation and removal of lignin (nearly 44.6% in initial material) took place. Both of the processes resulted in a lower yield of solids in comparison to that of the steam explosion experiment.

Table 4. Process steps used to produce different products from walnut shell and apricot seed shells.

Process Steps			Products	Ref.
Step-1	**Step-2**	**Enzymatic Hydrolysis**		
Walnut shell				
p-TsOH pretreatment	H_2O_2 pretreatment	Cellic CTec 2 (40 FPU/g cellulose)	Fermentable sugars	[53]
HNO_3 pretreatment		ZSL cellulose (40 FPU/g solid)	Fermentable sugars	[54]
Deep eutectic solvent		Cellulase from *Trichoderma viride* (Novozymes)	Lignin nano-particles Fermentable sugars	[55]
$NaClO_2$-CH_3COOH delignification	Alkaline extraction	Commercially available endo-1,4-β-xylanase	Xylooligosaccharides	[56]
Hydrothermal pretreatment	Organosolv delignification		Xylooligosaccharides Lignin Cellulose nanocrystals	[57]
Sequential organosolv delignification (3n)	Hydrothermal treatment		Lignin Cellulose nanocrystals	[58]
Apricot seed shells				
NH_4OH pretreatment	Phosphorylated in an aqueous solution		Biosorbent	[59]
"Sol-gel" technology. $(CuO)x^*(CoO)y^*(NiO)z^*(Fe_2O_3)k^*(MoO_3)m$/HSZ based catalyst			Nanocarbon	[60]
H3PO4/KOH carbonization			Lignin-derived activated carbon	[61]
NH_4OH pretreatment	Aminated using pyridine		Cellulose basic-ion exchangers	[62]
KOH carbonization			Nanocarbon	[63]

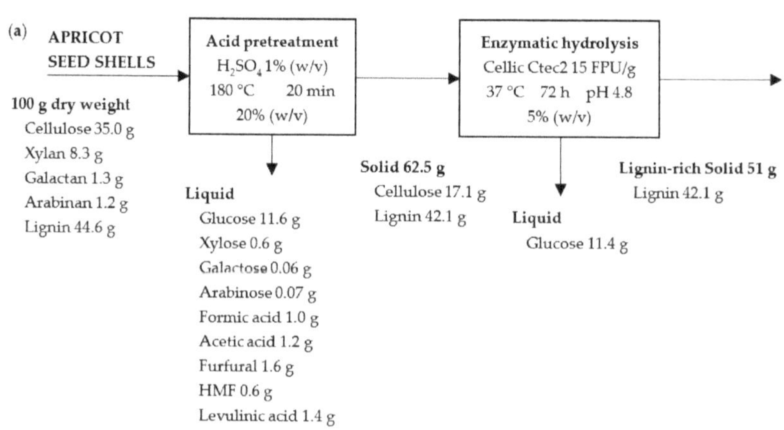

Figure 10. *Cont.*

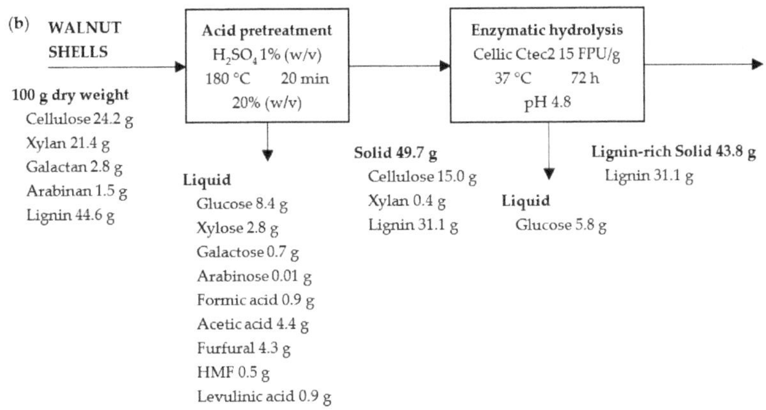

Figure 10. Mass balance for acid pretreatment and enzymatic hydrolysis of apricot seed shells (**a**) and walnut shells (**b**).

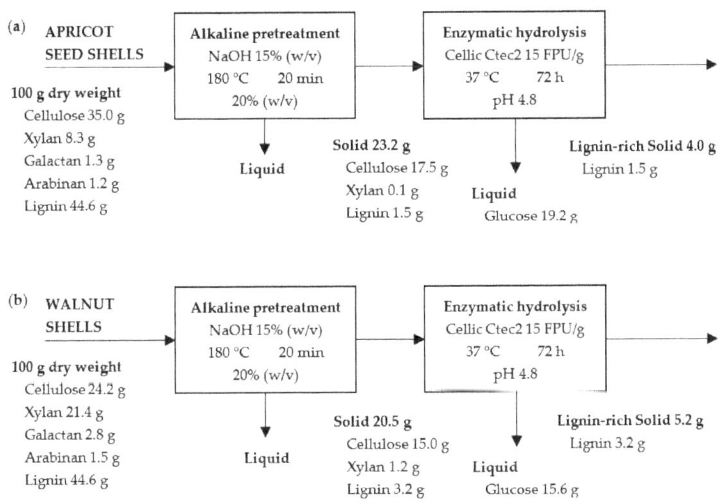

Figure 11. Mass balance for alkaline pretreatment and enzymatic hydrolysis of apricot seed shells (**a**) and walnut shells (**b**).

The second scheme (Figure 11) is characterized by the production of liquids with the most dissolved substances. In the first scheme (Figure 10), complete hemicellulose hydrolysis was achieved for the apricot seed shells, but for walnut shells, the resulting solid contains a small amount of xylose. At the same time, both of the solids are enriched in lignin. In the second scheme, the amount of lignin was reduced significantly for both the raw materials. The same happened regarding the amount of hemicellulose. In the third scheme (Figure 12), the maximum yield of the pretreated solid was obtained. The third scheme is characterized by maintaining the maximum amount of lignin and hemicellulosic sugars in the pretreated materials, resulting in the high solid yield (Figure 12).

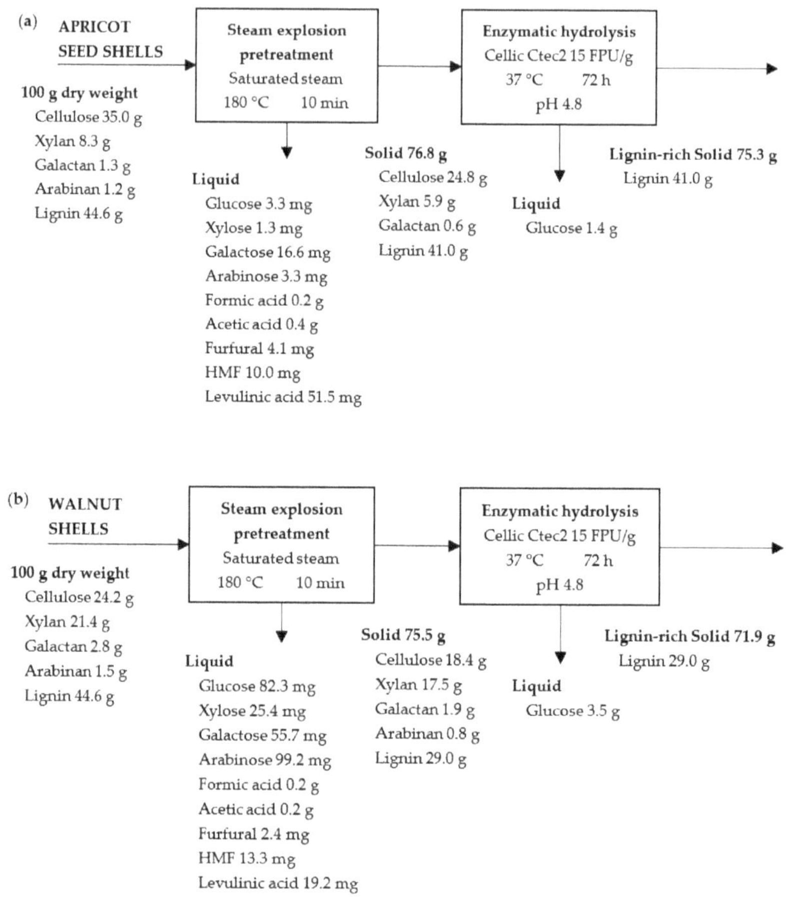

Figure 12. Mass balance for steam explosion pretreatment and enzymatic hydrolysis of apricot seed shells (**a**) and walnut shells (**b**).

The pretreated solids were submitted to enzymatic hydrolysis with an industrial cellulase Cellic CTec2 enzyme complex in the presence of β-glucanase. In the second stage, 11.4 g and 5.8 g of glucose after the acid pretreatment, 19.2 g and 15.6 g of glucose after the alkaline pretreatment, and 1.4 g and 3.5 g of glucose after the steam explosion pretreatment with the apricot seed shells and walnut shells, respectively, were generated. A complete cellulose conversion to glucose was not achieved, but the alkaline pretreatment with 15% NaOH was the most optimal type for the pretreatment of both of the raw materials, resulting in the highest glucose yields. The possible reason for this could be the removal of lignin and hemicellulosic sugars.

3. Materials and Methods

3.1. Materials

Shells of walnuts (*Juglans regia* L.) and shells of apricot seed (*Prunus armeniaca*) from the Odessa region of Ukraine were used as raw materials. The shells were milled. For this purpose, a laboratory hammer mill (Retsch, SM 100, Haan, Germany) was used. A fraction of material passed through a 1 mm screen was kept in a sealed plastic bag and used in the experiments.

Industrial cellulase Cellic CTec2 enzyme complex and β-glucanase were supplied by Novozymes A/S, Bagsværd, Denmark. The chemicals were purchased from Sigma-Aldrich (Darmstadt, Germany).

3.2. Biomass Pretreatment

The fraction of milled shells that passed through the 1 mm screen was used for the pretreatment experiments (dilute acid and alkali solution, as well as steam explosion), and the resulting pretreated solids were submitted to enzymatic hydrolysis. The pretreatments with 1% H_2SO_4 and 15% NaOH (*w*/*v*) were performed in a 1 L agitated tank reactor (Parr Instrument Company, Moline, IL, USA). The biomass loading inside the laboratory reactor was 20% (*w*/*v*). The targeted temperature was 180 °C, and the heating time was 40 min. As the targeted temperature was reached, the processing time was 20 min. Agitation at 300 rpm was applied. At the end of the acid and alkali pretreatment, we stopped heating it, and the reactor was cooled with tap water to room temperature, and the solids were filtered and washed with deionized water. The pretreated solids were oven dried at 50 °C for 48 h to reach a moisture content of 3%.

The steam explosion treatment of the shells without initial impregnation was carried out in a custom-built batch pilot unit in a 4 L reaction vessel. To perform the experiment, the vessel reactor was loaded with 500 g of sample, and then the reactor was heated with saturated steam to 180 °C. The process time was 10 min. Then, the reaction vessel was rapidly depressurized to an atmospheric pressure.

3.3. Enzymatic Hydrolysis

The pretreated solids were used as substrates for a further enzymatic hydrolysis. The prepared materials were hydrolyzed with an industrial cellulase Cellic CTec2 enzyme complex. The cellulase enzyme loading was exactly 15 FPU per 1 g of substrate. To provide the necessary β-glucosidase activity of cellulases, β-glucanase was added. The pH of the solution was adjusted to 4.8 with 0.05 M sodium citrate buffer, and then, the enzymes were added to the lignocellulosic substrate (5% *w*/*v* dry basis). During the experiment, the total working volume was 25 mL in 100 mL Erlenmeyer flasks. All the experiments were performed in triplicate reaction flasks. The flasks were kept at 50 °C for 72 h in an orbital shaker at 150 rpm (Certomat-R, B-Braun, Göttingen, Germany). To evaluate the efficiency of hydrolysis, samples of 1 mL were taken at 24, 48, and 72 h. The samples were centrifuged for 10 min at 10,000× *g* rpm using Sigma 1–14 Centrifuge (Osterode am Harz, Germany), and then used for the determination of glucose and xylose concentrations by high-performance liquid chromatography (HPLC, Agilent Technologies, Palo Alto, CA, USA).

3.4. Raw Material, Residual Solid, and Liquid Fraction (Hydrolysate) Characterization

The chemical composition of the biomass was investigated according to the methodology from NREL (National Renewable Energy Laboratory, Golden, CO, USA) [64–67]. The pretreated solids yields were determined gravimetrically. The liquid fraction was also analyzed and previously filtered through 0.45 μm nylon membranes. The monomer sugars, such as glucose, xylose, mannose, galactose, and arabinose, and inhibitor composition, such as acetic acid, formic acid, furfural, and hydroxymethylfurfural (HMF), of the liquid fraction after acid and steam explosion pretreatments were determined by HPLC using an Agilent Technologies 1200 series HPLC system (Santa Clara, CA, USA), which was equipped with ICSep ICE-COREGEL 87H3 column. The column was operating at 65 °C, and 5 mM sulfuric acid was used as the mobile phase (0.6 mL/min).

The amount of polysaccharide, acid-insoluble lignin, and acid-soluble lignin in the solids was also determined according to methodology from NREL. The hydrolysates after the pretreatment were also analyzed, and the amount of sugars were determined. Glucose and xylose recoveries were calculated as the ratio of sugars that remains in the pretreated solids and liquid fractions to sugars present in the raw material. Monomeric and oligomeric sugars in the liquid fractions after pretreatments were investigated by HPLC, as well as the amount of formic acid, acetic acid, furfural, hydroxymethylfurfural (HMF), and levulinic acid.

Glucose and xylose concentrations in the sample supernatant were determined by HPLC using the liquid chromatograph with a refractive index detector (CARBOSep CHO-782 Pb, Transgenomic, Inc., Omaha, NE, USA) with ultra-pure water as an eluent. The flow rate was 0.6 mL/min, and the column temperature was 70 °C.

Fourier transform infrared (FTIR) spectroscopy using a Perkin-Elmer Spectrum 100 FTIR spectrometer (Golden gate from Graseby Specac LTD, Kent, England) was used to evaluate structural changes in the solids. All of the materials were analyzed in the 4000–500 cm^{-1} range at a resolution of 4 cm^{-1}.

All of the experiments and determinations were carried out in triplicate. Relative standard deviations are below 5%.

3.5. *Temperature-Programmed Desorption Mass Experiments*

Temperature-programmed desorption spectrometry (TPD MS) was performed using a monopole mass spectrometer (MKh-7304A, Sumy, Ukraine) with electron impact ionization adapted to study pyrolysis kinetics [68–70]. About 8 mg of biomass samples were used for each experiment. TPD measurements were performed at a constant rate of 0.168 °C/s. Mass spectra, TPD curves, and the P/T pressure–temperature curves (P—the pressure of volatile pyrolysis products; T—temperature of the sample) were created and analyzed in a computer-based data acquisition and processing step.

4. Conclusions

The agricultural residues largely available in Ukraine such as apricot seed shells and walnut shells, which are underutilized, have a low cost, and lack applications, can be regarded as suitable renewable raw materials for multiproduct biorefinery. An evaluation of the influence of the pretreatments on raw materials shows that alkaline pretreatment is characterized by minimal solid recovery due to the removal of large amounts of hemicelluloses and lignin. Acid pretreatment is effective at solubilizing hemicellulose, while steam explosion has a less destructive effect on lignin and hemicellulose. The removal of the mentioned components exposes cellulose, resulting in an increase in its accessibility. Alkaline pretreatment allows us to achieve a high glucose yield. Both of the materials are a great alternative and prospective source of fermentable sugars, which indicates their potential in biofuel production (energy independence) for other bioproducts such as bioplastics. The results show that the alkali pretreatment is more advantageous, not only because, overall, a higher glucose yield is obtained, but also because it is industrially more feasible as lignin can be easily precipitated from the spent liquor and processed again into value-added products, as well as sodium hydroxide, which can be easily recovered and returned to the technological process. The proposed application represents an innovative method of disposal on the one hand, and the possibility of producing renewable compounds on the other side, thus contributing to the development of a circular bioeconomy.

Future work will focus on the conversion of sugars into valuable compounds as a result of a fermentation process; for these studies, attention will be paid especially to the formation of potential inhibitors of microorganism actions due to the conditions of the pretreatment, and a compromise between a high sugar yield and minimum degradation will be procured. In addition to sugars, a relevant future development will be the characterization of lignins in terms of average molecular weight, average molecular number, and polydispersity index, among others, with the general objective of taking full advantage of these renewable feedstocks.

Supplementary Materials: The following supporting information can be downloaded at: https://www.mdpi.com/article/10.3390/molecules28031455/s1, Figure S1: TPD curves for ions with m/z 154, 138, 136, 124, 110, 109, 108, 107, 94, 91, 78, 60, and 55 obtained via pyrolysis of the solid based on apricot seed shells after alkaline pretreatment; Figure S2: TPD curves for ions with m/z 154, 138, 136,

124, 110, 109, 108, 107, 94, 91, 78, 60, and 55 obtained via pyrolysis of the solid based on apricot seed shells after acid pretreatment; Figure S3: TPD curves for ions with m/z 154, 138, 136, 124, 110, 109, 108, 107, 94, 91, 78, 60, and 55 obtained via pyrolysis of the solid based on apricot seed shells after pretreatment with steam explosion; Figure S4: TPD curves for ions with m/z 154, 138, 136, 124, 110, 109, 108, 107, 94, 91, 78, 60, and 55 obtained via pyrolysis of the solid based on walnut shells after alkaline pretreatment; Figure S5: TPD curves for ions with m/z 154, 138, 136, 124, 110, 109, 108, 107, 94, 91, 78, 60, and 55 obtained via pyrolysis of the solid based on walnut shells after acid pretreatment; Figure S6: TPD curves for ions with m/z 154, 138, 136, 124, 110, 109, 108, 107, 94, 91, 78, 60, and 55 obtained via pyrolysis of the solid based on walnut shells after pretreatment with steam explosion.

Author Contributions: Conceptualization, V.H. and E.C.; methodology, J.M.R.-G. and M.G.; software, J.M.R.-G.; validation, V.H., J.M.R.-G. and A.M.V.; formal analysis, T.K. and B.P.; investigation, V.H., J.M.R.-G. and A.M.V.; resources, E.C.; data curation, V.H., T.K. and B.P.; writing—original draft preparation, V.H.; writing—review and editing, E.C.; visualization, T.K. and B.P.; supervision, E.C.; project administration, E.C.; funding acquisition, V.H. All authors have read and agreed to the published version of the manuscript.

Funding: This research was funded by CostAction (CA17128) (European Cooperation in Science and Technology). Junta de Andalucía (Postdoctoral researcher R-29/12/2020).

Institutional Review Board Statement: Not applicable.

Informed Consent Statement: Not applicable.

Data Availability Statement: Not applicable.

Acknowledgments: Vita Halysh is grateful to CostAction (CA17128) for the financial support to Short Term Scientific Mission, which is supported by COST (European Cooperation in Science and Technology). J.M. Romero-García expresses his gratitude to the Junta de Andalucía for financial support (Postdoctoral researcher R-29/12/2020).

Conflicts of Interest: The authors declare no conflict of interest.

References

1. Miller, R.G.; Sorrell, S.R. The future of oil supply. *Philos. Trans. A Math. Phys. Eng. Sci.* **2014**, *372*, 20130179. [CrossRef] [PubMed]
2. Bay, M.S.; Karimi, K.; Nasr Esfahany, M.; Kumar, R. Structural modification of pine and poplar wood by alkali pretreatment to improve ethanol production. *Ind. Crop. Prod.* **2020**, *152*, 112506. [CrossRef]
3. Grassi, M.C.B.; Pereira, G.A.G. Energy-cane and RenovaBio: Brazilian vectors to boost the development of Biofuels. *Ind. Crop. Prod.* **2019**, *129*, 201–205. [CrossRef]
4. Patni, N.; Pillai, S.G.; Dwivedi, A.H. Wheat as a promising substitute of corn for bioethanol production. *Procedia Eng.* **2013**, *51*, 355–362. [CrossRef]
5. Akao, S.; Yasutake, D.; Kondo, K.; Nagare, H.; Maeda, M.; Fujiwara, T. Effects of cultivation period on catch crop chemical composition and potential for bioenergy production. *Ind. Crop. Prod.* **2018**, *111*, 787–793. [CrossRef]
6. Chavan, S.; Gaikwad, A. Optimization of enzymatic hydrolysis of bamboo biomass for enhanced saccharification of cellulose through Taguchi orthogonal design. *J. Environ. Chem. Eng.* **2020**, *9*, 104807. [CrossRef]
7. Duruyurek, M.; Düşgün, C.; Gulhan, M.; Selamoglu, Z. Production of Bioethanol from Waste Potato. *Turkish J. Agric.-Food Sci. Tech.* **2015**, *3*, 331–334. [CrossRef]
8. Nashiruddin, N.I.; Mansor, A.F.; Rahman, R.A.; Ilias, R.M.; Yussof, H.W. Process parameter optimization of pretreated pineapple leaves fiber for enhancement of sugar recovery. *Ind. Crop. Prod.* **2020**, *152*, 112514. [CrossRef]
9. Kartel, M.; Galysh, V. New composite sorbents for Caesium and Strontium ions sorption. *Chem. J. Mold.* **2017**, *12*, 37–44. [CrossRef]
10. Deykun, I.; Halysh, V.; Barbash, V. Rapeseed straw as an alternative for pulping and papermaking. *Cellulose Chem. Technol.* **2018**, *52*, 833–839.
11. Halysh, V.; Trembus, I.; Deykun, I.; Ostapenko, A.; Nikolaichuk, A.; Ilnitska, G. Development of effective technique for the disposal of the Prunus Armeniaca seed shells. *East-Eur. J. Enterp. Technol.* **2018**, *1*, 4–9. [CrossRef]
12. Mahari, W.A.V.; Waiho, K.; Fazhan, H.; Necibi, M.C.; Hafsa, J.; Mrid, R.B.; Fal, S.; Arrousi, H.E.; Peng, W.; Tabatabaei, M.; et al. Progress in valorisation of agriculture, aquaculture and shellfish biomass into biochemicals and biomaterials towards sustainable bioeconomy. *Chemosphere* **2021**, *291*, 133036. [CrossRef]
13. Peinemann, J.C.; Pleissner, D. Continuous pretreatment, hydrolysis, and fermentation of organic residues for the production of biochemicals. *Bioresour. Technol.* **2020**, *295*, 122256. [CrossRef]
14. Gilna, P.; Lynd, L.R.; Mohnen, D.; Davis, M.F.; Davison, B.H. Progress in understanding and overcoming biomass recalcitrance: A BioEnergy Science Center (BESC) perspective. *Biotechnol. Biofuels* **2017**, *10*, 285. [CrossRef]

15. Zhang, Y.; Huang, M.; Su, J.; Hu, H.; Yang, M.; Huang, Z.; Chen, D.; Wu, J.; Feng, Z. Overcoming biomass recalcitrance by synergistic pretreatment of mechanical activation and metal salt for enhancing enzymatic conversion of lignocellulose. *Biotechnol. Biofuels* **2019**, *12*, 12. [CrossRef]
16. Meng, X.; Pu, Y.; Yoo, C.G.; Li, M.; Bali, G.; Park, D.Y.; Gjersing, E.; Davis, M.F.; Muchero, W.; Tuskan, G.A.; et al. An in-depth understanding of biomass recalcitrance using natural poplar variants as the feedstock. *Chem. Sus. Chem.* **2017**, *10*, 139–150. [CrossRef]
17. Kucharska, K.; Rybarczyk, P.; Hołowacz, I.; Łukajtis, R.; Glinka, M.; Kamiński, M. Pretreatment of lignocellulosic materials as substrates for fermentation processes. *Molecules* **2018**, *23*, 2937. [CrossRef]
18. Momayez, F.; Karimi, K.; Karimi, S.; Horváth, I.S. Efficient hydrolysis and ethanol production from rice straw by pretreatment with organic acids and effluent of biogas plant. *RSC Adv.* **2017**, *7*, 50537–50545. [CrossRef]
19. Xing, Y.; Yu, H.; Zhu, L.; Jiang, J. Efficient enzymatic hydrolysis of bamboo by pretreatment with steam explosion and alkaline peroxide. *BioResources* **2013**, *8*, 5392–5408. [CrossRef]
20. Cutrim, F.M.; Ramos, E.C.S.S.; Abreu, M.C.C.; Godinho, A.S.; Maciel, A.P.; Mendonça, C.J.S.; Cavalcante, K.S.B. A study of chemical composition and enzymatic hydrolysis of solid organic waste from Agrosilvopastoral systems. *J. Braz. Chem. Soc.* **2019**, *30*, 1955–1963. [CrossRef]
21. Kontogianni, N.; Barampouti, E.M.; Mai, S.; Malamis, D.; Loizidou, M. Effect of alkaline pretreatments on the enzymatic hydrolysis of wheat straw. *Environ. Sci. Pollut. Res.* **2019**, *26*, 35648–35656. [CrossRef] [PubMed]
22. De Carvalho, D.M.; Sevastyanova, O.; Penna, L.S.; de Silva, B.P.; Lindström, M.E.; Colodette, J.L. Assessment of chemical transformations in eucalyptus, sugarcane bagasse and straw during hydrothermal, dilute acid, and alkaline pretreatments. *Ind. Crop. Prod.* **2015**, *73*, 118–126. [CrossRef]
23. Romero-García, J.M.; Lama-Muñoz, A.; Rodríguez-Gutiérrez, G.; Moya, M.; Ruiz, E.; Fernández-Bolaños, J.; Castro, E. Obtaining sugars and natural antioxidants from olive leaves by steam-explosion. *Food Chem.* **2016**, *210*, 457–465. [CrossRef] [PubMed]
24. Li, G.; He, W.; Yuan, L. Aqueous ammonia pretreatment of sugar beet pulp for enhanced enzymatic hydrolysis. *Bioprocess Biosyst. Eng.* **2017**, *40*, 1603–1609. [CrossRef]
25. FAO—Food and Agriculture Organization. Food and Agriculture Data. Available online: https://www.fao.org/faostat/en/#home (accessed on 18 January 2023).
26. Halysh, V.; Trus, I.; Nikolaichuk, A.; Skiba, M.; Radovenchyk, I.; Deykun, I.; Vorobyova, V.; Vasylenko, I.; Sirenko, L. Spent biosorbents as additives in cement production. *J. Ecol. Eng.* **2020**, *21*, 131–138. [CrossRef]
27. Halysh, V.; Sevastyanova, O.; de Carvalho, D.M.; Riazanova, A.V.; Lindström, M.E.; Gomelya, M. Effect of oxidative treatment on composition and properties of sorbents prepared from sugarcane residue. *Ind. Crop. Prod.* **2019**, *139*, 111566. [CrossRef]
28. Trembus, I.V.; Trophimchuk, J.S.; Galysh, V.V. Preparation of pulp from sunflower stalks using peroxy acids. *Voprosy Khimii i Khimicheskoi Tekhnologii* **2018**, *2*, 122–127.
29. Trembus, I.; Hondovska, A.; Halysh, V.; Deykun, I.; Cheropkina, R. Feasible Technology for Agricultural Residues Utilization for the Obtaining of Value-Added Products. *Ecol. Eng. Environ. Technol.* **2022**, *2*, 107–112. [CrossRef]
30. Yu, Z.; Jameel, H.; Chang, H.M.; Park, S. The effect of delignification of forest biomass on enzymatic hydrolysis. *Bioresour. Technol.* **2011**, *102*, 9083–9089. [CrossRef]
31. Park, J.; Shin, H.; Yoo, S.; Zoppe, J.O.; Park, S. Delignification of lignocellulosic biomass and its effect on subsequent enzymatic hydrolysis. *BioResources* **2015**, *10*, 2732–2743. [CrossRef]
32. Corbett, D.B.; Kohan, N.; Machado, G.; Jing, C.; Nagardeolekar, A.; Bujanovic, B.M. Chemical composition of apricot pit shells and effect of hot-water extraction. *Energies* **2015**, *8*, 9640–9654. [CrossRef]
33. Queirós, C.; Cardoso, S.; Lourenço, A.; Ferreira, J.; Miranda, I.; Lourenço, M.; Pereira, H. Characterization of walnut, almond, and pine nut shells regarding chemical composition and extract composition. *Biomass Conv. Bioref.* **2020**, *10*, 175–188. [CrossRef]
34. Pirayesh, H.; Khazaeian, A.; Tabarsa, T. The potential for using walnut (*Juglans regia* L.) shell as a raw material for wood-based particleboard manufacturing. *Compos. Part B—Eng.* **2012**, *43*, 3276–3280. [CrossRef]
35. Wu, Z.; Hao, H.; Tu, Y.; Hu, Z.; Wei, F.; Liu, Y.; Zhou, Y.; Wang, Y.; Xie, G.; Gao, C.; et al. Diverse cell wall composition and varied biomass digestibility in wheat straw for bioenergy feedstock. *Biomass Bioenergy* **2014**, *70*, 347–355. [CrossRef]
36. Oka, D.; Kobayashi, K.; Isobe, N.; Ogawa, Y.; Yokoyama, T.; Kimura, S.; Kim, U.; Tokuyasu, K.; Wada, M. Enzymatic hydrolysis of wood with alkaline treatment. *J. Wood Sci.* **2013**, *59*, 484–488. [CrossRef]
37. Siddhu, M.A.H.; Li, W.; He, Y.; Liu, G.; Chen, C. Steam explosion pretreatment of rice straw to improve structural carbohydrates anaerobic digestibility for biomethanation. *Environ. Sci. Pollut. Res.* **2019**, *26*, 22189–22196. [CrossRef]
38. Benjamin, Y.; Cheng, H.; Görgens, J.F. Optimization of dilute sulfuric acid pretreatment to maximize combined sugar yield from sugarcane bagasse for ethanol production. *Appl. Biochem. Biotechnol.* **2014**, *172*, 610–630. [CrossRef]
39. Kim, J.S.; Lee, Y.Y.; Kim, T.H. A review on alkaline pretreatment technology for bioconversion of lignocellulosic biomass. *Bioresour. Technol.* **2016**, *199*, 42–48. [CrossRef]
40. Mafa, M.S.; Malgas, S.; Bhattacharya, A.; Rashamuse, K.; Pletschke, B.I. The effects of alkaline pretreatment on agricultural biomasses (corn cob and sweet sorghum bagasse) and their hydrolysis by a termite-derived enzyme cocktail. *Agronomy* **2020**, *10*, 1211. [CrossRef]
41. Herbaut, M.; Zoghlami, A.; Habrant, A.; Falourd, X.; Foucat, L.; Chabbert, B.; Paës, G. Multimodal analysis of pretreated biomass species highlights generic markers of lignocellulose recalcitrance. *Biotechnol. Biofuels* **2018**, *11*, 52. [CrossRef]

42. Steinbach, D.; Kruse, A.; Sauer, J.; Storz, J. Is Steam Explosion a Promising Pretreatment for Acid Hydrolysis of Lignocellulosic Biomass. *Processes* **2020**, *8*, 1626. [CrossRef]
43. Auxenfans, T.; Crônier, D.; Chabbert, B.; Paës, G. Understanding the structural and chemical changes of plant biomass following steam explosion pretreatment. *Biotechnol. Biofuels* **2017**, *10*, 36. [CrossRef] [PubMed]
44. He, Q.; Ziegler-Devin, I.; Chrusciel, L.; Ngwa Obame, S.; Hong, L.; Lu, X.; Brosse, N. Lignin-first integrated steam explosion process for green wood adhesive application. *ACS Sustain. Chem. Eng.* **2020**, *8*, 5380–5392. [CrossRef]
45. Kemppainen, K.; Inkinen, J.; Uusitalo, J.; Nakari-Setälä, T.; Siika-aho, M. Hot water extraction and steam explosion as pretreatments for ethanol production from spruce bark. *Biores Technol.* **2012**, *117*, 131–139. [CrossRef] [PubMed]
46. Wang, X.T.; Liu, L.S. Steam Explosion Pretreatment Technique and Application in Biomass Conversion. *Adv. Mater. Res.* **2010**, *113–116*, 525–528. [CrossRef]
47. Singh, J.; Suhag, M.; Dhaka, A. Augmented digestion of lignocellulose by steam explosion, acid and alkaline pretreatment methods: A review. *Carbohydr. Polym.* **2015**, *117*, 624–631. [CrossRef]
48. Díaz-Villanueva, M.; Cara-Corpas, C.; Ruiz-Ramos, E.; Romero-Pulido, I.; Castro-Galiano, E. Olive tree pruning as an agricultural residue for ethanol production. Fermentation of hydrolysates from dilute acid pretreatment. *Span. J. Agric. Res.* **2012**, *10*, 643–648. [CrossRef]
49. Baadhe, R.R.; Potumarthi, R.; Mekala, N.K. Influence of dilute acid and alkali pretreatment on reducing sugar production from corncobs by crude enzymatic method: A comparative study. *Bioresour. Technol.* **2014**, *162*, 213–217. [CrossRef]
50. Shimizu, F.L.; Monteiro, P.Q.; Ghiraldi, P.H.C.; Melati, R.B.; Pagnocca, F.C.; Souza, W.; Sant'Anna, C.; Brienzo, M. Acid, alkali and peroxide pretreatments increase the cellulose accessibility and glucose yield of banana pseudostem. *Ind. Crop. Prod.* **2018**, *115*, 62–68. [CrossRef]
51. Lebedev, A. *Mass Spectrometry in Organic Chemistry*; Binom: Moscow, Russia, 2003; 493p.
52. Palianytsia, B.; Kulik, T.; Dudik, O.; Cherniavska, T.; Tonkha, O. Study of the thermal decomposition of some components of biomass by desorption mass spectrometry. In Proceedings of the International Congress on Energy Efficiency and Energy Related Materials (ENEFM2013), Antalya, Turkey, 9–12 October 2013; pp. 19–25.
53. Zhu, J.; Jiao, N.; Li, H.; Xu, G.; Zhang, H.; Xu, Y. P-Toluenesulfonic Acid Combined with Hydrogen Peroxide-Assisted Pretreatment Improves the Production of Fermentable Sugars from Walnut (*Juglans regia* L.) Shells. *Bioresour. Technol.* **2022**, *355*, 127300. [CrossRef]
54. Tan, M.; Ma, L.; Rehman, M.S.U.; Ahmed, M.A.; Sajid, M.; Xu, X.; Sun, Y.; Cui, P.; Xu, J. Screening of Acidic and Alkaline Pretreatments for Walnut Shell and Corn Stover Biorefining using Two Way Heterogeneity Evaluation. *Renew. Energy* **2019**, *132*, 950–958. [CrossRef]
55. Li, H.; Liang, J.; Chen, L.; Ren, M.; Zhou, C. Utilization of Walnut Shell by Deep Eutectic Solvents: Enzymatic Digestion of Cellulose and Preparation of Lignin Nanoparticles. *Ind. Crop. Prod.* **2023**, *192*, 116034. [CrossRef]
56. Cebin, A.V.; Ralet, M.; Vigouroux, J.; Karača, S.; Martinić, A.; Komes, D.; Bonnin, E. Valorisation of Walnut Shell and Pea Pod as Novel Sources for the Production of Xylooligosaccharides. *Carbohydr. Polym.* **2021**, *263*, 117932. [CrossRef]
57. Morales, A.; Labidi, J.; Gullón, P. Hydrothermal Treatments of Walnut Shells: A Potential Pretreatment for Subsequent Product Obtaining. *Sci. Total Environ.* **2021**, *764*, 142800. [CrossRef]
58. Morales, A.; Labidi, J.; Gullón, P. Integral Valorisation of Walnut Shells Based on a Three-Step Sequential Delignification. *J. Environ. Manag.* **2022**, *310*, 114730. [CrossRef]
59. Yelatontsev, D. Production of Versatile Biosorbent Via Eco-Friendly Utilization of Non-Wood Biomass. *Chem. Eng. J.* **2023**, *451*, 138811. [CrossRef]
60. Yelatontsev, D.A.; Mukhachev, A.P. Synthesis and Properties of Ion Exchangers Derived from Non-Wood Cellulose. *ChemChemTech* **2020**, *63*, 88–95. [CrossRef]
61. Saha, D.; Taylor, B.; Alexander, N.; Joyce, D.F.; Faux, G.I.; Lin, Y.; Shteyn, V.; Orkoulas, G. One-Step Conversion of Agro-Wastes to Nanoporous Carbons: Role in Separation of Greenhouse Gases. *Bioresour. Technol.* **2018**, *256*, 232–240. [CrossRef]
62. Wang, L.; Feng, X.; Li, X.; Ma, H.; Wu, J.; Chen, Y.; Zhou, J. Valorization of Lignin: Application of Lignin-Derived Activated Carbon in Capacitors and Investigation of its Textural Properties and Electrochemical Performance. *Diam. Relat. Mat.* **2022**, *122*, 108791. [CrossRef]
63. Xolmirzayeva, H.N.; Fayzullayev, N.I. Obtaining Nanocarbon from Local Raw Materials and Studying its Textural and Sorption Properties. *Int. J. Eng. Trends Technol.* **2022**, *70*, 163–171. [CrossRef]
64. NREL—National Renewable Energy Laboratory. Determination of Structural Carbohydrates and Lignin in Biomass. Available online: https://www.nrel.gov/docs/gen/fy13/42618.pdf (accessed on 1 January 2022).
65. NREL—National Renewable Energy Laboratory. Ash in Biomass. Available online: https://www.nrel.gov/docs/gen/fy08/42622.pdf (accessed on 1 January 2022).
66. NREL—National Renewable Energy Laboratory. Extractives in Biomass. Available online: https://www.nrel.gov/docs/gen/fy08/42619.pdf (accessed on 1 January 2022).
67. NREL—National Renewable Energy Laboratory. Determination of Sugars, Byproducts, and Degradation Products in Liquid Fraction Process Samples. Available online: https://www.nrel.gov/docs/gen/fy08/42623.pdf (accessed on 1 January 2022).

68. Kulyk, K.; Palianytsia, B.; Alexander, J.D.; Azizova, L.; Borysenko, M.; Kartel, M.; Larsson, M.; Kulik, T. Kinetics of Valeric Acid Ketonization and Ketenization in Catalytic Pyrolysis on Nanosized SiO_2, γ-Al_2O_3, CeO_2/SiO_2, Al_2O_3/SiO_2 and TiO_2/SiO_2. *Chem. Phys. Chem.* **2017**, *18*, 1943–1955. [CrossRef] [PubMed]
69. Kulyk, K.; Borysenko, M.; Kulik, T.; Mikhalovska, L.; Alexander, J.D.; Palianytsia, B. Chemisorption and thermally induced transformations of polydimethylsiloxane on the surface of nanoscale silica and ceria/silica. *Polym. Degrad. Stab.* **2015**, *120*, 203–211. [CrossRef]
70. Kulik, T.; Palianytsia, B.; Larsson, M. Catalytic pyrolysis of aliphatic carboxylic acids into symmetric ketones over ceria-based catalysts: Kinetics, isotope effect and mechanism. *Catalysts* **2020**, *10*, 179. [CrossRef]

molecules

MDPI

Article

Analysis of Single-Step Pretreatments for Lignocellulosic Platform Isolation as the Basis of Biorefinery Design

Jhonny Alejandro Poveda-Giraldo, Maria Camila Garcia-Vallejo and Carlos Ariel Cardona Alzate *

Institute of Biotechnology and Agribusiness, Department of Chemical Engineering, Universidad Nacional de Colombia sede Manizales, Manizales 170001, Colombia

* Correspondence: ccardonaal@unal.edu.co; Tel.: +57-6-8879300 (ext. 50417); Fax: +57-6-8879300 (ext. 50452)

Abstract: Biorefinery feasibility is highly influenced by the early design of the best feedstock transformation pathway to obtain value-added products. Pretreatment has been identified as the critical stage in biorefinery design since proper pretreatment influences subsequent reaction, separation, and purification processes. However, many pretreatment analyses have focused on preserving and valorizing six-carbon sugars for future use in bioconversion processes, leaving aside fractions such as hemicellulose and lignin. To date, there has been no pretreatment systematization for the removal of lignocellulosic fractions. This work defines pretreatment efficacy through operational, economic, environmental, and social indicators. Thus, using the data reported in the literature, as well as the results of the simulation schemes, a multi-criteria weighting of the best-performing schemes for the isolation or removal of cellulose, hemicellulose, and lignin was carried out. As a main result, it was concluded that dilute acid is the most effective for cellulose isolation and hemicellulose removal for producing platform products based on six- and five-carbon sugars, respectively. Additionally, the kraft process is the best methodology for lignin removal and its future use in biorefineries. The results of this work help to elucidate a methodological systematization of the pretreatment efficacy in the design of biorefineries as an early feasibility stage considering sustainability aspects.

Keywords: biorefineries; pretreatment efficacy; lignocellulosic fractionation; heuristic analysis; techno-economic analysis; social impact

Citation: Poveda-Giraldo, J.A.; Garcia-Vallejo, M.C.; Cardona Alzate, C.A. Analysis of Single-Step Pretreatments for Lignocellulosic Platform Isolation as the Basis of Biorefinery Design. *Molecules* **2023**, *28*, 1278. https://doi.org/10.3390/molecules28031278

Academic Editors: Alejandro Rodriguez Pascual, Eduardo Espinosa Víctor, Carlos Martín and Dimitris P. Makris

Received: 29 November 2022
Revised: 19 January 2023
Accepted: 26 January 2023
Published: 28 January 2023

1. Introduction

The integral use of biomass for its subsequent conversion into high-value-added products is an approach associated with the biorefinery concept, where the product portfolio may comprise bulk and/or specialized chemical products [1]. The biorefinery concept starts from the analogy with crude oil refineries, where multiple products are generated from petroleum. To this end, several methodologies have been proposed for the conceptual design, optimization, and implementation of biorefineries [2]. Initially, a biorefinery should be understood as a network of facilities that integrates technologies, feedstocks, and equipment to transform biomass into products and energy [3]. However, other authors have proposed a broader definition specifying that *"a biorefinery is a complex system in which biomass is integrally processed or fractionated to obtain more than one product, including bioenergy, biofuels, chemicals, and high value-added compounds that can only be extracted from bio-based sources. The latter after a comprehensive study of the raw materials to be used and a sustainability analysis based on the latest state of the art technologies and approaches which include aspects of the three pillars of sustainability"* [1]. Therefore, the biorefinery design should include the evaluation of feedstocks, technologies and equipment, transformation routes, and products through technical, economic, environmental, and social analyses. The analysis of the dimensions is justified under the objectives of the biorefinery: (i) to maximize the value of the products; (ii) to reduce greenhouse gas emissions; (iii) to reduce dependence on non-renewable fuels; (iv) to stimulate rural development; and (iv) to promote the social welfare of the population [4]. The maximum use of biomass and the minimum production of waste, contributing economically, environmentally, and socially, is a sustainability challenge faced by the design of biorefineries. In this way, the objectives of biorefineries are framed in fulfilling the Sustainable Development Goals (SDGs) [5]. Additionally, different authors have established the direct relationship that the implementation of biorefineries has with the fulfillment of the 2030 agenda, framing biorefineries as a possible solution to multiple problems faced by society [6].

Different steps are involved during the biorefinery design by the early-stage approach: hierarchy, sequencing, and integration [7]. The hierarchization involves the global elements of the biorefinery, understood as feedstock flows, products, and transformation routes. The sequencing illustrates the logical order in which the unitary processes should be developed to maximize resource use. Finally, integration defines the possibilities of integrating raw materials, technologies, and products to obtain greater benefits [8]. Due to the diversity of biomass, social contexts, route and technology transformations, government policies, and local demand, biorefineries have become complex systems to analyze [9]. For this reason, a fundamental axis of biorefineries should be the contextualization of where the process will be developed. Therefore, preliminary localization studies are mandatory to benefit adjacent markets, considering the restrictions and identifying the possible economic areas that can be satisfied [10].

Biorefinery design involves different stages, starting with evaluating and selecting potential feedstocks. Then, the biomass is pretreated to recover or isolate its fractions to be further subjected to a combination of biological and/or chemical processes [1]. Based on the above and considering that biorefineries are established as a tool to promote the sustainability of production processes, biomass acquires relevance. Lignocellulosic biomass emerges as a fundamental axis to promote a sustainable society under the waste valorization transformed into value-added products [11]. Although biomass can mitigate certain pollution issues caused by fossil fuels, some authors have identified challenges related to availability, supply logistics, and conversion yields toward platform products [12]. The difficulties associated with the conversion processes show the importance of biomass conditioning, representing one of the most costly and relevant stages of the process [13]. Pretreatments involve physical and chemical changes in the biomass structure [11]. Physical pretreatments focus on increasing the surface area of the material to improve hydrolysis processes. Often, these structural changes are achieved by reducing particle size, increasing porosity, or altering structural regularity. For example, some authors suggest that using pretreatments such as ultrasound produces perforations in the biomass, increasing porosity and yield in enzymatic hydrolysis [14]. On the other hand, chemical pretreatments aim to generate ultrastructural and chemical modifications in the cell wall of the biomass. These modifications involve the fractionation of the polymers, which results in better accessibility by enzymes and better utilization of the lignocellulosic fractions [11]. The main components of biomass are cellulose, hemicellulose, and lignin coupled in a constitutive complex structure [15]. Currently, most applied pretreatments focus on recovering the cellulosic fraction for further conversion into biofuels [16]. However, in the design of biorefineries, pretreatments must seek the greatest benefit from all available lignocellulosic fractions to obtain value-added products from each one or their possible combination. For example, after diluted acid pretreatments, the xylose-rich fraction could be dehydrated for furfural production and used for agrochemical and solvent applications. Some studies have also studied the xylose valorization for xylitol production from steam-exploded corn straw due to the nutritional properties of this sugar alcohol [17]. Similarly, kraft pretreatment allows obtaining a black liquor with soluble lignin that can be fractionated for vanillin production. Pretreatments such as steam explosions make it possible to obtain concentrated cellulose with multiple valorization possibilities, such as the production of biofuels, bioproducts, and biosurfactants.

The main purpose of biomass pretreatment should be the access of biocatalysts to plant polysaccharides to be converted into platform products [18]. Additionally, some authors have reported that an adequate pretreatment should (i) decrease the enzymatic load necessary for the hydrolysis processes, (ii) avoid the loss of valuable fractions to platform products, (iii) minimize the generation of inhibitory compounds for the hydrolysis or fermentation processes, and (iv) allow the recovery of other fractions (such as lignin and hemicellulose) that can be converted into valuable co-products [18]. Thus, pretreatments become relevant in promoting the availability of lignocellulosic fractions and increasing the productivity of the process. For example, it has been reported that adequate pretreatment can increase the rate of enzymatic hydrolysis of lignocellulosic materials from 3 to 10 times [19]. Although different pretreatment performance studies have been performed, mostly focusing on cellulose isolation for biofuel production, no design factors have been involved in the scale-up stage focused on biorefineries. There are no in-depth technical, economic, and environmental assessments considering operational and scaling factors. In addition, there currently needs to be a clear systematization of pretreatments with applications in biorefinery design since effective pretreatment improves the performance of reaction and downstream processes. Therefore, the pretreatment processes in biorefineries should consider the integral valorization of lignocellulosic fractions, maximizing the isolation of fractions in both branched chains or platform products for further processing. This work aims to analyze the sustainability of pretreatment efficacy and its influence on the design of production processes in biorefinery schemes. Therefore, pretreatments for lignocellulosic isolation were assessed considering technical, economic, environmental, and social perspectives based on ten sustainable indicators.

2. Results and Discussion

2.1. Pretreatment Screening

Tables 1–3 show the possible pretreatments that would remove or isolate each lignocellulosic fraction for future use in biorefinery schemes. It should be noted that the hemicellulose and lignin fractions are removed together in the liquor in most pretreatments. Hemicellulose is usually hydrolyzed to form five-carbon oligomers and monosaccharides such as xylose and arabinose, and a small six-carbon fraction such as galactose, glucose, mannose, and 4-O-methyl-d-glucuronic acid [20]. Therefore, this gives rise to the synthesis of by-products with industrial attraction due to their future processing for producing ethanol, lactic acid, and sugar alcohols such as xylitol, sorbitol, mannitol, ethylene glycol, and glycerol, among others. Unlike cellulose, hemicellulose has an amorphous and slightly random structure with low thermal and chemical stability, achieving more than 75% removal in aqueous hydrolysis at high temperatures [21]. This overview of hemicellulose hydrolysis and its by-products makes it a valuable candidate for the biofuel, food, and biomaterials industry. On the other hand, lignin is easily hydrolyzed by alkaline agents at high temperatures due to its amorphous and cross-linked structure together with hemicellulose [22]. Lignin pretreatments are focused on its isolation, and the degree of polymerization decreases, modifying the three-dimensional structure as it becomes less complex [23]. However, the aromatic structure of lignin, with functional groups including aliphatic, phenolic hydroxyl, and carbonyl groups, make it a potential raw material in the food, pharmaceutical, and perfume industries.

Initially, biorefineries were proposed to produce biofuels through the efficient use of biomass, focusing on cellulose valorization through enzymatic hydrolysis and further fermentation [24]. Therefore, pretreatments focused on cellulose preservation or isolation should constitute an essential part of the process design since some physicochemical, structural, and compositional factors of the biomass must be overcome. The molecular configuration and chain length cause cellulose to have a high degree of polymerization, which hinders enzymatic and microbial attack [25]. In addition, the cellulose structure has a large number of intramolecular and intermolecular hydrogen bonds due to the hydroxyl groups forming a superstructure (supramolecular structure) and hindering easy hydrolysis compared to hemicellulose and lignin [26]. That fraction of cellulose that can be removed after pretreatment belongs to the amorphous fraction of the polymer, resulting in soluble and insoluble cellulose molecules in the form of polymers and oligomers [27]. For example, it has been reported that after dilute acid pretreatment, the insoluble cellulose molecules have a degree of polymerization of 100-20,000, while the soluble one of 2-12 [28], or values of 500-1500 after kraft pulping [29]. Pretreatments have been described as alternatives to decrease the crystallinity while increasing the surface area, promoting the adsorption capacity of enzymes to the substrate [30].

The early selection of pretreatments was performed through the removal index (R_I). Although many criteria should be considered for pretreatment assessment, it is possible to identify some schemes that are not optimal for removing specific lignocellulosic fractions. It should also be noted that the operating conditions, summarized in the severity factor (Log (R_0)), drastically influence the R_I. For example, removals of 14.6–31.3%, 26.2–77.9%, and 29.7–93.8% can be achieved for cellulose, hemicellulose, and lignin, respectively, when the severity factor varies between 3.1 and 4.4 for SO_2-catalyzed steam explosion pretreatment [31]. Even hemicellulose removals of 19.1–26.3% and lignin removals of 66.9–71.4% have been achieved when small severity factor variations of 0.2–0.7 were performed in recycled aqueous ammonia expansion (RAAE) pretreatments [32]. The information condensed in Tables 1–3 was based on a comprehensive review of the best single pretreatment schemes for lignocellulosic biomass. For cellulose isolation, it can be observed that wet air oxidation, diluted acid, and kraft pretreatments are the most effective in preserving the biopolymer, with values around 90–93% (or isolations of 7–10%). In contrast, ammonia fiber expansion (AFEX) and organosolv have been postulated for cellulose preservation as well as hemicellulose and lignin removal [33]; however, more than 19% of cellulose is hydrolyzed into the liquor as losses. Pretreatment screening should be based on the removal of the lignocellulosic fraction coupled with the accessibility degree (A_I) for further valorization. Although dilute acid and kraft pretreatment have cellulose removals of 10%, their A_I is 57–62%, which can hinder bioprocesses that are inhibited by the presence of undesired compounds. For example, after dilute acid pretreatment, the water-insoluble solid (WIS) is constituted by cellulose and a large composition of lignin (cellulignin solids) that must be removed, as it inhibits bioconversions to ethanol or other bioproducts [34]. However, bioprocesses have been studied for the joint valorization of cellulose and hemicellulose-based substrates to avoid disposing of pentoses as waste, being the kraft process a candidate for these systems as it preserves 90% and 60% of the initial biopolymer respectively. For example, Mishra and Ghosh studied the fermentation of glucose and xylose after kans grass biomass fractionation with sulfuric acid for bioethanol production through a co-culture based on *Scheffersomyces shehatae* and *Zymomonas mobilis* [35]. Therefore, based on the R_I and A_I indexes, the kraft, organosolv, ionic liquid, dilute acid, RAAE, and wet air oxidation pretreatments were selected for cellulose isolation in the WIS.

Due to different operational (i.e., feed ratio, acid catalyst addition, temperature, residence time) and structural factors of the feedstock (i.e., recalcitrance, crystallinity, porosity, degree of polymerization), the pretreatments are not selective to a single fraction; instead, more than two fractions can be obtained in the hydrolyzed liquor. For example, acidification of biomass with SO_2-catalyzed steam explosion and dilute acid has effectively decreased the crystallinity and degree of polymerization of lignocellulosic biomass while removing hemicellulose and small amounts of lignin [36]. Therefore, pretreatments can be repeated in lignocellulosic isolation schemes, such as kraft, which removes lignin while preserving a large amount of cellulose. For hemicellulose removal, steam explosion and diluted acid pretreatments are the best schemes for obtaining five-carbon platform products based on the R_I and A_I indexes. In steam explosion pretreatments, large amounts of xylose are generated, followed by aliphatic acids and furanic derivatives from the hemicellulose [37]. Aliphatic acids such as acetic acid are derived from the hydrolysis of acetyl groups, while other acids such as formic and levulinic acids are products of catalyzed thermochemical degradation. Acetic acid has been reported to decrease the pH of the liquor, stimulating the acid-catalyzed hydrolysis of the other components, and some of the pentose sugars are subsequently dehydrated to furfural and hydroxymethylfurfural (HMF) [34]. From

Table 2, steam explosion, liquid hot water (LHW), organosolv, ionic liquids, dilute acid, and ammonia recycled percolation (ARP) schemes were selected as the best pretreatments for hemicellulose removal. On the other hand, lignin is usually insoluble in aqueous solutions and can be removed from the liquor, together with the hydrolyzed hemicellulose, through thermochemical treatments. Delignification reactions involve the cleavage of the non-phenolic β-O-4 bond and the α-O-4 phenolic bond [38]. Therefore, there is no selective pretreatment to lignin removal; instead, it is necessary to remove it from the liquor with hydrolyzed pentoses through other treatments. Normally, acidification steps are implemented by adding CO_2 and inorganic acids, which promotes lignin precipitation. This precipitation can be explained by the colloidal nature of lignin, specifically as a hydrocolloid that precipitates at low pH due to the protonation effect of the acid groups of the lignin structure [39]. Based on the data reported in Table 3, the best lignin removal schemes were alkali, kraft, LHW, ionic liquids, RAAE, and organosolv pretreatments.

Table 1. Pretreatment screening for cellulose isolation.

Pretreatment	Raw Material	Operating Conditions			R_I (%)			A_I (%) ***	Reference
		Temperature (°C)	Pressure (bar)	Log (R_0)	Cellulose	Hemicellulose	Lignin		
Kraft	*Eucalyptus globulus*	155	-	3.09	10.0	39.4	84.6	62.0	[40]
Organosolv	Wheat straw	160	-	3.37	19.6	93.4	62.5	77.9	[41]
Ionic liquids	Corn stover	160	-	4.02	15.5	81.5	69.2	75.4	[42]
Diluted acid	Bamboo green	180	-	3.83	9.9	98.8	16.6	57.7	[43]
RAAE *	Corn stalks	85	20.4	0.67	10.3	14.1	71.4	42.8	[32]
Wet air oxidation	Rice husk	195	5.0	3.79	7.1	75.5	97.3	86.4	[44]
AFEX **	Corn stover	130	44.8	2.06	27.8	34.6	23.5	29.1	[45]
Biological	Corn stalks	28	-	2.66	57.0	41.0	11.0	14.1	[46]
Biological	Switchgrass	28	-	2.29	22.0	14	24	81	[47]

* Recycled aqueous ammonia expansion; ** ammonia fiber expansion; *** calculated using Equation (3).

Table 2. Pretreatment screening for hemicellulose isolation.

Pretreatment	Raw Material	Operating Conditions			R_I (%)			A_I (%) ***	Reference
		Temperature (°C)	Pressure (bar)	Log (R_0)	Cellulose	Hemicellulose	Lignin		
Steam explosion	Sugarcane bagasse	195	-	3.67	2.3	81.7	12.1	92.8	[37]
LHW *	Bermuda grass	170	-	3.84	29.8	88.8	33.8	68.2	[48]
Organosolv	Wheat straw	160	-	3.34	19.6	93.4	62.5	58.9	[41]
Ionic liquids	Switchgrass	160	-	4.02	15.5	81.5	69.2	57.6	[42]
Diluted acid	Bamboo green	180	-	3.83	9.9	98.8	16.6	86.8	[43]
ARP **	Corn stover	170	23.0	3.06	39.8	63.3	80.4	39.9	[49]
Biological	Hardwood	28	-	2.28	15.8	17.9	3.0	93.6	[47]
Biological	Wheat straw	28	-	2.48	16	94	49	67.5	[50]

* Liquid hot water ** ammonia recycled percolation; *** calculated using Equation (2).

Table 3. Pretreatment screening for lignin isolation.

Pretreatment	Raw Material	Operating Conditions			R_I (%)			A_I (%) ****	Reference
		Temperature (°C)	Pressure (bar)	Log (R_0)	Cellulose	Hemicellulose	Lignin		
Alkali	*Eucalyptus camaldulensis*	150	-	3.25	6.7	33.3	63.6	79.9	[51]
Kraft	*Eucalyptus globulus*	165	-	3.69	17.2	51.0	97.4	65.9	[40]
LHW *	Wheat straw	190	-	3.95	41.8	92.3	64.9	32.9	[52]
Ionic liquids	Corn stover	140	-	3.43	49.4	56.9	94.4	46.8	[53]
RAAE **	Corn stalks	85	20.4	0.67	10.3	14.2	71.4	87.7	[32]
Organosolv	Wheat straw	160	-	3.37	19.6	93.4	62.5	43.5	[41]
ARP ***	Corn stover	170	23.0	3.06	39.8	63.3	80.4	79.9	[49]
Biological	Bamboo culms	60	-	2.28	8.8		53.3	95.6	[54]
Biological	Corn stover	28	-	2.66	58	51	64	45.5	[55]

* Liquid hot water; ** recycled aqueous ammonia expansion; *** ammonia recycled percolation; **** calculated using Equation (2).

2.2. By-Product and Inhibitors Formation

After pretreatment, by-products are obtained from the hydrolysis of macromolecules, especially hemicellulose and lignin, including five- and six-carbon sugars in the form of oligomers and monomers, aliphatic acids, furans, and phenolic compounds, among others. Each by-product can be used in other thermochemical and biochemical processes as raw material (platform product) to obtain value-added products or have an inhibitory effect in bioconversions. In hemicellulose hydrolysis pretreatments, mainly xylose and xylo-oligomers are produced from the cleavage of the glycosidic bond in the xylan chain, as well as small amounts of glucose, galactose, and arabinose [31]. Based on different bioconversion processes, xylooligosaccharides can be used as a platform product in the functional food, pharmaceutical, and chemical industries. Although xylooligosaccharides are not digestible by humans, they favor the proliferation of healthy microorganisms due to their prebiotic nature [56]. In general, oligosaccharides based on glucose, xylose, fructose, and galactose have been shown to be a potential partial substitute for sugars promoting the growth of bacterial flora and also for their antioxidant and antiallergic capacities [57]. Different pretreatments have been proposed for producing xylo-oligomers, highlighting LHW or autohydrolysis, acid, and alkaline hydrolysis [58]. On the other hand, xylose is an important industrial product as it can be used to produce xylitol, used in the food industry as a sugar substitute to produce diabetic-friendly or low-calorie products. Panjiar et al. have studied the production of biosurfactants from xylose-rich hydrolysates to reduce surface tension in oil-water mixtures in the food, pharmaceutical, and petrochemical industries [59]. The production of bioethanol [60] and xylonic acid, used as an additive in the food industry and precursor for producing green solvents and copolyamides [61], has also been explored.

The rich phenolic content of lignin makes it promising to use this heteropolymer in different industrial processes of food, pharmacy, and perfumery. Solubilized or precipitated lignin has been studied to produce vanillin as an additive in the chemical industry [62]. Other studies have focused on using lignin in producing new materials, such as phenolic resin, epoxy, and polyurethane, as emulsifying and carbon agents, among others [23]. The thermostability and mechanical properties of lignin mixed with thermoplastics such as polypropylene have been demonstrated [63]. Substituting phenol for lignin in phenol-formaldehyde synthesis has also been studied [64]. However, the main drawback is the high heterogeneity of the molecule and an efficient isolation method with high purity. Depending on the pretreatment employed, the structure of lignin may vary in terms of molecular weight, phenolic distribution, polydispersity, inter structural bonds, among others. As an alternative, fractionation through membrane filtration, solvents coupled with microwaves, or sequential precipitation has been studied [22]. Pretreatments with acid catalyst addition promote the formation of low molecular weight phenolic and aromatic compounds that can act as inhibitors in bioconversion processes, affecting both microbial growth and overall yield. This inhibition can be given to specific functional groups, as they can interfere with the cell membrane influencing its microbial function [34].

Furfural and HMF are also produced by the dehydration of monosaccharides at high temperatures. Furfural has gained importance as a solvent or as a precursor for the production of pesticides. In contrast, HMF has been widely used as an additive in petroleum-based polymer blends, such as polyester and polyurethane. Furfural and HMF inhibit yeast growth and decrease the yield and productivity of ethanol fermentations. However, anaerobic fermentations have been performed in recombinant transformants with xylose as substrate followed by the addition of furfural, leading to increased ethanol production due to the reduction of furfural to furfuryl alcohol to decrease the production of undesirable by-products such as xylitol [65]. Alkaline pretreatment has greater potential for fermentative processes as less inhibitory substances such as HMF and furfural are produced than acid pretreatment [66]. Other degradation by-products, such as aliphatic acids (i.e., levulinic acid, acetic acid, formic acid), can serve as end products or platform products for pharmaceuticals, plasticizers, and other additives. However, inhibition of these acids has been observed at concentrations higher than 0.1 M, as they can pass through the cell wall and dissociate, lowering the intracellular pH and leading to death [34,67].

2.3. Techno-Economic Analysis

The results of the technical analysis of the selected pretreatment schemes are summarized in Table 4. Regarding the technical aspect, the cellulose isolation pretreatment schemes showed no significant difference; only small yield changes for organosolv and ionic liquid were observed. Regarding hemicellulose, the yields were calculated based on the production of oligomers and five-carbon monosaccharides, with the diluted acid, organosolv, and LHW being the best schemes. It was observed that the ionic liquid was the worst scheme in terms of yield since the hydrolyzed liquor was submitted to ionic liquid recovery stages, preventing a good recovery of the hemicellulose carbohydrates. For the lignin removal yield, it was observed that the kraft process allows obtaining a greater amount of the heteropolymer, which was to be expected since it is the main methodology for obtaining soluble lignin in the black liquor. Likewise, in terms of energy, it can be seen in Table 4 that the schemes with the highest energy demand were the RAAE, ionic liquid, and organosolv pretreatments. This behavior is justified by using energy-demanding equipment such as distillation towers for solvent recovery in organosolv and pressurized equipment such as RAAE, among others. On the other hand, the pretreatment schemes with the lowest energy demand were LHW, dilute acid, and steam explosion. These pretreatments are characterized by short reaction times and low reagent demand. Thus, these pretreatments allow the isolation of the lignocellulosic fractions without high energy demands. However, some of these pretreatments require specific designs for their proper operation. For example, although steam explosion is not energy-demanding, it requires an operational design material resistant to high pressures or the impact between the high-pressure steam and the biomass, or the dilute acid requires a specific alloy construction that withstands the acidic nature of the medium and does not corrode the reactor and its accessories. Therefore, an economic analysis of the operating and investment costs of each pretreatment scheme is essential.

Table 4. Technical analysis of pretreatments to produce different lignocellulosic fractions.

Lignocellulosic Fraction	Pretreatment	Yield (kg 100 kg^{-1}) ****	Utilities		
			Cooling Water (ton h^{-1})	Steam (ton h^{-1})	Electricity (kW)
Cellulose	Wet oxidation	20.9	4220	59.6 ***	147.6
	RAAE	20.2	5260	547.7 *	8589.0
	Organosolv	18.1	22,200	190.4 * 25.8 **	226.3
	Diluted acid	20.3	1580	21.2 ***	9.4
	Kraft	20.2	1580	33.3 **	151.8
	Ionic liquid	18.9	3720	247.3 * 7.8 **	361.9
Hemicellulose	Steam explosion	9.9	161	1.1 ***	61.20
	LHW	11.5	1500	18.4 **	111.8
	Ionic liquid	0.7	327	247.3 ** 7.8 **	361.9
	Organosolv	12.2	22,100	25.4 **	308.9
	Diluted acid	12.8	1540	21.2 ***	109.3
	ARP	8.1	-	4.1 **	105.1

Table 4. *Cont.*

Lignocellulosic Fraction	Pretreatment	Yield (kg 100 kg^{-1}) ****	Utilities		
			Cooling Water (ton h^{-1})	Steam (ton h^{-1})	Electricity (kW)
Lignin	Alkali	14.2	5860	69.4 **	236.8
	Kraft	21.6	1220	16.9 **	122.8
	Ionic liquid	3.4	244	1.3 * 7.4 ***	264.3
	LHW	13.9	136	4.1 ***	64.3
	Organosolv	13.9	22,100	190.2 * 25.4 **	308.9
	RAAE	15.9	5260	547.7 *	8589.0

* Low-pressure steam; ** medium-pressure steam; *** high-pressure steam; **** yield based on the isolated lignocellulosic fraction per initial raw material (50 tons d^{-1}). For cellulose, this is the six-carbon fraction of the WIS. For hemicellulose, this corresponds to the oligomers and monosaccharides of five carbons in the hydrolysate. For lignin, this is the solubilized lignin. See Equations (5)–(7).

Table 5 summarizes the economic results regarding investment (CapEx) and operating (OpEx) costs. In terms of CapEx, the most expensive pretreatment schemes were organosolv and RAAE, which can be explained by their use of complex equipment such as the distillation system (i.e., the organosolv scheme), which includes the use of the column, condenser, reboiler, reflux pump, and collection tank. In contrast, the most economical processes were LHW, dilute acid, and kraft. These processes were characterized by using few operating units and mild operating conditions, resulting in the lowest OpEx. On the other hand, the pretreatments with the highest OpEx were the ionic liquid and RAAE processes. Expensive reagents characterize the ionic liquid process, and its recirculation is crucial. To reduce the OpEx in the ionic liquid schemes, recirculation of the ionic liquid was proposed by adding anti-solvent to the reactive mixture in a 1:1 by weight ratio followed by centrifugation and evaporation to remove the anti-solvent from the ionic liquid for its subsequent recirculation [68]. This process allowed more than 10% decrease in OpEx associated with the demand for reagents. Finally, the organosolv and RAAE pretreatments showed a high demand for OpEx due to the need for utilities during the process, specifically electricity and steam.

Table 5. Economic assessment of pretreatments to produce different lignocellulosic fractions.

Lignocellulosic Fraction	Pretreatment	CapEx (M-USD)	OpEx (M-USD Year^{-1})				
			Raw Materials	Utilities	Depreciation	Others *	Total
Cellulose	Wet oxidation	2.46	0.63	6.76	0.58	0.16	8.13
	RAAE	6.65	1296.46	48.23	1.57	0.41	1346.67
	Organosolv	3.73	40.54	25.28	0.88	0.24	66.93
	Diluted acid	0.81	0.79	2.83	0.19	0.06	3.88
	Kraft	2.25	764.03	3.73	0.53	0.15	768.44
	Ionic liquid	1.96	6603.88	18.67	0.46	0.13	6623.14
Hemicellulose	Steam explosion	0.63	0.44	0.75	0.15	0.05	1.39
	LHW	0.88	0.41	2.60	0.21	0.06	3.28
	Ionic liquid	2.45	6603.88	18.66	0.58	0.16	6623.27
	Organosolv	4.09	40.76	15.80	0.97	0.26	67.30
	Diluted acid	0.96	0.79	2.83	0.23	0.07	3.92
	ARP	1.12	0.89	0.38	0.26	0.08	1.61
Lignin	Alkali	1.62	41.60	8.19	0.38	0.11	50.28
	Kraft	0.99	1.20	2.38	0.23	0.07	3.89
	Ionic liquid	2.15	197.47	1.51	0.51	0.14	199.63
	LHW	0.60	0.37	0.96	0.14	0.05	1.52
	Organosolv	4.09	40.54	25.29	0.97	0.26	67.05
	RAAE	6.45	1181.08	48.23	1.52	0.40	1231.23

* Maintenance and labor costs.

The pretreatment selection should involve operational aspects and the overall performance of the process. It is also necessary to identify scaling factors or technology development level since, in many cases, its implementation on an industrial scale depends on technical, economic, and environmental design factors. Therefore, the most appropriate selection of the isolation scheme should involve the technology readiness level (TRL) since readiness defines whether the technology is applicable at the industrial or pilot scale, as well as whether it is in the innovation and development stage. Figure 1 shows the TRL results for the selected processing schemes for each lignocellulosic fraction. It can be observed that the high capital cost of pretreatments, such as RAAE, limits investors from using them on an industrial scale. Pretreatments such as organosolv involve solvents that increase the energy requirements and operating costs of the process, decreasing its economic viability. The economic feasibility of organosolv schemes has been widely reported for lignin extraction at the pilot scale [69]. The kraft and alkaline processes are used at the industrial scale for removing lignin from woody biomass for pulp and paper production. Although dilute acid pretreatment has been widely discussed in biotechnological applications, its implementation at the industrial scale has been slowed down. These delays are largely due to the generation and accumulation of inhibitory compounds and the high corrosion of processing units and piping [69]. Therefore, TRL must be analyzed from an operational and design perspective with techno-energetic and economic considerations.

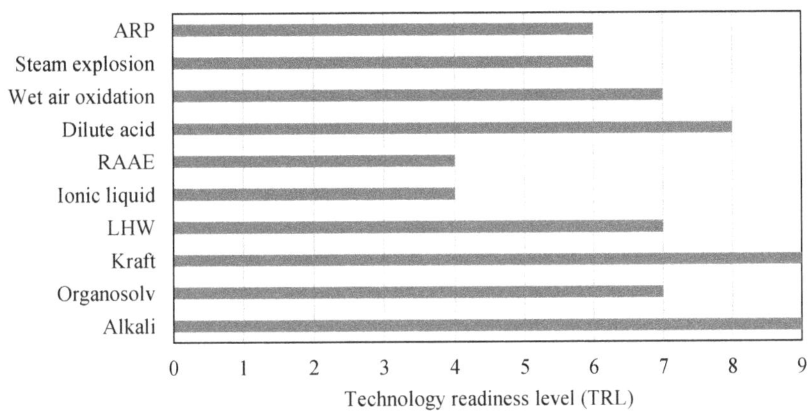

Figure 1. Technology readiness level of the selected lignocellulosic pretreatments.

2.4. Environmental Analysis

The potential environmental impact of the pretreatment schemes during the isolation of the cellulose fraction is shown in Figure 2. It can be seen that all pretreatments presented similar impacts in the categories analyzed. Therefore, positive environmental impacts were observed, given that their values are negative, interpreted as environmental relief. Additionally, it can be seen that in the photochemical oxidation potential (PCOP) category, the organosolv pretreatment generated less impact because, during pretreatment, there were no considerable emissions of polluting gases, dust, or smoke. Although there could have been impacts related to human toxicity by ingestion (HTPI), there was no negative environmental impact since more than 95% of solvents are recirculated to the process. On the other hand, the RAAE pretreatment is not considered in Figure 2 since its potential presented considerable differences concerning the other pretreatments. The RAAE generated a harmful environmental impact, with 106 PEI kg^{-1} of product. This impact was reflected in higher rates in the HTPI and terrestrial toxicity potential (TTP) categories since they involve effects on humans either by ingestion, inhalation or dermal exposure to ammonia

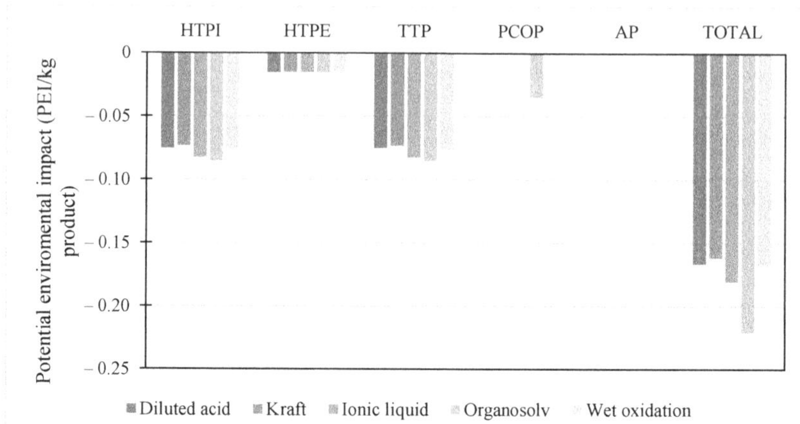

Figure 2. Potential environmental impact of pretreatment schemes for cellulose isolation.

The environmental impact of the pretreatment schemes for the isolation of the hemicellulose fraction is presented in Figure 3. The pretreatment schemes that exhibited the greatest environmental benefit were the ionic liquid and steam explosion in the human impact categories and the greenhouse gas emissions. Therefore, the organosolv pretreatment, although it contemplates using solvents, does not present pollutant gas emissions due to the recovery or recirculation of the solvent. This analysis also applies to the ionic liquid pretreatment, with the recirculation of the main process reagent. On the other hand, the ARP pretreatment showed the lowest favorable index due to the use of ammonia during the reactive process. Different authors have previously reported the effects of ammonia on the ecosystem and human health [70]. Thus, organosolv pretreatment and steam explosion are the best environmental schemes for hemicellulose recovery.

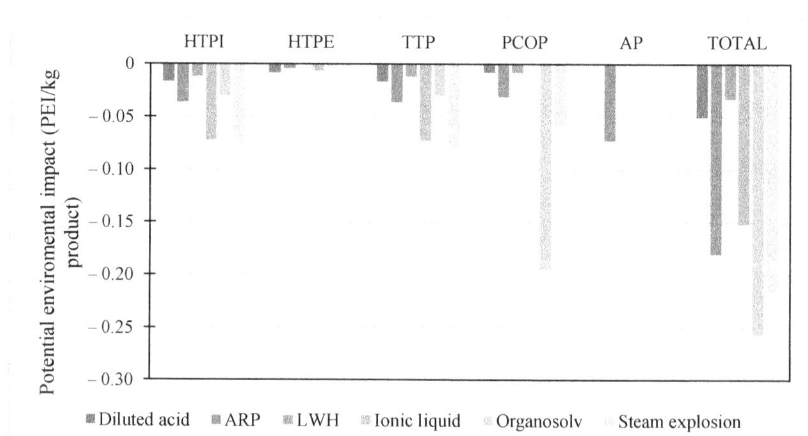

Figure 3. Potential environmental impact of pretreatment schemes for hemicellulose isolation.

The potential environmental impact generated by the pretreatment schemes for the isolation of the lignin fraction is shown in Figure 4. Similar to the pretreatments for cellulose isolation and hemicellulose removal, the schemes used for lignin fractionation did not express any harmful environmental impact, except for the RAAE pretreatment concerning human toxicity. The RAAE pretreatment is the only pretreatment that shows harmful environmental impacts for pretreatments focused on cellulose and lignin recovery.

This phenomenon is associated with the high ammonia concentration in the reactive mixture. These concentrations reach values of more than 50% v/v. Therefore, downstream ammonia removal becomes a complex process, considering subsequent separation stages. In addition, the lignin obtained from pretreatments such as RAAE cannot be used in food-type production schemes because the contaminant traces cannot be eliminated.. The impact associated with using hazardous substances during the process triggers complications in the development of pretreatment since it requires the installation of extensive and rigorous safety protocols.

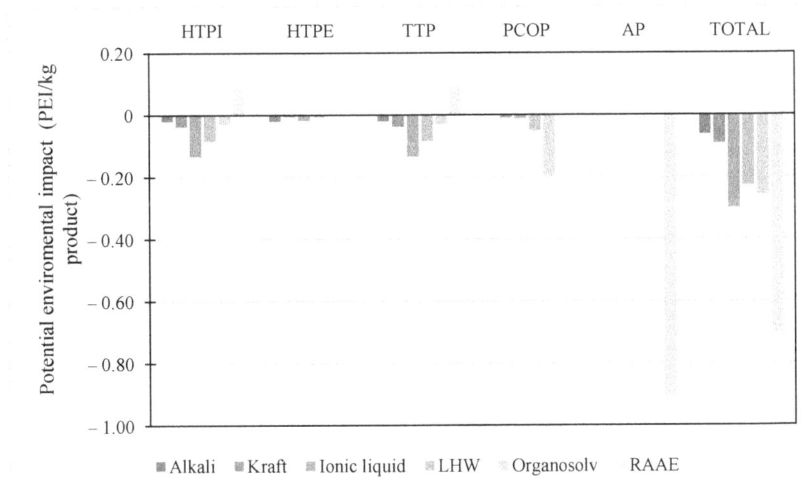

Figure 4. Potential environmental impact of pretreatment schemes for lignin isolation.

2.5. Social Analysis

For the social analysis, a comparison of indicators contextualized by the capacity in Colombia was performed. These ratios considered statistics at the national scale for certain economic sectors. Based on water consumption in pretreatment schemes for cooling and processing, two indicators were calculated: (i) water for the industrial sector, where Colombia demanded 3.73×10^9 m^3 in 2019, and (ii) total water available in the country, with reported values of 2145×10^9 m^3 in 2019 (AQUASTAT). Likewise, indicators referring to the demand for electricity from the processing units were analyzed with the national electricity capacity of 72,824 GWh in 2021 and the consumption of diesel (in energy terms) for the generation of thermal energy from the different steam qualities of the pretreatment schemes with the amount of energy provided by Colombia in 2021 (2.40×10^{11} MJ) (UPME). For the latter analysis, a boiler thermal efficiency of 80% was assumed. Table 6 shows the results of the indicators analyzed for each lignocellulosic fraction. It can be seen that for all indicators, except for the level of industrial water use (withdrawal), the values are very small since they are contextualized to the country. Therefore, the comparison and analysis were carried out between the different schemes and not based on a risk scale. From the perspective of the level of industrial water use (withdrawal), it can be observed that the organosolv pretreatment, in the three lignocellulosic fractions targeted, consumes the most water in the industrial sector. The addition of the dilute solvent during hydrolysis, the cooler prior to filtration of the hydrolysate, and the condensation of the solvent in the recovery stage represents the greatest social risk for its implementation. It was expected that pretreatments such as steam explosion would not greatly affect industrial water use (renewable) since small feed ratios between steam and biomass are required. Regarding electricity energy demand, the RAAE, LHW, and kraft pretreatments are the most at risk for cellulose isolation and removal of hemicellulose and lignin. Finally, from the perspective of fossil fuel extraction, there was no significant impact on the national diesel data. This shows that there is no great dependence on this fuel in boilers for steam production, and it does not represent any risk due to competition in its use. The low consumption of process water in steam explosion means that low levels of steam and diesel are required, making it the scheme with the lowest risk.

Table 6. Social indicator results of access to material resources adapted to the Colombian context.

Lignocellulosic Fraction	Pretreatment	FWU_{sector} (%)	$FWU_{country}$ (%)	ED (%)	EF (%)
Cellulose	Wet oxidation	1.00	1.74×10^{-3}	2.03×10^{-4}	1.71×10^{-5}
	RAAE	3.27	0.32×10^{-3}	0.01	1.53×10^{-4}
	Organosolv	5.20	9.05×10^{-3}	3.11×10^{-4}	6.06×10^{-5}
	Diluted acid	0.37	6.51×10^{-4}	1.29×10^{-4}	6.09×10^{-6}
	Kraft	5.62	0.10×10^{-3}	2.08×10^{-4}	9.56×10^{-6}
	Ionic liquid	0.13	2.21×10^{-4}	4.97×10^{-4}	7.14×10^{-5}
Hemicellulose	Steam explosion	0.04	6.63×10^{-5}	2.03×10^{-4}	3.07×10^{-7}
	LHW	0.36	6.19×10^{-4}	0.01	5.27×10^{-6}
	Ionic liquids	0.13	2.18×10^{-4}	3.11×10^{-4}	7.14×10^{-5}
	Organosolv	5.20	9.04×10^{-3}	1.29×10^{-4}	6.05×10^{-5}
	Diluted acid	0.36	6.33×10^{-4}	2.08×10^{-4}	6.09×10^{-6}
	ARP	0.21	9.47×10^{-7}	4.97×10^{-4}	1.17×10^{-6}
Lignin	Alkali	1.39	2.42×10^{-3}	2.03×10^{-4}	1.99×10^{-5}
	Kraft	0.29	5.00×10^{-4}	0.01	4.85×10^{-6}
	Ionic liquid	0.06	9.98×10^{-5}	3.11×10^{-4}	2.48×10^{-6}
	LHW	0.03	5.60×10^{-5}	1.29×10^{-4}	1.18×10^{-6}
	Organosolv	5.19	9.03×10^{-3}	2.08×10^{-4}	6.05×10^{-5}
	RAAE	1.46	2.55×10^{-3}	4.97×10^{-4}	1.53×10^{-4}

FWU_{sector}: level of industrial water use (withdrawal), $FWU_{country}$: level of industrial water use (renewable), ED: energy demand, EF: extraction of fossil fuels.

2.6. Pretreatment Efficacy

Based on the results of each evaluative indicator, an overall assessment of pretreatment efficacy was carried out by weighing the heuristic analysis. Heuristic analyses have been described as subjective analyses based on theoretical, experimental, or simulated results. Some investigations have shown satisfactory results of raw material selection [71] or transformation routes [72] through manual weightings. Therefore, the efficacy calculation contemplated the weighting of two types of ratings (see Table 7): (i) weighting by equal assignment of 10% for each indicator (ω_1); and (ii) weighting by statistical analysis (ω_2). As can be seen, despite the use of statistical weighting, where there is greater relevance to economic design indicators appealing to investors, there are no differences between the efficacy results compared to equal distribution. On the other hand, Table 7 shows that dilute acid, wet air oxidation or organosolv pretreatment is recommended to isolate the cellulose fraction. It should be noted that after each pretreatment, it is necessary to rinse the WIS to remove the greatest amount of unwanted soluble compounds or the excess of initial chemical reagents. In biorefinery schemes, the cellulose should be as clean as possible since degradation compounds such as furans or phenolics inhibit the bioconversion processes. It is recommended to perform neutralization stages with lime to neutralize and detoxify the solid stream, whose reaction product, such as calcium sulfate or gypsum, has been used as a co-product in biorefineries based on lignocellulosic biomass [73]. On the other hand, to remove hemicellulose in five-carbon sugar platform products, LHW, steam explosion, or dilute acid should be used, while to remove lignin heteropolymer, kraft, alkali, or LHW pretreatment is recommended. Many authors have previously reported that pretreatments are efficient when there is the direct formation of sugars or after enzymatic processes with the least loss of sugars, the formation of inhibitory compounds is limited, and energy demand and operating costs are minimized [74]. Other research has focused on the fact that pretreatments should remove lignin and hemicellulose, as well as reduce cellulose crystallinity and increase biomass porosity [27]. Although these definitions have already been discussed, many focus on the maximum cellulose utilization for bioconversion processes aimed at biofuels, leaving aside the other lignocellulosic fractions. The present work focuses on a more exhaustive assessment involving more evaluative indicators of operability, energy demands, profitability, and environmental and social impact. Therefore, a first approximation of the sustainability of pretreatment in biorefinery designs is given through technical, economic, environmental, and social assessment pillars. For example, under controlled conditions, dilute acid is sustainable because it generates a large amount of hydrolyzed five-carbon sugars while preserving cellulose, its TRL is high for industrial application, it has low energy demand, low investment and operating costs, a positive environmental impact, and good indexes of access to material resources in the social sphere.

The single-step pretreatment efficacy results give a tentative route to the future reaction and downstream processes that should be involved in the early design of biorefineries. Although the heuristic analysis provides design support, there are certain challenges to be considered in the biorefinery approach. (i) Insufficient separation of hemicellulose and lignin in the hydrolyzed liquor requires additional steps for lignin precipitation through acids, whose concentration can alter the soluble carbohydrates to form degradation compounds. (ii) In pretreatments that are acid-catalyzed, such as dilute acid, organosolv, or steam explosion, extra design factors must be considered due to the high temperatures used together with the corrosive action of the acids. Therefore, hot acid corrosion-resistant alloys must be used for reactor design, which makes the cost of the reactor, exchangers, filters, and piping an important element of the CapEx as well as possible maintenance costs. (iii) The main demand for process water in biotechnological processes comes from the pretreatment stage due to the high feed ratios needed to hydrolyze the lignocellulosic matrix, which sometimes makes it necessary to add costly chemical reagents that are difficult to recover and that alter the environmental impact of the waste streams. However, processes such as organosolv and kraft have been implemented at high scales considering the recovery of reagents. For example, green liquor causticization of the kraft process after the black liquor incineration. (iv) High energy use in the first stage of the process (pretreatment), such as organosolv. Therefore, it is proposed to contemplate energy integration stages to reduce the demand and costs associated with thermal energy.

During biorefinery design, the integral use of raw materials to produce high-value-added compounds is essential. Moreover, the objective is to minimize the environmental impact and costs associated with waste generation and maximize profits through the production of more compounds. Therefore, the design of pretreatment schemes through efficacy should involve using the other lignocellulosic fractions that were not considered a target product. For example, alkali pretreatment for lignin removal was studied without contemplating the future utilization of cellulose-rich WIS, followed by unhydrolyzed hemicellulose. Therefore, the pretreatment efficacy study may involve additional schemes sequentially to utilize each fraction best. Since pretreatment is the critical stage in the design of biorefineries, as explained by the onion diagram, it is required to optimize the processes to maximize the production of platform products. Different authors have studied sequential pretreatments to improve operationally future processing steps, such as enzymatic hydrolysis [75], buffering problems in the LHW [76], and decrease inhibitory compounds and residence times [77].

Table 7. Pretreatment efficacy by heuristic analysis.

Lignocellulosic Fraction	Pretreatment	Evaluative Indicator											
		R_I	A_I	By-Products	Inhibitors	TRL_I	E_I	CapEx	OpEx	PEI	SI_I	Total Using ω_1	Total Using ω_2
		Weight Factor (ω_1 %) *											
		10	10	10	10	10	10	10	10	10	10	100	-
		Weight Factor (ω_2 %) **											
		8	10	8	10	12	11	12	12	15	3	-	100
Cellulose	Wet oxidation	9	6	5	8	10	5	4	7	6	8	6.8	6.7
	RAAE	8	8	6	7	8	1	3	2	0	8	5.1	4.5
	Organosolv	9	8	8	6	4	3	5	6	9	6	6.3	6.3
	Diluted acid	9	6	7	4	9	7	8	8	6	8	7.2	7.1
	Kraft	9	4	7	7	4	6	5	3	6	6	5.8	5.6
	Ionic liquid	9	9	4	7	8	2	4	1	7	8	5.9	5.6
Hemicellulose	LHW	7	8	8	5	8	8	7	8	8	7	7.7	7.6
	Organosolv	6	7	7	5	3	3	9	8	6	6	6.4	6.2
	Ionic liquids	8	6	7	3	5	1	7	6	6	8	5.7	5.5
	Dilute acid	7	4	8	5	8	9	7	8	7	7	7.1	7.1
	ARP	7	7	4	6	8	8	7	6	7	7	6.8	6.9
	Steam explosion	7	7	9	8	8	8	8	8	8	7	7.5	7.5
Lignin	Alkali	6	8	7	9	10	5	8	6	7	8	7.4	7.4
	Organosolv	6	4	6	6	8	2	4	5	8	8	5.7	5.6
	Kraft	10	7	7	9	10	6	8	8	7	6	7.7	7.8
	LHW	7	3	7	7	8	8	8	8	8	8	7.2	7.2
	Ionic liquid	9	5	7	8	4	8	5	3	8	6	6.4	6.3
	RAAE	9	9	6	8	4	1	2	1	7	8	5.6	5.2

* Weighting using equal distribution of 10%; ** weighting using statistical analysis.

3. Methodology

The pretreatment assessment in biorefinery schemes was performed based on indicators involving literature data and simulation results. This assessment began with selecting a processing objective: to identify pretreatments that best isolate each lignocellulosic fraction separately for further valorization. A screening of pretreatment schemes was performed considering composition and fraction removal constraints. Afterward, ten indicators were described and assessed, involving technical, economic, environmental, and social aspects. Finally, each indicator was scored based on data from the literature and simulation depending on what is most suitable in the final biorefinery design to improve the sustainability of the process. The best pretreatment schemes were selected based on the efficacy results performed by the heuristic analysis.

3.1. Pretreatment Screening

A literature review was performed to determine the best pretreatments to isolate each lignocellulosic fraction individually, either as a water-insoluble solid (WIS) or in the pretreatment liquor, considering thermal, chemical, and thermochemical processes. The screening involved more than 100 updated review and research-type papers (articles from 2018 to 2022) concerning pretreatments with experimental and simulation sections. Biomass containing a lignocellulosic compositional range of 25% < cellulose < 50%, 20% < hemicellulose < 40%, and 10% < lignin < 35% was chosen to restrict the analysis scope to comparable raw materials by composition. After the screening, six pretreatments were selected that best removed or isolated the fractions of interest based on the removal index (R_I) (see Equation (1)), resulting in a total of 18 schemes: six for cellulose, six for hemicellulose, and six for lignin. Pretreatments with removals to liquor higher than 60% or conservations higher than 80% in the WIS were considered.

$$\text{Removal } (R_I) = \left(1 - \frac{\text{Pretreated fraction}_{\text{dry basis}}}{\text{Raw feedstock}_{\text{dry basis}}}\right) \times 100\% \quad (1)$$

3.2. Efficacy Assessment

The pretreatment efficacy in biorefineries was assessed based on ten evaluative indicators to obtain the isolated lignocellulosic fraction or a platform product from the fraction of interest. The indicators were classified as technical (indicators (i)–(vi)), economical (indicators (vii) and (viii)), environmental (indicators (ix)), and social (indicator (x)). (i) Lignocellulosic fraction removal (R_I). This factor involves the lignocellulosic fractional amount isolated in the liquor or in the WIS after pretreatment (see Equation (1)). For lignin and hemicellulose, high individual removals are desired in the hydrolysate, while cellulose should be minimized. (ii) Accessibility index (A_I). During pretreatment, isolating only one fraction without altering the others is impossible. Therefore, the A_I represents the accessibility rate of the isolated fraction affected by the remaining fractions or the contamination by undesired fractions. For valorization in the hydrolysate (see Equation (2)), high R_I of undesired fractions lead to liquor contamination (leading to low A_I), whereas the fraction valorization in the WIS (see Equation (3)) requires high R_I of undesired fractions (leading to high A_I). (iii) Formation of by-products or hydrolyzed products (HP_I). Many oligomers, monosaccharides, and carboxylic acids are produced from the hydrolysis of the representative biopolymer biomass [78], which can be considered platform products in biorefineries for further processing. Thus, higher by-product formation implies higher valorization proposals to improve biorefinery profitabilities. (iv) Formation of inhibitory products (IP_I). This indicator involves the number of degradation products inhibitory to biochemical processes, such as furan compounds, and their removal is essential [79]. Thus, minimal IP_I values are desired for pretreatments. (v) Technology readiness level (TRL_I). Many of the pretreatments are applied at laboratory or pilot scale since their scaling up is complex due to design and operability factors, cost, and energy demand. Therefore, the TRL_I demonstrates the progress or applicable extent of research and scaling up of the technology, as shown in Table 8. (vi) Energy demand (E_I). Utilities, such as steam, cooling water, and electricity, can be expressed as annual operating expenditures (OpEx). Therefore, those processes where energy demand is minimized are the most promising for scale-up. (vii) Capital cost (CapEx) and (viii) OpEx. Higher technological complexity of processing units and sub-units increases the investment cost, decreasing its attractiveness for investors. Therefore, these indicators relate to the total investment cost of the pretreatment system and the operating costs described as raw material, utilities, labor, maintenance, and depreciation, which must be minimal for economic feasibility. (ix) Potential environmental impact (PEI). Many pretreatments involve chemical additives that are harmful to both humans and the environment, as well as corrosive, and may damage the durability of the processing units. Therefore, the PEI considers the environmental impact of pretreatment that should be minimized. (x) Social impact (SI_I). This indicator involves comparing energy and water consumption to national or regional availability, providing a perspective of the level of risk to the communities surrounding the biorefineries that involve the pretreatment schemes. The energy and water demand of the process must be minimized so as not to affect the energy grid and water sources in the country.

Table 8. Indicator scale for the technology readiness level indicator (TRL_I).

Scale	Description	Group
1	Fundamental research	
2	Technology formulation	Research
3	Applied research (proof of concept)	
4	Small-scale development (laboratory scale)	
5	Scale-up development (pilot scale)	Development
6	Full-scale development	
7	System validated in simulation	
8	System validated in real life	Innovation
9	Commercial application	

$$\text{Accessibility index in the liquor } (A_I) = 100\% - (0.5\ R_I \text{ of undesired fraction}_1 + 0.5\ R_I \text{ of undesired fraction}_2) \quad (2)$$

$$\text{Accessibility index in the WIS } (A_I) = 0.5\ R_I \text{ of undesired fraction}_1 + 0.5\ R_I \text{ of undesired fraction}_2 \quad (3)$$

Pretreatment efficacy was assessed using a quantitative approach to heuristic analysis methodologies [71]. This score was calculated manually based on literature data and used to estimate the values of indicators (i)–(v) and as inputs for the simulation schemes, such as for indicators (vi)–(x). The assessment considered a rating from 1 to 10 for each efficacy indicator, where 1 represents the lowest score and 10 the highest. For the TRL_I indicator, normalizations to a scale of ten were performed according to the data in Table 8. The ratings were weighted to obtain a comprehensive indicator of each pretreatment based on the relevance of each parameter (weight factor) using Equation (4), where ω_i is the weight factor and I_i the evaluative indicator. The weight factor was calculated as the ratio of the specific variability ranges for each indicator over the best-case scenario, as suggested elsewhere for multi-criteria decisions in biorefineries [80].

$$\text{Efficacy} = \sum_i^n \omega_i \times I_i \quad (4)$$

3.3. Simulation Procedure

For the assessment of the E_I, CapEx, OpEx, PEI, and SI_I indicators, the pretreatment schemes were simulated in Aspen Plus v9.0 software (Aspen Technologies, Inc., USA), considering production yields, removals, operating conditions, and feed ratios described in the literature. A processing flow rate of 50 tons d^{-1} of rice husk and a simulation scope up to filtration or separation of liquor and WIS fractions were assumed. The raw material characterization is presented in Table 9. The chemical and thermodynamic properties of cellulose, hemicellulose and lignin were specified based on that reported by the National Research Energy Laboratory (NREL) [81]. The properties and chemical equilibria of the liquid and vapor phases were estimated through the Non-Random Two Liquids (NRTL) thermodynamic method and the Soave–Redlich–Kwong equation of state, respectively.

Table 9. Physicochemical characterization of rice husk.

Parameter	Mass Composition (g 100 g^{-1}) on a Dry Basis
Initial moisture	12.01
Cellulose	29.34
Hemicellulose	15.02
Lignin	29.14
Total extract	7.86
Fats	3.80
Protein	1.29
Pectin	13.55
Ash	18.52

3.3.1. Techno-Economic Assessment

The overall pretreatment yields were calculated for each lignocellulosic fraction based on the initial raw material flow rate. For the cellulose yield, the six-carbon total content in the WIS was considered, as shown in Equation (5). The oligomers and monosaccharides of five carbons were used for hemicellulose yield (see Equation (6)). Meanwhile, Equation (7) involves the solubilized lignin in the liquor fraction. On the other hand, utility requirements were calculated for steam, cooling water, and electricity. For the CapEx analysis, the direct cost of the processing equipment was estimated using the Aspen Process Economic Analyzer v9.0 software (Aspen Technologies, Inc., USA) and based on the mass and energy balances of the simulations. The CapEx was calculated considering the sum of the direct cost with mechanical, civil, and instrumentation work as well as piping and electrical wiring. The OpEx involves raw materials, utilities, maintenance, depreciation, and labor costs. The raw material cost includes the feedstock and chemical reagents costs (see Table 10). For utilities, values of 7.89 USD ton^{-1}, 8.07 USD ton^{-1}, 8.15 USD ton^{-1}, and 0.1 USD kWh^{-1} were used for low-pressure steam, medium-pressure steam, high-pressure steam, and electricity, respectively. The cooling and process water cost was calculated using a Chemical Engineering Plant Cost Index (CEPCI) of 701.4 for 2021. The maintenance was calculated as 6% of the CapEx and the depreciation using an interest rate of 17% based on the straight-line method. Finally, the labor cost was estimated considering a wage of 5.21 USD h^{-1} for an eight-hour shift per day.

$$\text{Yield}_{\text{cellulose}} = \frac{\text{Cellulose}_{\text{WIS}}}{\text{Raw feedstock}} \times 100\% \tag{5}$$

$$\text{Yield}_{\text{hemicellulose}} = \frac{\left(\text{Oligomers}_{C_5} + \text{Monosaccharides}_{C_5}\right)_{\text{hydrolizate}}}{\text{Raw feedstock}} \times 100\% \tag{6}$$

$$\text{Yield}_{\text{Lignin}} = \frac{\text{Lignin}_{\text{hydrolyzate}}}{\text{Raw feedstock}} \times 100\% \tag{7}$$

Table 10. Feedstock and chemical reagent cost.

Input	Cost (USD ton^{-1})	Reference
Rice husk	20	Colombian regional market
Sodium carbonate	234	Means of Alibaba *
Ammonia	450	
Ethanol	863	Colombian regional market
Sulfuric acid	94	Means of Alibaba *
Sodium hydroxide	450	
Sodium sulfide	350	
Ionic liquid	13,500	

* Cost calculated as a mean of www.alibaba.com (accessed on 20 November 2022).

3.3.2. Environmental Assessment

The environmental analysis was carried out based on the indicators established by the Environmental Protection Agency (EPA). The Waste Reduction Algorithm (WAR) software (Environmental Protection Agency, USA) calculated the potential environmental impact (PEI), considering only the pretreatment stage as the control volume. Thus, the PEI was calculated as the difference between the environmental impact generated by the process input and output streams through five impact categories: terrestrial toxicity potential (TTP), human toxicity potential by inhalation or dermal exposure (HTPE), human toxicity potential by ingestion (HTPI), smog formation or photochemical oxidation potential (PCOP), and acidification potential or acid rain (AP).

3.3.3. Social Assessment

Social analysis is considered one of the three fundamental pillars in determining the sustainability of a process. Therefore, this work proposed an analysis to determine the social impact of implementing pretreatment schemes for rice husks related to the local community. It is important to point out that other categories associated with workers, value chain agents, and consumers were excluded from the analysis since they could not be identified and evaluated. For example, a life cycle analysis that includes the agronomic and transport stages of rice husks is not carried out, hindering the analysis of employees in the value chain. Therefore, the scope of the social analysis also involved pretreatment schemes from raw material intake to filtration for obtaining the WIS and the hydrolysate. In this work, it was assumed that two operators would be working in the pretreatment stage; therefore, an analysis of employment generated was not performed since it would be constant for all the schemes. It has previously been reported that two employees can be assumed for each processing section, involving raw material and reagent reception, pretreatment, reaction, and separation [82]. Table 11 summarizes the stakeholders, subcategories and indicators used to evaluate the social impact of the different rice husk pretreatment systems. All indicators were normalized using statistics and information derived from the industrial sector in the Colombian context. Additionally, the methodology for the social analysis was carried out following the Product Social Life Cycle Assessment (PSILCA) database developed by GreenDelta [83].

Table 11. Social indicators used to evaluate the social impact of pretreatment schemes.

Stakeholder	Subcategory	Indicator	Unit
Local community	Access to material resources	Level of industrial water use (withdrawal)	%
		Level of industrial water use (renewable)	%
		Energy demand	%
		Extraction of fossil fuels	%

Stakeholder: Local Community

The social impact caused by the pretreatment schemes was evaluated by considering a subcategory related to the use of natural resources, energy demand, and fossil fuel extraction. Thus, the first two indicators associated with using natural resources aim to assess the level of water use in the industrial sector and the renewable available in the country. For the water use in the industrial sector, the flow used in each scheme (cooling water and process water) and the total water used at the national scale in the industrial sector are correlated. Furthermore, for the renewable water available in the country, the flow used in each scheme is related to the total water available at the national scale. The AQUASTAT database from the Food and Agriculture Organization of the United Nations (FAO) was used to calculate these indicators [84]. On the other hand, the energy demand indicator relates the energy of each scheme with the national energy demand associated with the industrial sector. For the calculation of this indicator, the annual reports of the Mining-Energy Planning Unit (UPME) were used [85]. Finally, the fossil fuel extraction indicator is understood as the diesel energy (fuel for a steam boiler) required based on the thermal demand of low, medium, and high-pressure steams in the pretreatment schemes. This indicator also considers the UPME annual reports. The equations for calculating the social indicators are summarized in Table 12.

Table 12. Social indicators used to evaluate the social impact.

Indicator	Equation
Level of industrial water use (withdrawal)	$FWU_{sector} = \frac{W_{process} + W_{cooling}}{W_{withdrawal} \text{ by industry sector in Colombia}}$
Level of industrial water use (renewable)	$FWU_{country} = \frac{W_{process} + W_{cooling}}{W_{renewable} \text{ in Colombia}}$
Energy demand	$ED = \frac{\text{Energy demand in process}}{\text{Energy demand in Colombia}}$
Extraction of fossil fuels	$EF = \frac{\text{Dieselenergy}}{\text{National energy demand (diesel)}}$

4. Conclusions

This work demonstrated the definition of pretreatment efficacy in biorefinery schemes assessed through ten sustainable indicators involving operational considerations, cost-effectiveness, environmental impact, and social issues, described as follows. (i) lignocellulosic fraction removal; (ii) accessibility index; (iii) formation of by-products or hydrolyzed products; (iv) formation of inhibitory compounds; (v) technology readiness level; (vi) energy demand; (vii) capital costs; (viii) operating costs; (ix) environmental impact; and (x) social impact. Through an in-depth literature review and process simulation, it was possible to identify the best pretreatment schemes for individual cellulose isolation and the removal of hemicellulose and lignin. It was concluded that at a preliminary analysis, the best pretreatments for cellulose isolation in the WIS are dilute acid, wet air oxidation, or organosolv. On the other hand, if biorefineries are planned to valorize the hemicellulose fraction, it is recommended to implement LHW, steam explosion, or dilute acid pretreatments. In lignin-based biorefineries, it is proposed to use kraft, alkali, or LHW pretreatments. As a main result, an approximation of sustainable pretreatments in the Colombian context is described due to the assessment of techno-energetic, economic, environmental, and social indicators.. These results would help future work on designing complex biorefineries to choose the best pretreatment scheme as it drastically influences the reaction and downstream processes for product separation.

Author Contributions: Conceptualization, J.A.P.-G. and M.C.G.-V.; Methodology, J.A.P.-G. and M.C.G.-V.; Software, Simulation and Validation, J.A.P.-G.; Formal Analysis, J.A.P.-G. and M.C.G.-V.; Writing—Original Draft Preparation, J.A.P.-G. and M.C.G.-V.; Writing—Review and Editing, J.A.P.-G. and C.A.C.A.; Supervision, C.A.C.A.; Project administration and Funding acquisition, C.A.C.A. All authors have read and agreed to the published version of the manuscript.

Funding: This research was funded by the "Ministerio de Ciencia, Tecnología e Innovación (Minciencias)", Colombia. (Contract number: 80740-903-2020).

Institutional Review Board Statement: Not applicable.

Informed Consent Statement: Not applicable.

Data Availability Statement: No new data were created or analyzed in this study. Data sharing is not applicable to this article.

Acknowledgments: This work was supported by the project entitled "Impulsando el desarrollo de biosurfactantes a través de su ciclo de vida sistemático" from the Ministerio de Ciencia, Tecnología en Innovación (Minciencias) with the contract number 80740-903-2020.

Conflicts of Interest: The authors declare no conflict of interest.

References

1. Moncada, B.J.; Aristizábal, M.V.; Cardona, A.C.A. Design strategies for sustainable biorefineries. *Biochem. Eng. J.* **2016**, *116*, 122–134. [CrossRef]
2. Smith, R. *Chemical Process: Design and Integration*, 1st ed.; McGraw-Hill: Manchester, UK, 2005. [CrossRef]
3. Cherubini, F. The biorefinery concept: Using biomass instead of oil for producing energy and chemicals. *Energy Convers. Manag.* **2010**, *51*, 1412–1421. [CrossRef]
4. Aristizábal, M.V.; Gómez, P.Á.; Cardona, A.C.A. Biorefineries based on coffee cut-stems and sugarcane bagasse: Furan-based compounds and alkanes as interesting products. *Bioresour. Technol.* **2015**, *196*, 480–489. [CrossRef] [PubMed]
5. JSolarte-Toro, C.; Alzate, C.A.C. Biorefineries as the base for accomplishing the sustainable development goals (SDGs) and the transition to bioeconomy: Technical aspects, challenges and perspectives. *Bioresour. Technol.* **2021**, *340*, 125626. [CrossRef]
6. Bautista, S.; Enjolras, M.; Narvaez, P.; Camargo, M.; Morel, L. Biodiesel-triple bottom line (TBL): A new hierarchical sustainability assessment framework of principles criteria & indicators (PC&I) for biodiesel production. Part II-validation. *Ecol. Indic.* **2016**, *69*, 803–817. [CrossRef]
7. Alzate, C.A.C.; Aristiza, V.; Botero, J.M. *Biorefineries: Design and Analysis*; CRC Press, Taylor & Francis Group: Boca Raton, FL, USA, 2019. [CrossRef]
8. Gutiérrez, C.D.B.; Serna, D.L.R.; Alzate, C.A.C. A comprehensive review on the implementation of the biorefinery concept in biodiesel production plants. *Biofuel Res. J.* **2017**, *4*, 691–703. [CrossRef]
9. Aristizábal-Marulanda, V.; Alzate, C.A.C. Methods for designing and assessing biorefineries: Review. *Biofuels Bioprod. Biorefining* **2019**, *13*, 789–8082019. [CrossRef]
10. Berndes, G.; Hoogwijk, M.; van den Broek, R. The contribution of biomass in the future global energy supply: A review of 17 studies. *Biomass Bioenergy* **2003**, *25*, 1–28. [CrossRef]
11. Galbe, M.; Wallberg, O. Pretreatment for biorefineries: A review of common methods for efficient utilisation of lignocellulosic materials. *Biotechnol. Biofuels* **2019**, *12*, 294. [CrossRef]
12. Chandel, A.K.; Garlapati, V.K.; Singh, A.K.; Antunes, F.A.F.; da Silva, S.S. The path forward for lignocellulose biorefineries: Bottlenecks, solutions, and perspective on commercialization. *Bioresour. Technol.* **2018**, *264*, 370–381. [CrossRef]
13. Mosier, N.; Wyman, C.; Dale, B.; Elander, R.; Lee, Y.Y.; Holtzapple, M.; Ladisch, M. Features of promising technologies for pretreatment of lignocellulosic biomass. *Bioresour. Technol.* **2005**, *96*, 673–686. [CrossRef] [PubMed]
14. Easson, M.W.; Condon, B.; Dien, B.S.; Iten, L.; Slopek, R.; Yoshioka-Tarver, M.; Lambert, A.; Smith, J. The application of ultrasound in the enzymatic hydrolysis of switchgrass. *Appl. Biochem. Biotechnol.* **2011**, *165*, 1322–1331. [CrossRef] [PubMed]
15. Gupta, R.; Khasa, Y.P.; Kuhad, R.C. Evaluation of pretreatment methods in improving the enzymatic saccharification of cellulosic materials. *Carbohydr. Polym.* **2011**, *84*, 1103–1109. [CrossRef]
16. Lopez-Hidalgo, A.M.; Magaña, G.; Rodriguez, F.; de Leon-Rodriguez, A.; Sanchez, A. Co-production of ethanol-hydrogen by genetically engineered Escherichia coli in sustainable biorefineries for lignocellulosic ethanol production. *Chem. Eng. J.* **2021**, *406*, 126829. [CrossRef]
17. Wang, W.; Ling, H.; Zhao, H. Steam explosion pretreatment of corn straw on xylose recovery and xylitol production using hydrolysate without detoxification. *Process Biochem.* **2015**, *50*, 1623–1628. [CrossRef]
18. Silveira, M.H.; Morais, A.R.; da Costa Lopes, A.M.; Olekszyszen, D.N.; Bogel-Łukasik, R.; Andreaus, J.; Pereira Ramos, L. Current Pretreatment Technologies for the Development of Cellulosic Ethanol and Biorefineries. *ChemSusChem* **2015**, *8*, 3366–3390. [CrossRef]
19. Aden, A.; Ruth, M.; Ibsen, K.; Jechura, J.; Neeves, K.; Sheehan, J.; Wallace, B.; Montague, L.; Slayton, A.; Lukas, J. Lignocellulosic Biomass to Ethanol Process Design and Economics Utilizing Co-Current Dilute Acid Prehydrolysis and Enzymatic Hydrolysis for Corn Stover. National Renewable Energy Laboratory. 2002. Available online: http://www.osti.gov/bridge (accessed on 21 October 2022).
20. Brunner, G. Processing of Biomass with Hydrothermal and Supercritical Water. In *Supercritical Fluid Science and Technology*; Elsevier: Amsterdam, The Netherlands, 2014; Volume 5, pp. 395–509. [CrossRef]

21. Li, L.; Rowbotham, J.S.; Greenwell, H.C.; Dyer, P.W. An Introduction to Pyrolysis and Catalytic Pyrolysis: Versatile Techniques for Biomass Conversion. In *New and Future Developments in Catalysis: Catalytic Biomass Conversion*; Elsevier B.V.: Oxford, UK, 2013; pp. 173–208. [CrossRef]
22. Ribca, I.; Jawerth, M.E.; Brett, C.J.; Lawoko, M.; Schwartzkopf, M.; Chumakov, A.; Roth, S.V.; Johansson, M. Exploring the Effects of Different Cross-Linkers on Lignin-Based Thermoset Properties and Morphologies. *ACS Sustain. Chem. Eng.* **2021**, *9*, 1692–1702. [CrossRef]
23. Poveda-Giraldo, J.A.; Solarte-Toro, J.C.; Alzate, C.A.C. The potential use of lignin as a platform product in biorefineries: A review. *Renew. Sustain. Energy Rev.* **2021**, *138*, 110688. [CrossRef]
24. Demirbas, M.F. Biorefineries for biofuel upgrading: A critical review. *Appl. Energy* **2009**, *86* (Suppl. S1), S151–S161. [CrossRef]
25. Karimi, K.; Taherzadeh, M.J. A critical review on analysis in pretreatment of lignocelluloses: Degree of polymerization, adsorption/desorption, and accessibility. *Bioresour. Technol.* **2016**, *203*, 348–356. [CrossRef]
26. Nishiyama, Y. Structure and properties of the cellulose microfibril. *J. Wood Sci.* **2009**, *55*, 241–249. [CrossRef]
27. Kumar, P.; Barrett, D.M.; Delwiche, M.J.; Stroeve, P. Methods for pretreatment of lignocellulosic biomass for efficient hydrolysis and biofuel production. *Ind. Eng. Chem. Res.* **2009**, *48*, 3713–3729. [CrossRef]
28. Zhang, Y.H.P.; Lynd, L.R. Toward an aggregated understanding of enzymatic hydrolysis of cellulose: Noncomplexed cellulase systems. *Biotechnol. Bioeng.* **2004**, *88*, 797–824. [CrossRef] [PubMed]
29. Santos, E.O.; Silva, A.M.S.; Fragoso, W.D.; Pasquini, C.; Pimentel, M.F. Determination of degree of polymerization of insulating paper using near infrared spectroscopy and multivariate calibration. *Vib. Spectrosc.* **2010**, *52*, 154–157. [CrossRef]
30. Harris, D.; DeBolt, S. Relative crystallinity of plant biomass: Studies on assembly, adaptation and acclimation. *PLoS ONE* **2008**, *3*, e2897. [CrossRef] [PubMed]
31. Montipó, S.; Roslander, C.; Camassola, M.; Galbe, M.; Wallberg, O. Steam Pretreatment of Rice Hulls to Release Fermentable Saccharides: An Approach to Improve Recovery of (Hemi)Cellulosic Sugars Through Multivariate Design. *Rice Sci.* **2021**, *28*, 501–510. [CrossRef]
32. Zhang, C.; Pang, F.; Li, B.; Xue, S.; Kang, Y. Recycled aqueous ammonia expansion (RAAE) pretreatment to improve enzymatic digestibility of corn stalks. *Bioresour. Technol.* **2013**, *138*, 314–320. [CrossRef]
33. Bals, B.; Rogers, C.; Jin, M.; Balan, V.; Dale, B. Evaluation of Ammonia Fibre Expansion (AFEX) Pretreatment for Enzymatic Hydrolysis of Switchgrass Harvested in Different Seasons and Locations. *Biotechnol. Biofuels* **2010**, *3*, 1. [CrossRef]
34. Jönsson, L.J.; Alriksson, B.; Nilvebrant, N.O. Bioconversion of lignocellulose: Inhibitors and detoxification. *Biotechnol. Biofuels* **2013**, *6*, 16. [CrossRef] [PubMed]
35. Mishra, A.; Ghosh, S. Saccharification of kans grass biomass by a novel fractional hydrolysis method followed by co-culture fermentation for bioethanol production. *Renew. Energy* **2020**, *146*, 750–759. [CrossRef]
36. Kumar, R.; Mago, G.; Balan, V.; Wyman, C.E. Physical and chemical characterizations of corn stover and poplar solids resulting from leading pretreatment technologies. *Bioresour. Technol.* **2009**, *100*, 3948–3962. [CrossRef] [PubMed]
37. Fockink, D.H.; Sánchez, J.H.; Ramos, L.P. Comprehensive analysis of sugarcane bagasse steam explosion using autocatalysis and dilute acid hydrolysis (H3PO4 and H2SO4) at equivalent combined severity factors. *Ind. Crops Prod.* **2018**, *123*, 563–572. [CrossRef]
38. Ibrahim, M.N.M.; Chuah, S.B.; Rosli, W.D.W. Characterization of lignin precipitated from the soda black liquor of oil pal empty fruit bunch fibers by various mineral acids. *ASEAN J. Sci. Technol. Dev.* **2004**, *21*, 57–67. [CrossRef]
39. Santos, P.S.B.; Erdocia, X.; Gatto, D.A.; Labidi, J. Characterisation of Kraft lignin separated by gradient acid precipitation. *Ind. Crops Prod.* **2014**, *55*, 149–154. [CrossRef]
40. Monrroy, M.; García, J.R.; Mendonça, R.T.; Baeza, J.; Freer, J. Kraft pulping of eucalyptus globulus as a pretreatment for bioethanol production by simultaneous saccharification and fermentation. *J. Chil. Chem. Soc.* **2012**, *57*, 1113–1117. [CrossRef]
41. ISalapa; Katsimpouras, C.; Topakas, E.; Sidiras, D. Organosolv pretreatment of wheat straw for efficient ethanol production using various solvents. *Biomass Bioenergy* **2017**, *100*, 10–16. [CrossRef]
42. Li, C.; Knierim, B.; Manisseri, C.; Arora, R.; Scheller, H.V.; Auer, M.; Vogel, K.P.; Simmons, B.A.; Singh, S. Comparison of dilute acid and ionic liquid pretreatment of switchgrass: Biomass recalcitrance, delignification and enzymatic saccharification. *Bioresour. Technol.* **2010**, *101*, 4900–4906. [CrossRef]
43. Li, Z.; Jiang, Z.; Fei, B.; Cai, Z.; Pan, X. Comparison of bamboo green, timber and yellow in sulfite, sulfuric acid and sodium hydroxide pretreatments for enzymatic saccharification. *Bioresour. Technol.* **2014**, *151*, 91–99. [CrossRef]
44. Banerjee, S.; Sen, R.; Pandey, R.A.; Chakrabarti, T.; Satpute, D.; Giri, B.S.; Mudliar, S. Evaluation of wet air oxidation as a pretreatment strategy for bioethanol production from rice husk and process optimization. *Biomass Bioenergy* **2009**, *33*, 1680–1686. [CrossRef]
45. Lau, M.W.; Gunawan, C.; Dale, B.E. The impacts of pretreatment on the fermentability of pretreated lignocellulosic biomass: A comparative evaluation between ammonia fiber expansion and dilute acid pretreatment. *Biotechnol. Biofuels* **2009**, *2*, 30. [CrossRef]
46. Du, W.; Yu, H.; Song, L.; Zhang, J.; Weng, C.; Ma, F.; Zhang, X. The promoting effect of byproducts from Irpex lacteus on subsequent enzymatic hydrolysis of bio-pretreated cornstalks. *Biotechnol. Biofuels* **2011**, *4*, 37. [CrossRef] [PubMed]
47. Wan, C.; Li, Y. Effectiveness of microbial pretreatment by Ceriporiopsis subvermispora on different biomass feedstocks. *Bioresour. Technol.* **2011**, *102*, 7507–7512. [CrossRef] [PubMed]

48. Lee, J.M.; Jameel, H.; Venditti, R.A. A comparison of the autohydrolysis and ammonia fiber explosion (AFEX) pretreatments on the subsequent enzymatic hydrolysis of coastal Bermuda grass. *Bioresour. Technol.* **2010**, *101*, 5449–5458. [CrossRef] [PubMed]
49. Tae, H.K.; Lee, Y.Y. Pretreatment and fractionation of corn stover by ammonia recycle percolation process. *Bioresour. Technol.* **2005**, *96*, 2007–2013. [CrossRef]
50. Cianchetta, S.; di Maggio, B.; Burzi, P.L.; Galletti, S. Evaluation of selected white-rot fungal isolates for improving the sugar yield from wheat straw. *Appl. Biochem. Biotechnol.* **2014**, *173*, 609–623. [CrossRef]
51. Oka, D.; Kobayashi, K.; Isobe, N.; Ogawa, Y.; Yokoyama, T.; Kimura, S.; Kim, U.J.; Tokuyasu, K.; Wada, M. Enzymatic hydrolysis of wood with alkaline treatment. *J. Wood Sci.* **2013**, *59*, 484–488. [CrossRef]
52. Ertas, M.; Han, Q.; Jameel, H.; Chang, H. Enzymatic hydrolysis of autohydrolyzed wheat straw followed by refining to produce fermentable sugars. *Bioresour. Technol.* **2014**, *152*, 259–266. [CrossRef] [PubMed]
53. Gao, X.; Kumar, R.; Singh, S.; Simmons, B.A.; Balan, V.; Dale, B.E.; Wyman, C.E. Comparison of enzymatic reactivity of corn stover solids prepared by dilute acid, AFEX^{TM}, and ionic liquid pretreatments. *Biotechnol. Biofuels* **2014**, *7*, 71. [CrossRef]
54. Suhara, H.; Kodama, S.; Kamei, I.; Maekawa, N.; Meguro, S. Screening of selective lignin-degrading basidiomycetes and biological pretreatment for enzymatic hydrolysis of bamboo culms. *Int. Biodeterior. Biodegradation* **2012**, *75*, 176–180. [CrossRef]
55. Song, L.; Yu, H.; Ma, F.; Zhang, X. Biological Pretreatment under Non-sterile Conditions for Enzymatic Hydrolysis of Corn Stover. *BioResources* **2013**, *8*, 3802–3816. [CrossRef]
56. de Freitas, C.; Carmona, E.; Brienzo, M. Xylooligosaccharides production process from lignocellulosic biomass and bioactive effects. *Bioact. Carbohydr. Diet. Fibre* **2019**, *18*, 100184. [CrossRef]
57. Samanta, A.K.; Jayapal, N.; Jayaram, C.; Roy, S.; Kolte, A.P.; Senani, S.; Sridhar, M. Xylooligosaccharides as prebiotics from agricultural by-products: Production and applications. *Bioact. Carbohydr. Diet. Fibre* **2015**, *5*, 62–71. [CrossRef]
58. Otieno, D.O.; Ahring, B.K. The potential for oligosaccharide production from the hemicellulose fraction of biomasses through pretreatment processes: Xylooligosaccharides (XOS), arabinooligosaccharides (AOS), and mannooligosaccharides (MOS). *Carbohydr. Res.* **2012**, *360*, 84–92. [CrossRef] [PubMed]
59. Panjiar, N.; Mattam, A.J.; Jose, S.; Gandham, S.; Velankar, H.R. Valorization of xylose-rich hydrolysate from rice straw, an agroresidue, through biosurfactant production by the soil bacterium Serratia nematodiphila. *Sci. Total Environ.* **2020**, *729*, 138933. [CrossRef] [PubMed]
60. Nosrati-Ghods, N.; Harrison, S.T.L.; Isafiade, A.J.; Tai, S.L. Analysis of ethanol production from xylose using Pichia stipitis in microaerobic conditions through experimental observations and kinetic modelling. *Biochem. Eng. J.* **2020**, *164*, 107754. [CrossRef]
61. Bondar, M.; da Fonseca, M.M.R.; Cesário, M.T. Xylonic acid production from xylose by Paraburkholderia sacchari. *Biochem. Eng. J.* **2021**, *170*, 107982. [CrossRef]
62. Poveda-Giraldo, J.A.; Cardona, C.A. Biorefinery potential of Eucalyptus grandis to produce phenolic compounds and biogas. *Can. J. For. Res.* **2020**, *51*, 89–100. [CrossRef]
63. Cazacu, G.; Pascu, M.C.; Profire, L.; Kowarski, A.I.; Mihaes, M.; Vasile, C. Lignin role in a complex polyolefin blend. *Ind. Crops Prod.* **2004**, *20*, 261–273. [CrossRef]
64. Tejado, A.; Peña, C.; Labidi, J.; Echeverria, J.M.; Mondragon, I. Physico-chemical characterization of lignins from different sources for use in phenol-formaldehyde resin synthesis. *Bioresour. Technol.* **2007**, *98*, 1655–1663. [CrossRef]
65. Wahlbom, C.F.; Hahn-Hägerdal, B. Furfural, 5-hydroxymethyl furfural, and acetoin act as external electron acceptors during anaerobic fermentation of xylose in recombinant Saccharomyces cerevisiae. *Biotechnol. Bioeng.* **2002**, *78*, 172–178. [CrossRef]
66. Mathew, A.K.; Abraham, A.; Mallapureddy, K.K.; Sukumaran, R.K. Lignocellulosic Biorefinery Wastes, or Resources? In *Waste Biorefinery: Potential and Perspectives*; Elsevier: Oxford, UK, 2018; pp. 267–297. [CrossRef]
67. Pampulha, M.E.; Loureiro-Dias, M.C. Combined effect of acetic acid, pH and ethanol on intracellular pH of fermenting yeast. *Appl. Microbiol. Biotechnol.* **1989**, *31*, 547–550. [CrossRef]
68. Chuetor, S.; Panakkal, E.J.; Ruensodsai, T.; Cheenkachorn, K.; Kirdponpattara, S.; Cheng, Y.S.; Sriariyanun, M. Improvement of enzymatic saccharification and ethanol production from rice straw using recycled ionic liquid: The effect of anti-solvent mixture. *Bioengineering* **2022**, *9*, 115. [CrossRef] [PubMed]
69. Yu, L.; Seabright, K.; Bajaj, I.; Keffer, D.J.; Alonso, D.M.; Hsieh, C.T.; Li, M.; Chen, H.; Gandomi, Y.A.; Maravelias, C.T.; et al. Performance and Economic Analysis of Organosolv Softwood and Herbaceous Lignins to Activated Carbons as Electrode Materials in Supercapacitors. *Front. Energy Res.* **2022**, *10*, 849949. [CrossRef]
70. Li, Y.; Pan, L.; Zeng, X.; Zhang, R.; Li, X.; Li, J.; Xing, H.; Bao, J. Ammonia exposure causes the imbalance of the gut-brain axis by altering gene networks associated with oxidative metabolism, inflammation and apoptosis. *Ecotoxicol. Environ. Saf.* **2021**, *224*, 112668. [CrossRef] [PubMed]
71. Arias, W.A.A.; Pineda, J.C.; Toro, O.J.S.; Alzate, C.A.C. Comparative analysis of different amylaceous raw materials for obtaining fuel alcohol. In Proceedings of the XXIV Congreso Colombiano de Ingeniería Química, Cali, Colombia, 24–26 September 2007.
72. Alzate, C.C.; Rodríguez, M.M.; Suárez, J.Q. Selection of appropriate technologies for fuel ethanol production. *Ing. De Recur. Nat. Ambiente* **2004**, *2*, 48–55.
73. Serna-loaiza, S.; Carmona-garcia, E.; Cardona, C.A. Potential raw materials for biorefineries to ensure food security: The Cocoyam case. *Ind. Crops Prod.* **2018**, *126*, 92–102. [CrossRef]
74. Win, N.N.; Weinwurm, F.; Friedl, A. Investigation of organosolv and hot-compressed water pretreatments of rice straw. *Biomass Convers. Biorefin.* **2016**, *6*, 355–364. [CrossRef]

75. Hoşgün, E.Z.; Ay, S.B.; Bozan, B. Effect of sequential pretreatment combinations on the composition and enzymatic hydrolysis of hazelnut shells. *Prep. Biochem. Biotechnol.* **2020**, *51*, 570–579. [CrossRef]
76. Tang, W.; Wu, X.; Huang, C.; Huang, C.; Lai, C.; Yong, Q. Enhancing enzymatic digestibility of waste wheat straw by presoaking to reduce the ash-influencing effect on autohydrolysis. *Biotechnol. Biofuels* **2019**, *12*, 222. [CrossRef]
77. Mood, S.H.; Golfeshan, A.H.; Tabatabaei, M.; Jouzani, G.S.; Najafi, G.H.; Gholami, M.; Ardjmand, M. Lignocellulosic biomass to bioethanol, a comprehensive review with a focus on pretreatment. *Renew. Sustain. Energy Rev.* **2013**, *27*, 77–93. [CrossRef]
78. Shrotri, A.; Kobayashi, H.; Fukuoka, A. Catalytic Conversion of Structural Carbohydrates and Lignin to Chemicals. *Adv. Catal.* **2017**, *60*, 59–123. [CrossRef]
79. Deshavath, N.N.; Dasu, V.V.; Goud, V.; Rao, P.S. Development of dilute sulfuric acid pretreatment method for the enhancement of xylose fermentability. *Biocatal. Agric. Biotechnol.* **2017**, *11*, 224–230. [CrossRef]
80. Martinkus, N.; Latta, G.; Rijkhoff, S.A.; Mueller, D.; Hoard, S.; Sasatani, D.; Pierobon, F.; Wolcott, M. A multi-criteria decision support tool for biorefinery siting: Using economic, environmental, and social metrics for a refined siting analysis. *Biomass Bioenergy* **2019**, *128*, 105330. [CrossRef]
81. Wooley, R.J.; Putsche, V. *Development of an ASPEN PLUS Physical Property Database for Biofuels Components*; NREL/MP-425-20685; National Renewable Energy Lab: Golden, CO, USA, 1996.
82. Peters, M.; Timmerhaus, K.; West, R. *Plant Design and Economics for Chemical Engineers*, 5th ed.; McGraw-Hill: New York, NY, USA, 2003.
83. Maister, K.; di Noi, C.; Ciroth, A.; Srocka, M. *PSILCA Database v.3 Documentation*; PSILCA: Berlin, Germany, 2020.
84. Food and Agriculture Organization of the United Nations. AQUASTAT. FAO. 2021. Available online: https://tableau.apps.fao.org/views/ReviewDashboard-v1/country_dashboard?%3Aembed=y&%3AisGuestRedirectFromVizportal=y (accessed on 16 November 2022).
85. Martínez, W.A.; Rojas, M.P.; García, J.S.; Martínez, J.F.; Rodríguez, R.A. *Proyección Demanda Energía Eléctrica Gas Natural y Combustibles Líquidos 2022–2036*; UPME: Bogotá, Colombia, 2021.

molecules

MDPI

Article

Alkaline Modification of *Arabica*-Coffee and *Theobroma*-Cocoa Agroindustrial Waste for Effective Removal of Pb(II) from Aqueous Solutions

Carmencita Lavado-Meza [1,*], Leonel De la Cruz-Cerrón [2], Yvan J.O. Asencios [3], Francielle Candian Firmino Marcos [4] and Juan Z. Dávalos-Prado [5,*]

[1] Escuela Profesional de Ingeniería Ambiental, Universidad Nacional Intercultural de la Selva Central Juan Santos Atahualpa, Chanchamayo 12856, Peru
[2] Facultad de Ingeniería, Universidad Continental, Huancayo 12000, Peru
[3] Institute of Marine Science, Federal University of São Paulo, Santos 11030-100, SP, Brazil
[4] Escola Politecnica, Department of Chemical Engineering, Universidade de São Paulo, Av. Prof. Luciano Gualberto, t. 3, 380, São Paulo 05508-010, SP, Brazil
[5] Instituto de Química Física "Rocasolano", CSIC, 28006 Madrid, Spain
* Correspondence: authors: clavado@uniscjsa.edu.pe (C.L.-M.); jdavalos@iqfr.csic.es (J.Z.D.-P.)

Abstract: *Arabica*-coffee and *Theobroma*-cocoa agroindustrial wastes were treated with NaOH and characterized to efficiently remove Pb(II) from the aqueous media. The maximum Pb(II) adsorption capacities, q_{max}, of *Arabica*-coffee (WCAM) and *Theobroma*-cocoa (WCTM) biosorbents (q_{max} = 303.0 and 223.1 mg·g^{-1}, respectively) were almost twice that of the corresponding untreated wastes and were higher than those of other similar agro-industrial biosorbents reported in the literature. Structural, chemical, and morphological characterization were performed by FT-IR, SEM/EDX, and point of zero charge (pH_{PZC}) measurements. Both the WCAM and WCTM biosorbents showed typical uneven and rough cracked surfaces including the OH, C=O, COH, and C-O-C functional adsorbing groups. The optimal Pb(II) adsorption, reaching a high removal efficiency %R (>90%), occurred at a pH between 4 and 5 with a biosorbent dose of 2 g·L^{-1}. The experimental data for Pb(II) adsorption on WACM and WCTM were well fitted with the Langmuir-isotherm and pseudo-second order kinetic models. These indicated that Pb(II) adsorption is a chemisorption process with the presence of a monolayer mechanism. In addition, the deduced thermodynamic parameters showed the endothermic ($\Delta H^0 > 0$), feasible, and spontaneous ($\Delta G^0 < 0$) nature of the adsorption processes studied.

Keywords: biosorption; Pb(II) removal; agroindustrial waste; heavy metals

Citation: Lavado-Meza, C.; De la Cruz-Cerrón, L.; Asencios, Y.J.; Marcos, F.C.F.; Dávalos-Prado, J.Z. Alkaline Modification of *Arabica*-Coffee and *Theobroma*-Cocoa Agroindustrial Waste for Effective Removal of Pb(II) from Aqueous Solutions. *Molecules* **2023**, *28*, 683. https://doi.org/10.3390/molecules28020683

Academic Editor: Xiaomin Xu

Received: 23 November 2022
Revised: 20 December 2022
Accepted: 28 December 2022
Published: 10 January 2023

1. Introduction

Effluents from industrial activities such as smelting, mining, painting, tanning, etc. are causing severe environmental pollution by depositing heavy metals, particularly in aquatic ecosystems [1]. These metals are highly toxic, are not degradable, and can accumulate in living organisms and affect many of their vital functions. Lead (Pb) is the second most toxic metal and can adversely affect the nervous, digestive, and reproductive systems and can even cause death [2,3]. For these reasons and for preventive purposes, for example, the World Health Organization (WHO) has established the permissible limit of Pb in drinking water at 0.01 mg L^{-1} [4].

Various methods are used to remove heavy metals, such as Pb, from wastewater: coagulation–flocculation, liquid–liquid extraction, ion exchange, and electrochemical treatment. However, these methods have disadvantages such as the high operating costs, long operating times, and the generation of a large volume of toxic sludge [5,6]. In this context, the removal of contaminants using biological materials (biosorbents), such as algae, cyanobacteria, fungi, or particularly agroindustrial wastes has become an economical, ecological, and promising alternative method compared with the conventional methods mentioned above [7–10]. These biosorbents can be modified to improve the adsorption capacity, structural stability, and reusability. The modification can be carried out by chemical reagents (chemical modification), physical calcination, or grinding methods [11]. In particular, Calero et al. [12], Moyo et al. [13], Petrović et al. [14], and Ye and Yu [15], among others, reported that chemical modification, with NaOH, of biosorbent-precursors improved the removal capacity of Pb.

Peru is an important producer of coffee and cocoa worldwide, with annual productions of 136 and 218 Kt, respectively [16]. The processing of coffee cherries and cocoa pods generates wastes at approximately 80% (coffee) [17] and 70% (cocoa) [18,19] of the total weight of the product, constituting a serious environmental problem for their producing regions [20]. However, these agro-industrial residues could be used for the effective cleaning of aqueous ecosystems of the surrounding crops, in which we have evidence that there is contamination with heavy metals such as Pb.

In this work, we significantly improved the absorption capacity of Pb(II) by means of alkaline modification (with NaOH) to coffee- and cocoa-untreated wastes [21].

2. Results and Discussion

1.1. Effect of Alkaline Treatment

After treatment with NaOH, both the *arabica*-coffee and *theobroma*-cocoa biosorbents lost weight, 40.2% and 38.4%, respectively (Table 1). This loss may be due to the fact that, during NaOH treatment, the hydrolysis reactions that take place would cause a high dissolution of organic compounds from the biomass and, therefore, its considerable disintegration [12,22].

Table 1. Physical–chemical characteristics of untreated and alkaline treated *Arabica*-coffee and *Theobroma*-cocoa wastes.

	WAC Coffe Waste	WTC Cocoa Waste	WACM Coffe Waste	WTCM Cocoa Waste
	Untreated		Alkaline Treated [a]	
Point of Zero Charge, pH_{PZC}	4.8	6	6	6.8
Acid titrable sites (mmol g^{-1})	2.8×10^{-2}	1.97×10^{-2}	0	0
Basic titrable sites (mmol g^{-1})	2.12×10^{-2}	1.84×10^{-2}	2.97×10^{-2}	2.63×10^{-2}
% Biomass loss due to treatment			40.2	38.4

[a] 0.1 M NaOH.

The basic titrable sites on both WACM- and WTCM-treated biosorbents were almost 1.4 times higher than that in the respective untreated precursors WAC and WTC (see Table 1). According to Santos et al. [22], and Bulgariu and Bulgariu [23], among others, the NaOH treatment provides, due to the hydrolysis reactions, the formation of more carboxylic (-COO^-) and hydroxyl (-OH) groups (basic titrable sites), both in undissociated as dissociated forms, that improve the Pb-binding properties of WACM and WTCM biosorbents. It is interesting to mention the absence of acid-titrable sites on the surface of each treated biomass (See Table 1).

The point of zero charge, pH_{PZC}, values for WACM and WTCM were higher than those for WAC and WTC, respectively (Table 1). This result indicates an increase in the surface basicity of the treated biosorbents, which is consistent with the increase in the concentration of its basic titrable sites, described above. A similar feature was reported by Blázquez et al. [24] for olive stone biomass modified with NaOH.

An interesting consequence of the alkaline treatment of the studied biomasses is the considerable increase in the Pb(II) removal capacity. Thus, taking into account optimal conditions, described below (Section 1.3), the Pb(II) adsorption capacity q_e of WACM and WTCM were almost three times higher than of corresponding non-treated WAC and WTC biomasses (see Figure 1). Mangwandi et al. [6], Ren et al. [11], and Gupta et al. [25], among others, reported that chemical modification of a biomass considerably improves its adsorption capacity of heavy metals.

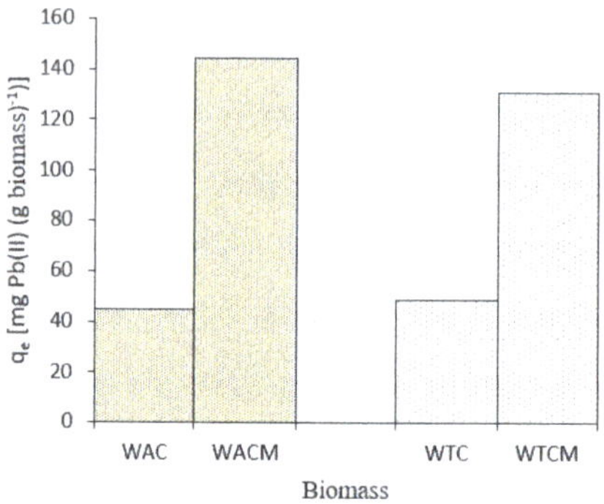

Figure 1. Pb(II) adsorption capacity q_e for *Arabica*-coffee and *Theobroma*-cocoa wastes, untreated (WAC and WTC) and treated (WACM and WTCM) biosorbents. Initial Pb(II) concentration, C_0 = 48.42 mg L^{-1}, biosorbent dose = 2 g L^{-1}, pH 4.

1.2. SEM/EDX and FTIR Analysis

SEM micrographs of the WACM and WTCM biosorbents before and after Pb(II) was loaded are shown in Figure 2. Typical uneven and rough surface morphologies are observed. Before adsorption, the images show more porous and less compact structures, than that with Pb(II) loaded. The results of the EDX spectra (Figure 3) confirmed that Pb(II) is adsorbed on the surface of both Pb-loaded biosorbents. Furthermore, the disappearance of the Na peak on these samples (Figure 3 right) can be attributed to ionic exchange with Pb(II). Similar morphologies were reported by Jaihan et al. [5] in papaya peels loaded with Pb(II).

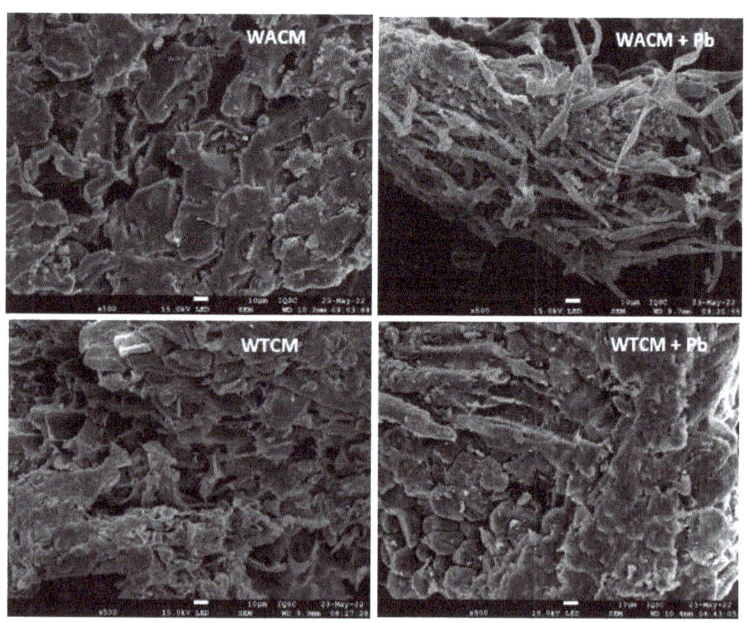

Figure 2. SEM of WACM and WTCM before (**left**) and after (**right**) Pb(II) biosorption. pH = 4, C_0 = 130.8 mg L^{-1}, T = 293 K, t = 120 min.

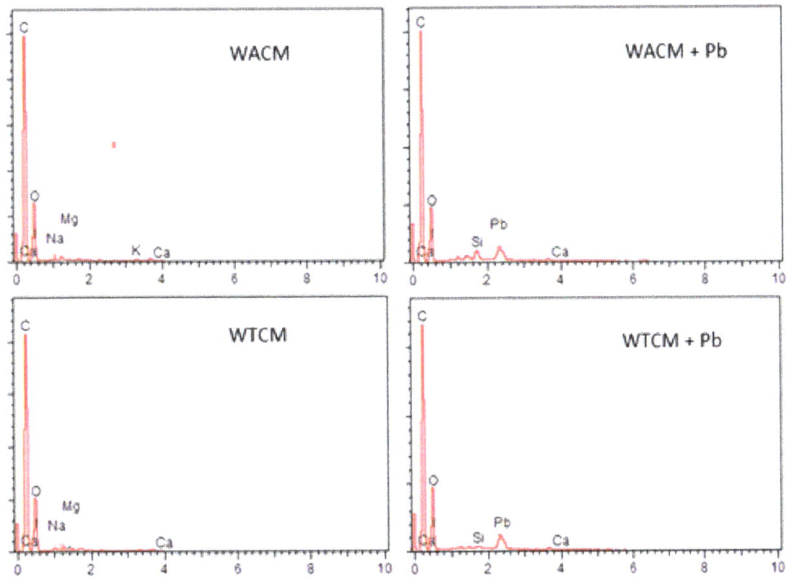

Figure 3. EDX spectra of WACM and WTCM before (**left**) and after Pb adsorption (**right**). pH = 4, C_0 = 130.8 mg L^{-1}, T = 293 K, t = 120 min.

Figure 4 shows the FTIR spectra of WACM and WTCM before and after Pb(II) sorption. The FTIR spectra of the unloaded-Pb samples show the positions of the peaks and absorption bands (in parentheses for WTCM) at the following:

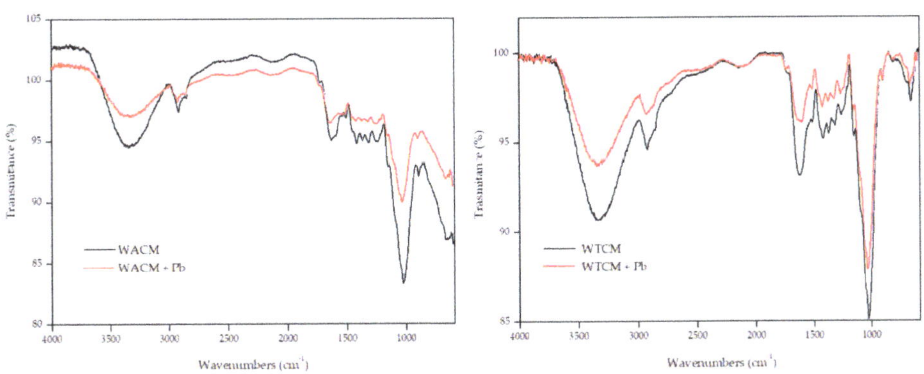

Figure 4. FTIR spectra before (black) and after (red) Pb(II) adsorption by WACM (**left**) and WTCM (**right**). pH 4, C_0 = 130.8 mg L^{-1}, T = 293 K, t = 120 min.

(1) 3339.3 (3335.7) cm^{-1}, assignable to typical -OH bond stretching vibrations in samples such as cellulose, lignin, or water [26–29];

(2) 2919.9 (2910.9) cm^{-1}, assignable to the symmetric stretching of the C-H bonds of aliphatic acids [30];

(3) 1636.2 (1621.6) cm^{-1} assignable to the asymmetric stretching of the double bond of C=O carbonyl groups [29];

(4) 1420.4 (1420.4) cm^{-1}, assignable to the stretching C-OH and C=O groups of carboxylates [29,31];

(5) 1019.1 (1028.8) cm^{-1}, characteristic of C-O-C stretching in polysaccharides [32];

The FTIR spectra after Pb(II) was loaded show changes in the intensity and position of some peaks and bands with respect to those of the clean samples. Thus, for both biosorbents, the positions of the peaks or bands 1, 3, and 4 are displaced with respect to the clean sample values at $\Delta_1 = -13.3$ (-9.7), $\Delta_3 = -4.9$ (-23.2), $\Delta_4 = -8.5$ (-3.6) cm^{-1}. These results indicate that the OH, C=O, and C-O groups would be involved in the biosorption of Pb(II). A similar behavior was reported, among others, by Barka et al. [33] and Mahyoob et al. [29] in the removal of Pb(II) by biomasses such as cladodes of prickly pear or olive tree leaves.

1.3. Adsorption Experiments

1.3.1. Influence of pH Solution

The pH plays an important role in the adsorption process and provides necessary information on the adsorption–desorption mechanisms. The effect of pH was studied in the range of 2 to 5 (Figure 5) since Pb precipitates, at pH > 5, into $Pb(OH)_2$ [27]. The Pb(II) adsorption capacity q_e, was very low for an acidic medium close to pH 2. It is due to a competing effect between H^+ and Pb(II) ions for fill surface active sites [30,34]. For pH > 2, q_e increased greatly, reaching values of 227.1 and 214.3 mg g^{-1} at pH 4 for WACM and WTCM, respectively. With a further increase at pH 5, the q_e values showed improvements by almost 17% for WACM and by only 2% for WTCM. Accordingly, when pH increases, the repulsive interactions between H^+ and Pb(II) ions decrease, facilitating access of Pb(II) to the surface adsorption sites.

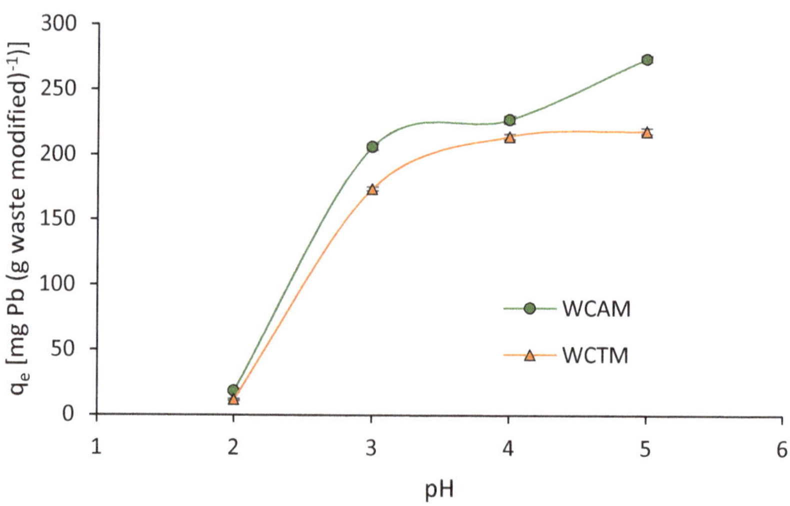

Figure 5. Influence of pH on the Pb(II) adsorption capacity, q_e, for T = 293 K, adsorption time = 120 min, biosorbent dose = 2 g L^{-1}, C_0 = 130.8 mg L^{-1}.

1.3.2. Influence of Biomass Dosage

Figure 6 depicts the Pb(II) removal efficiency, %R, as a function of the biomass dosage. A significant increase in %R is observed for both the WACM and WTCM biosorbents, reaching almost 55% and 60%, respectively, for a biomass dosage of 2 g L^{-1}. A further increase in the dosage produces a slight increase in %R for WTCM but a rapid decrease for WACM. The latter trend is due to the agglomeration of the WACM biomass, observed during the experimentation, which would reduce the effective surface area available for the interaction between Pb(II) and the biosorbent [35].

For both the WACM and WTCM biosorbents, the dose of 2 g L^{-1} was selected as the optimal value, which would provide a suitable surface area for the efficient removal of Pb(II).

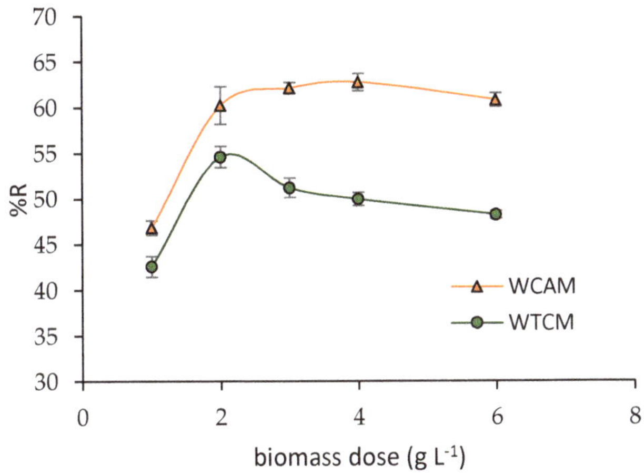

Figure 6. Effect of biomass dosage. T = 293 K, adsorption time = 120 min, pH = 4, C_0 = 130.8 mg L^{-1}.

1.3.3. Influence of Initial Pb(II) ion Concentration, C_0

The effect of the initial Pb(II) ion concentration C_0 on the adsorption capacity q_e and removal efficiency %R was studied in the range of 5.6 to 130.8 mg L^{-1} (See Figure 7). For both the WACM and WTCM biosorbents, the highest %R values (>92%) were obtained for low C_0 concentrations (in the range of 5.6 to 40 mg L^{-1}), where the ratio of surface active sites to the free Pb(II) ions is high, resulting in rapid adsorption [36]. For high C_0 concentrations, %R decreased until almost 60% was reached at C_0 = 130.8 mg L^{-1}. The diminishing %R with increasing C_0 is attributed to the saturation of the available adsorption sites [32,36].

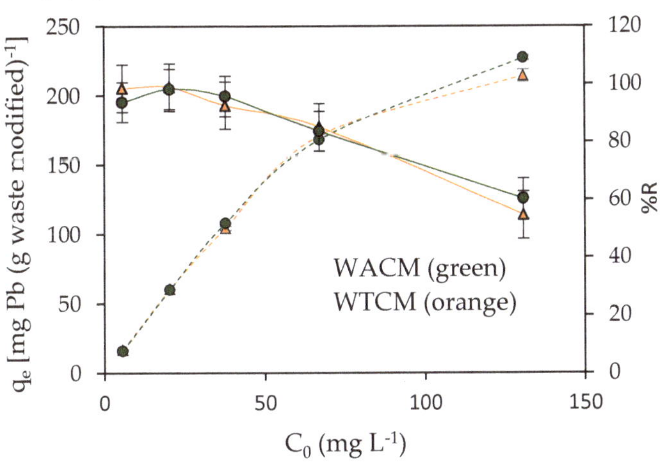

Figure 7. Effect of the initial Pb(II) concentration C_0 on q_e (dotted lines) and %R (full lines). t = 60 min, T = 293 K, pH 4, biosorbent dose = 2 g L^{-1}.

On the other hand, the Pb(II) adsorption capacity q_e of both the WACM and WTCM biosorbents showed an opposite trend to %R, since it increases with increasing C_0. This result can be explained considering that for a given amount of adsorbent, an increase in the amount of Pb(II) ions, in solution, produces a concentration gradient that drives a greater interaction with the active binding sites of the biosorbent [37,38].

1.4. Adsorption Isotherms

The adsorption isotherms were studied in a range of initial Pb(II) concentrations C_0 between 5.6 and 130.8 mg L^{-1}, at pH 4 and pH 5, and T = 293 K and t = 120 min. The results are depicted in Figure 8, where Pb(II) adsorption capacity q_e vs. C_e concentration of Pb(II) in equilibrium is presented. These data were fitted to two very common isotherm models [39,40]: (i) Langmuir model, which assumes solute sorption in monolayers with a homogeneous sorption energy; (ii) Freundlich model, which assumes multilayer sorption, with heterogeneous sorption energies. The adjustment parameters with both models were obtained by fitting the corresponding experimental data [C_e/q_e vs. C_e] and [log(q_e) vs. log(C_e)].

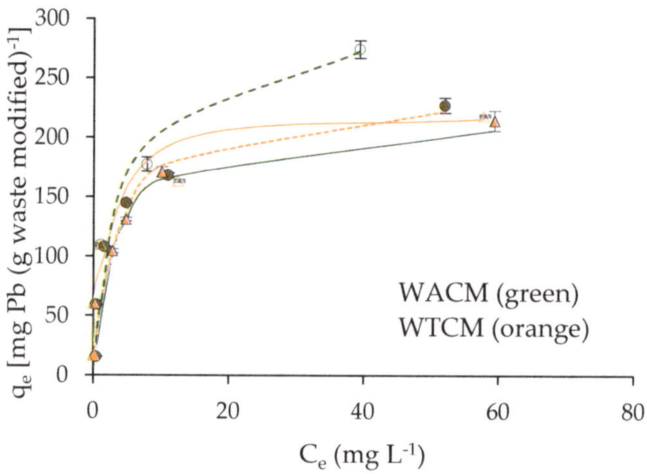

Figure 8. WACM and WTCM adsorption isotherms fitted to Langmuir model. For pH 4 (continuous lines) and pH 5 (dotted lines); biosorbent dose = 2 g L^{-1}, t = 120 min, T = 293 K.

We can see (Table 2) that adsorption isotherms are fitted better with Langmuir (R^2 close to 1) than the Freundlich model ($R^2 \leq 0.87$). From the first model and, for both pH 4 and pH 5, lower K_L and higher maximum sorption capacity q_{max} values are obtained for WACM than for WTCM. At pH 5, q_{max} = 303.0 mg g^{-1} for WACM is almost 21% higher than the value at pH 4, while for WTCM, q_{max} = 223.1 mg g^{-1} is practically the same at both pHs.

Table 2. Isothermal parameters for Pb(II) adsorption on WACM and WTCM, adjustment to Langmuir and Freundlich models.

	Parameters	WACM pH = 4	WACM pH = 5	WTCM pH = 4	WTCM pH = 5
Langmuir model	K_L ($L \cdot mg^{-1}$)	0.32	0.22	0.45	0.61
	q_{max} ($mg \cdot g^{-1}$)	238.1	303.0	222.2	223.1
	R^2	~1	0.98	~1	~1
Freundlich model	K_F ($mg \cdot g^{-1}$ $L^{(1/n)} \cdot mg^{-(1/n)}$)	58.47	56.78	66.08	71.69
	n_F	2.39	2.04	2.80	3.11
	R^2	0.73	0.72	0.87	0.79

q_{max} values of agro-industrial wastes, with alkaline treatment, are consigned in Table 3. We can note that q_{max} of WACM and WTCM are among the highest. On the other hand, it is important to mention that that q_{max} value of the treated biosorbent is almost twice that of its corresponding untreated precursor [21].

Table 3. Comparative table of the maximum adsorption capacity q_{max} of Pb(II), for biosorbents with alkaline treatment.

Biosorbent Wastes	q_{max} (mg g^{-1})	Reference
Apricot shells	37.37	[35]
Mangifera indica seed shells	59.25	[13]
Olive tree pruning	121.60	[12]
Grape pomace	137	[14]
Moringa oleifera tree leaves	209.54	[36]
Theobroma cacao; WTCM	223.1	This work
Arabica coffee; WACM	303.0	This work

1.5. Kinetic of Biosorption

The kinetic study is of great importance for the practical and effective use of biosorbents in the industry [41]. In this work, the kinetic studies were carried out by varying the adsorption time from 0 to 180 min, at pH 4; C_0 = 48.42 and 130.8 mg L^{-1}; biosorbent dosage = 2 g L^{-1}; and T = 293 K.

The experimental kinetic data were modeled using three adsorption kinetic models (Table 4): pseudo first-order model, pseudo second-order model, and ploting q_t vs. $t^{1/2}$ (Weber and Morris model). The parameters obtained after the non-linear adjustments, including correlation coefficient R^2, are consigned in Table 4. Figure 9 shows the non-linear fit of the pseudo second-order equation to the kinetic data.

Table 4. Kinetic parameters of Pb(II) adsorption on WACM and WTCM biosorbents.

Model	Parameters	WACM		WTCM	
		C_0 = 130.8 mg L^{-1}	C_0 = 48.42 mg L^{-1}	C_0 = 130.8 mg L^{-1}	C_0 = 48.42 mg L^{-1}
Pseudo 1st order	k_1(min^{-1})	0.075	0.063	0.075	0.043
	$q_{e,cal}$ (mg g^{-1}) [a]	206.62	132.93	206.62	124.92
	R^2	0.94	0.84	0.94	0.87
Pseudo 2nd order	k_2 (g mg^{-1} min^{-1})	0.0003	0.0006	0.0004	0.0004
	$q_{e,cal}$ (mg g^{-1}) [a]	249.53	147.98	229.99	141.23
	h	20.99	13.14	21.16	7.98
	R^2	0.94	0.93	0.97	0.92
Weber and Morris model	$k_{d,I}$ (mg g^{-1} $min^{-1/2}$) [b]	75.4	10.9	56.2	10,3
	R^2	0.94	0.96	0.90	0.99
	$k_{d,II}$ (mg g^{-1} $min^{-1/2}$) [b]	10.7	0.4	12.4	0.7
	R^2	0.95	1	0.89	1
	$k_{d,III}$ (mg g^{-1} $min^{-1/2}$) [b]	0.12	-	0.58	-
	R^2	1	-	1	-

[a] Calculated adsorption capacity. [b] Intraparticle diffusion rate constant.

For both the WACM and WTCM biosorbents, a better correlation ($R^2 \approx 1$) is obtained with *pseudo*-second-order than the first-order adjustment models. This result indicates that Pb(II) adsorption is a chemisorption process [42]. We can note that the calculated adsorption capacities $q_{e,cal}$ are close to those determined experimentally and, for a given C_0 concentration, greater for WACM than for WTCM. The adsorption rates (k_2, rate constant adsorption and h, initial adsorption rate) are comparable for both biosorbents.

The q_t vs. $t^{0.5}$ data, for both WACM and WTCM biosorbents, are depicted in Figure 10 and fitted with the intra-particle diffusion Weber–Morris model. According to the k_d intra-particle diffusion rate constants (in mg g^{-1} $min^{-1/2}$, Table 4), we can distinguish three parts: The first part shows rapid growth of q_t at time t ($k_{d,I}$ > 10.3), particularly at high initial Pb(II) concentrations (C_0 = 130.8 mg L^{-1}), where $k_{d,I}$ can reach values up to seven times higher than for those at low C_0 concentrations (e.g., 48.4 mg L^{-1}). These results would indicate the rapid absorption of Pb(II) ions on the surface of the biosorbents. The second part shows slower growth of q_t with t (0.16 < $k_{d,II}$ < 12.4), which would be related to a gradual sorption process, where Pb(II) sorbed would fill the biosorbent pores;

this part would be related to the diffusion of Pb(II) inside the biosorbent (intraparticle diffusion) [24]. Finally, the 3rd part shows that q_t is practically constant with very low $k_{d,III}$ values ($k_{d,III} < 0.6$). It indicates that the equilibrium between Pb(II) ions in the solution and the sorbent surface is reached.

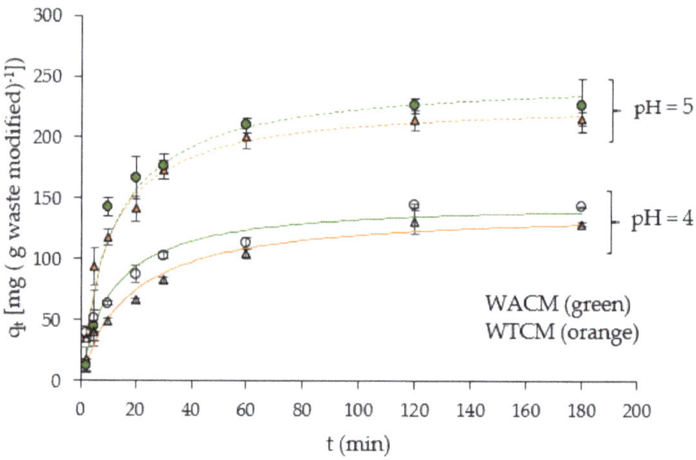

Figure 9. q_t vs. time t. C_0 = 48.4 (continuous lines) and 130.8 mg L^{-1} (dotted lines); dose = 2 g L^{-1}, T = 293 K. q_t = amount Pb(II) removed per mass unit of biosorbent at time t.

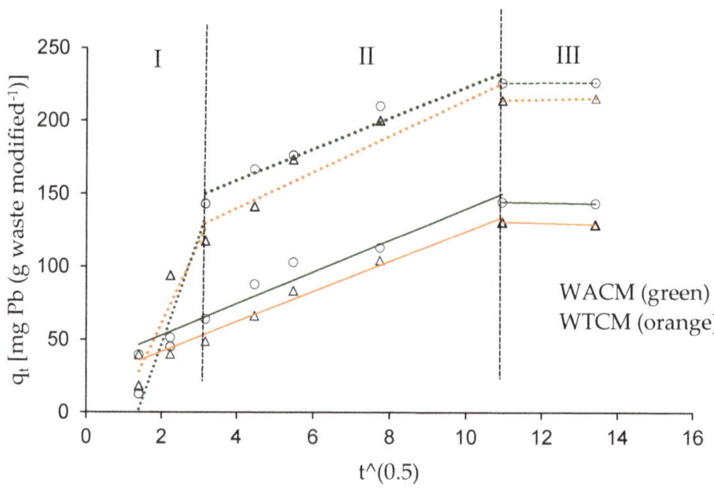

Figure 10. Weber–Morris plots of Pb(II) adsorption on WACM and WTCM. C_0 = 48.4 (continuous lines) and 130.8 mg L^{-1} (dotted lines).

1.6. Biosorption Thermodynamics

ΔG^0 was calculated from Equations (4) and (5). The plot lnK_c vs. 1/T (Equation (7)), depicted in Figure 11, was fitted using the least squares method, aiming to calculate the ΔH^0 and ΔS^0 values. The results are consigned in Table 5 and indicate that the Pb(II) adsorption process on both the WACM and WTCM biosorbents is: (i) feasible, spontaneous (positive ΔG^0 values), and more favorable with increasing temperature; (ii) endothermic by nature (positive ΔH^0 values); and (iii) a process with increasing randomness (positive ΔS^0 values) at the solid–liquid interface [43].

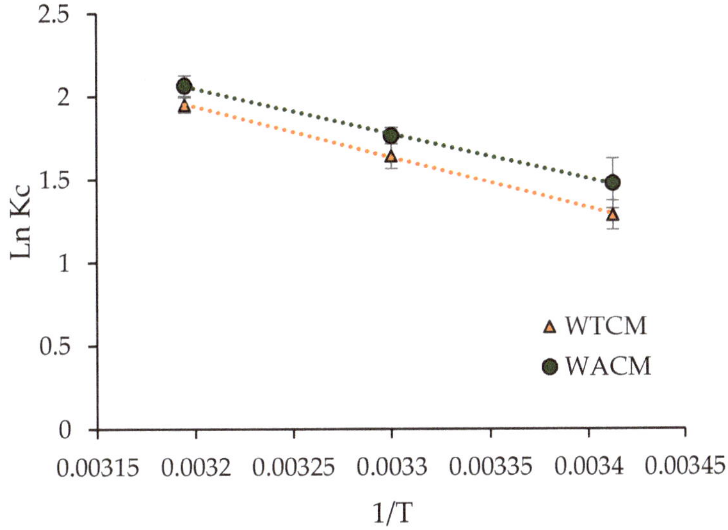

Figure 11. lnK_c vs. 1/T plot for Pb(II) biosorption onto WTCM and WACM at pH = 4, biosorbent dosages = 2 g L^{-1}, t = 120 min, C_0 = 130.8 mg L^{-1}.

Table 5. Thermodynamic parameters for Pb(II) adsorption on WACM and WTCM.

	ΔH^0 (kJ mol^{-1})	ΔS^0 (J mol^{-1} K^{-1})	ΔG^0 (kJ mol^{-1})		
			293 K	**303 K**	**313 K**
WACM	22.5	88.9	−35.90	−44.42	−53.70
WTCM	25.4	97.2	−31.26	−41.32	−50.70

Similar results were reported by Song et al. [27], Morosanu et al. [31], Milojkovic et al. [44], among others, for Pb(II) removal by *Auricularia auricular* spent substrate, rapeseed biomass, and *Myriophyllum spicatum* and its compost, respectively. However, a feasible, spontaneous, but exothermic process of Pb(II) adsorption was also reported by Petrović et al. [14] when removing Pb(II) with a hydrochar of grape pomace. Mahyoob et al. [29] determined that Pb(II) adsorption on co-processed olive tree leaves was an exothermic process with decreasing randomness (negative ΔS^0).

2. Materials and Methods

2.1. Preparation of Biosorbents

Arabica-coffee (WAC) and *Theobroma*-cocoa (WTC) waste were obtained from the Satipo and Chanchamayo provinces, respectively, located at Junín, Perú. Both samples were previously washed with abundant distilled water, dried in an oven at 60 °C for 48 h, and finally ground. WAC and WTC were treated with a 0.1 M NaOH solution in a solid–liquid ratio of 1:10 (g biomass: mL solution) for 24 h with constant stirring at 300 rpm. Once treated, both products were filtered and washed with abundant deionized water and then were dried in an oven at 60 °C for 48 h, and finally, both treated samples (WACM and WTCM) were ground again and homogenized with a 70 mesh sieve.

The alkaline treatment of precursor wastes produced a biomass loss. The percentage loss was determined as the difference between the initial sample weight (before treatment) and the final sample weight (after treatment)

All chemical reagents used in this work were of analytical grade.

2.2. Biosorbent Characterization

- The point of zero charge (pH_{PZC}) study was evaluated according to the methodology reported by do-Nascimento et al. [45]. A mixture of 0.05 g of biomass with 50 mL of an aqueous solution under different initial pHs (pH_0) ranging from 1 to 12 was prepared. The acid dilutions were prepared from a 1 M HCl solution, while basic dilutions were from 1 M NaOH. After 24 h of equilibrium, the final pHs (pH_f) were measured.
- The concentrations of the acid and basic groups (or acid/basic titrable sites) on the surface of WACM and WTCM were determined using the Boehm method reported by Aygun et al. [46]. For acid titrable sites (between brackets for basic sites), mixtures of 0.25 g (0.5 g) of the biosorbent with 50 mL of a standardized 0.05 M NaOH (0.1 M HCl) solution were prepared. All the mixtures were shaken, at room temperature, for 24 h at 100 rpm, and then, for each mixture, 20 mL (10 mL) of the supernatant liquid was pipetted and excess acid (base) was adequately titrated using bromocresol blue or phenolphtalien) as an indicator.
- Fourier transform infrared (FTIR, SHIMADZU IR Affinity) spectroscopy, over a spectral range of 4000 to 500 cm^{-1} was used to characterize the functional groups present on the surface of WACM and WTCM before and after Pb(II) biosorption.
- Morphological and elemental analysis on the surface of biosorbents were performed by Scanning Electron Microscopy (SEM) coupled with EDX (Energy Dispersive X-rays spectroscopy) (LEO 440 model).

2.3. Adsorption Experiments

Batch experiments were carried out using $Pb(NO_3)_2$ solution, with varying Pb(II) concentrations between 5.63 to 130.8 mg L^{-1}. The dose of the modified biomass was varied in the range of 0.5 to 6 g L^{-1} and adjusted to a pH in the range of 2.0 to 5.0 (using LAQUA PH1200) by adding 0.01 M HNO_3 or 0.01 M NaOH. The solutions, at room temperature, were stirred to 150 rpm for 120 min, and the samples were taken at certain time intervals.

The Pb(II) concentrations, before and after adsorption, were evaluated using an Atomic Absorption Spectrophotometer (SHIMADZU-AAS 6800). All adsorption experiments were replicated three times, and the results were averaged.

Pb(II) adsorption capacity q_e(in mg g^{-1}) and removal efficiency (%R) were determined using Equations (1) and (2), respectively [9].

$$q_e = \frac{(C_0 - C_e) \times V}{m} \quad (1)$$

$$\%R = \frac{(C_0 - C_e)}{C_0} \times 100 \quad (2)$$

where C_0 and C_e (in mg·L^{-1}) are the initial and equilibrium final Pb(II) concentrations, respectively, and V (in L) is the volume of solution and *m* (in g) is the biosorbent mass.

The adsorption isotherms and kinetic data were obtained by contacting a biomass dose of 2 g L^{-1} with different Pb(II) concentrations at pH 4 (also pH 5 for isotherms). The experimental data were adjusted to the corresponding adsorption models described in Table 6 (isotherms) and Table 7 (kinetic).

Table 6. Adsorption isotherm models.

Model	Equation	Parameters
Langmuir	$\frac{C_e}{q_e} = \frac{1}{q_{mx}k_L} + \frac{C_e}{q_{mx}}$	q_e (mg g^{-1}): adsorption capacity C_e (mg L^{-1}): adsorbate concentration in equilibrium q_{max} (mg g^{-1}): maximum sorption capacity k_L (L mg^{-1}): Langmuir constant related to the affinity between sorbent and sorbate
Freundlich	$lnq_e = lnk_F + \frac{1}{n}lnC_e$	k_F (mg g^{-1} $L^{(1/n)}$.$mg^{-(1/n)}$): equilibrium constant n: constant related to the affinity between sorbent and sorbate

Table 7. Kinetic adsorption models.

Model	Equation	Parameters
Pseudo-first order	$q_t = q_e\left(1 - e^{-K_1 t}\right)$	q_e (mg g^{-1}): adsorption capacity q_t (mg g^{-1}): amount of Pb(II) retained per unit biomass at time t. k_1 (min^{-1}): first-order kinetic constant
Pseudo-second order	$q_t = q_e \frac{q_e K_2 t}{1+q_e K_2 t}$ $h = k_2 q_e^2$	k_2 (g (mg min)$^{-1}$): rate constant adsorption h (mg(g min)$^{-1}$): initial adsorption rate k_d (mg g^{-1} $min^{-1/2}$): intraparticle diffusion rate constant
Weber and Morris	$q_t = k_d t^{1/2} + B$	B (mg g^{-1}): constant related to the thickness of the adsorbent boundary layer

2.4. Thermodynamic Parameters

This study was carried out by varying the temperature from 293 to 313 K (20 to 40 °C) using a biosorbent dose of 2 g L^{-1}, an initial Pb(II) concentration C_0 = 130.8 mg L^{-1}, and a contact time 120 min at pH 4. Parameters such as free energy change $\Delta G°$, enthalpy change $\Delta H°$, and entropy change, $\Delta S°$ for the adsorption process studied were calculated using Equations (3)–(6) [47,48]:

$$K_c = \frac{C_{es}}{C_e} \quad (3)$$

$$\Delta G^o = -RTLnK_c \quad (4)$$

$$\Delta G^o = \Delta H^o - T\Delta S^o \quad (5)$$

$$lnK_c = \frac{\Delta S^o}{R} - \frac{\Delta H^o}{RT} \quad (6)$$

where R is the ideal gas constant (8.314 J mol^{-1} K^{-1}); T is the absolute temperature of the solution; K_c is the thermodynamic equilibrium constant; and C_{es} and C_e are Pb(II) concentrations at equilibrium, respectively, in the biosorbent and solution.

3. Conclusions

Arabica-coffee (WACM) and *Theobroma*-cocoa (WTCM) biosorbents were chemically modified, with 0.1 M NaOH to improve their Pb(II) adsorption capacities an aqueous medium. After treatment, both WACM and WTCM biosorbents lost weight 40.2% and 38.4%, respectively.

The point of zero charge, pH_{PZC} values at 6 and 6.8, for WACM and WTCM, respectively, were higher than those for the corresponding WAC and WTC untreated samples. These measurements were consistent with the concentration of basic titrable sites, which was almost 1.4 times higher for treated than untreated biosorbents. The basic sites would be associated with the OH, C=O, COH, and C-O-C functional groups, which were identified by FTIR measurements.

SEM/EDX analyses showed typical uneven and rough surface morphologies, more porous and less compact for clean than for Pb(II)-loaded biosorbents.

Both the WACM and WTCM biosorbents reached high Pb(II) removal efficiency %R values (>90%) for a biomass dosage of 2 g L^{-1} at pH between 4 and 5, with initial Pb(II) concentrations C_0 in the range of 5 to 40 mg L^{-1}. For these conditions, the Pb(II) adsorption capacity q_e of WACM (and WTCM) was almost three times higher than that of the corresponding untreated biosorbent.

The adsorption isotherm data were well fitted with the Langmuir model (monolayer adsorption mechanism), which provided, at pH 5, maximum Pb(II) adsorption capacities, q_{max}, equal to 303.0 and 223.1 mg g^{-1} for WACM and WTCM, respectively. These values are i) twice those corresponding to untreated samples and ii) higher than for other similar alkaline-treated biosorbents reported in the literature.

The adsorption kinetic data were well fitted with the *pseudo*-second-order model, indicating that the Pb(II) adsorption on WACM and WTCM was a chemisorption process.

This process, according to our thermodynamic results, can be characterized as endothermic ($\Delta H^0 > 0$), feasible, and spontaneous ($\Delta G^0 < 0$) and with an increasing randomness ($\Delta S^0 > 0$) at the solid–liquid interface.

Author Contributions: C.L.-M. and L.D.l.C.-C.: experiments, data analysis, writing—original draft. Y.J.A. and F.C.F.M.: experimental characterization, paper review. J.Z.D.-P.: data analysis, writing of the final version. All authors have read and agreed to the published version of the manuscript.

Funding: This work was funded by Universidad Nacional Intercultural Juan Santos Atahualpa (UNICJSA), Project 207-2020-UNISCJSA.

Institutional Review Board Statement: Not applicable.

Informed Consent Statement: Not applicable.

Data Availability Statement: Not applicable.

Conflicts of Interest: The authors declare no conflict of interest.

Sample Availability: Not applicable.

References

1. Li, X.; Bing, J.; Zhang, J.; Guo, L.; Deng, Z.; Wang, D.; Liu, L. Ecological risk assessment and sources identification of heavy metals in surface sediments of a river–reservoir system. *Sci. Total Environ.* **2022**, *842*, 156683. [CrossRef] [PubMed]
2. Collin, S.; Baskar, A.; Geevarghese, D.M.; Ali, M.N.V.S.; Bahubali, P.; Choudhary, R.; Lvov, V.; Tovar, G.I.; Senatov, F.; Koppala, S.; et al. Bioaccumulation of lead (Pb) and its effects in plants: A review. *J. Hazard. Lett.* **2022**, *3*, 100064. [CrossRef]
3. O'Connor, D.; Hou, D.; Ok, Y.S.; Lanphear, B.P. The effects of iniquitous lead exposure on health. *Nat. Sustain.* **2020**, *3*, 77–79. [CrossRef]
4. Taguchi, R.; Seki, H.; Maruyama, H. Biosorption of Pb and Cd onto Polygonum sachalinense. *Colloids Surfaces A Physicochem. Eng. Asp.* **2022**, *650*, 129210. [CrossRef]
5. Jaihan, W.; Mohdee, V.; Sanongraj, S.; Pancharoen, U.; Nootong, K. Biosorption of lead (II) from aqueous solution using Cellulose-based Bio-adsorbents prepared from unripe papaya (Carica papaya) peel waste: Removal Efficiency, Thermodynamics, kinetics and isotherm analysis. *Arab. J. Chem.* **2022**, *15*, 103883. [CrossRef]
6. Mangwandi, C.; Kurniawan, T.A.; Albadarin, A.B. Comparative biosorption of chromium (VI) using chemically modified date pits (CM-DP) and olive stone (CM-OS): Kinetics, isotherms and influence of co-existing ions. *Chem. Eng. Res. Des.* **2020**, *156*, 251–262. [CrossRef]
7. Brazesh, B.; Mousavi, S.M.; Zarei, M.; Ghaedi, M.; Bahrani, S.; Hashemi, S.A. Biosorption. *Interface Sci. Technol.* **2021**, *33*, 587–628.
8. Cheng, S.Y.; Show, P.L.; Lau, B.F.; Chang, J.S.; Ling, T.C. New Prospects for Modified Algae in Heavy Metal Adsorption. *Trends Biotechnol.* **2019**, *37*, 1255–1268. [CrossRef]
9. Basu, A.; Ali, S.S.; Hossain, S.K.S.; Asif, M. A Review of the Dynamic Mathematical Modeling of Heavy Metal Removal with the Biosorption Process. *Processes* **2022**, *10*, 1154. [CrossRef]
10. Syeda, H.I.; Sultan, I.; Razavi, K.S.; Yap, P.S. Biosorption of heavy metals from aqueous solution by various chemically modified agricultural wastes: A review. *J. Water Process Eng.* **2022**, *46*, 102446. [CrossRef]
11. Ren, B.; Jin, Y.; Zhao, L.; Cui, C.; Song, X. Enhanced Cr(VI) adsorption using chemically modified dormant Aspergillus niger spores: Process and mechanisms. *J. Environ. Chem. Eng.* **2020**, *10*, 106955. [CrossRef]
12. Calero, M.; Pérez, A.; Blázquez, G.; Ronda, A.; Martín-lara, M.A. Characterization of chemically modified biosorbents from olive tree pruning for the biosorption of lead. *Ecol. Eng.* **2013**, *58*, 344–354. [CrossRef]
13. Moyo, M.; Pakade, V.E.; Modise, S.J. Biosorption of lead(II) by chemically modified Mangifera indica seed shells: Adsorbent preparation, characterization and performance assessment. Process Saf. *Environ. Prot.* **2017**, *111*, 40–51. [CrossRef]
14. Petrović, J.T.; Stojanović, M.D.; Milojković, J.V.; Petrović, M.S.; Šoštarić, T.D.; Laušević, M.D.; Mihajlović, M.L. Alkali modified hydrochar of grape pomace as a perspective adsorbent of Pb2+ from aqueous solution. *J. Environ. Manag.* **2016**, *182*, 292–300. [CrossRef] [PubMed]
15. Ye, H.; Yu, Z. Adsorption of Pb(II) onto Modified Rice Bran. *Nat. Resour.* **2010**, *01*, 104–109. [CrossRef]
16. Vásquez, Z.S.; de Carvalho Neto, D.P.; Pereira, G.V.M.; Vandenberghe, L.P.S.; de Oliveira, P.Z.; Tiburcio, P.B.; Rogez, H.L.G.; Góes Neto, A.; Soccol, C.R. Biotechnological approaches for cocoa waste management: A review. *Waste Manag.* **2019**, *90*, 72–83. [CrossRef] [PubMed]
17. Informe Mensual del "Valor Bruto de la Producción Agropecuaria" Diciembre 2021. Available online: https://cdn.www.gob.pe/uploads/document/file/3282594/Informe%20Mensual%20del%20%22Valor%20Bruto%20de%20la%20Producci%C3%B3n%20Agropecuaria%22%20Diciembre%202021.pdf?v=1655825207 (accessed on 15 December 2022).
18. Anastopoulos, I.; Karamesouti, M.; Mitropoulos, A.C.; Kyzas, G.Z. A review for coffee adsorbents. *J. Mol. Liq.* **2017**, *229*, 555–565. [CrossRef]

19. Mendoza Martinez, C.L.; Saari, J.; Melo, Y.; Cardoso, M.; de Almeida, G.M.; Vakkilainen, E. Evaluation of thermochemical routes for the valorization of solid coffee residues to produce biofuels: A Brazilian case. *Renew. Sustain. Energy Rev.* **2021**, *137*, 110585. [CrossRef]
20. Park, S.J.; Yang, H.K. Ultra-fast synthesis of carbon dots using the wasted coffee residues for environmental remediation. *Curr. Appl. Phys.* **2022**, *36*, 9–15. [CrossRef]
21. Lavado-Meza, C.; De la Cruz-Cerrón, L.; Cisneros-Santos, G.; De la Cruz, A.H.; Angeles-Suazo, J.; Dávalos-Prado, J.Z. Arabica-coffee and teobroma-cocoa agro-industrial waste biosorbents, for Pb(II) removal in aqueous solutions. *Environ. Sci. Pollut. Res.* 2022. [CrossRef]
22. Santos, S.C.R.; Ungureanu, G.; Volf, I.; Boaventura, R.A.R.; Botelho, C.M.S. Macroalgae Biomass as Sorbent for Metal Ions. In *Biomass as Renewable Raw Material to Obtain Bioproducts of High-Tech Value*; Elsevier: Amsterdam, The Netherlands, 2018; pp. 69–112.
23. Bulgariu, L.; Bulgariu, D. Enhancing Biosorption Characteristics of Marine Green Algae (Ulva lactuca) for Heavy Metals Removal by Alkaline Treatment. *J. Bioprocess. Biotech.* **2014**, *04*, 1–8. [CrossRef]
24. Blázquez, G.; Calero, M.; Ronda, A.; Tenorio, G.; Martín-Lara, M.A. Study of kinetics in the biosorption of lead onto native and chemically treated olive stone. *J. Ind. Eng. Chem.* **2014**, *20*, 2754–2760. [CrossRef]
25. Gupta, N.K.; Gupta, A.; Ramteke, P.; Sahoo, H.; Sengupta, A. Biosorption-a green method for the preconcentration of rare earth elements (REEs) from waste solutions: A review. *J. Mol. Liq.* **2019**, *274*, 148–164. [CrossRef]
26. Basu, M.; Guha, A.K.; Ray, L. Adsorption of Lead on Cucumber Peel. *J. Clean. Prod.* **2017**, *151*, 603–615. [CrossRef]
27. Song, T.; Yu, S.; Wang, X.; Teng, C.; Bai, X.; Liang, J.; Dong, L.; Ouyang, F.; Qu, J.; Jin, Y. Biosorption of Lead(II) from Aqueous Solution by Sodium Hydroxide Modified Auricularia auricular Spent Substrate: Isotherms, Kinetics, and Mechanisms. *Water. Air. Soil Pollut.* **2017**, *228*, 1–17. [CrossRef]
28. Taşar, Ş.; Kaya, F.; Özer, A. Biosorption of lead(II) ions from aqueous solution by peanut shells: Equilibrium, thermodynamic and kinetic studies. *J. Environ. Chem. Eng.* **2014**, *2*, 1018–1026. [CrossRef]
29. Mahyoob, W.; Alakayleh, Z.; Abu Hajar, H.A.; Al-Mawla, L.; Altwaiq, A.M.; Al-Remawi, M.; Al-Akayleh, F. A novel co-processed olive tree leaves biomass for lead adsorption from contaminated water. *J. Contam. Hydrol.* **2022**, *248*, 104025. [CrossRef]
30. Fawzy, M.; Nasr, M.; Abdel-Rahman, A.M.; Hosny, G.; Odhafa, B.R. Techno-economic and environmental approaches of Cd2+ adsorption by olive leaves (*Olea europaea* L.) waste. *Int. J. Phytoremediat.* **2019**, *21*, 1205–1214. [CrossRef]
31. Morosanu, I.; Teodosiu, C.; Paduraru, C.; Ibanescu, D.; Tofan, L. Biosorption of lead ions from aqueous effluents by rapeseed biomass. *N. Biotechnol.* **2017**, *39*, 110–124. [CrossRef]
32. Amin, M.T.; Alazba, A.A.; Shafiq, M. Removal of Copper and Lead using Banana Biochar in Batch Adsorption Systems: Isotherms and Kinetic Studies. *Arab. J. Sci. Eng.* **2018**, *43*, 5711–5722. [CrossRef]
33. Barka, N.; Abdennouri, M.; El Makhfouk, M.; Qourzal, S. Biosorption characteristics of cadmium and lead onto eco-friendly dried cactus (Opuntia ficus indica) cladodes. *J. Environ. Chem. Eng.* **2013**, *1*, 144–149. [CrossRef]
34. Lavado-Meza, C.; Asencios, O.Y.J.; Cisneros-Santos, G.; Unchupaico-Payano, I. Revista Mexicana de Ingeniería Química. *Rev. Mex. Ing. Química* **2021**, *2*, 941–954. [CrossRef]
35. Šoštarić, T.D.; Petrović, M.S.; Pastor, F.T.; Lončarević, D.R.; Petrović, J.T.; Milojković, J.V.; Stojanović, M.D. Study of heavy metals biosorption on native and alkali-treated apricot shells and its application in wastewater treatment. *J. Mol. Liq.* **2018**, *259*, 340–349. [CrossRef]
36. Reddy, D.H.K.; Harinath, Y.; Seshaiah, K.; Reddy, A.V.R. Biosorption of Pb(II) from aqueous solutions using chemically modified Moringa oleifera tree leaves. *Chem. Eng. J.* **2010**, *162*, 626–634. [CrossRef]
37. Reddy, D.H.K.; Seshaiah, K.; Reddy, A.V.R.; Lee, S.M. Optimization of Cd(II), Cu(II) and Ni(II) biosorption by chemically modified Moringa oleifera leaves powder. *Carbohydr. Polym.* **2012**, *88*, 1077–1086. [CrossRef]
38. Alver, E.; Metin, A.Ü.; Brouers, F. Methylene blue adsorption on magnetic alginate/rice husk bio-composite. *Int. J. Biol. Macromol.* **2020**, *154*, 104–113. [CrossRef]
39. Gonçalves, A.C.; Schwantes, D.; Campagnolo, M.A.; Dragunski, D.C.; Tarley, C.R.T.; Dos Santos Silva, A.K. Removal of toxic metals using endocarp of açaí berry as biosorbent. *Water Sci. Technol.* **2018**, *77*, 1547–1557. [CrossRef]
40. Ayawei, N.; Ebelegi, A.N.; Wankasi, D. Modelling and Interpretation of Adsorption Isotherms. *J. Chem.* **2017**, *2017*, 3039817. [CrossRef]
41. Lavado-Meza, C.; Sun-Kou, M.R.; Castro-Arroyo, T.K.; Bonilla-Mancilla, H.D. Biosorción de plomo (II) en solución acuosa con biomasa de los cladodios de la tuna (opuntia ficus indica). *Rev. Colomb. Quim.* **2020**, *49*, 36–46. [CrossRef]
42. Oliveira, M.R.F.; Abreu, K.V.; Romão, A.L.E.; Davi, D.M.V.; Magalhães, C.E.C.; Carrilho, E.N.V.; Alves, C.R. Carnauba (Copernicia prunifera) palm tree biomass as adsorbent for Pb(II) and Cd(II) from water medium. *Environ. Sci. Pollut. Res.* **2021**, *28*, 18941–18952. [CrossRef]
43. Tavana, M.; Pahlavanzadeh, H.; Zarei, M.J. The novel usage of dead biomass of green algae of Schizomeris leibleinii for biosorption of copper(II) from aqueous solutions: Equilibrium, kinetics and thermodynamics. *J. Environ. Chem. Eng.* **2020**, *8*, 104272. [CrossRef]
44. Milojković, J.V.; Mihajlović, M.L.; Stojanović, M.D.; Lopičić, Z.R.; Petrović, M.S.; Šoštarić, T.D.; Ristić, M.D. Pb(II) removal from aqueous solution by Myriophyllum spicatum and its compost: Equilibrium, kinetic and thermodynamic study. *J. Chem. Technol. Biotechnol.* **2014**, *89*, 662–670. [CrossRef]

45. do Nascimento, J.M.; de Oliveira, J.D.; Leite, S.G.F. Chemical characterization of biomass flour of the babassu coconut mesocarp (Orbignya speciosa) during biosorption process of copper ions. *Environ. Technol. Innov.* **2019**, *16*, 100440. [CrossRef]
46. Aygun, A.; Yenisoy-Karakaş, S.; Duman, I. Production of granular activated carbon from fruit stones and nutshells and evaluation of their physical, chemical and adsorption properties. *Microporous Mesoporous Mater.* **2003**, *66*, 189–195. [CrossRef]
47. Shooto, N.D.; Thabede, P.M.; Bhila, B.; Moloto, H.; Naidoo, E.B. Lead ions and methylene blue dye removal from aqueous solution by mibuna beans (velvet beans) adsorbents. *J. Environ. Chem. Eng.* **2020**, *8*, 103557. [CrossRef]
48. Kushwaha, S.; Suhas; Chaudhary, M.; Tyagi, I.; Bhutiani, R.; Goscianska, J.; Ahmed, J.; Manila; Chaudhary, S. Utilization of Phyllanthus emblica fruit stone as a Potential Biomaterial for Sustainable Remediation of Lead and Cadmium Ions from Aqueous Solutions. *Molecules* **2022**, *27*, 3355. [CrossRef] [PubMed]

molecules

Article

Co-Fermentation of Glucose–Xylose Mixtures from Agroindustrial Residues by Ethanologenic *Escherichia coli*: A Study on the Lack of Carbon Catabolite Repression in Strain MS04

Estefanía Sierra-Ibarra [1], Alejandra Vargas-Tah [1], Cessna L. Moss-Acosta [1], Berenice Trujillo-Martínez [1], Eliseo R. Molina-Vázquez [1], Alberto Rosas-Aburto [2], Ángeles Valdivia-López [2], Martín G. Hernández-Luna [2], Eduardo Vivaldo-Lima [2] and Alfredo Martínez [1,*]

[1] Departamento de Ingeniería Celular y Biocatálisis, Instituto de Biotecnología, Universidad Nacional Autónoma de Mexico. Av. Universidad 2001, Col. Chamilpa, Cuernavaca 62210, Mexico
[2] Departamento de Ingeniería Química, Facultad de Química, Universidad Nacional Autónoma de Mexico, Ciudad de Mexico 04510, Mexico
* Correspondence: alfredo.martinez@ibt.unam.mx; Tel.: +52-7773291601

Citation: Sierra-Ibarra, E.; Vargas-Tah, A.; Moss-Acosta, C.L.; Trujillo-Martínez, B.; Molina-Vázquez, E.R.; Rosas-Aburto, A.; Valdivia-López, Á.; Hernández-Luna, M.G.; Vivaldo-Lima, E.; Martínez, A. Co-Fermentation of Glucose–Xylose Mixtures from Agroindustrial Residues by Ethanologenic *Escherichia coli*: A Study on the Lack of Carbon Catabolite Repression in Strain MS04. *Molecules* **2022**, *27*, 8941. https://doi.org/10.3390/molecules27248941

Academic Editors: Alejandro Rodriguez Pascual, Eduardo Espinosa Víctor and Carlos Martín

Received: 28 October 2022
Accepted: 13 December 2022
Published: 15 December 2022

Abstract: The production of biofuels, such as bioethanol from lignocellulosic biomass, is an important task within the sustainable energy concept. Understanding the metabolism of ethanologenic microorganisms for the consumption of sugar mixtures contained in lignocellulosic hydrolysates could allow the improvement of the fermentation process. In this study, the ethanologenic strain *Escherichia coli* MS04 was used to ferment hydrolysates from five different lignocellulosic agroindustrial wastes, which contained different glucose and xylose concentrations. The volumetric rates of glucose and xylose consumption and ethanol production depend on the initial concentration of glucose and xylose, concentrations of inhibitors, and the positive effect of acetate in the fermentation to ethanol. Ethanol yields above 80% and productivities up to 1.85 g_{EtOH}/Lh were obtained. Furthermore, in all evaluations, a simultaneous co-consumption of glucose and xylose was observed. The effect of deleting the *xylR* regulator was studied, concluding that it plays an important role in the metabolism of monosaccharides and in xylose consumption. Moreover, the importance of acetate was confirmed for the ethanologenic strain, showing the positive effect of acetate on the co-consumption rates of glucose and xylose in cultivation media and hydrolysates containing sugar mixtures.

Keywords: lignocellulosic hydrolysates; *Escherichia coli*; bioethanol; monosaccharides co-consumption; catabolite repression

1. Introduction

Ethanol is an important commodity in transportation and industry, as it is considered a sustainable, renewable and eco-friendly energy source [1]. Unlike oil-derived fossil fuels, ethanol can be produced from renewable lignocellulosic biomass through microbial fermentation; particularly, the synthesis of ethanol using slurries from agroindustrial residues as culture media has additional economic and environmental benefits [2]. Therefore, several methodologies have been applied for the pretreatment of these feedstocks to obtain slurries enriched with fermentable sugars [3]. It is important to note that lignocellulose has three main components: cellulose, hemicellulose, and lignin. These components are arranged into macrofibrils that confer structural stability to the plant cell wall [4]. When these fibrils are pretreated with diluted acid, they release syrups containing fermentable sugars, mainly glucose and xylose.

Lignocellulose is found in agricultural waste, which is considered a major global issue as it is underutilized and produced in huge quantities, making it a contaminant of difficult disposal. Latin American countries, such as Mexico and Colombia, are known for their variety of agricultural and timber products, making them major producers of agricultural waste [5]. In Mexico, the production of tequila and mezcal generates more than 360,000 metric tons of agave bagasse per year [6]. Otherwise, 42 million tons of corn stover are generated each year [7], making these residues some of the most relevant in the country, as they have high potential to be used as feedstock in the synthesis of value-added products. Timber is another important industry in Mexico. For example, the production of teak wood constitutes 29,000 ha of planted trees around the country (https://www.eleconomista.com.mx/opinion/Plantaciones-forestales-comerciales-de-teca-en-Mexico-II-20190110-0137.html (accessed on 20 October 2022)); thus, the residues that this industry produces in Mexico are also significant. Meanwhile, Colombia is one of the most important producers of coffee in the world [8], where spent coffee grounds constitute the main byproduct, with potential to be converted into ethanol, biodiesel and other high-value biorefinery products [9]. Finally, in recent decades, the production of barley has increased to approximately 140 million tons per year due to the rise in craft breweries and its use for animal and human consumption [10], making the development of processes for its waste valorization an attractive venture.

The residues from the agroindustrial crops mentioned above contain different concentrations of fermentable sugars. Nevertheless, their use as feedstocks to produce value-added products such as ethanol represents an important economic and environmental opportunity for the mentioned countries. Hence, the study of ethanologenic fermentation of hydrolysates obtained under different treatment conditions and with different concentrations of sugars and inhibitors is relevant for the development and improvement of specific production processes for each agricultural waste. However, wild-type ethanologenic microorganisms are not able to efficiently ferment all the sugars contained in the hydrolysate slurries, which limits the use of lignocellulosic residues for ethanol production [11]. Therefore, metabolically engineered ethanologenic *Escherichia coli* strains that produce ethanol from glucose or xylose have previously been developed. These strains achieved yields of up to 90% when cultured in mineral media or in certain lignocellulosic hydrolysates [7]. Furthermore, unlike other common ethanologenic microorganisms such as yeast, *E. coli* can consume different hexoses and pentoses such as glucose, mannose and galactose, and xylose and arabinose. Nonetheless, *E. coli* preferentially consumes glucose over other sugars; therefore, the fermentation of pentoses, such as xylose, becomes inefficient and is often not completed [12]. The preference in glucose consumption is due to the phosphoenolpyruvate-dependent glucose phosphotransferase system (PTS) [13]. This system is involved in the regulation of several cellular processes, such as carbon catabolite repression, and is part of a global regulatory network controlling the capability of cells to find, select, transport and metabolize several types of carbon sources [14].

On the other hand, the transport and metabolism of xylose in *E. coli* are mediated by two major transcriptional units, *xylFGHR* and *xylAB*, whose expression is governed by promoters activated by xylose and repressed by glucose [15]. However, the xylose operon regulatory protein *xylR* has a weak non-sugar regulated promoter [16]. Moreover, it has been shown that in the presence of xylose, *xylR* forms a dimmer that activates transcription of the transcriptional units mentioned above [17]. It has also been reported that in wild-type and ethanologenic strains of *E. coli*, two-point mutations on the *xylR* gene release the carbon catabolic repression of glucose and arabinose over xylose, since they provide a higher binding affinity to the DNA of promoter regions for the xylose catabolic operons [16].

In this work, the metabolic engineered *E. coli* strain MS04 was used for ethanologenic fermentation of lignocellulosic syrups from different agricultural wastes. Fermentations in simulated hydrolysates were also performed as controls. Co-fermentation of hexoses and pentoses, glucose and xylose, as well as ethanol production, were evaluated in *E. coli* MS04 and a mutant of this strain. This work discusses the simultaneous consumption of glucose and xylose by *E. coli* MS04 in different lignocellulosic hydrolysates and mineral media, while also analyzing potential causes for the lack of catabolite repression in the strain.

2. Results and Discussion

2.1. Co-Consumption of Glucose and Xylose by E. coli MS04 in Simulated Media and Plant Hydrolysates Containing High Glucose Concentration

E. coli MS04 was used for ethanol production in laboratory-simulated hydrolysates (LSH) with the so-called AM1 mineral medium (see Section 3) supplemented with glucose and xylose (Figure 1a), as well as agroindustrial lignocellulosic hydrolysates (ALH), such as teak wood residues (Figure 1b) and agave bagasse (Figure 1c). Given that enzymatic saccharification was performed after pretreatment and before fermentation, the glucose concentration of the slurries was higher than that of xylose. Kinetic and stoichiometric parameters are shown in Table 1.

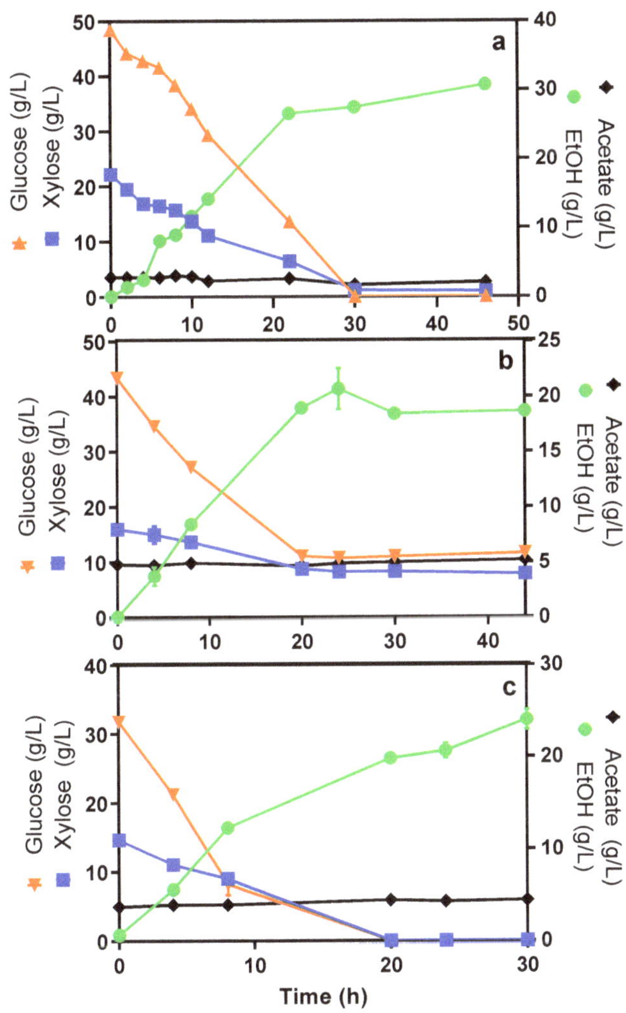

Figure 1. Glucose and xylose consumption and ethanol production by *E. coli* MS04 in LSH and ALH with syrups containing relatively high glucose concentrations and lower amounts of xylose. (**a**) Mineral medium, (**b**) teak wood, (**c**) agave bagasse.

Table 1. Ethanol yields and volumetric parameters for the fermentation of hydrolysates with relatively high glucose concentrations (values in parenthesis indicate standard error from duplicates).

Parameter	Control LSH	Teak Wood	Agave Bagasse
X_{max} (g_{DCW}/L)	1.69	ND	ND
μ (h^{-1})	0.16	ND	ND
Y_{EtOH}(%)	88	92 (5)	81 (5)
Q_{glc} (g_{glu}/Lh)	1.61	0.72 (0.00)	1.59 (0.02)
Q_{xyl} (g_{xyl}/Lh)	0.69	0.18 (0.02)	0.73 (0.01)
Q_{EtOH} (g_{EtOH}/Lh)	0.92	0.42 (0.02)	0.96 (0.05)

Xmax: maximum cell mass concentration, DCW: dry cell weight, ND: not determined, μ: specific growth rate, Y_{EtOH}: ethanol yield from consumed sugars; Q_{glc}: volumetric consumption rate of glucose, Q_{xyl}: volumetric consumption rate for xylose, Q_{EtOH}: volumetric productivity of ethanol.

Although both ALH and LSH have different sugar concentrations, it is interesting to note that both sugars were consumed simultaneously. In all the evaluated hydrolysates, most of the sugars were consumed between 5 and 20 h after the start of the experiments (Figure 1b,c). Depending on the hydrolysate, the volumetric consumption rate of glucose (Q_{glc}) was 1.5 to 4 times higher than the consumption rate for xylose (Q_{xyl}). In addition, the volumetric productivity of ethanol (Q_{EtOH}) also shows a broad range of values (Table 1). Therefore, the variations in volumetric consumption and production rates appear to be determined by the nature of the residue and the initial concentration of fermentable monosaccharides.

The fermentation of the teak wood hydrolysate showed the lowest volumetric sugar consumption rates, as 25% and 53% of the initial glucose and xylose, respectively, remained at the end of the tests (Figure 1b). These low sugar consumption and ethanol productivity rates were due to the relatively high amounts of phenolic-derived compounds such as vanillin, vanillic acid, guaiacol and catechol [18] contained in these slurries (12 g/L), which are toxic to *E. coli* in concentrations above 1 g/L [19]. On the other hand, agave bagasse hydrolysates have been previously fermented into ethanol by yeasts, achieving productivities between 0.5 and 7.2 g_{EtOH}/Lh [20,21]; however, in such studies, there were low consumption rates of pentoses. Similar results were reported for the fermentation of pine hydrolysates amended with yeast extract and fermented with the ethanologenic *E. coli* KO11, where a Q_p of 0.73 g/Lh was achieved without a complete xylose depletion [22]. Meanwhile, the use of ethanologenic *E. coli* MS04 in agave bagasse hydrolysate has shown yields and productivities of 81.6–85.3% and 0.68–1.2 g_{EtOH}/Lh for treatments with ionic liquids and organosolv, respectively [23]. These values are in accordance with those obtained in the present work (Table 1). Additionally, in both studies, a simultaneous consumption of glucose and xylose was observed, which in turn enhanced the efficiency of the fermentation process.

As shown in Figure 1, despite the differences in biomass treatment conditions and the varying contents of glucose and xylose in the hydrolysates, the co-consumption of hexoses and pentoses was noteworthy in all three experiments. Furthermore, in the case of teak wood hydrolysates, the higher amount of available glucose at the initial fermentation times and the content of inhibitory amounts of phenolic compounds did not seriously hinder the consumption of sugars.

2.2. Co-Consumption of Glucose and Xylose by E. coli MS04 in Simulated and Plant Hydrolysates Containing High Xylose Concentration

Low-glucose-concentration syrups obtained from the thermochemical hydrolysis of corn stover (Figure 2b), barley straw (Figure 2c) and spent coffee grounds (Figure 2d) were used as media for ethanol production with *E. coli* MS04. It is important to note that, except for the one pertaining to the coffee grounds, the used slurries were not saccharified before the fermentation process. These results show that the simultaneous consumption of both monosaccharides was also observed when using lignocellulosic slurries with higher xylose concentrations.

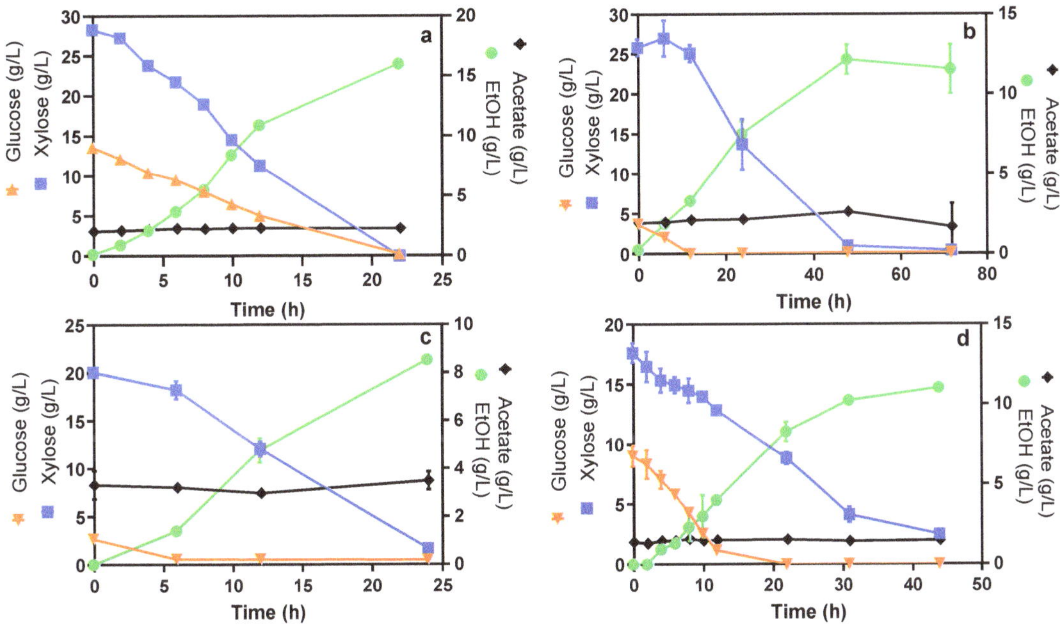

Figure 2. Glucose and xylose consumption and ethanol production by *E. coli* MS04 in LSH and ALH with higher proportions of xylose. (**a**) Mineral medium, (**b**) corn stover, (**c**) barley straw, (**d**) spent coffee grounds.

Regarding spent coffee grounds (Figure 2d), this biomass has been used as feedstock for biodiesel production due to its chemical composition and desirable lipid profile [24]; however, the relatively high proportion of carbohydrates contained in this residue (approximately 13% and 42% of cellulose and hemicellulose, respectively) also makes it a suitable source for bioethanol production [9]. For example, in previous studies, ethanol production from spent coffee grounds hydrolysates was performed by using the yeast *S. cerevisiae* as a biocatalyst, achieving Y_{EtOH} and Q_{EtOH} up to 91% and 1.0 g/Lh, respectively [25], from the consumption of glucose. This productivity rate is on average 2.6 times higher than that reported in our study (Table 2). Nevertheless, an advantage of using *E. coli* MS04 as an ethanol-producing microorganism is that it allows the use of hexoses and pentoses as substrate. Furthermore, prior to fermentation, an extraction of oils and phenolic lignin derivatives from the spent coffee ground hydrolysates could be a suitable strategy to improve ethanol productivity in cultures with *E. coli* [26].

Table 2. Ethanol yields and volumetric parameters for the fermentation of hydrolysates with relatively high xylose concentrations (values in parenthesis indicate standard deviation from triplicates).

Parameter	Control LSH	Corn Stover	Barley Straw
X_{max} (g_{DCW}/L)	1.61	ND	ND
μ (h^{-1})	0.19	ND	ND
Y_{EtOH}(%)	75	81 (1)	82 (4)
Q_{glc} (g_{glc}/Lh)	0.61	0.31 (0.01)	0.35 (0.02)
Q_{xyl} (g_{xyl}/Lh)	1.29	0.52 (0.02)	0.77 (0.03)
Q_{EtOH} (g_{ETOH}/Lh)	0.72	0.25 (0.02)	0.36 (0.01)

Fermentation of hydrolysates from corn stover (Figure 2b) and barley straw (Figure 2c) did not show a clear co-consumption of both monosaccharides because their initial glucose concentrations were under 4 g/L. This issue makes the analysis difficult because, as reported in previous studies, under limited-oxygen conditions, low glucose concentrations trigger high transcriptional levels of *ptsG*, a gene that is directly involved in catabolic repression, in turn affecting metabolic cellular regulation [27]. In addition, the glucose uptake rate is limited by glucose concentration [28], which explains the lower Q_{glc} for these syrups (Table 2). Meanwhile, similarly to sugar consumption rates, Q_{EtOH} seems to be negatively affected by the initial sugar concentration, being 0.5 to 4 times lower than other reports for ethanol production from corn and barley [29,30]. Furthermore, a Y_{EtOH} above 80% demonstrates that low glucose concentration affects the kinetics of the process but not the yields.

As expected, xylose uptake rates were higher than glucose rates in media with low glucose concentrations (Table 2) since sugar consumption is less regulated in these cases [29]. The only exception to this is the case of coffee grounds, where Q_{glc} was higher than Q_{xyl} (Table 2). From these results, we can infer that the kinetics of sugar consumption and ethanol production not only depend on the proportion of sugars, but also on the hydrolysate composition. This statement is reinforced when comparing the results for control cultures in mineral media (Tables 1 and 2), where the cell growth expressed as X_{max} and μ is similar, and Q_{EtOH} is only slightly lower for the medium with more xylose, despite the differences in sugar consumption rates.

Taking into consideration the previously discussed results, this work aims to gain further insight into what causes the co-consumption of xylose and glucose with the ethanologenic *E. coli* MS04.

2.3. *The Lack of Acetate in the Mineral Media Negatively Affects Sugar Consumption*

One of the most studied inhibitory compounds contained in ALH is acetate, which is released during hemicellulose hydrolysis in concentrations between 1.5 and 13 g/L [31]. Nevertheless, some ethanologenic *E. coli* strains have shown tolerance and improved growth under certain concentrations of acetate; specifically, *E. coli* strain MS04 grows at higher rates when at least 2 g/L of acetate is present in the culture medium, and it is able to tolerate acetate concentrations of up to 10 g/L [32]. In addition, *E. coli* can co-consume mixtures of glucose, arabinose and xylose at a higher rate in the presence of acetate than when it is absent [33]. Accordingly, we studied the co-consumption of these three monosaccharides (at initial concentrations of 25 g/L of each sugar) and ethanol production by *E. coli* MS04 in mineral media without acetate (WOA) and with acetate (WA) at an initial concentration of 2 g/L.

The curves in Figure 3b,c show the positive effect that acetate has on the rates of sugar consumption, ethanol production and in maximum cell concentration (Table 3). In medium WA, glucose was completely consumed after 48 h, with minor amounts of xylose and arabinose remaining after this elapsed fermentation time, but these pentoses were completely consumed after 72 h (Figure 3b). Although the glucose consumption rate was 50% higher than the xylose and arabinose consumption rate (Table 3), it is important to note that there was simultaneous uptake of the three sugars and that a high ethanol yield was observed. On the contrary, the fermentation in the medium WOA showed lower and more variable sugar consumption (Figure 3c). Therefore, after 48 h, 6%, 39% and 22% of the initial concentrations of glucose, xylose and arabinose remained in the culture media, respectively. In addition, Q_{EtOH} and X_{max} (Table 3) were also 94% and 72% lower compared to the medium WA. Regarding μ values, the rate of cell growth does not seem to be affected by the presence of acetate.

The results described above demonstrate that under non-aerated conditions, moderate concentrations of acetate improve the consumption of sugars and the production of ethanol by *E. coli* MS04. This is due in part to the strain's necessity for acetyl CoA generation as the *pflB* gene was deleted [32].

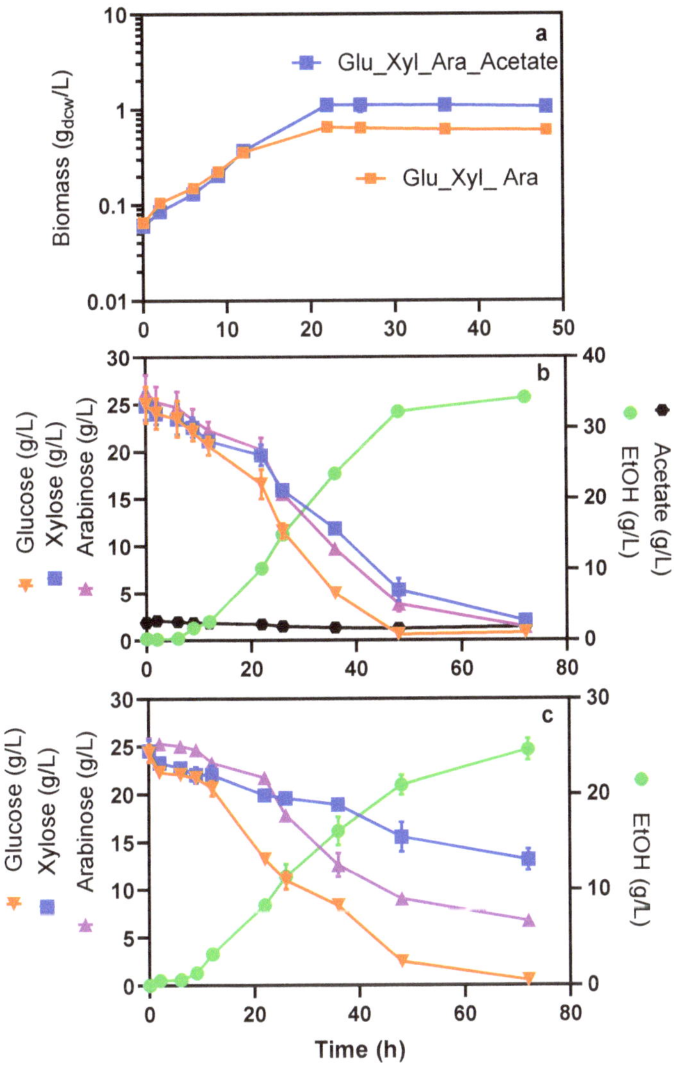

Figure 3. Kinetics of cell growth (**a**); glucose, xylose and arabinose consumption (25 g/L each at initial time), and ethanol production by *E. coli* MS04 in mineral media WA 2 g/L (**b**) and WOA (**c**).

Table 3. Ethanol yields and volumetric parameters for fermentation of 25 g/L (each) glucose–xylose–arabinose mixtures by *E. coli* MS04 in mineral media WA 2 g/L and WOA (values in parenthesis indicate standard error from duplicates).

Parameter	WA	WOA
μ (h^{-1})	0.14 (0.00)	0.13 (0.00)
X_{max} (g_{DCW}/L)	1.12 (0.09)	0.65 (0.06)
Y_{EtOH} (%)	95 (5)	90 (2)
Q_{glc} (g_{glu}/Lh)	0.51 (0.04)	0.33 (0.01)
Q_{xyl} (g_{xyl}/Lh)	0.32(0.02)	0.16 (0.00)
Q_{ara} (g_{ara}/Lh)	0.35 (0.02)	0.25 (0.01)
Q_{EtOH} (g_{ETOH}/Lh)	0.66 (0.01)	0.34 (0.02)

In relation to the simultaneous consumption of sugar mixtures, the curves in Figure 3b,c show that a lack of acetate does not release catabolic repression. Furthermore, the established hierarchy for sugars preferably consumed by *E. coli* is clearly noticeable in the fermentation WOA, where glucose was consumed at the highest rate, followed by arabinose and then xylose (Table 3). These results concur with the model of carbon catabolite repression in *E. coli*, which states that glucose is preferentially consumed, followed by arabinose and finally xylose. This is because the arabinose operon regulates xylose metabolism, leading to a favorable consumption of arabinose over xylose [34]. This behavior suggests that the acetate contained in the evaluated hydrolysates could be partly responsible for the lack of regulation of sugar uptake. Still, other regulatory mechanisms induce this phenomenon, making the consumption of sugars by *E. coli* MS04 strain a complex metabolic regulation system.

2.4. Xylose Consumption Is Governed by the XylR Activator

In *E. coli*, XylR is a transcriptional activator for xylose uptake. Therefore, in the presence of xylose, it promotes the transcription of the *xylFGH* and *xylAB* operons, which are required for xylose metabolism [17]. Although *xylFGH* was deleted from *E. coli* MS04 strain genome, previous studies performed by our group have demonstrated that the MS04 parental strain also internalizes xylose through the galactitol transporter GatC while *xylAB* genes are transcriptionally active [35]. Hence, we studied the function of *xylR* in the co-utilization of sugars by *E. coli* MS04, its possible relationship to the lack of carbon catabolite repression and the combined effect with acetate. To this end, the *xylR* gene was deleted from the *E. coli* MS04 genome (*E. coli* MS04 Δ*xylR*) and fermentations were performed with this new strain. In these tests, glucose–xylose mixtures at initial concentrations of 25 g/L for each were fermented in two different mediums, one with 5 g/L of acetate and another without acetate. Figure 4 shows the ethanol yield and kinetic parameters obtained for these fermentations.

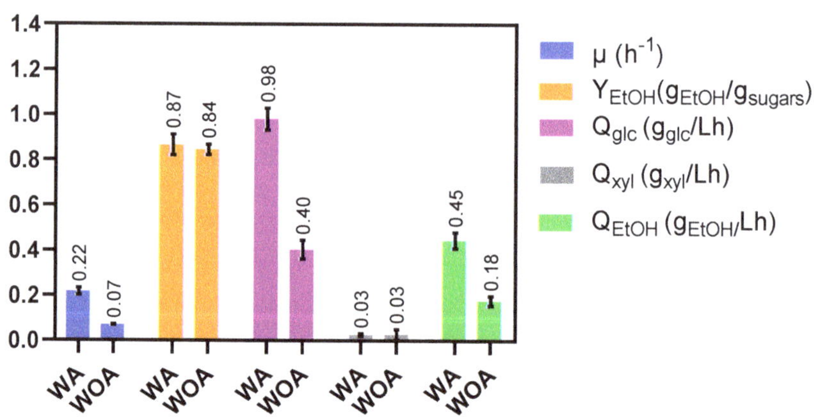

Figure 4. Ethanol yield and kinetic parameters for fermentation of 25 g/L (each) glucose-xylose mixtures by *E. coli* MS04 Δ*xylR* in mineral media with acetate (WA, 5 g/L) and without acetate (WOA) (bars indicate standard deviation from triplicate experiments).

These results show that xylose consumption was null in both cultures, demonstrating the need of XylR for xylose metabolism in *E. coli*. Unlike MS04 (Table 3), the μ for the MS04 Δ*xylR* strain (Figure 4) was three times lower when there was no acetate in the culture medium, demonstrating the importance of xylose consumption for cell growth even in media containing glucose. Although Y_{EtOH} values were similar, glucose consumption and ethanol production rates were about 2.5 times lower for the fermentations WOA. Additionally, Q_{glc} in the presence of acetate was double compared to that observed for

MS04 (Table 3 and Figure 4), suggesting that in the absence of an active xylose metabolism pathway, all the catabolic energy is directed towards glucose consumption.

The results obtained for *E. coli* MS04 Δ*xylR* demonstrate that, independently of the sugars consumed, *XylR* plays an essential role in xylose metabolism, and also highlight the importance of acetate for sugar consumption and ethanol production rates. Nevertheless, these results do not explain the lack of carbon catabolite repression in *E. coli* MS04. Therefore, an exhaustive review of the sequence of MS04 strain genome was performed, with the objective of finding possible gene deletions, changes or single-nucleotide polymorphism (a variation at a single position in a DNA sequence in the genome) that could help explain the co-utilization of sugars. Moreover, the sequence of *xylR* was carefully analyzed, as it has previously been reported that single-nucleotide polymorphisms in this gene void the preference for glucose over other monosaccharides [36]. Nevertheless, no changes were found on the MS04 *xylR* sequence. On the other hand, the results presented in this work show a clear co-fermentation of sugar mixtures under different media composition and lignocellulosic hydrolysates; therefore, additional studies are needed to reveal the causes for the lack of carbon catabolite repression in *E. coli* MS04.

3. Materials and Methods

3.1. *Strains*

3.1.1. *Escherichia coli* MS04

Escherichia coli strain MS04 (MG1655: Δ*pflB*, Δ*adhE*, Δ*frdA*, Δ*xylFGH*, *gatC*-S184L, Δ*midarpA*, Δreg 27.3 kb, Δ*ldhA*) [7] was used as a biocatalyst for ethanol production in mineral medium and lignocellulosic hydrolysates.

3.1.2. *Escherichia coli* MS04 ΔxylR

Deletion of *xylR* gene from *E. coli* MS04 was carried out using the phage transduction methodology. *Escherichia coli* strain MS04 is kanamycin (Km)-resistant due to a mutation in the FRT sequence. The Km-resistant cassette placed in the locus *xylFGH* was replaced by a chloramphenicol (Cm)-resistant cassette. This was performed by a P1 transduction event as previously reported [37]. An *E. coli* MG1655 strain Δ*xylFGH*::*Cm*^{R} constructed in our laboratory (unpublished data) was used as the donor strain. The mutant MS04 Δ*xylFGH*::*Cm*^{R} was selected in LB plates supplemented with Cm at 30 μg/mL, and Km sensitivity was verified. The cassette resistance replacement was confirmed by PCR. The resulting strain was used for a second P1 transduction process. Strain JW3541-2 (BW25113 Δ*xylR*::Km^R) from the Keio collection was used as the donor strain [38]. Furthermore, the mutant MS04 Δ*xylFGH*::Cm^R Δ*xylR*::Km^R was selected in LB plates and supplemented with Cm and Km at 30 μg/mL each. Finally, the elimination of *xylR* gene in the MS04 background was confirmed by PCR.

3.2. *Culture Media*

3.2.1. Laboratory-Simulated Hydrolysates

Control fermentations and the study of acetate effect with strain MS04, and cultures with MS04 Δ*xylR*, were performed in AM1 mineral medium [39]. The composition of the AM1 medium was 2.63 g/L $(NH_4)_2HPO_4$, 0.87 g/L $NH_4H_2PO_4$, 1 mL/L $MgSO_47H_2O$ (1 M), 1 mL/L KCl (2 M), 1 mL/L betaine HCl (1 M) and 1.5 mL/L trace elements. The medium was supplemented with 0.1 g/L sodium citrate and different concentrations of xylose, glucose and arabinose, and sodium acetate was added as needed. The trace element solution contains per liter 1.6 g $FeCl_3$, 0.2 g $CoCl_2{\cdot}6H_2O$, 0.1 g $CuCl_2$, 0.2 g $ZnCl_2{\cdot}4H_2O$, 0.2 g Na_2MoO_4, 0.05 g H_3BO_3 and 0.33 g $MnCl_2{\cdot}4H_2O$. When required, 30 μg/L of Km or Cm was used for inoculum development.

3.2.2. Lignocellulosic Hydrolysates

Teak Wood and Agave Bagasse Hydrolysates

Hydrolysates of teak wood and agave bagasse residues were prepared according to a previously reported methodology [40]. The procedure includes a thermochemical hydrolysis performed in a 5 L parr-type reactor containing a gas–liquid–solid system of 18% *w*/*w* biomass powder, 7% *w*/*w* SO_2 and water. Typical reaction conditions were operated at 140 °C and 450 rpm for 90 min. The enzymatic saccharification of slurries obtained from thermochemical pretreatments of teak wood and agave bagasse was performed with a commercial cellulase cocktail (43 FPU g^{-1}, 100 $U_{xylanase}$ g^{-1}) (NEO Biotech Co., Ltd., Xi'an, China) in a set of six 0.2 L (working volume) mini-reactors fitted with a peg mixer [6]. Enough sodium citrate was added to obtain a final concentration of 50 mM, while the slurry pH was adjusted to 4.8 by adding 10 N KOH, and 15 FPU/g_{glucan} of the cellulase cocktail was supplemented for saccharification.

Barley Straw and Corn Stover Hydrolysates

The hydrolysates from barley straw and corn stover were obtained through thermochemical hydrolysis carried out with a dry biomass charge of 15% *w*/*w*, 1% *w*/*w* H_2SO_4, and at 121 °C for 30 min (holding time). The hydrolysates were used as culture medium without additional pretreatment steps.

Spent Coffee Grounds Hydrolysates

Spent coffee grounds hydrolysates were obtained through a thermochemical pretreatment carried out with 15% *w*/*w* dry biomass, 1% *w*/*w* H_2SO_4, and at 121 °C for 30 min (holding time). After pretreatment, enzymatic saccharification was performed with commercial cellulase complex GC220 (Genencor International, Rochester, NY, USA) and β-glucosidase NS50010 (Novozymes, Copenhagen, Denmark) at 15 FPU/g_{glucan} and 30 UCB/$g_{dry\ biomass}$, respectively. The saccharification process was performed at 50 °C and pH 4.5 for 48 h.

3.3. Fermentation in Simulated and Agroindustrial Lignocellulosic Hydrolysates

Fermentations for both laboratory-simulated hydrolysates (LSH) and agroindustrial lignocellulosic hydrolysates (ALH) were carried out by using the ethanologenic *E. coli* strain MS04 [32] or its derivatives. The preinoculum was prepared by growing cells from a cryovial (1:1 glycerol 80% and strain at OD_{600} ~2.0 in mineral media) in test tubes containing 4 mL of mineral medium with 10 g/L of glucose or xylose. Afterwards, the cells were incubated for 12 h at 37 °C and 300 rpm. The inoculum was prepared by transferring the content of the test tubes to 200 mL fermenters containing AM1 mineral media, 20 g/L of glucose or xylose and 2 g/L of acetate. The inoculum was grown for ~24 h at 37 °C, 150 rpm and pH 7 (controlled with 2N KOH), until OD_{600} between 1.5 and 2 was reached. The cells were harvested by centrifugation (4 °C, 10 min, 4300× *g*), resuspended in fresh mineral media and then used to inoculate the fermenters with LSH or ALH at an initial OD_{600} ~0.5.

Fermentations were performed in the same bioreactor system used for saccharification. Before inoculation, the pH of the vessels containing ALH was adjusted to 7 with 8 N KOH and concentrated solutions of betaine (as osmoprotectant) to a final concentration of 1 mM. Then, phosphate buffer, citric acid and Km were added until final concentrations of 1 mM, 5 mM, 100 mg/L and 30 μg/L, respectively. The cultures were incubated at 37 °C, 150 rpm and pH 7 (controlled with 2 N KOH) until complete substrate depletion or consumption was achieved.

3.4. Biomass, Glucose, Xylose, Arabinose, Acetate and Ethanol Determinations

Growth in LSH was determined spectrophotometrically using an optical density of 600 nm (DU 70, Beckman Instruments, Inc. Fullerton, CA, USA). The values given by the spectrophotometer were converted to dry cell weight (DCW) by using a calibration curve, which indicates that 1 optical density at 600 nm = 0.37 g_{DCW}/L. The samples were centrifuged, and the cell-free culture broth was frozen for its subsequent analysis. Xylose, arabinose and acetate concentrations were measured by high-performance liquid chromatography (Waters U6K, Millipore Co., Milford, MA, USA) using an Aminex HPX-87H ion exclusion column (300 mm × 7.8 mm; Bio-Rad Laboratories, Hercules, CA, USA) at 45 °C, while using 5.0 mM H_2SO_4 solution as the mobile phase (0.5 mL/min), a photodiode array detector at 210 nm (Model 996, Waters, Millipore Co) and a refractive index detector (Model 2410, Waters, Millipore Co., Milford, MA, USA). Glucose concentration in the culture medium was measured with a biochemical analyzer (YSI model 2700, YSI Inc., Yellow Springs, OH, USA). Finally, as described by Fernandez-Sandoval et al. [32], the ethanol produced from fermentations was quantified through gas chromatography (Agilent, 6850 series GC System, Wilmington, DE, USA).

3.5. *Calculation of Kinetic and Stoichiometric Parameters*

The kinetic and stoichiometric parameters for ethanol fermentations were calculated according to Equations (1)–(4) [19].

3.5.1. Specific Growth Rate (μ)

In the case of experiments with hydrolysates, due to the syrups' color and/or the presence of solids, the optical density was not measured. Thus, only the μ for control experiments was calculated during the exponential cell growth using Equation (1):

$$\mu = \frac{\ln\left(\frac{X}{X_0}\right)}{t - t_0} \quad (1)$$

where X (g_{DCW}/L) is the biomass concentration at the end of exponential cell growth, X_0 (g_{DCW}/L) is the initial biomass concentration, t (h) is the time elapsed at the end of the exponential cell growth and t_0 (h) is the initial time.

3.5.2. Ethanol Yield (Y_{EtOH})

The ethanol yield from total sugars was calculated based on a theoretical yield of 0.51 g_{EtOH}/g_{sugar} according to Equation (2).

$$Y_{EtOH} = \frac{EtOH_f \quad EtOH_0}{(S_{t0} - S_{tf}) \times 0.51} \times 100\% \quad (2)$$

where $EtOH_f$ (g_{EtOH}/L) is the maximum ethanol concentration, $EtOH_0$ (g_{EtOH}/L) is the initial ethanol concentration, S_{T0} (g_{sugars}/L) is the sum of initial sugar concentration and S_{to} and S_{tf} (g_{sugars}/L) are the sum of sugar concentration at the start and end of the fermentations.

3.5.3. Volumetric Sugar Consumption Rate (Q_S)

The volumetric consumption rates for individual sugars, i.e., glucose (glc), arabinose (ara) or xylose (xyl), were determined by Equation (3).

$$Q_S = \frac{S_0 - S_f}{t_f - t_0} \quad (3)$$

where S_0 (g_{sugar}/L) is the initial sugar (glc, ara or xyl) concentration, and S_f (g_{sugar}/L) is the sugar concentration (glc, ara or xyl) at the end of the fermentation.

3.5.4. Volumetric Productivity of Ethanol (Q_{EtOH})

The volumetric ethanol production rate was calculated using Equation (4).

$$Q_{EtOH} = \frac{EtOH_f - EtOH_0}{t_{EtOH} - t_0} \quad (4)$$

where $EtOH_f$ (g_{EtOH}/L) is the maximum ethanol concentration, $EtOH_0$ (g_{EtOH}/L) is the initial ethanol concentration, t_{EtOH} is the elapsed time (h) at the maximum ethanol concentration and t_0 is the initial time (h).

4. Conclusions

The ethanologenic *Escherichia coli* strain MS04 was able to efficiently produce ethanol from several lignocellulosic hydrolysates, which were obtained from different pretreatments and saccharification processes. Despite the differences in biomass treatment conditions and the varying contents of glucose and xylose, a clear co-fermentation of hexoses and pentoses in sugar mixtures under different media compositions and lignocellulosic hydrolysates was shown, indicating a partial lack of carbon catabolite repression. Even the presence of phenolic compounds that inhibit *E. coli* growth and fermentation did not seriously hinder the co-consumption of sugars and conversion into ethanol. The volumetric rates of sugar consumption and ethanol production depend on the proportion of initial glucose and xylose, concentrations of inhibitors and a positive effect of acetate (generated in the hydrolysates from the deacetylation of hemicellulose). Furthermore, the deletion of the *xylR* gene involved in xylose metabolism confirms its essential role in xylose consumption. These results give a first insight into the metabolism of monosaccharides in *E. coli* MS04 from a phenomenological perspective, but additional studies are needed to reveal the causes for the lack of carbon catabolite repression in *E. coli* MS04.

Author Contributions: Conceptualization: E.S.-I. and A.M.; methodology: E.S.-I., A.V.-T., C.L.M.-A., B.T.-M. and E.R.M.-V.; validation: Á.V.-L. and E.V.-L.; formal analysis: E.S.-I., A.R.-A. and A.M.; investigation: E.S.-I., A.V.-T. and A.M.; resources: M.G.H.-L., E.V.-L. and A.M.; writing—original draft preparation: E.S.-I. and A.M.; writing—review and editing: A.V.-T., E.V.-L. and A.M.; supervision: A.R.-A., Á.V.-L., M.G.H.-L. and A.M.; project administration: E.V.-L. and A.M.; funding acquisition: E.V.-L., M.G.H.-L. and A.M. All authors have read and agreed to the published version of the manuscript.

Funding: This research was funded by Universidad Nacional Autónoma de México (UNAM)—PAPIIT-DGAPA-UNAM Grant IV100119 and IG10122.

Institutional Review Board Statement: Not applicable.

Informed Consent Statement: Not applicable.

Data Availability Statement: The data presented in this study are available on request from the corresponding author.

Acknowledgments: We thank Georgina Hernández-Chavez, Patricia Bustos-Arcos, Omar Arriaga-Pérez, Arturo Ocádiz-Ramírez and Servando Aguirre-Cruz for their technical support. This work was supported by Universidad Nacional Autónoma de México (UNAM)—PAPIIT-DGAPA-UNAM Grant IV100119 and IG10122. E.S.-I. held a scholarship from DGAPA-UNAM.

Conflicts of Interest: The authors declare no conflict of interest.

Sample Availability: Samples of the compounds are available from the authors.

References

1. Sebayang, A.; Masjuki, H.; Ong, H.C.; Dharma, S.; Silitonga, A.; Mahlia, T.; Aditiya, H. A perspective on bioethanol production from biomass as alternative fuel for spark ignition engine. *RSC Adv.* **2016**, *6*, 14964–14992. [CrossRef]
2. Balat, M.; Balat, H. Recent trends in global production and utilization of bio-ethanol fuel. *Appl. Energy* **2009**, *86*, 2273–2282. [CrossRef]
3. Kumar, P.; Barrett, D.; Delwiche, M.; Stroeve, P. Methods for pretreatment of lignocellulosic biomass for efficient hydrolysis and biofuel production. *Ind. Eng. Chem. Res.* **2009**, *48*, 3713–3729. [CrossRef]
4. Zhao, X.; Zhang, L.; Liu, D. Biomass recalcitrance. Part II: Fundamentals of different pre-treatments to increase the enzymatic digestibility of lignocellulose. *Biofuels Bioprod. Biorefin.* **2012**, *6*, 561–579. [CrossRef]
5. Moya, R.; Tenorio, C.; Oporto, G. Short rotation wood crops in Latin American: A review on status and potential uses as biofuel. *Energies* **2019**, *12*, 705. [CrossRef]

6. Caspeta, L.; Caro-Bermúdez, M.A.; Ponce-Noyola, T.; Martinez, A. Enzymatic hydrolysis at high-solids loadings for the conversion of agave bagasse to fuel ethanol. *Appl. Energy* **2014**, *113*, 277–286. [CrossRef]
7. Vargas-Tah, A.; Moss-Acosta, C.L.; Trujillo-Martinez, B.; Tiessen, A.; Lozoya-Gloria, E.; Orencio-Trejo, M.; Gosset, G.; Martinez, A. Non-severe thermochemical hydrolysis of stover from white corn and sequential enzymatic saccharification and fermentation to ethanol. *Bioresour. Technol.* **2015**, *198*, 611–618. [CrossRef]
8. Guhl, A. Coffee production intensification and landscape change in Colombia, 1970–2002. In *Land-Change Science in the Tropics: Changing Agricultural Landscapes*; Millington, A., Jepson, W., Eds.; Springer: Boston, MA, USA, 2008; pp. 93–116.
9. Karmee, S.K. A spent coffee grounds based biorefinery for the production of biofuels, biopolymers, antioxidants and biocomposites. *J. Waste Manag.* **2018**, *72*, 240–254. [CrossRef]
10. Zhou, M.X. Barley Production and Consumption. In *Genetics and Improvement of Barley Malt Quality*; Zhang, G., Li, C., Eds.; Springer: Berlin/Heidelberg, Germany, 2010; pp. 1–17.
11. Zaldivar, J.; Nielsen, J.; Olsson, L. Fuel ethanol production from lignocellulose: A challenge for metabolic engineering and process integration. *Appl. Microbiol. Biotechnol.* **2001**, *56*, 17–34. [CrossRef]
12. Nichols, N.; Dien, B.; Bothast, R. Use of catabolite repression mutants for fermentation of sugar mixtures to ethanol. *Appl. Microbiol. Biotechnol.* **2001**, *56*, 120–125. [CrossRef]
13. Postma, P.; Lengeler, J.; Jacobson, G. Phosphoenolpyruvate: Carbohydrate phosphotransferase systems of bacteria. *Microbiol. Rev.* **1993**, *57*, 543–594. [CrossRef] [PubMed]
14. Hernández-Montalvo, V.; Martínez, A.; Hernández-Chavez, G.; Bolivar, F.; Valle, F.; Gosset, G. Expression of *galP* and *glk* in an *Escherichia coli* PTS mutant restores glucose transport and increases glycolytic flux to fermentation products. *Biotechnol. Bioeng.* **2003**, *83*, 687–694. [CrossRef] [PubMed]
15. Kim, H.-H.; So, J.-H.; Shin, J.-H.; Rhee, I.-K. Role of *xylR* gene on the construction of xylose-inducible expression vectors using *xylF* promoter of *Escherichia coli*. *J. Korean Soc. Appl. Biol. Chem.* **2010**, *53*, 790–797. [CrossRef]
16. Sievert, C.; Nieves, L.M.; Panyon, L.A.; Loeffler, T.; Morris, C.; Cartwright, R.A.; Wang, X. Experimental evolution reveals an effective avenue to release catabolite repression via mutations in *XylR*. *Proc. Natl. Acad. Sci. USA* **2017**, *114*, 7349–7354. [CrossRef] [PubMed]
17. Song, S.; Park, C. Organization and regulation of the D-xylose operons in *Escherichia coli* K-12: *XylR* acts as a transcriptional activator. *J. Bacteriol.* **1997**, *179*, 7025–7032. [CrossRef] [PubMed]
18. Jönsson, L.; Palmqvist, E.; Nilvebrant, N.-O.; Hahn-Hägerdal, B. Detoxification of wood hydrolysates with laccase and peroxidase from the white-rot fungus *Trametes versicolor*. *Appl. Microbiol. Biotechnol.* **1998**, *49*, 691–697. [CrossRef]
19. Sierra-Ibarra, E.; Alcaraz-Cienfuegos, J.; Vargas-Tah, A.; Rosas-Aburto, A.; Valdivia-López, Á.; Hernández-Luna, M.G.; Vivaldo-Lima, E.; Martinez, A. Ethanol production by *Escherichia coli* from detoxified lignocellulosic teak wood hydrolysates with high concentration of phenolic compounds. *J. Ind. Microbiol. Biotechnol.* **2021**, *49*, kuab077. [CrossRef]
20. Rios-González, L.J.; Morales-Martínez, T.K.; Rodríguez-Flores, M.F.; Rodríguez-De la Garza, J.A.; Castillo-Quiroz, D.; Castro-Montoya, A.J.; Martinez, A. Autohydrolysis pretreatment assessment in ethanol production from agave bagasse. *Bioresour. Technol.* **2017**, *242*, 184–190. [CrossRef]
21. Saucedo-Luna, J.; Castro-Montoya, A.J.; Martinez-Pacheco, M.M.; Sosa-Aguirre, C.R.; Campos-Garcia, J. Efficient chemical and enzymatic saccharification of the lignocellulosic residue from *Agave tequilana* bagasse to produce ethanol by *Pichia caribbica*. *J. Ind. Microbiol. Biotechnol.* **2011**, *38*, 725–732. [CrossRef]
22. Barbosa, M.F.; Beck, M.J.; Fein, J.E.; Potts, D.; Ingram, L.O. Efficient fermentation of *Pinus* sp. acid hydrolysates by an ethanologenic strain of *Escherichia coli*. *Appl. Environ. Microbiol.* **1992**, *58*, 1382–1384. [CrossRef]
23. Pérez-Pimienta, J.A.; Vargas-Tah, A.; López-Ortega, K.M.; Medina-López, Y.N.; Mendoza-Pérez, J.A.; Avila, S.; Singh, S.; Simmons, B.A.; Loaces, I.; Martinez, A. Sequential enzymatic saccharification and fermentation of ionic liquid and organosolv pretreated agave bagasse for ethanol production. *Bioresour. Technol.* **2017**, *225*, 191–198. [CrossRef] [PubMed]
24. Ballesteros, L.F.; Teixeira, J.A.; Mussatto, S.I. Chemical, functional, and structural properties of spent coffee grounds and coffee silverskin. *Food Bioprocess Technol.* **2014**, *7*, 3493–3503. [CrossRef]
25. Atabani, A.E.; Al-Muhtaseb, A.a.H.; Kumar, G.; Saratale, G.D.; Aslam, M.; Khan, H.A.; Said, Z.; Mahmoud, E. Valorization of spent coffee grounds into biofuels and value-added products: Pathway towards integrated bio-refinery. *Fuel* **2019**, *254*, 115640. [CrossRef]
26. Battista, F.; Barampouti, E.M.; Mai, S.; Bolzonella, D.; Malamis, D.; Moustakas, K.; Loizidou, M. Added-value molecules recovery and biofuels production from spent coffee grounds. *Renew. Sustain. Energy Rev.* **2020**, *131*, 110007. [CrossRef]
27. Negrete, A.; Ng, W.-I.; Shiloach, J. Glucose uptake regulation in *E. coli* by the small RNA SgrS: Comparative analysis of *E. coli* K-12 (JM109 and MG1655) and *E. coli* B (BL21). *Microb. Cell Factories* **2010**, *9*, 75. [CrossRef] [PubMed]
28. Neubauer, P.; Häggström, L.; Enfors, S.O. Influence of substrate oscillations on acetate formation and growth yield in *Escherichia coli* glucose limited fed-batch cultivations. *Biotechnol. Bioeng.* **1995**, *47*, 139–146. [CrossRef] [PubMed]
29. Saha, B.C.; Qureshi, N.; Kennedy, G.J.; Cotta, M.A. Enhancement of xylose utilization from corn stover by a recombinant *Escherichia coli* strain for ethanol production. *Bioresour. Technol.* **2015**, *190*, 182–188. [CrossRef]
30. Saha, B.C.; Cotta, M.A. Comparison of pretreatment strategies for enzymatic saccharification and fermentation of barley straw to ethanol. *N. Biotechnol.* **2010**, *27*, 10–16. [CrossRef]

31. Fernández-Sandoval, M.T.; Galíndez-Mayer, J.; Moss-Acosta, C.L.; Gosset, G.; Martinez, A. Volumetric oxygen transfer coefficient as a means of improving volumetric ethanol productivity and a criterion for scaling up ethanol production with *Escherichia coli*. *J. Chem. Technol. Biotechnol.* **2017**, *92*, 981–989. [CrossRef]
32. Fernández-Sandoval, M.; Huerta-Beristain, G.; Trujillo-Martinez, B.; Bustos, P.; González, V.; Bolivar, F.; Gosset, G.; Martinez, A. Laboratory metabolic evolution improves acetate tolerance and growth on acetate of ethanologenic *Escherichia coli* under non-aerated conditions in glucose-mineral medium. *Appl. Microbiol. Biotechnol.* **2012**, *96*, 1291–1300. [CrossRef]
33. Xia, T.; Eiteman, M.A.; Altman, E. Simultaneous utilization of glucose, xylose and arabinose in the presence of acetate by a consortium of *Escherichia coli* strains. *Microb. Cell Factories* **2012**, *11*, 77. [CrossRef] [PubMed]
34. Fox, K.J.; Prather, K.L.J. Carbon catabolite repression relaxation in *Escherichia coli*: Global and sugar-specific methods for glucose and secondary sugar co-utilization. *Curr. Opin. Chem. Eng.* **2020**, *30*, 9–16. [CrossRef]
35. Utrilla, J.; Licona-Cassani, C.; Marcellin, E.; Gosset, G.; Nielsen, L.K.; Martinez, A. Engineering and adaptive evolution of *Escherichia coli* for D-lactate fermentation reveals GatC as a xylose transporter. *Metab. Eng.* **2012**, *14*, 469–476. [CrossRef] [PubMed]
36. Heo, J.M.; Kim, H.J.; Lee, S.J. Efficient anaerobic consumption of D-xylose by *E. coli* BL21 (DE3) via *xylR* adaptive mutation. *BMC Microbiol.* **2021**, *21*, 332. [CrossRef]
37. Thomason, L.C.; Costantino, N.; Court, D.L. *E. coli* Genome Manipulation by P1 Transduction. *Curr. Protoc. Mol. Biol.* **2007**, *79*, 1171–1178. [CrossRef]
38. Baba, T.; Ara, T.; Hasegawa, M.; Takai, Y.; Okumura, Y.; Baba, M.; Datsenko, K.A.; Tomita, M.; Wanner, B.L.; Mori, H. Construction of *Escherichia coli* K-12 in-frame, single-gene knockout mutants: The Keio collection. *Mol. Syst. Biol.* **2006**, *2*, 2006. [CrossRef]
39. Martinez, A.; Grabar, T.B.; Shanmugam, K.T.; Yomano, L.P.; York, S.W.; Ingram, L.O. Low salt medium for lactate and ethanol production by recombinant *Escherichia coli* B. *Biotechnol. Lett.* **2007**, *29*, 397–404. [CrossRef]
40. Hernández Luna, M.G.; Vivaldo Lima, E.; Alcaraz Cienfuegos, J.; Valdivia López, M.A. Proceso de Tratamiento ácido en Fase de Gas de Materiales Lignocelulosicos. MX Publication No. WO 2018/004327 A1, 1 April 2018.

molecules

MDPI

Article

Fractionation of Raw and Parboiled Rice Husks with Deep Eutectic Solvents and Characterization of the Extracted Lignins towards a Circular Economy Perspective

Chiara Allegretti [1], Emanuela Bellinetto [1], Paola D'Arrigo [1,2,*], Monica Ferro [1], Gianmarco Griffini [1], Letizia Anna Maria Rossato [1], Eleonora Ruffini [1], Luca Schiavi [3], Stefano Serra [2,*], Alberto Strini [3] and Stefano Turri [1,†]

1 Department of Chemistry, Materials and Chemical Engineering "Giulio Natta", Politecnico di Milano, p.zza L. da Vinci 32, 20133 Milano, Italy
2 Istituto di Scienze e Tecnologie Chimiche "Giulio Natta", Consiglio Nazionale delle Ricerche (SCITEC-CNR), Via Luigi Mancinelli 7, 20131 Milano, Italy
3 Istituto per le Tecnologie della Costruzione, Consiglio Nazionale delle Ricerche (ITC-CNR), Via Lombardia 49, 20098 San Giuliano Milanese, Italy
* Correspondence: paola.darrigo@polimi.it (P.D.); stefano.serra@cnr.it (S.S.); Tel.: +39-2-23993075 (P.D.); +39-2-23993076 (S.S.)
† The authors are listed in alphabetical order.

Citation: Allegretti, C.; Bellinetto, E.; D'Arrigo, P.; Ferro, M.; Griffini, G.; Rossato, L.A.M.; Ruffini, E.; Schiavi, L.; Serra, S.; Strini, A.; et al. Fractionation of Raw and Parboiled Rice Husks with Deep Eutectic Solvents and Characterization of the Extracted Lignins towards a Circular Economy Perspective. *Molecules* **2022**, 27, 8879. https://doi.org/10.3390/molecules27248879

Academic Editors: Carlos Martín, Alejandro Rodriguez Pascual and Eduardo Espinosa Víctor

Received: 19 October 2022
Accepted: 5 December 2022
Published: 14 December 2022

Abstract: In the present work, rice husks (RHs), which, worldwide, represent one of the most abundant agricultural wastes in terms of their quantity, have been treated and fractionated in order to allow for their complete valorization. RHs coming from the raw and parboiled rice production have been submitted at first to a hydrothermal pretreatment followed by a deep eutectic solvent fractionation, allowing for the separation of the different components by means of an environmentally friendly process. The lignins obtained from raw and parboiled RHs have been thoroughly characterized and showed similar physico-chemical characteristics, indicating that the parboiling process does not introduce obvious lignin alterations. In addition, a preliminary evaluation of the potentiality of such lignin fractions as precursors of cement water reducers has provided encouraging results. A fermentation-based optional preprocess has also been investigated. However, both raw and parboiled RHs demonstrated a poor performance as a microbiological growth substrate, even in submerged fermentation using cellulose-degrading fungi. The described methodology appears to be a promising strategy for the valorization of these important waste biomasses coming from the rice industry towards a circular economy perspective.

Keywords: deep eutectic solvent (DES); lignocellulosic biomass fractionation; raw rice husk; parboiled rice husk; parboiling process; lignin; cement plasticizers; sustainability; circular economy

1. Introduction

Rice (*Oriza sativa*) is one of the most important food crops, with about 755 million tons of rough rice (also known as paddy rice) produced worldwide in 2019 [1] and a >25% production increase in the 2000–2020 timeframe [2]. Its importance is undoubtedly established when it is taken into account that it feeds around three billion people all over the world.

Raw rice is composed by an external hull and the brown rice grain, which are separated during the dehusking process. Rice husks (or hull, RH), deriving from paddy rice dehusking, represent a very important source of waste biomass and are one of the most abundant by-products in the rice production line. Considering that ca. 20–25% of paddy rice mass is constituted by the outer husk [3], it is possible to estimate a global annual rice husk production of >150 million tons, currently used mainly for power generation [4].

Parboiled (partially boiled) rice is obtained from rough rice by a hydrothermal process constituted by a hydration stage followed by gelatinization and drying. Because of its several nutritional and physical benefits, parboiled rice possesses between two and three times the economic value of white rice and its importance on the global market is currently increasing (about 20% of the produced rice is treated by the parboiling process at the present time) [5,6]. The parboiling process is carried out mainly on rough rice that then must undergo the hulling process, thus leading to the parboiled rice husk as a by-product. As a consequence, this latter constitutes a significant part of the global rice husk production with solid forecasts of a high growth in the coming years.

For these reasons, the exploitation of both a raw rice husk (rRH) and parboiled rice husk (pRH) as lignocellulosic renewable feedstock constitutes a very important objective in a circular economy perspective, where the valorization of bio-agricultural wastes as a secondary raw material is largely preferable to their incineration for heat generation. Globally, this is particularly important in Asian countries where a rice cultivation is the primary food source and, regionally, in local realities where rice is a traditional source (e.g., in Lombardy in Italy). Up to now, RHs are instead mainly exploited as an energy source in rice processing plants and as bedding for farm animals. Therefore, the availability of this huge quantity of wastes and their low cost have pushed the current research towards the investigation and development of new sustainable possible applications for such materials.

Deep eutectic solvents (DESs) are a promising class of new solvents in the context of sustainable process development that can be prepared from a broad range of components, which are often from biobased sources [7]. DESs are characterized by a low toxicity and the typically high renewability and recyclability of the constituents [8]. Their wide solvating power modulation capabilities make them a first-choice medium for setting up separation processes, with particular reference to the fractionation of agro-industrial lignocellulosic wastes [9–11]. Moreover, it is even more interesting that the potential use of reactive DESs (RDESs, [12,13]) could allow for the exploitation of the same DES formulation as a separation medium (for lignocellulosic fractionation) and reaction medium (for lignin derivatization), even possibly in one-pot processes, enhancing the sustainability of the process [14].

Thus, we have focused on the set-up of a cascade multistep process, reported in Figure 1, composed of a preprocess in hot water and in an autoclave, followed by a DES-mediated treatment of the rRH and pRH biomasses, in order to deconstruct the natural lignocellulosic skeleton that usually hampers the further exploitation of the different components of RHs. The fractionation of these waste materials in two main fractions, i.e., a cellulose-enriched fraction and lignin constitutes the key-point for the successive development of valuable applications. The methodology based on the DES treatment was previously successfully applied by the authors to another common agri-food biomass, i.e., Brewer's Spent Grain (BSG) [15]. The main aim of the present work is to study the DES-mediated fractionation process focusing on RHs, and to compare and characterize in detail both the composition and properties of the lignin fractions recovered from rRH and pRH. The latter is particularly interesting because of its aforementioned increase in commercial interest. Indeed, such a detailed characterization should represent a strategic milestone for the successful incorporation of lignin, as a macromolecular precursor, into bio-based polymers of a high added value [16–18], which have found an application in a variety of industrial [19–22] and technological [23–28] fields.

Moreover, in this work, the potentialities of these materials were also demonstrated in the field of cement water reducers, the latter being an important component of all modern concrete formulations. Lignin is a well-known component of traditional water reducers and has been widely used for this purpose for a long time [29]. In recent years, several research efforts aimed at the study of renewable high-performance water reducers inspired a renewed interest in advanced formulations based on lignin derivatives [30–32]. Any process resulting in lignin as a by-product is thus a potential resource for the development of sustainable concrete water reducers. For such reasons, the obtained lignins were here assessed for this specific application by rheological measurements on cement pastes. The results clearly indicate a water reducing performance equivalent to that of a commercial soda lignin (Protobind 1000), thus demonstrating the potential of the approach proposed herein.

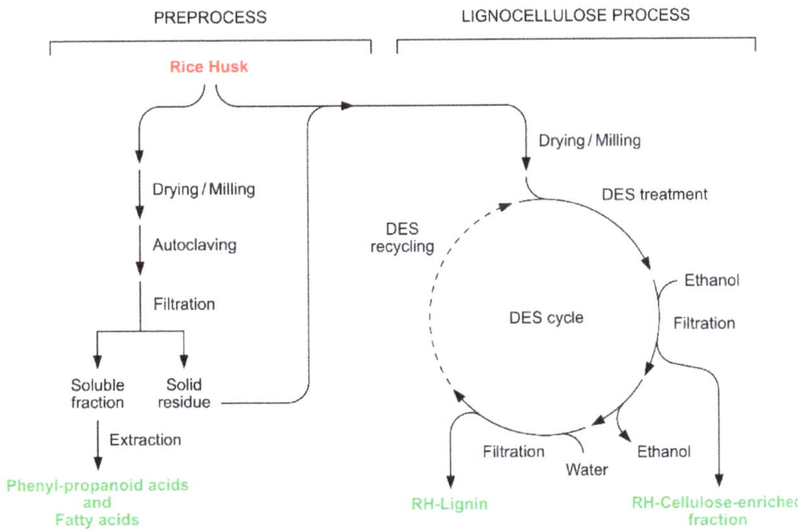

Figure 1. Complete fractionation process of raw and parboiled rice husks constituted by a hydrothermal preprocess and a DES-mediated lignocellulose process.

A fermentation-based pretreatment was also studied, aiming to pretreat the biomass before the fractionation steps. The rationale behind this approach was to verify whether the soluble saccharides components of RHs could be exploitable as substrates in industrial fermentation processes. The expected effect of the pretreatment was the partial deconstruction of the lignocellulose matrix with the reduction in the saccharides' content. However, the limited results obtained indicated that both raw and parboiled RHs are poor performers as a microbiological growth media, possibly due to their relatively low oligosaccharide and high silica contents.

2. Results and Discussion

In this work, two rice husk waste biomasses were selected from raw and parboiled rice production, respectively (see Section 3.1 for details). The present work aims to fractionate and compare them in terms of the composition and properties with the special aim of providing fundamental information for the exploitation of the different fractions, focusing in particular on the isolated lignins. Moreover, a deep comparison between rRH and pRH has not been presented until now.

The two samples of the starting biomasses employed in the present work are illustrated in Figure 2. As it is evident from the picture, pRH (on the right) appeared to be slightly darker than rRH (on the left). As soon as they were received, these biomasses, even if they have been provided in a quite dried form, were put in a ventilated oven (60 °C for 24 h) and finely ground with an electric mixer.

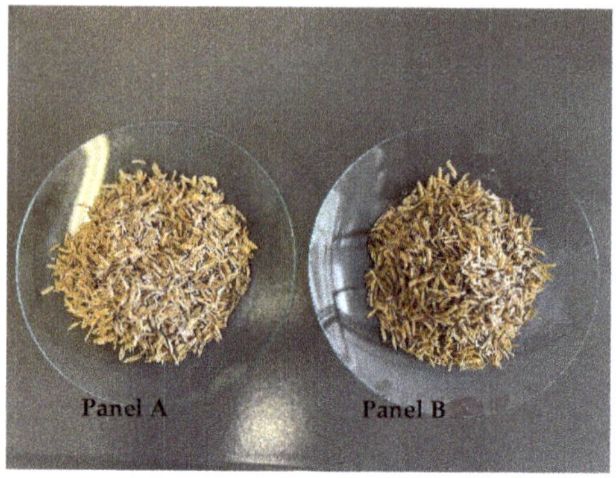

Figure 2. Picture of raw rice husk (**Panel A**) and parboiled rice husk (**Panel B**).

2.1. Rice Husks Composition

As most lignocellulosic biomasses, RHs are an intricate material where all the components are closely related to each other. For that reason, in order to separate and quantify the four main constituents of rRH and pRH of this study (hemicellulose, cellulose, silica, and lignin), a multistep process based on a classical method of fractionation by successive water and acid treatments was performed (see Section 3.3 for details) [33]. The composition of the presently studied biomasses is reported in Table 1 and appears to be in agreement with those reported in the literature [34,35].

Table 1. Composition of the studied RHs biomasses.

Sample	Hemicellulose (% *w/w*)	Cellulose (% *w/w*)	Silica (% *w/w*)	Lignin (% *w/w*)	Soluble Fraction (% *w/w*)
rRH	19.5	43.2	12.9	19.2	5.20
pRH	27.3	36.7	12.4	18.5	5.10

Moreover, the detailed composition, as monosaccharides components, of the isolated hemicellulose fraction (reported in Table 2) has been determined by GC-MS analysis after the total hydrolysis, reduction, and acetylation of the samples following a well-known procedure (see Section 3.3.1 for experimental details) [36]. The two main components of RHs hemicellulose were arabinose and xylose, with a low amount of galactose and traces of rhamnose and fucose. The detected glucose residues were probably due to a contamination of the cellulose fraction that appeared to be nearly completely eliminated after the pretreatment of the biomass performed in the fractionation process (the resulting residual glucose was 0.2–0.4%).

Table 2. Relative abundance of the monosaccharides deriving from RHs hemicellulose hydrolysis.

RH	Rhamnose	Fucose	Arabinose	Xylose	Mannose	Glucose	Galactose
rRH	0.4	0.5	49.0	44.9	-	1.2	4.0
pRH	0.2	0.5	22.6	69.2	-	4.9	2.6

2.2. Rice Husks Preprocess

In our previous work, we described an efficient procedure for the water-mediated extraction of Brewer's Spent Grain (BSG) [15]. Accordingly, the treatment of BSG with water at a high temperature allowed for the removal of a considerable (25–30% *w/w*) fraction of this waste material. Hence, we checked the effectiveness of the same procedure when applied to RHs.

2.2.1. Hydro-Thermal Preprocess

We performed a hydro-thermal treatment of rRH and pRH by autoclaving the aqueous suspensions of the latter materials at 121 °C for 20 min. The two suspensions were then filtered, leading to solid residues which were indicated as rRH_T (raw rice husk-treated) and pRH_T (parboiled rice husk-treated), whereas the obtained aqueous extracts were concentrated at a reduced pressure and the weights of the resulting residues were compared to those of the parent RHs samples. We observed that the described procedure allowed for the extraction of only 2 and 2.8% (*w*/*w*) of rRH and of pRH, respectively. This water-mediated extraction gave results comparable with those described in Table 1, proving that rRH and pRH contained a very low amount of soluble sugars. In order to further characterize these soluble fractions, the extracts were further treated with ethyl acetate. The organic solvent soluble fractions were made up of a mixture of fatty acids (~0.05% *w*/*w* total biomass) and phenylpropanoid metabolites such as cinnamic acid (~0.74% *w*/*w* total biomass) and ferulic acid (0.13% *w*/*w* total biomass).

2.2.2. Microbiological Preprocess

As discussed above, the results of the hydro-thermal preprocess have confirmed that both rRH and pRH are waste materials of a very difficult valorization, as they are devoid of water-soluble components such as sugars, starch, or proteins. In addition, the biomass of the RHs was almost unaffected by our thermal treatment. As a consequence, we evaluated whether a biological pre-treatment could break up the lignocellulosic structure of the RHs. To this end, we evaluated the potential of some selected fungal strains in the submerged fermentation of the RHs samples. More specifically, we singled out the filamentous fungi *Myceliophthora thermophila*, *Rhizomucor pusillus*, and *Trichoderma viride*. The latter microorganisms have been widely employed in the industry for the production of hydrolytic enzymes such as cellulases, amylases, pectinases, and chitinases [37–40]. In addition, the first two species are thermophiles with two different optimal growth temperatures. Therefore, the fermentation of the three identical samples of pRH with *M. thermophila*, *R. pusillus*, and *T. viride* were performed at 45, 35, and 24 °C, respectively. The pRH biomasses recovered after the fermentation indicated a weight loss of the 18, 19, and 14%, respectively. These results clearly demonstrate that all the tested strains utilized pRH as a carbon source for their growth, regardless of the temperature fermentation. Despite this fact, none of the experiments displayed a weight loss superior to 19%, thus confirming the difficult degradability of RHs, even when using cellulose-degrading fungi. Interestingly, for all the trials, we observed an initial luxuriant growth followed by the transformation of the formed filamentous biomass into a slurry. The microscopic view of the biomasses showed the presence of a very short hypha, most likely deriving by the grinding of the fungal filaments. We supposed that RH, with its high silica content, was not a suitable substrate for a submerged fermentation. The shaking, necessary for oxygenation, damaged the fungal hypha, thus reducing their activity. As a consequence of the described results, the microbiological preprocessing of this specific waste was considered to be unsuitable and thus excluded from the complete fractionation process depicted in Figure 1.

2.3. DES-Mediated Lignocellulose Process

The subsequent step of the biomass treatment after the previously described pre-process was constituted of a DES-mediated fractionation process. Different DESs have been prepared and tested in this work. They have been obtained by mixing and heating a mixture of the two components: a hydrogen bond acceptor (HBA) such as choline chloride (ChCl) and betaine glycine (BetG), and a hydrogen bond donor (HBD) composed by acetic acid and L-lactic acid. These two HBDs have been selected since it is known that the presence of the carboxylic group in an HBD enhances the lignin yields during the lignocellulosic destructuration [41,42]. The composition and density of the DESs are reported in Table 3. Typically, DESs had a higher density than water and those containing L-lactic acid (DESs 2 and 4) resulted in being denser than those with acetic acid (DESs 1 and 3), likely because of the presence of the hydroxyl group in the HBD molecules, which should allow for the formation of stronger hydrogen bonds.

Table 3. List, composition, and density (measured at 18 °C) of prepared DESs.

DES Number	Composition DES HBA/HBD	Molar Ratio (HBA/HBD)	Density of Pure DES (g/cm^3)
1	Choline chloride/Acetic acid	1/2	1.10
2	Choline chloride/L-Lactic acid	1/5	1.18
3	Betaine Glycine/Acetic acid	1/2	1.11
4	Betaine Glycine/L-Lactic acid	1/5	1.20

Following the scheme of the process reported in Figure 1, dried and finely milled treated biomasses (rRH_T and pRH_T) were submitted to a fractionation. For comparison, the native biomasses (rRH and pRH) were also treated in the same way. The biomasses were then suspended in the different DESs at 120 °C under magnetic stirring for 24 h. After cooling, ethanol was gradually added, leading to the precipitation of the first fractions, the cellulose-enriched fractions, which were then further separated in order to quantify the silica content. To this end, we slightly modified the Yoshida procedure [43], consisting in the oxidative degradation of all the organic fractions (see Section 3.5.3 for details). The cellulose-enriched fractions were heated at a reflux with a mixture of concentrated nitric acid, sulfuric acid, and perchloric acid, and the resulting silica suspensions allowed for the isolation of the silica samples. Their weights were employed to calculate the silica content of the cellulose-enriched fractions (given as a *w*/*w* biomass percentage and reported in Figure 3). The silica content was quantified at around 13%, almost identical in all the samples, confirming that the treatments performed on the RHs did not affect their SiO_2 content.

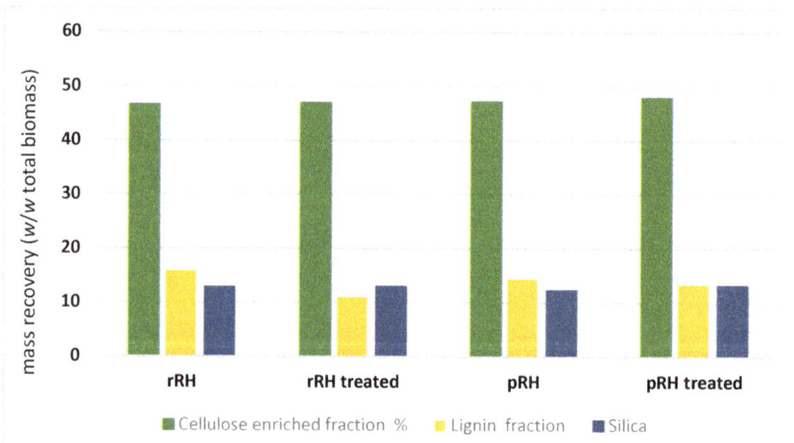

Figure 3. Data of mass recovery of cellulose-enriched fraction (in green), lignin (in yellow), and silica (in blue) during the DES 2-mediated fractionation process of raw rice husk (rRH), raw treated rice husk, parboiled rice husk (pRH), and parboiled treated rice husk.

The filtrates were then concentrated to eliminate ethanol and were treated with water acting as an anti-solvent, to induce the precipitation of the lignins. After the centrifugation, filtration, and solvent evaporation, the final solid fractions (rRH-Lignin, rRH_T-Lignin, pRH-Lignin, and pRH_T-Lignin) were recovered and quantified.

After the examination of the results obtained with the four DESs in terms of the handling of the reaction mixtures and the final mass recovery, DES 2 (choline chloride/L-lactic acid, 1/5) was selected for this work, since it gave the higher yields in terms of the fractions recovery and lignin recovery. The data of the quantitative recovery of the different fractions are reported in the Supplementary Materials (Table S1) for all the tested DESs, whereas the mass recovery results obtained on the DES 2-mediated fractions for the four RHs biomasses have been illustrated more precisely in Figure 3.

As it appears clearly from Figure 3, the weight percentages of the three fractions were almost identical for the four fractionated RHs. The cellulose-enriched fraction accounted for 47% of the total biomass in the three cases, lignin for 15.8 to 11% (for rRH and rRH_T) and to 14.5 to 15.5% (for pRH and pRH_T), and silica at around 13% in all cases. The results clearly prove that both the parboiling process and the autoclave treatment do not influence so much the percentages of the three main fractions and do not induce substantial differences in the natural biomass fractionation.

2.4. Lignin Characterization

Lignins derived from the complete fractionation process (using DES 2) of both raw and parboiled rice husks (rRH-Lignin and pRH-Lignin, respectively) were subjected to an extensive characterization to evaluate their properties and potential differences. A well-known commercial soda lignin (Protobind 1000, indicated from now on as Protobind) was used as the reference [44].

2.4.1. Solvent Solubilization

One of the key features in the lignin's characterization and valorization is represented by the solubilization studies, which can afford the starting point in the set-up of the best conditions for lignin processing. In fact, it should be stressed that one of the main issues concerning lignin exploitation is its low solubility in the most frequently used organic solvents. In this work, the solubilization screening has been performed in six organic solvents and in water. The selected solvents have been: *tert*-butyl methyl ether (MTBE), *n*-butyl acetate (BuOAc), ethyl acetate (EtOAc), 2-butanone (MEK), methanol (MeOH), and tetrahydrofuran (THF). Figure 4 shows the obtained extraction yields for rRH- (in pale yellow), rRH_T- (in dark yellow), pRH- (in orange), pRH_T-Lignins (in brown), and Protobind (in grey) for comparison, as a w/w percentage of the solubilized lignin vs. total lignin.

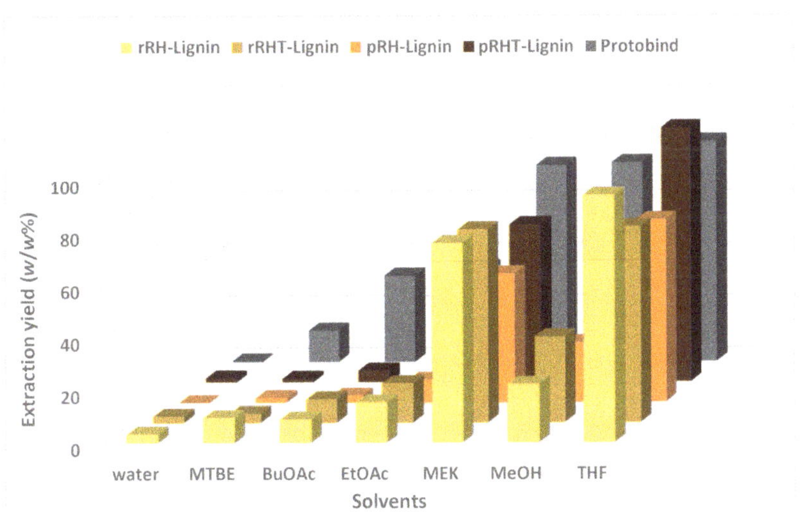

Figure 4. Extraction yield of the commercial technical lignin Protobind (in grey), and the lignins extracted from raw (in pale yellow), raw treated (dark yellow), parboiled (in orange), and parboiled treated (in brown) rice husks. The data are reported as the percentage (w/w) of solubilized fraction vs. total fraction in 6 different organic solvents and water.

As expected, the four RH-lignins are quite insoluble in water, similar to what was observed in the reference material (Protobind). When considering the solubility response to organic solvents, RH-lignins show a substantial similarity in their solubility: a very poor solubility (less than 10% *w/w*) in MTBE and BuOAc, and a slightly higher solubility in EtOAc and MeOH in a range between 10 and 30% *w/w*. The best solvents appear to be MEK and THF, as for the reference material (Protobind), where the value of the solubility ranges from 50 up to 99% *w/w*. It should be noted that the lignins from rRHs seemed to be slightly more soluble in all the solvents, except THF, if compared with the pRHs. Moreover, the solubility of the RH_T-lignins appeared to be higher than the RH-lignins in all the organic solvents (except THF), indicating that the biomass pretreatment could generally positively affect the solubility characteristic of the recovered lignins. This aspect appears as an interesting outcome and suggests a potential straightforward exploitation of such RH-recovered lignins in line with the similar exploitation routes typical of commercial lignin materials.

2.4.2. Molar Mass Distribution

The evaluation of the molar mass distribution in the lignin recovered fractions is a fundamental tool in order to successfully exploit these biomasses. To that end, GPC analyses were performed to determine the molecular weight and the molecular weight distribution of the extracted lignins. In Table 4, the number average molecular weight (M_n), the weight average molecular weight (M_w), and the polydispersity index ($Đ$) of all the RH-lignin samples and of the reference lignin Protobind 1000 are reported.

Table 4. Number average molecular weight (M_n), weight average molecular weight (M_w), and polydispersity index ($Đ$) of all examined lignins (samples were eluted after acetylation; reported values are relative to polystyrene standards).

Sample	M_n (g/mol)	M_w (g/mol)	$Đ$
rRH-Lignin	1380	5330	3.86
rRH_T-Lignin	1360	5195	3.82
pRH-Lignin	1320	3930	2.98
pRH_T-Lignin	1310	3860	2.95
Protobind 1000	830	2800	3.37

The RH-Lignins showed a higher M_n and M_w and a similar $Đ$, with respect to the Protobind lignin. Especially, in all RH-Lignins, a comparable M_n was observed. In contrast to this, the M_w of both rRH-Lignins was found to be higher compared to both pRH-Lignins, resulting in a higher $Đ$. No differences were observed between the treated and untreated lignins, indicating that the pretreatment did not affect the molecular weight and the molecular weight distribution of these lignins.

These results are in accordance with what was previously discussed in terms of solubility. Indeed, on average, due to their higher M_n and M_w, RHs-lignins resulted in being slightly less soluble in organic solvents compared to the Protobind lignin.

2.4.3. Sugar Content

The precipitated lignin fractions were washed three times after the water-mediated precipitation from the DES solutions in order to eliminate potential DES residues, which could contaminate the lignins and provide a sticky response of the final products. The sugar quantification, reported in Table 5, was performed with an already established procedure using the bicinchoninic assay method with some modifications [15,45]. Although they were not further purified after the precipitation, all the lignins possessed a very low carbohydrate content similar to the Protobind value; it accounted for less than 0.4% (*w/w*) of the free reducing sugars. Furthermore, after the hydrolysis, the rRH- and pRH-Lignins showed the presence of a certain quantity of complex sugars (from 5.7 to 9.1%), which was also in this case substantially lower than the Protobind value (13%). However, pRH_T-Lignin exhibited quite the same value as Protobind. This aspect has to be considered in view of the further applications of these RH-Lignins in the preparation of cement water reducers.

Table 5. Results of the total reducing sugar quantification.

Sample	Reducing Sugars/Biomass (*w*/*w*%)	Reducing Sugars/Biomass after Hydrolysis (*w*/*w*%)
rRH-Lignin	0.19	5.7
rRH_T-Lignin	0.51	9.2
pRH-Lignin	0.31	9.1
pRH_T-Lignin	0.38	14
Protobind 1000	0.34	13

2.4.4. ^{13}C CP-MAS NMR

The solid-state ^{13}C cross-polarization magic angle spinning (CP-MAS) NMR spectroscopy was used to define the fingerprint of each lignin and to highlight the structural differences among the examined samples. Figure 5 shows the superposition of the ^{13}C CP-MAS NMR spectra of the commercial Protobind lignin with the extracted lignin samples. At first, the RHs samples had an almost overlapping profile but clearly differed from the Protobind reference sample (see Figure 5 Panel A). The region between 210 and 190 ppm corresponded to the non-conjugated carbonyl groups C=O of aldehydes. In this range of chemical shifts, there were no substantial differences among the spectra, whereas in the aromatic region (160–100 ppm), Protobind showed a lower signal intensity at 175.7 ppm compared to the other spectra. This indicated a lower presence of both $-CO_2H$ carboxylic and $-CO_2R$ ester groups. In addition, the Protobind spectrum showed another signal which was less intense than the other samples around 115 ppm, which very likely could be attributed to a lower content of the guaiacyl units.

However, the main spectral differences were between 100 and 0 ppm (Figure 5 in Panel B). Between 65 and 75, Protobind presented two more peaks at 74.5 and 63.7 ppm: peaks in such a region are typical of the β-O-4 structural moiety with the –OH group at the α-position. On the other hand, the extracted lignin samples showed at 67.4 ppm a signal which was not present in the Protobind lignin. This region (62–75 ppm) is diagnostic of β-O-4 structures. In the alkyl part of the spectra, the intensity of the $O-CH_3$ peak at 55.9 ppm was comparable in all the samples. Lastly, the most significant differences were at around 30 ppm (corresponding to the alkyl CH_2) and around 20 ppm (typical of the methyl groups). This analysis could be a very important tool in order to compare the lignins extracted from different biomasses.

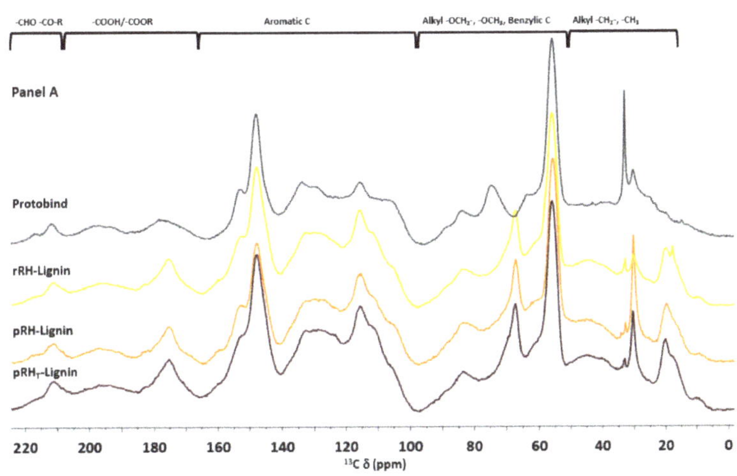

Figure 5. *Cont.*

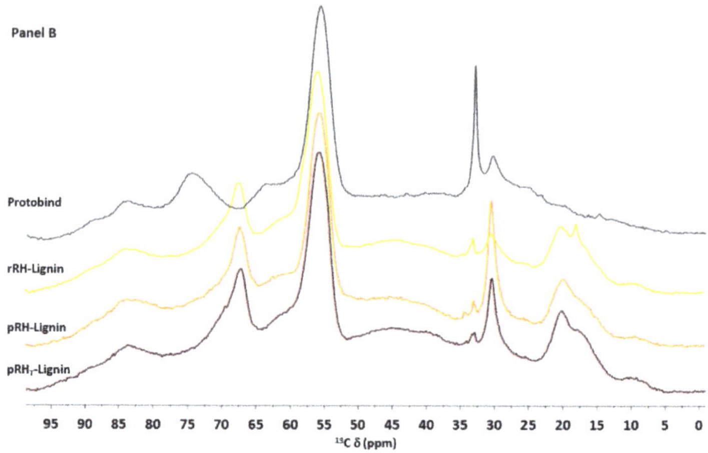

Figure 5. Solid state ^{13}C CP-MAS NMR spectra of Protobind lignin (in grey), rRH-Lignin (in yellow), pRH-Lignin (in orange), pRH$_T$-Lignin (in brown). (**Panel A**) the whole spectrum; (**Panel B**) zoom of 0–100 ppm region.

2.4.5. Total Phenolic Content

The total phenolic content was determined in the recovered lignins from RHs using a modified Folin–Ciocalteu (FC) assay (see Section 3.6.5 for details). In this protocol, the samples were initially fully solubilized in DMSO before being incubated with the specific redox reagent (FC reagent). The phenolic content results are summarized in Table 6 as the vanillin equivalents (mmol/g of dry lignin), using the data from Protobind as the reference.

Table 6. Results of the determination of phenolic hydroxyl groups expressed as vanillin equivalents/g of lignin sample. Estimated standard errors ± 0.1 mmol/g vanillin equivalent (1σ, from calibration data).

Sample	Vanillin Equivalent Content (mmol/g)
rRH-Lignin	1.6
rRH$_T$-Lignin	1.3
pRH-Lignin	1.6
pRH$_T$-Lignin	1.7
Protobind 1000	3.1

All four lignins from RHs show quite the same content of vanillin equivalents (around 1.3–1.6 mmol/g), which is half the value found in Protobind. In addition, also in this case it appears that the parboiled treatment does not affect the majority of the functionalities of the phenolic lignin.

2.4.6. Hydroxyl Groups Quantification by ^{31}P NMR

The different hydroxyl groups present in the recovered lignins were assessed and identified through high resolution ^{31}P NMR spectroscopy. This technique allows for the quantification of the different hydroxyl groups present in the lignin backbone after derivatization with 2-chloro-4,4,5,5-tetramethyl-1,3,2-dioxaphospholane [46,47]. The high natural abundance of the ^{31}P nucleus allows us to obtain well-resolved NMR signals. Their attribution to the different OH groups in the sample could be performed because the chemical shifts are largely dependent on the surrounding chemical environment of the derivatized hydroxyls. The spectra are reported in Figure 6.

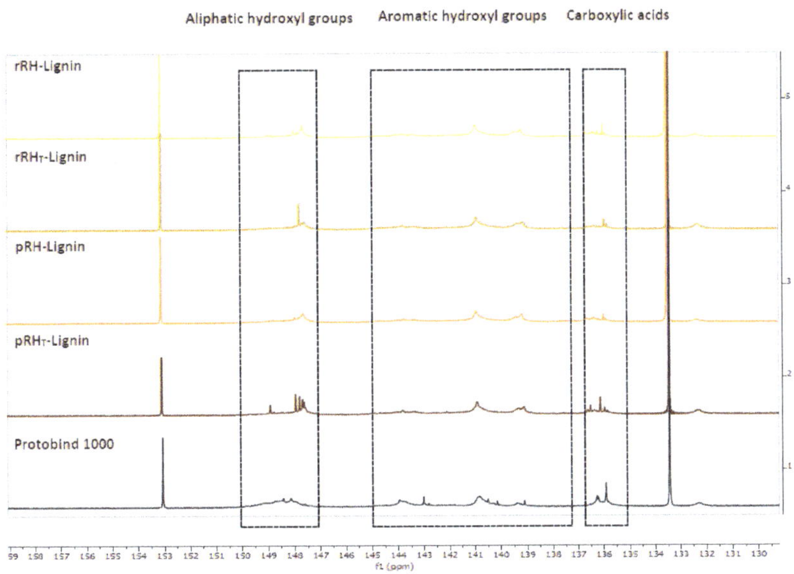

Figure 6. ^{31}P NMR Spectra of isolated Lignins and Protobind as reference.

The signals of the RH-Lignins and Protobind spectra were integrated to differentiate the hydroxyl groups: the signals from 150 to 147 ppm were associated with the aliphatic hydroxyl groups, the signals from 145 to 138 ppm were related to the aromatic hydroxyl groups, and, lastly, the signals centered at 136 ppm were attributed to the carboxylic acid residues. The peak integration of these three chemical shift regions led to the quantification of the total hydroxyl groups expressed in mmol of the functional group per g of dry lignin, as reported in Table 7 below.

Table 7. Detailed hydroxyl/carboxyl quantification by ^{31}P NMR (as mmol of functional group per g of dry lignin).

Sample	–OH Aliphatic (mmol/g)	–OH Aromatic (mmol/g)	–COOH (mmol/g)
rRH-Lignin	2.42	7.12	1.43
rRH$_T$-Lignin	2.48	7.10	1.02
pRH-Lignin	3.01	7.77	1.42
pRH$_T$-Lignin	3.08	7.49	1.36
Protobind 1000	3.00	4.67	1.42

Additionally, in this case, the ^{31}P NMR results showed that the three RH-Lignins had quite the same amount of hydroxyl and carboxyl groups in their composition. In particular, when compared with Protobind, the lignins extracted from RHs appeared to have quite the same quantity of aliphatic OH and carboxylic groups, whereas the main differences were showed in the aromatic hydroxyl groups, which appeared to be one and a half times higher than in Protobind.

2.4.7. Fourier-Transform Infrared Spectroscopy

Fourier-transform infrared (FTIR) spectroscopy was carried out to further study the chemical composition of the RH-Lignins. The absorption spectra obtained are shown in Figure 7, where the FTIR spectrum of Protobind is also presented for comparison purposes. In general, no noticeable differences among the recovered lignins could be observed. All the lignin fractions showed a broad absorption band in the region of 3400–3200 cm^{-1}, related to the stretching vibrations of the aliphatic and phenolic OH groups, and the signals in the region of 3050–2800 cm^{-1} were associated with the CH bond stretching in the methyl and methylene groups. The major differences between the RH-Lignins and Protobind lignin were found in the fingerprint region. In particular, while Protobind showed a single signal located at 1710 cm^{-1} attributable to the stretching vibration of C=O in unconjugated ketones, carbonyl, and ester groups, all four RH-Lignins were characterized by an additional peak at 1740 cm^{-1}. This signal was associated with C=O stretching vibrations likely resulting from the presence of different carbohydrate species entrapped in the recovered RH-Lignin fractions, as also evident from Table 5. This result was also confirmed by the signal detectable at 1650 cm^{-1} for such materials, related to conjugated carbonyl/carboxyl stretching in C=O moieties [48].

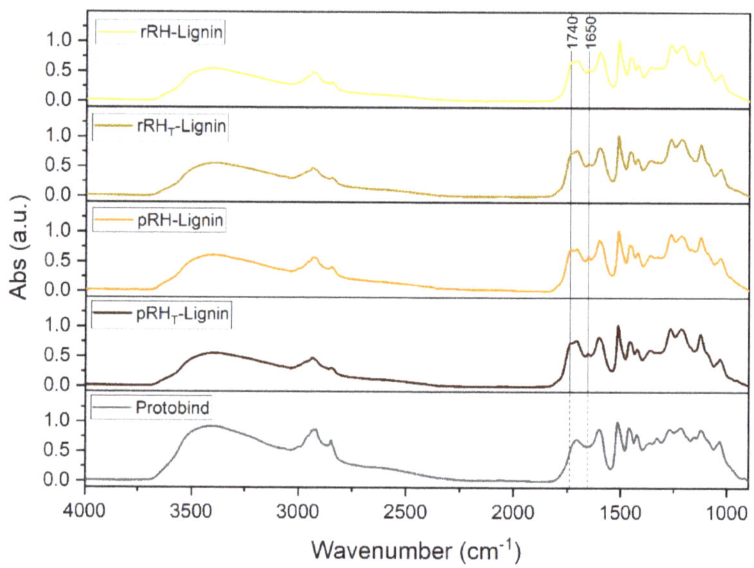

Figure 7. FT-IR spectra of rRH-Lignin (in pale yellow), rRH$_T$-Lignin (in dark yellow), pRH-Lignin (in orange), pRH$_T$-Lignin (in brown), and Protobind (in grey).

2.4.8. Thermal Behavior

The glass transition temperatures (Tg) of all the lignin samples were assessed by means of differential scanning calorimetry (DSC). The thermograms reported in Figure 8 showed that the Tg of the recovered RH-Lignins were slightly higher compared to that of the reference lignin material (Protobind). In particular, rRH- and pRH-Lignins were characterized by the same Tg, close to 160 °C. This value was higher compared to the Tg of the rRH$_T$- and pRH$_T$-Lignin, which instead were detected at around 157 and 149 °C, respectively, meaning that the treatment has slightly affected the molecular structure and thus the macromolecular mobility of such a lignin fraction. The behavior is perfectly in accordance with the results of both GPC analyses and ^{31}P NMR. Especially, the GPC analyses showed a higher Mw and Mn of the RH-Lignins. Along the same lines, ^{31}P NMR allowed us to detect a higher abundance of OH aromatic functionalities, which can induce the higher intra/intermolecular hydrogen bonding capability, ultimately resulting in a reduced molecular motion (viz., reduced free volume and higher Tg).

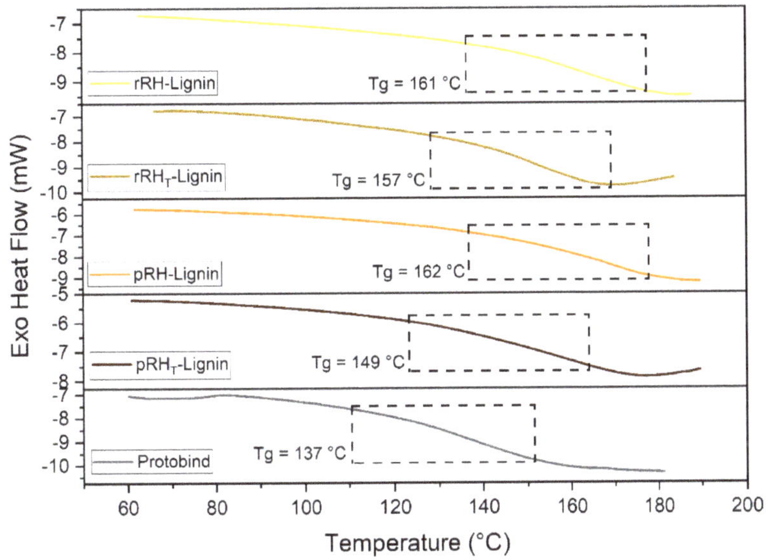

Figure 8. DSC scans (second heating ramp) of rRH-Lignin (in pale yellow), rRH$_T$-Lignin (in dark yellow), pRH-Lignin (in orange), pRH$_T$-Lignin (in brown), and Protobind (in grey).

To study the thermal stability behavior of the lignin samples, thermogravimetric analysis (TGA) was performed as well. The mass loss traces as a function of the temperature of the RH-Lignins and of Protobind are reported in Figure 9, where the mass derivative (DTG) is also shown.

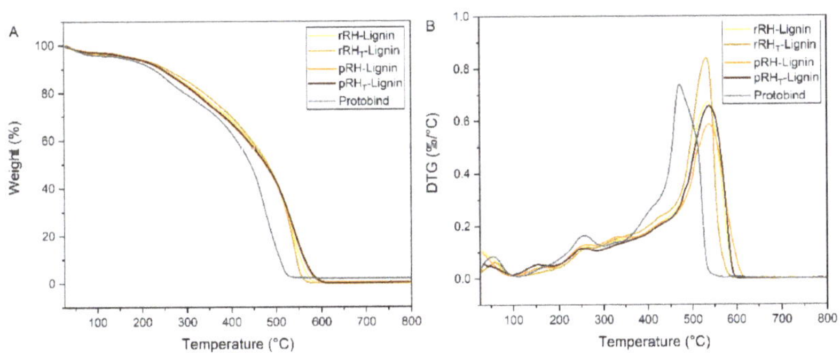

Figure 9. (**A**) TGA and (**B**) DTG traces of rRH-Lignin (in pale yellow), rRH$_T$-Lignin (in dark yellow), pRH-Lignin (in orange), pRH$_T$-Lignin (in brown), and Protobind (in grey). Analyses were conducted in air.

As it can be observed, all RH-Lignin fractions exhibited a slightly improved thermal response as compared with the reference material (Protobind) over the entire temperature range which was investigated. This can be associated with the slightly higher molecular weights (and characteristic thermal transitions) found in RH-Lignins, which result in a higher stability towards thermo-oxidative degradation. In particular, a relatively monotonic mass loss response is observed, where a single broad degradation event can be detected. This behavior is reflected in the temperatures at which 10% and 50% mass losses are registered, which appear in line with Protobind (Table 8).

Table 8. Characteristic mass loss temperatures (10% mass loss, 50% mass loss, and maximum degradation rate) for the analyzed lignins. The reference Protobind system is also reported.

Sample	$T_{10\%}$ (°C)	$T_{50\%}$ (°C)	T_{DTGmax} (°C)
rRH-Lignin	247	480	532
rRH_T-Lignin	256	482	531
pRH-Lignin	245	477	538
pRH_T-Lignin	243	486	538
Protobind 1000	227	444	472

2.5. Water Reduction in Cement Pastes

Natural lignins are a very interesting resource for the development of sustainable traditional cement water reducers. Moreover, high-performance lignin-based cement water reducers can also be developed through different types of lignin derivatization [49–52]. Typically, lignosulfonates show better water-reducing capabilities than natural lignins, but even the latter have been successfully adopted as a starting point in the development of high-performance experimental cement plasticizers. This was obtained both in the presence [53] and in the absence of sulfonation steps [49]. The DES-mediated fractionation of the lignocellulosic biomass is a mild process that results in the production of non-derivatized lignins comparable to those obtained by organosolv or soda processes. The lignins (rRH- and pRH-Lignins) obtained in the present study were thus evaluated using a commercial soda lignin (Protobind) as a reference. The measured water reduction capabilities, made by cement pastes (a blend of water and cement powder), are reported in Figure 10.

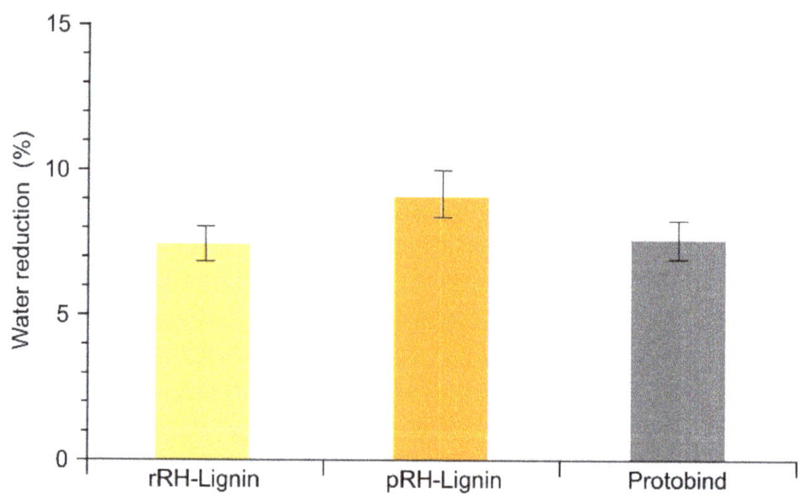

Figure 10. Measured water reduction capability for cement pastes containing 0.2% wt. of lignin. Error bars are the confidence interval (1σ) of the inverse linear regression.

The results indicated a comparable water reduction activity for the lignins derived from the rRH and pRH samples. Both lignins demonstrated an equivalent performance in comparison to the commercial soda lignin Protobind. The reported data are in a good agreement with a previous study based on BSG as a lignocellulosic starting material [15]. Indeed, the obtained results indicate that the lignin derived from the proposed process is comparable to a typical industrial soda lignin and that it constitutes a suitable basis for the development of sustainable, high performance concrete water reducers. Moreover, these results indicate that the lignin derived from the pRH is comparable to that of rRH.

3. Materials and Methods

3.1. Materials and General Methods

The rice husk samples (from *Oriza sativa*) were kindly provided by Riso Scotti S.p.A. (Pavia, Italy) as raw rice husk derived from the processing of *japonica* (90%) and *indica* (10%) rice varieties and parboiled rice husk derived from *indica* (75%) and *japonica* (25%) varieties. Protobind 1000 (a mixed wheat straw/sarkanda grass lignin from the soda pulping of non-woody biomass) was provided by Tanovis (Alpnach, Switzerland).

All air- and moisture-sensitive reactions were carried out using dry solvents and under a static atmosphere of nitrogen. Choline chloride (C0329), betaine glycine (B0455) and L-lactic acid (L0165), D-(+)-Galactose, and L-(−)-fucose were purchased from TCI (Milano, Italy), whereas (L)-(+)-rhamnose, L-(+)-arabinose, D-(+)-xylose, D-(+)-mannose, D-(+)-glucose, and the other reagents and the employed solvents, used without a further purification, were obtained from Merck (Merck Life Science S.R.L., Milan, Italy).

Thin layer chromatography (TLC) Merck silica gel 60 F_{254} plates (Merck Millipore, Milan, Italy) were used for analytical TLC.

The identification and the quantification of the low molecular weight aromatic compounds obtained in the pre-treatment step were carried out using gas chromatography coupled with mass spectrometry (GC-MS). The GC-MS apparatus used is an Agilent GC System 7890A, with an inert MSD with a Triple-Axis Detector 7975C. The gas carrier was helium at a flux of 1.18 mL/min. The separation was performed on a DB-5MS column (30 m × 250 μm × 0.25 μm, Phenomenex, Bologna, Italy) with a temperature program of 50 °C (1 min) to 280 °C at 10 °C/min, 280 °C at 15 min (total run time 39 min). A solvent delay of 4 min was selected. The samples were dissolved in methanol or acetone in a concentration of around 0.5–1 mg/mL.

3.2. Microorganisms and Growth Media

Myceliophthora thermophila (CBS 866.85) and *Rhizomucor pusillus* (CBS 354.68) were purchased from the CBS-KNAW collection (Utrecht, The Netherlands). *Trichoderma viride* (DSM 63065) was purchased from the DSMZ GmbH collection (Braunschweig, Germany).

The microbial inoculum for the RH pretreatment consisted of the suspension of the suitable fungal spores in distilled water. The latter suspensions were prepared adding sterile water to the suitable fungal culture, grown on an agar slant, followed by the scrubbing of the sporulated surface.

Trace elements solution: $FeCl_3$ (50 mM), $CaCl_2$ (20 mM), $MnCl_2$ (10 mM), $ZnSO_4$ (10 mM), $CoCl_2$ (2 mM), $CuCl_2$ (2 mM), $NiCl_2$ (2 mM), Na_2MoO_4 (2 mM), Na_2SeO_3 (2 mM), and H_3BO_3 (2 mM). All inorganic salts were purchased from Carlo Erba (Milano, Italy).

3.3. Determination of Rice Husks Composition

The rRH and pRH compositions have been determined with a known multistep procedure with minor modifications [33]. The RH (1 g, value a) was suspended and stirred in deionized water (150 mL) at 100 °C for 1 h. Then, after filtration, the solid was washed by deionized water (300 mL), dried in an oven at 80 °C, and weighted (value b). The solid was treated with 1 N of H_2SO_4 (150 mL) at 100 °C for 1 h. Then, the suspension was filtered again, washed with water, and the solid was dried and weighted (value c). The solid was mixed with 72% H_2SO_4 (10 mL) for 4 h at r.t. and treated with 1 N of H_2SO_4 (150 mL) under a reflux for 1 h. After cooling, the solid was isolated by filtration, dried, and weighted (value d). The final residue was then calcinated in an oven at 600 °C for 6 h and the residue has been quantified (value e). The fractions of the different components were quantified with the following equations:

$$\text{hemicellulose (\%)} = 100 \times [(b - c)/a];$$

$$\text{cellulose (\%)} = 100 \times [(c - d)/a];$$

$$\text{lignin (\%)} = 100 \times [(d - e)/a].$$

3.3.1. Determination and Quantification of Hemicellulose Components

The procedure was performed according to Foster et al., with minor modifications [36]. The monosaccharides fraction obtained from the hemicellulose hydrolysis (obtained in Section 3.3) was dissolved in deionized water (30 mL) and was treated with $NaBH_4$ (1 g, 26.4 mmol) stirring at r.t. for 2 h. Then, the reaction was quenched by the careful addition of glacial acetic acid (10 mL), keeping the temperature under 30 °C by external cooling. The solvent (water) and the excess of acetic acid were removed by evaporation under a reduced pressure and the obtained powder was treated with pyridine (30 mL) and acetic anhydride (30 mL) stirring at a reflux for 1 h. Hence, the reaction was concentrated to the dryness under a reduced pressure and the residue was partitioned between ethyl acetate (70 mL) and water (100 mL). The aqueous phase was extracted again with ethyl acetate (50 mL) and the combined organic phases were washed in turn with saturated $NaHCO_3$ aq. (100 mL) and with brine (100 mL). The resulting solution was dried (Na_2SO_4) and concentrated in vacuo. The residue contained the alditol acetates of the hemicellulose monosaccharides, whose relative composition was determined by GC-MS analysis. GC-MS analyses for the determination of sugar composition in hemicellulose were performed on an HP-6890 gas chromatograph equipped with a 5973 mass detector and using an HP-5MS column (30 m × 0.25 mm, 0.25 μm film thickness; Hewlett Packard, Palo Alto, CA, USA). The temperature program was the following: 120 °C (3 min)—12 °C/min—195 °C (10 min)—12 °C/min—300 °C (10 min); carrier gas: He; constant flow 1 mL/min; and split ratio: 1/30. The reference standards of the alditole acetates were prepared starting from the corresponding monosaccharides, following the reduction/acetylation protocol described above. The retention times of the alditole acetates are given below for each monosaccharide derivative: rhamnose 14.03 min; fucose 14.30 min; arabinose 14.46 min; xylose 14.88 min; mannose 21.55 min; glucose 21.65 min; and galactose 21.96 min.

3.4. *Rice Husks Preprocess*

3.4.1. Hydro-Thermal Preprocess

A sample of rice husk (30 g) was suspended in distilled water (1 L) and was heated in an autoclave at 121 °C for 20 min. After cooling, the solid residue was removed by filtration and the aqueous phase was concentrated under a reduced pressure until it reached complete dryness. The crude extracts of rRH and pRH were 2.1% and 2.8% *w*/*w* of the starting biomass, respectively. They were further fractionated by extraction with ethyl acetate and filtrated on a short silica gel column. The so-obtained solvent soluble fractions were derivatized as previously described and then analyzed by GC-MS [54]. Briefly, a mixture of 25 μL of pyridine, 250 μL of dioxane, and 75 μL of silylation mixture composed of N,O-bis(trimethylsilyl)trifluoroacetamide (BSTFA, Sigma T6381) with trimethylchlorosilane (TMC, Sigma T6381) was incubated with 1 mg of the sample heated in a thermomixer (1.5 mL vial Eppendorf Thermomixer Comfort) at 70 °C and 600 rpm for 30 min. At the end, 100 μL of the mixture were withdrawn, added to 100 μL of dioxane, and analysed by GC-MS with the same apparatus and analytical method reported in Section 3.1. The identification of the compounds was performed by means of an NIST 2008 mass spectral library search and then the selected peaks were confirmed with the known standards (comparing both the mass spectrum and chromatographic coordinate).

3.4.2. Microbiological Preprocess

A sample of ground pRH (25 g) was added to a 1 L flask which contained distilled water (250 mL), yeast extract (0.3 g), NH_4Cl (0.8 g), and the microelement solution (4 mL). The flask was sealed with a cellulose plug and was sterilized at 121 °C for 15 min. Then, the obtained mixture was inoculated with the selected strain and the microbial growth was performed according to the experimental conditions indicated below.

Myceliophthora thermophila (CBS 866.85): 45 °C, 6 days, 130 rpm
Rhizomucor pusillus (CBS 354.68): 35 °C, 8 days, 130 rpm
Trichoderma viride (DSM 63065): 24 °C, 8 days, 130 rpm

Hence, the microbial transformation was interrupted by filtration on a narrow mesh metallic filter. The recovered RH was washed with deionized water and was dried (ventilated oven, 60 °C, 24 h).

According to this procedure, the difference between the weight of the recovered pRH and the weight of the untreated sample takes into account of the sum of the husk soluble fractions and of the husk organic fraction depleted by the fungal strain during its growth process. Overall, the fraction between this weight difference over the corresponding RH initial weight (given as *w*/*w* percentages) indicates the degradative capability of the strain. The obtained results are listed below:

Myceliophthora thermophila: 18%; *Rhizomucor pusillus*: 19%; and *Trichoderma viride*: 14%.

3.5. Rice Husks Lignocellulose Process

3.5.1. Preparation of DESs

The DESs were prepared in a closed flask by mixing the anhydrous hydrogen bond acceptor (HBA) with the hydrogen bond donor (HBD) in the determined molar ratio (see Section 2.3) and was stirred at 120 °C for 4 h until the liquid phase appeared to be completely homogeneous and clear. The products were then dried under a vacuum and stored at room temperature in a desiccator in the presence of anhydrous calcium chloride until further use. The ^{1}H NMR spectrum of the selected DES 2 (choline chloride/L-lactic acid 1/5) for the full fractionation is reported in the Supplementary Materials (Figure S1).

3.5.2. DES-Mediated Lignocellulose Process

The rice husk (25 g) was suspended in a DES of choline chloride/L-lactic acid 1/5 (250 mL) at 130 °C in a round-bottom flask under magnetic stirring for 24 h. After cooling, ethanol (500 mL) was then added gradually over 2 h in order to precipitate the cellulose-enriched fraction. The solid particulate was separated by centrifugation and filtration, washed many times with ethanol, and dried to give a final solid RH-Cellulose-enriched fraction (~15 g) with a yield of 59–61% (*w*/*w* initial biomass). This fraction was further fractionated in order to quantify the silica content (see Section 3.5.3). The silica content accounted for 12.8% (*w*/*w* biomass).

The filtrate after the ethanol treatment was then concentrated by the rotary evaporation under the vacuum to eliminate the solvent. Water (500 mL) was added, and the suspension was stirred for 24 h at 4 °C. The obtained precipitate was then centrifuged, filtered, and washed three times for 1 h with a solution of water/ethanol 9/1. After the centrifugation, filtration, and solvent evaporation, the final fraction (RH-Lignin, ~3 g) was recovered with a final yield of 10% (*w*/*w* initial biomass).

3.5.3. Determination of Silica Content in Cellulose-Enriched Fractions

The cellulose-enriched fractions, derived from DES-mediated fractionation, were analyzed in order to determine their silica contents, using the Yoshida procedure based on the complete oxidation of the organic materials by heating with a mixture of strong acids with some modifications. We employed an acid mixture with the following composition: HNO_3 aq. (65% *w*/*v*) 75 mL, H_2SO_4 aq. (96% *w*/*v*) 15 mL, and $HClO_4$ aq. (70% *w*/*v*) 30 mL.

Each sample of the cellulose-enriched fractions (about 2 g, dry weight) was placed in a round bottomed flask containing a stirring bar and equipped with a condenser. The fractions were treated with 20 mL of the acid mixture and were allowed to predigest under a fume hood for 3 h. Then, the obtained brown slurries were heated at reflux under stirring for six h. The resulting colorless silica suspensions were ice-cooled, diluted with 60 mL of distilled water, and centrifuged (9000 rpm, 4 °C). The collected silica samples were washed with distilled water (2 × 25 mL), with acetone (25 mL), and were then dried at a reduced pressure. The weights of the obtained silica samples were employed to calculate the silica content of the cellulose-enriched fractions and thus of the starting rice husk samples (both given as a *w*/*w* percentage).

3.6. Lignin Characterization

3.6.1. Solvent Solubilization Determination

The lignin solvent's solubility was determined by treating 1 g of the analyzed lignin with 10 mL of the different solvents under stirring at 400 rpm. Each test was carried out overnight at room temperature. The suspension was then filtered and the solvent was evaporated at a reduced pressure, and the final residue was dried until a constant weight was achieved prior to the quantification.

3.6.2. Molar Mass Determination

A Waters 510 HPLC system equipped with a refractive index detector was used for the GPC analyses. Tetrahydrofuran (THF) was used as the eluent. The analyzed lignin sample (volume 200 μL, concentration 1 mg/mL in THF) was injected into a system of the three columns connected in series (Ultrastyragel HR, Waters − dimensions 7.8 mm × 300 mm) and the analysis was performed at 30 °C at a flow rate of 0.5 mL/min. The calibration was performed against polystyrene standards in the 10^2-10^4 g/mol molecular weight range. The samples have been acetylated to allow for a complete solubility in the THF eluent. The estimation of the number-average and weight-average molecular weights of the obtained lignin fractions was performed excluding the signals related to the solvent (THF) and the solvent stabilizer (butylated hydroxytoluene), visible at long elution times (>29.5 min).

3.6.3. Sugar Content Determination

The total sugars quantification was performed with a previously described method based on a bicinchoninic assay [15,44].

3.6.4. ^{13}C CP-MAS NMR Analysis

Solid state NMR experiments were carried out on a Bruker NEO 500 MHz spectrometer (Bruker, Billerica, MA, USA) equipped with an i-probe solid state probe. The samples (150 mg) were packed into 4 mm zirconia rotors and sealed with Kel-F caps. All the NMR spectra were recorded at 298 K. The ^{13}C CP-MAS NMR spectra were performed at 125 MHz. The spin-rate was kept at 8 KHz. The 90-pulse width was 3.2 μs, the relaxation delay was 4 s, the acquisition number was 17,000, and the contact time was 1.5 ms. The spectra were obtained using 1024 data points in the time domain, zero-filled, and Fourier transformed. All the data were processed using the MestreNova 6.0.2 software (Mestrelab Research, Santiago de Compostela, Spain). An adamantane standard was used as the external referencing standard to calibrate the ^{13}C chemical shifts.

3.6.5. Folin–Ciocalteu Analysis

The total phenolic content of the lignins was determined by a modified Folin–Ciocalteu (FC) protocol with some modifications to the sample preparation step, as previously described [54,55]. Briefly, the samples were dissolved in DMSO with a final concentration of 2 mg/mL. For each determination, 5 μL of the working solution (or the standard solution) were then mixed with 120 μL of deionized water, 125 μL of FC reagent (Sigma 47641), and kept for 6 min at room temperature after 30 s of vortex stirring. Then, after the addition of 1.25 mL of 5% sodium carbonate and mixing, the vial was incubated in a thermoshaker at 40 °C for 30 min. The reaction mixture absorbance was measured using a UV–Vis spectrophotometer (Jasco V-560) equipped with a temperature-controlled cuvette holder and a thermostatic water bath (Haake K10, Karlsruhe, Germany). All the spectrophotometric measurements were carried out at 760 nm, 25 °C, using a 1 cm optical path cuvette and deionized water as the blank sample. Vanillin was chosen as the reference standard. The calibration curve was constructed with nine different vanillin solutions in DMSO with a concentration in the range 0–500 μg/mL (see Supplementary Materials Figure S2). Each FC assay determination was carried out in triplicate.

3.6.6. ^{31}P NMR Analysis

^{31}P NMR spectroscopic analyses were recorded on a Bruker Instrument AVANCE400 spectrometer (Milano, Italy). The acquisition and data treatment were performed with Bruker TopSpin 3.2 software (Milano, Italy). The spectra were collected at 29 °C with a 4 s acquisition time, 5 s relaxation delay, and 256 scans. Prior to the analysis, the samples were dried for 24 h under a vacuum and then derivatized according to the following procedure.

The sample (40 mg) was completely dissolved in 300 µL of *N,N*-dimethylformamide. To this solution, the following components were added: 200 µL of dry pyridine, 100 µL of solution of an internal standard (10 mg of Endo-*N*-hydroxy-5-norbornene-2,3-dicarboximide (Sigma 226378) dissolved in 0.5 mL of a mixture of pyridine and $CDCl_3$ 1.6:1 v/v), 50 µL of a relaxation agent solution (5.7 mg of chromium (III) acetylacetonate (Sigma 574082) dissolved in 0.5 mL of a mixture of pyridine and $CDCl_3$ 1.6:1 v/v), 100 µL of 2-chloro-4,4,5,5-tetramethyl-1,3,2-dioxaphospholane (Sigma 447536), and, at the end, 200 µL of $CDCl_3$. The solution was centrifuged and/or filtered if necessary. All the chemical shifts reported were related to the reaction product of the phosphorylating agent with water, which gave a signal at 132.2 ppm.

3.6.7. Fourier-Transform Infrared Spectroscopy Analysis

The FTIR spectra were collected with a Nicolet Nexus 760 FTIR spectrophotometer. The samples were prepared by pressing the lignin powders with KBr powder to obtain thin discs. The spectra were recorded at room temperature, in air, in transmission mode (64 accumulated scans at a resolution of 4 cm^{-1}) in the 4000–1000 cm^{-1} wavenumber range.

3.6.8. Differential Scanning Calorimetry Analysis

DSC analysis was employed to investigate the thermal transitions in the lignin samples. Measurements were carried out on 10–15 mg samples by means of a Mettler-Toledo DSC 823e instrument. Three runs (heating/cooling/heating) were performed: from 25 °C to 150 °C to remove the water from the samples, from 150 °C to 25 °C, and from 25 °C to 200 °C, at a scan rate of 20 °C/min under a nitrogen flux. The glass transition temperature (Tg) of the samples was evaluated as the inflection point in the second heating run.

3.6.9. Thermogravimetric Analysis

TGA was performed on all the lignin samples (~15 mg) by means of a Q500 TGA system (TA Instruments, Milan, Italy) from an ambient temperature to 800 °C, at a scan rate of 10 °C/min in air.

3.7. Determination of Water Reduction in Cement Pastes

The complete description of the laboratory method for the preliminary estimation of the water reduction capabilities of the fractionated lignins was reported in detail in a previous work [15]. Briefly, the analysis is based on the evaluation of the relative Bingham dynamic yield torque of cement pastes containing 0.2% of lignin (w/w relative to dry cement) at a different water/cement (w/c) ratio [56].

By measuring the dynamic yield torque of the cement pastes at various w/c ratios, it is possible to calculate by inverse regression the w/c ratio needed to obtain a predetermined reference torque value and thus to determine the water reduction capacity for each lignin under study. In this study, the reference Bingham dynamic yield torque was obtained with a pure cement paste with a 0.45 w/c ratio. The actual water reduction obtained for each lignin is given by the following equation

$$W_R = 100 \cdot (1 - \frac{W_L}{W_{ref}})$$

where W_R is the water reduction (%), W_L is the measured w/c ratio at the reference yield torque for the given lignin, and W_{ref} is the reference w/c ratio (0.45). This procedure allows us to measure the effective water reduction capability of each lignin in the preparation of cement pastes in the given conditions. In the present work, the water reduction capability was measured for rRH-Lignin, for pRH-Lignin and, as a reference, for a well-known commercial soda lignin (Protobind).

4. Conclusions

Among the agri-food cultivations, rice has a huge social and economic importance. For this reason, the exploitation of waste residues coming from its production possesses an important interest for the scientific community and for the society.

In this work, we demonstrated a multistep lignocellulosic fractionation process of raw and parboiled rice husks by means of DESs based on non-toxic, renewable components. The fractionation allowed us to recover a cellulose/silica fraction, precipitated by the addition of ethanol, and a lignin-based fraction, precipitated by the addition of water.

The lignin fractions obtained from both raw and parboiled rice husks were fully characterized and compared for the first time. The results show similar physico-chemical properties between the two recovered lignins, indicating that the parboiling process does not alter significantly the lignin characteristics. Moreover, because of the mild separation conditions allowed by the DES-mediated process, the resulting lignins were underivatized and demonstrated a limited degradation.

Both lignins were studied as a potential base for the production of cement water reducers, the latter being an important application field for lignin derivatives. The obtained results confirmed the substantial equivalence of the two lignins for the target application and indicated that both are comparable to a commercial soda lignin taken as a reference benchmark.

A preprocess step based on fermentation was also studied. However, the obtained results demonstrated the limited degradability of both raw and parboiled RHs, even using cellulose-degrading fungi. A possible cause for such behavior was found in the relatively high silica content in both RHs (about 13%).

These preliminary data show, at a 25 g lab-scale, the potentialities of DES-mediated fractionation processes in the lignocellulosic fractionation of RHs. This protocol is a promising base for a future implementation on a larger scale. Considering the very large volume of the worldwide rice production and the environmental friendliness of DES-based processes, the reported results constitute an encouraging step towards the exploitation of RH wastes in line with the strategic objectives of the circular economy.

Supplementary Materials: The following supporting information can be downloaded at: https://www.mdpi.com/article/10.3390/molecules27248879/s1, Figure S1: ^{1}H NMR spectrum of DES 2 choline chloride/L-lactic acid (1:5 mol/mol); Figure S2: Calibration curve of Folin–Ciocalteu phenol titration with vanillin.; Table S1: Yields of mass recovery in the different tested DESs.

Author Contributions: Conceptualization, P.D., S.S. and A.S.; methodology, P.D., A.S., L.S. and S.S.; investigation, C.A., E.B., M.F., G.G., L.A.M.R., E.R., L.S., S.S. and A.S.; resources, P.D., G.G., S.S., A.S. and S.T.; writing—original draft preparation, P.D., G.G., S.S. and A.S.; writing—review and editing, C.A., P.D., M.F., G.G., L.A.M.R., S.S., A.S., E.R. and S.T.; funding acquisition, P.D., G.G., S.S. and S.T. All authors have read and agreed to the published version of the manuscript.

Funding: This research project has been partially funded by Regione Lombardia and Fondazione Cariplo (grant number 2018-1739, project: POLISTE) and by the European Union's Horizon 2020 Research and Innovation Programme (grant Agreement no. 952941, project: BIOMAC). L.R. is a Ph.D. student of the Research Doctorate Program in Chemical Engineering and Industrial Chemistry at Politecnico di Milano.

Institutional Review Board Statement: Not applicable.

Informed Consent Statement: Not applicable.

Data Availability Statement: Data are contained within the article and the Supplementary Materials.

Acknowledgments: The authors gratefully acknowledge Francesco Gatti (Politecnico of Milano) for the scientific discussions and his valuable help. The authors also gratefully thank Riso Scotti S.p.A. (Pavia, Italy) for providing the rice husk samples.

Conflicts of Interest: The authors declare no conflict of interest.

References

1. FAO. *World Food and Agriculture—Statistical Yearbook 2021*; FAO: Rome, Italy, 2021. [CrossRef]
2. FAO. FAOSTAT Database. Available online: www.fao.org/faostat (accessed on 31 May 2022).
3. Buggenhout, J.; Brijs, K.; Celus, I.; Delcour, J.A. The breakage susceptibility of raw and parboiled rice: A review. *J. Food Eng.* **2013**, *117*, 304–315. [CrossRef]
4. Tan, B.L.; Norhaizan, M.E. *Rice By-Products: Phytochemicals and Food Products Application*; Springer Nature: Cham, Switzerland, 2020. [CrossRef]
5. Balbinoti, T.C.V.; Nicolin, D.J.; Jorge, L.M.D.; Jorge, R.M.M. Parboiled Rice and Parboiling Process. *Food Eng. Rev.* **2018**, *10*, 165–185. [CrossRef]
6. Shimizu, N.; Kimura, T. Quality Evaluation of Parboiled Rice with Physical Properties. *Food Sci. Technol. Res.* **2001**, *7*, 57–63. [CrossRef]
7. Liu, J.; Li, X.; Row, K.H. Development of deep eutectic solvents for sustainable chemistry. *J. Mol. Liq.* **2022**, *362*, 119654. [CrossRef]
8. Isci, A.; Kaltschmitt, M. Recovery and recycling of deep eutectic solvents in biomass conversions: A review. *Biomass Convers. Biorefinery* **2022**, *12*, 197–226. [CrossRef]
9. Gong, L.; Zha, J.; Pan, L.; Ma, C.; He, Y.C. Highly efficient conversion of sunflower stalk-hydrolysate to furfural by sunflower stalk residue-derived carbonaceous solid acid in deep eutectic solvent/organic solvent system. *Bioresour. Technol.* **2022**, *351*, 126945. [CrossRef]
10. Wu, M.; Gong, L.; Ma, C.; He, Y.-C. Enhanced enzymatic saccharification of sorghum straw by effective delignification via combined pretreatment with alkali extraction and deep eutectic solvent soaking. *Bioresour. Technol.* **2021**, *340*, 125695. [CrossRef]
11. Lin, G.; Tang, Q.; Huang, H.; Yu, J.; Li, Z.; Ding, B. Process optimization and comprehensive utilization of recyclable deep eutectic solvent for the production of ramie cellulose fibers. *Cellulose* **2022**, *29*, 3689–3701. [CrossRef]
12. Zhu, Y.; Yu, Z.; Zhu, J.; Zhang, Y.; Ren, X.; Jiang, F. Developing flame-retardant lignocellulosic nanofibrils through reactive deep eutectic solvent treatment for thermal insulation. *Chem. Eng. J.* **2022**, *445*, 136748. [CrossRef]
13. Allegretti, C.; Gatti, F.G.; Marzorati, S.; Rossato, L.A.M.; Serra, S.; Strini, A.; D'Arrigo, P. Reactive Deep Eutectic Solvents (RDESs): A New Tool for Phospholipase D-Catalyzed Preparation of Phospholipids. *Catalysts* **2021**, *11*, 655. [CrossRef]
14. Zhou, M.; Fakayode, O.A.; Yagoub, A.A.; Ji, Q.H.; Zhou, C.S. Lignin fractionation from lignocellulosic biomass using deep eutectic solvents and its valorization. *Renew. Sustain. Energy Rev.* **2022**, *156*, 111986. [CrossRef]
15. Allegretti, C.; Bellinetto, E.; D'Arrigo, P.; Griffini, G.; Marzorati, S.; Rossato, L.A.M.; Ruffini, E.; Schiavi, L.; Serra, S.; Strini, A.; et al. Towards a Complete Exploitation of Brewers' Spent Grain from a Circular Economy Perspective. *Fermentation* **2022**, *8*, 151. [CrossRef]
16. de Baynast, H.; Tribot, A.; Niez, B.; Audonnet, F.; Badel, E.; Cesar, G.; Dussap, C.-G.; Gastaldi, E.; Massacrier, L.; Michaud, P.; et al. Effects of Kraft lignin and corn cob agro-residue on the properties of injected-moulded biocomposites. *Ind. Crops Prod.* **2022**, *177*, 114421. [CrossRef]
17. Fiorani, G.; Crestini, C.; Selva, M.; Perosa, A. Advancements and Complexities in the Conversion of Lignocellulose Into Chemicals and Materials. *Front. Chem.* **2020**, *8*, 797. [CrossRef]
18. Tanase-Opedal, M.; Espinosa, E.; Rodríguez, A.; Chinga-Carrasco, G. Lignin: A Biopolymer from Forestry Biomass for Biocomposites and 3D Printing. *Materials* **2019**, *12*, 3006. [CrossRef]
19. Carlos de Haro, J.; Magagnin, L.; Turri, S.; Griffini, G. Lignin-Based Anticorrosion Coatings for the Protection of Aluminum Surfaces. *ACS Sustain. Chem. Eng.* **2019**, *7*, 6213–6222. [CrossRef]
20. de Haro, J.C.; Allegretti, C.; Smit, A.T.; Turri, S.; D'Arrigo, P.; Griffini, G. Biobased Polyurethane Coatings with High Biomass Content: Tailored Properties by Lignin Selection. *ACS Sustain. Chem. Eng.* **2019**, *7*, 11700–11711. [CrossRef]
21. Tribot, A.; Amer, G.; Abdou Alio, M.; de Baynast, H.; Delattre, C.; Pons, A.; Mathias, J.-D.; Callois, J.-M.; Vial, C.; Michaud, P.; et al. Wood-lignin: Supply, extraction processes and use as bio-based material. *Eur. Polym. J.* **2019**, *112*, 228–240. [CrossRef]
22. Wang, H.; Pu, Y.; Ragauskas, A.; Yang, B. From lignin to valuable products-strategies, challenges, and prospects. *Bioresour. Technol.* **2019**, *271*, 449–461. [CrossRef]
23. de Haro, J.C.; Tatsi, E.; Fagiolari, L.; Bonomo, M.; Barolo, C.; Turri, S.; Bella, F.; Griffini, G. Lignin-Based Polymer Electrolyte Membranes for Sustainable Aqueous Dye-Sensitized Solar Cells. *ACS Sustain. Chem. Eng.* **2021**, *9*, 8550–8560. [CrossRef]
24. Trano, S.; Corsini, F.; Pascuzzi, G.; Giove, E.; Fagiolari, L.; Amici, J.; Francia, C.; Turri, S.; Bodoardo, S.; Griffini, G.; et al. Lignin as Polymer Electrolyte Precursor for Stable and Sustainable Potassium Batteries. *ChemSusChem* **2022**, *15*, e202200294. [CrossRef] [PubMed]
25. Zhao, B.; Borghei, M.; Zou, T.; Wang, L.; Johansson, L.-S.; Majoinen, J.; Sipponen, M.H.; Österberg, M.; Mattos, B.D.; Rojas, O.J. Lignin-Based Porous Supraparticles for Carbon Capture. *ACS Nano* **2021**, *15*, 6774–6786. [CrossRef] [PubMed]

26. Österberg, M.; Sipponen, M.H.; Mattos, B.D.; Rojas, O.J. Spherical lignin particles: A review on their sustainability and applications. *Green Chem.* **2020**, *22*, 2712–2733. [CrossRef]
27. Momayez, F.; Hedenström, M.; Stagge, S.; Jönsson, L.J.; Martín, C. Valorization of hydrolysis lignin from a spruce-based biorefinery by applying γ-valerolactone treatment. *Bioresour. Technol.* **2022**, *359*, 127466. [CrossRef] [PubMed]
28. Giannì, P.; Lange, H.; Crestini, C. Functionalized Organosolv Lignins Suitable for Modifications of Hard Surfaces. *ACS Sustain. Chem. Eng.* **2020**, *8*, 7628–7638. [CrossRef] [PubMed]
29. Edmeades, R.M.; Hewlett, P.C. Cement admixtures. In *Lea's Chemistry of Cement and Concrete*, 4th ed.; Hewlett, P.C., Ed.; Butterworth-Heinemann: Oxford, UK, 1998; pp. 841–905.
30. Lou, H.M.; Lai, H.R.; Wang, M.X.; Pang, Y.X.; Yang, D.J.; Qiu, X.Q.; Wang, B.; Zhang, H.B. Preparation of Lignin-Based Superplasticizer by Graft Sulfonation and Investigation of the Dispersive Performance and Mechanism in a Cementitious System. *Ind. Eng. Chem. Res.* **2013**, *52*, 16101–16109. [CrossRef]
31. Zheng, T.; Zheng, D.F.; Qiu, X.Q.; Yang, D.J.; Fan, L.; Zheng, J.M. A novel branched claw-shape lignin-based polycarboxylate superplasticizer: Preparation, performance and mechanism. *Cem. Concr. Res.* **2019**, *119*, 89–101. [CrossRef]
32. Gupta, C.; Nadelman, E.; Washburn, N.R.; Kurtis, K.E. Lignopolymer Superplasticizers for Low-CO_2 Cements. *Acs Sustain. Chem. Eng.* **2017**, *5*, 4041–4049. [CrossRef]
33. Okur, M.; Eslek Koyuncu, D.D. Investigation of pretreatment parameters in the delignification of paddy husks with deep eutectic solvents. *Biomass Bioenergy* **2020**, *142*, 105811. [CrossRef]
34. Jackson, M.G. Review article: The alkali treatment of straws. *Anim. Feed. Sci. Technol.* **1977**, *2*, 105–130. [CrossRef]
35. Xiao, B.; Sun, X.F.; Sun, R. Chemical, structural, and thermal characterizations of alkali-soluble lignins and hemicelluloses, and cellulose from maize stems, rye straw, and rice straw. *Polym. Degrad. Stab.* **2001**, *74*, 307–319. [CrossRef]
36. Foster, C.E.; Martin, T.M.; Pauly, M. Comprehensive Compositional Analysis of Plant Cell Walls (*Lignocellulosic biomass*) Part I: Lignin. *J. Vis. Exp.* **2010**, *37*, e1745. [CrossRef] [PubMed]
37. Singh, B. *Myceliophthora thermophila* syn. *Sporotrichum thermophile*: A thermophilic mould of biotechnological potential. *Crit. Rev. Biotechnol.* **2016**, *36*, 59–69. [CrossRef] [PubMed]
38. Hu, K.Y.; Zhang, Z.C.; Wang, F.; Fan, Y.J.; Li, J.H.; Liu, L.; Wang, J. Optimization of the Hydrolysis Condition of Pretreated Corn Stover using *Trichoderma viride* Broth based on Orthogonal Design and Principal Component Analysis. *Bioresources* **2018**, *13*, 383–398. [CrossRef]
39. Adams, P.R. Extracellular amylase activities of *Rhizomucor pusillus* and *Humicola lanuginosa* at initial stages of growth. *Mycopathologia* **1994**, *128*, 139–141. [CrossRef]
40. Henriksson, G.; Akin, D.E.; Slomczynski, D.; Eriksson, K.-E.L. Production of highly efficient enzymes for flax retting by *Rhizomucor pusillus*. *J. Biotechnol.* **1999**, *68*, 115–123. [CrossRef]
41. Hong, S.; Shen, X.-J.; Xue, Z.; Sun, Z.; Yuan, T.-Q. Structure–function relationships of deep eutectic solvents for lignin extraction and chemical transformation. *Green Chem.* **2020**, *22*, 7219–7232. [CrossRef]
42. Tan, Y.T.; Ngoh, G.C.; Chua, A.S.M. Effect of functional groups in acid constituent of deep eutectic solvent for extraction of reactive lignin. *Bioresour. Technol.* **2019**, *281*, 359–366. [CrossRef]
43. Yoshida, S.; Institute, I.R.R. *Laboratory Manual for Physiological Studies of Rice*; International Rice Research Institute: Los Baños, Philippines, 1976.
44. Allegretti, C.; Boumezgane, O.; Rossato, L.; Strini, A.; Troquet, J.; Turri, S.; Griffini, G.; D'Arrigo, P. Tuning Lignin Characteristics by Fractionation: A Versatile Approach Based on Solvent Extraction and Membrane-Assisted Ultrafiltration. *Molecules* **2020**, *25*, 2893. [CrossRef]
45. D'Arrigo, P.; Allegretti, C.; Tamborini, S.; Formantici, C.; Galante, Y.; Pollegioni, L.; Mele, A. Single-batch, homogeneous phase depolymerization of cellulose catalyzed by a monocomponent endocellulase in ionic liquid [BMIM][Cl]. *J. Mol. Catal. B Enzym.* **2014**, *106*, 76–80. [CrossRef]
46. Meng, X.; Crestini, C.; Ben, H.; Hao, N.; Pu, Y.; Ragauskas, A.J.; Argyropoulos, D.S. Determination of hydroxyl groups in biorefinery resources via quantitative ^{31}P NMR spectroscopy. *Nat. Protoc.* **2019**, *14*, 2627–2647. [CrossRef] [PubMed]
47. Archipov, Y.; Argyropoulos, D.S.; Bolker, H.I.; Heitner, C. ^{31}P NMR Spectroscopy in Wood Chemistry. I. Model Compounds. *J. Wood Chem. Technol.* **1991**, *11*, 137–157. [CrossRef]
48. Faix, O. Fourier Transform Infrared Spectroscopy. In *Methods in Lignin Chemistry*; Lin, S.Y., Dence, C.W., Eds.; Springer: Berlin/Heidelberg, Germany, 1992; pp. 83–109. [CrossRef]
49. Kalliola, A.; Vehmas, T.; Liitia, T.; Tamminen, T. Alkali-O_2 oxidized lignin—A bio-based concrete plasticizer. *Ind. Crops Prod.* **2015**, *74*, 150–157. [CrossRef]
50. He, W.M.; Fatehi, P. Preparation of sulfomethylated softwood kraft lignin as a dispersant for cement admixture. *Rsc. Adv.* **2015**, *5*, 47031–47039. [CrossRef]
51. Takahashi, S.; Hosoya, S.; Hattori, M.; Morimoto, M.; Uraki, Y.; Yamada, T. Performance of Softwood Soda-Anthraquinone Lignin-Polyethylene Glycol Derivatives as Water-Reducing Admixture for Concrete. *J. Wood Chem. Technol.* **2015**, *35*, 348–354. [CrossRef]
52. Ji, D.; Luo, Z.Y.; He, M.; Shi, Y.J.; Gu, X.L. Effect of both grafting and blending modifications on the performance of lignosulphonate-modified sulphanilic acid-phenol-formaldehyde condensates. *Cem. Concr. Res.* **2012**, *42*, 1199–1206. [CrossRef]

53. Li, S.Y.; Li, Z.Q.; Zhang, Y.D.; Liu, C.; Yu, G.; Li, B.; Mu, X.D.; Peng, H. Preparation of Concrete Water Reducer via Fractionation and Modification of Lignin Extracted from Pine Wood by Formic Acid. *Acs Sustain. Chem. Eng.* **2017**, *5*, 4214–4222. [CrossRef]
54. Allegretti, C.; Fontanay, S.; Rischka, K.; Strini, A.; Troquet, J.; Turri, S.; Griffini, G.; D'Arrigo, P. Two-Step Fractionation of a Model Technical Lignin by Combined Organic Solvent Extraction and Membrane Ultrafiltration. *ACS Omega* **2019**, *4*, 4615–4626. [CrossRef]
55. Allegretti, C.; Fontanay, S.; Krauke, Y.; Luebbert, M.; Strini, A.; Troquet, J.; Turri, S.; Griffini, G.; D'Arrigo, P. Fractionation of Soda Pulp Lignin in Aqueous Solvent through Membrane-Assisted Ultrafiltration. *ACS Sustain. Chem. Eng.* **2018**, *6*, 9056–9064. [CrossRef]
56. Haist, M.; Link, J.; Nicia, D.; Leinitz, S.; Baumert, C.; von Bronk, T.; Cotardo, D.; Pirharati, M.E.; Fataei, S.; Garrecht, H.; et al. Interlaboratory study on rheological properties of cement pastes and reference substances: Comparability of measurements performed with different rheometers and measurement geometries. *Mater. Struct.* **2020**, *53*, 92. [CrossRef]

molecules MDPI

Review

Bioethanol Production from Lignocellulosic Biomass—Challenges and Solutions

Magdalena Broda [1,*], Daniel J. Yelle [2] and Katarzyna Serwańska [3,4]

1 Department of Wood Science and Thermal Techniques, Faculty of Forestry and Wood Technology, Poznań University of Life Sciences, Wojska Polskiego 28, 60-637 Poznan, Poland
2 Forest Biopolymer Science and Engineering, Forest Products Laboratory, USDA Forest Service, One Gifford Pinchot Drive, Madison, WI 53726, USA
3 Department of Animal Anatomy, Faculty of Veterinary Medicine and Animal Sciences, Poznan University of Life Sciences, Wojska Polskiego 71c, 60-625 Poznan, Poland
4 Department of Sports Dietetics, Poznan University of Physical Education, 61-871 Poznan, Poland
* Correspondence: magdalena.broda@up.poznan.pl

Abstract: Regarding the limited resources for fossil fuels and increasing global energy demands, greenhouse gas emissions, and climate change, there is a need to find alternative energy sources that are sustainable, environmentally friendly, renewable, and economically viable. In the last several decades, interest in second-generation bioethanol production from non-food lignocellulosic biomass in the form of organic residues rapidly increased because of its abundance, renewability, and low cost. Bioethanol production fits into the strategy of a circular economy and zero waste plans, and using ethanol as an alternative fuel gives the world economy a chance to become independent of the petrochemical industry, providing energy security and environmental safety. However, the conversion of biomass into ethanol is a challenging and multi-stage process because of the variation in the biochemical composition of biomass and the recalcitrance of lignin, the aromatic component of lignocellulose. Therefore, the commercial production of cellulosic ethanol has not yet become well-received commercially, being hampered by high research and production costs, and substantial effort is needed to make it more widespread and profitable. This review summarises the state of the art in bioethanol production from lignocellulosic biomass, highlights the most challenging steps of the process, including pretreatment stages required to fragment biomass components and further enzymatic hydrolysis and fermentation, presents the most recent technological advances to overcome the challenges and high costs, and discusses future perspectives of second-generation biorefineries.

Keywords: bioethanol; ethanol; lignocellulose; lignocellulosic materials; lignocellulosic biomass; lignocellulosic complex; fermentation; biomass utilisation; biofuel; green fuel; biorefinery

check for updates

Citation: Broda, M.; Yelle, D.J.; Serwańska, K. Bioethanol Production from Lignocellulosic Biomass—Challenges and Solutions. *Molecules* **2022**, *27*, 8717. https://doi.org/10.3390/molecules27248717

Academic Editors: Alejandro Rodriguez Pascual, Eduardo Espinosa Víctor and Carlos Martín

Received: 3 November 2022
Accepted: 7 December 2022
Published: 9 December 2022

1. Introduction

Concerning the continuously increasing global demand for energy, fossil fuel resources on our planet are anticipated to become depleted within the next several decades, endangering worldwide energy security. More importantly, the combustion of fossil fuels contributes to CO_2 emissions and hence global warming, a rise in sea levels, urban pollution, and loss of biodiversity, constituting a threat to the global environment. Therefore, the energy transition to low-carbon-intensity fuels becomes necessary to tackle climate change [1,2].

All these negative environmental, social, political, and energy security concerns of the current world has boosted interest in alternative energy sources, including biofuels. However, although alternative energy sources hold the key to solving the three critical global problems, i.e., energy demand and security and climate change (Figure 1), the transition from fossil fuels to more sustainable energy resources require a high initial investment and innovative technologies. Therefore, employing an energy mix of fossil fuels, biofuels, and renewable energy sources seems to be a good starting strategy to switch to solely sustainable resources in the near future [1–4].

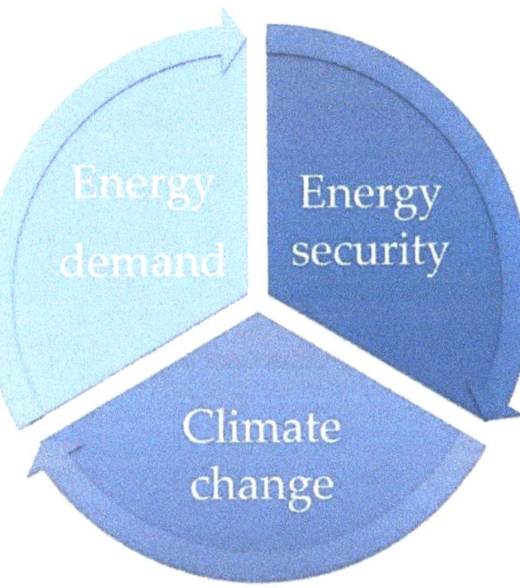

Figure 1. Three main reasons to develop the production of biofuels.

Biofuels emerged as a promising alternative to fossil fuels [5–7]. Among them, bioethanol is one of the most attractive as it can substitute gasoline [8–11]. As a result, several countries, including the USA, Brazil, China, Canada, India, Thailand, Argentina, and many EU members, have already proclaimed commitments to reducing their dependence on fossil fuels towards developing bioethanol production. However, industrial-scale bioethanol production still faces a severe challenge of suitable feedstock acquisition and economic viability while environmentally friendly production technology [1,12,13].

Bioethanol can be produced from a variety of renewable materials rich in carbohydrates, which can be hydrolysed to fermentable sugars and converted to ethanol. Three main feedstock types can be used for bioethanol production: sucrose and starch crops, such as cereals, sugarcane, corn, and others similar (the first-generation bioethanol) [14–16], lignocellulosic biomass (the second-generation bioethanol) [12,17,18], and microalgae (the third-generation bioethanol) [19–21].

The first-generation bioethanol constitutes the majority of over 27,000 million gallons (over 102,060 million litres) of bioethanol produced worldwide (status as of 2021), with the United States of America and Brazil being the indisputable leaders producing almost 85% of the global output mainly from corn and sugarcane, respectively (Figure 2) [22]. On the other hand, France and Germany are the leading bioethanol producers in Europe (Table 1). The primary feedstock for bioethanol production is wheat (in Belgium, Germany, France, and the UK), corn (in Central Europe, the Netherlands, and Spain), sugar beets (in France, Germany, the UK, the Czech Republic, Belgium, and Austria), as well as beet pulp or concentrated juice (in Austria and Belgium) [23]. However, the increasing bioethanol production levels, along with the growing population, raise concerns over the long-term sustainability of first-generation bioethanol, including a threat to global food and feed security, demand for land and water resources, and potential contamination of soil with distillation residues, which prompts intensive research on alternatives, such as second- and third-generation bioethanol production technologies [14,24,25].

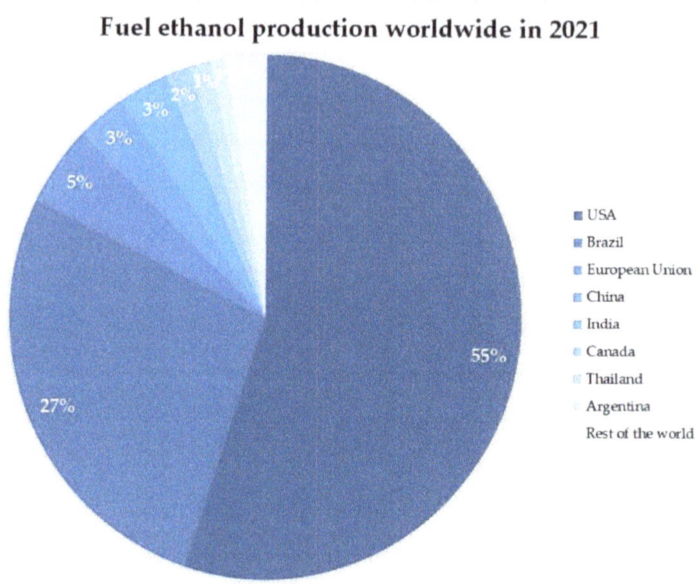

Figure 2. World's top leaders in bioethanol production in 2021 (based on [22]).

Table 1. The main fuel ethanol producers in the European Union (in millions of litres per year) [23].

Country/Calendar Year	2014 ^{r}	2015 ^{r}	2016 ^{r}	2017 ^{r}	2018 ^{r}	2019 ^{e}	2020 ^{e}	2021 ^{f}
France	1018	1039	987	1000	1138	1299	1049	1095
Germany	920	870	882	810	799	676	875	950
Hungary	456	591	633	633	645	689	639	640
Netherlands	519	563	443	532	563	570	538	570
Spain	454	494	328	377	522	547	487	480
Belgium	557	557	570	620	646	620	380	380
Poland	181	214	241	258	259	286	277	285
Austria	230	223	224	235	251	254	241	255
United Kingdom	329	538	658	684	443	190	127	190
Total	**5190**	**5165**	**5159**	**5373**	**5497**	**5281**	**4747**	**5000**

^{r} = revised/^{e} = estimated/^{f} = forecasted EU FAS Posts. Source: EU FAS Posts (based on [23]).

Over the past few decades, the experience gained from first-generation bioethanol production has paved the way for new technologies enabling the utilisation of more sustainable feedstock without adverse effects on food supplies and the environment. Second-generation biorefineries are based on widely available lignocellulosic biomass generated from various sectors (see Section 2.1) that are not directly used as food, and new technologies to convert the biomass into ethanol and other valuable co-products have been continuously developed. Such an approach has the potential to meet energy demands sustainably in an economically viable and environmentally safe way [13,14].

The existing or planned second-generation biorefineries in the US with their production capacity are listed in Table 2. Total cellulosic ethanol production for 2022 in Brazil is estimated at 55 million litres, with an increase of 15 million litres compared to 2021 [26]. In the European Union, there are only a few advanced biofuel plants producing second-generation bioethanol at a commercial scale (Table 3); several others, based on sawdust, forest residues, cereal straw, and by-products from cellulose production, are planned to be opened soon in Finland, Norway, Slovakia, Romania, and Austria [23]. However, commercial production of second-generation bioethanol still represents only an insignificant share of total ethanol production worldwide.

Table 2. Second-generation or mixed biorefineries in the US (state as of 2022) [27]; MGY–million gallons per year; biorefineries under construction are in italics.

Company	City	State	Feedstock	Production Capacity (MGY)	Under Construction (MGY)
NewEnergyBlue LLC	Mason City	IA	Cellulosic Biomass	-	20
Project LIBERTY	Emmetsburg	IA	Cellulosic Biomass	25	-
VERBIO North America Corp.	Nevada	IA	Corn/Cellulosic Biomass	-	60
Quad County Corn Processors	Galva	IA	Corn/Cellulosic Biomass	38	-
Ace Ethanol LLC	Stanley	WI	Corn/Cellulosic Biomass	54	-
POET Biorefining-Iowa Falls LLC	Iowa Falls	IA	Corn/Cellulosic Biomass	115	-
Louis Dreyfus Grand Junction LLC	Grand Junction	IA	Corn/Cellulosic Biomass	125	-
POET Biorefining-Shell Rock LLC	Shell Rock	IA	Corn/Cellulosic Biomass	140	-
PureField Ingredients LLC	Russell	KS	Corn/Sorghum/Cellul. Biomass	55	-
Pelican Acquisition LLC	Stockton	CA	Corn/Sorghum/Cellul. Biomass	60	-
ELEMENT LLC	Colwich	KS	Corn/Sorghum/Cellul. Biomass	70	-
LanzaTech Freedom Pines Fuels LLC	Soperton	GA	Industrial Off-Gases/Biomass/Biogas	-	10
Total	-	-	-	682	90

Table 3. The operational or close to operational advanced biofuel plants in the EU that produce second-generation bioethanol commercially (state as of 2022, based on [23]).

Country	Feedstock	Capacity (Million Litres Per Year)	Year of Opening
Finland	Sawdust	10	2018
Italy	Biomass	28	2020
Austria	Wood sugar	30	2020
Romania	Wheat straw	65	2021
Bulgaria	Corn stover	50	2021
Total	-	183	-

Based on: EU FAS Posts [23].

Second-generation bioethanol has gained increased interest from governments, large companies, and academic research over the past two decades since it represents an attractive renewable alternative to diminishing fossil fuels. Its production is also widely accepted by the general public as it is perceived as non-competitive with the food and feed market and can help mitigate climate change. However, the commercial production of cellulosic ethanol is still in its infancy, being hampered by the high cost of research and production, and tremendous efforts are required to make it more widespread and profitable [17,28,29].

In this paper, we present the state of the art in bioethanol production from lignocellulosic biomass, discuss the most challenging stages of the process, highlight the up-to-date solutions and technological advances that can increase the efficiency of fuel ethanol yield and reduce production costs, and debate future perspectives of second-generation biorefineries.

2. Production of Bioethanol from Lignocellulosic Biomass

Bioethanol production from lignocellulosic biomass is a complex and lengthy process. It includes several steps from resources to end products, such as sourcing of raw materials (lignocellulosic biomass) and their transportation, biomass pretreatment, saccharification, fermentation and ethanol dehydration, products and by-products management, plus all other resources necessary for the production process, including labour, machinery, utilities, and chemicals (Figure 3).

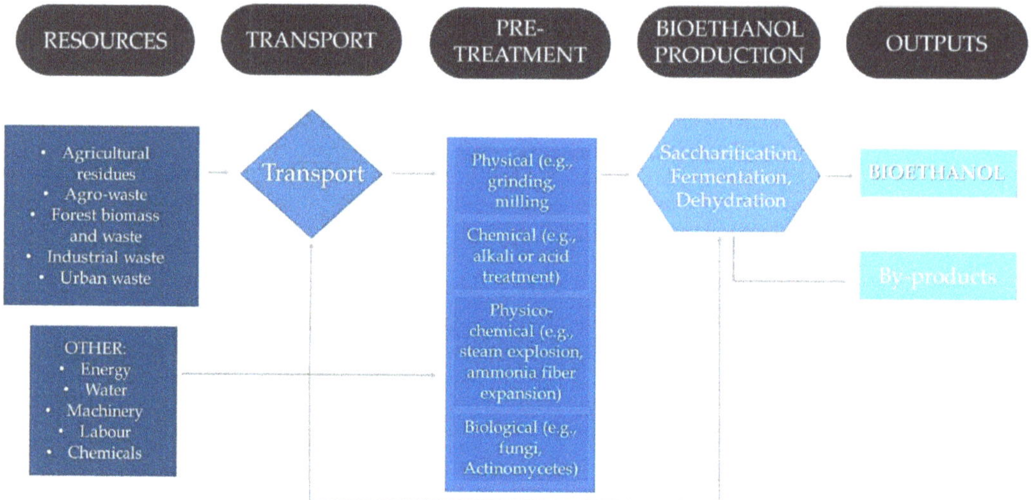

Figure 3. Five elements of bioethanol production, including resources, transportation, all stages of processing and conversion technologies with associated inputs, and final products (based on [3,30]).

Although biomass transportation and the utilisation of other resources such as energy, chemicals, labour, machinery, and water are essential and often challenging elements in ethanol production, we omit them in this paper to focus primarily on the processes directly related to the conversion of lignocellulosic biomass into bioethanol, namely pretreatment, hydrolysis, and bioethanol release, and on the associated challenges and new solutions.

2.1. Lignocellulose Resources

Lignocellulosic biomass (or lignocellulose) refers to all plant dry matter (biomass) on Earth, which is the most abundant renewable raw material. Due to the presence of fermentable components (carbohydrates, cellulose, and hemicelluloses), lignocellulose is one of the alternative feedstocks for bioethanol production [1].

Lignocellulose resources can be broadly classified into three main groups: virgin biomass, energy crops, and waste biomass (Figure 4). Virgin biomass comprises all naturally growing terrestrial plants, including herbaceous plants (annual, biennial, and perennial plants) and woody plants (trees, bushes, and dwarf shrubs), as well as aquatic plants (e.g., water hyacinth, water fern, water lettuce, and duckweed). Energy crops include perennial grasses and other dedicated energy crops that produce a high yield of lignocellulosic biomass (e.g., switchgrass, giant reed, elephant grass, and miscanthus). Waste biomass is a low-value by-product of different industrial sectors such as agriculture (bagasse, cereal straws, stover, and husks), forestry (branches from dead trees, pruning, and thinning residues), and wood and paper production (bark, sawdust, and wood chips). It also includes an organic portion of municipal solid wastes [1,31,32].

Each group of lignocellulose sources has some potential to serve as a raw material for bioethanol production. However, their usability depends on the polysaccharide content that varies between the type of biomass, plant species, and individual parts of the plant. Generally, lignocellulosic biomass is the worldwide most abundant feedstock for ethanol production and has numerous advantages: it is cost-efficient, it is readily available, it does not interfere with food and feed production, it does not require any extra land, and it provides a continuous and reliable supply. Additionally, its utilisation for bioethanol production lessens the problem of waste biomass management and fits in with a sustainable, environmentally-friendly, zero-waste circular economy [1,13,33].

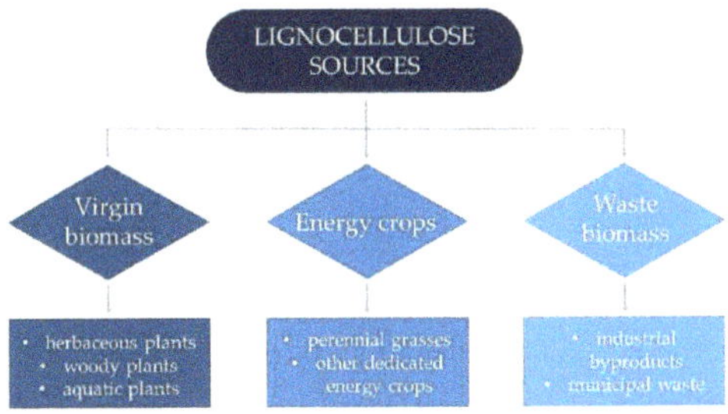

Figure 4. Sources of lignocellulosic biomass for bioethanol production.

The Structure of the Lignocellulosic Complex

Lignocellulose is the main structural component of plant cell walls. It is composed of three different polymers: polysaccharides (cellulose and hemicelluloses) and lignin, a complex aromatic polymer synthesised via radicals of hydroxycinnamyl alcohols (Figure 5). The amount of individual polymers vary depending on the biomass origin (e.g., plant species and part of the plant). However, cellulose and hemicelluloses usually constitute about two-thirds of its total dry mass [1,34].

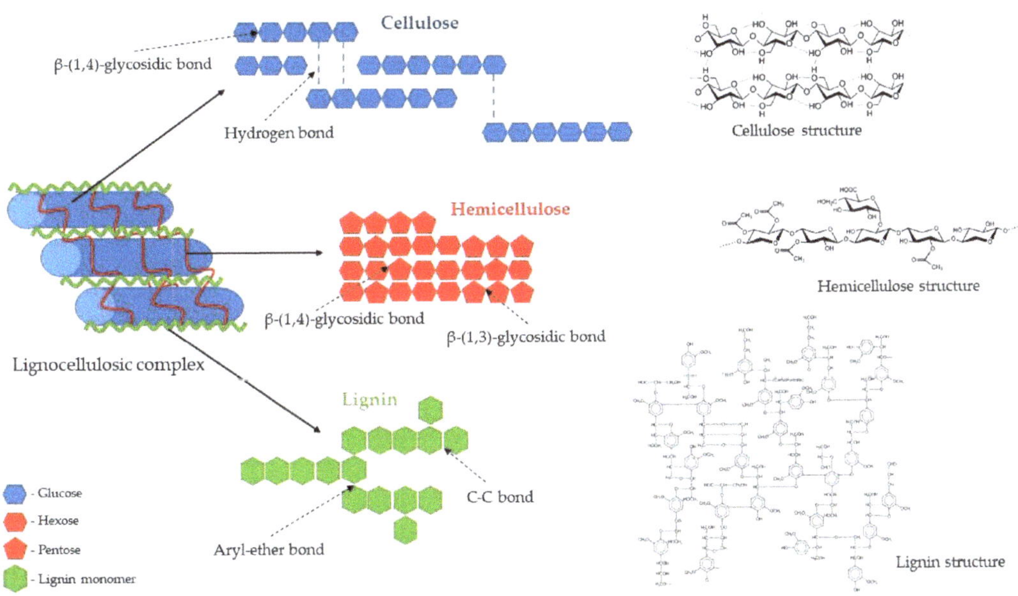

Figure 5. A schematic structure of the lignocellulosic complex (based on [35,36]).

Cellulose is the main structural polysaccharide of the plant cell wall, providing it with high tensile strength and rigidity. Its amount usually ranges from 30% to 50% of the dry weight of lignocellulosic biomass. Regardless of its origin, cellulose is generally a highly crystalline and a high-molecular-weight polymer with a strong tendency to form high-crystalline fibres. A cellulose molecule is a long-chain straight linear homopolysaccharide with an average molecular weight of about 100,000 Da. It consists of β-D-glucopyranose units linked by β-1,4-glycosidic bonds, with the disaccharide cellobiose, made of two β-glucose molecules connected by a β(1→4) bond, as its repeating unit (Figure 5). Owing to the presence of reactive hydroxyl groups at C6, C2, and C3 of glucopyranose units, an extensive network of intra- and inter-chain hydrogen bonds is formed. It induces crystalline structure and facilitates the organisation of the individual cellulose chains into bundles (microfibrils) that are additionally stabilised by van der Waals interactions (hydrophobic interactions), leading to a fibrous state. Highly ordered crystalline regions in cellulose are interspersed with disordered amorphous regions. The explicit cellulose structure makes it insoluble in water and resistant to depolymerisation [1,37,38].

Hemicelluloses are, collectively, the second most abundant polysaccharide of the plant cell wall, which accounts for 15–30% of the lignocellulosic dry mass. They are complex polymers consisting of short linear and highly branched heteropolysaccharides (Figure 5) with an average molecular weight of about 30,000 Da, consisting of various sugar units, both pentoses (β–D–xylose and α–L–arabinose) and hexoses (β–D–glucose, α–D–galactose, and β–D–mannose), as well as uronic acids (α–D–glucuronic, α–D–galacturonic, and α–D–4–*O*–methylgalacturonic acid), and minor amounts of α–L–rhamnose and α–L–fructose. Among heteropolymers present in hemicelluloses are the most common xylan and glucomannan, as well as glucuronoxylan, arabinoxylan, and xyloglucan. This composition makes hemicelluloses' structure random and amorphous, or only partially crystalline. Hemicelluloses are embedded in the plant cell walls, and their segments bind with adjacent cellulose microfibrils through hydrogen bonding and van der Waals interactions. This cellulose–hemicellulose network acts as a load-bearing element that strengthens the cell wall [1,39–41].

Lignin is the third principal component of lignocellulosic biomass, constituting about 15–30% of its dry mass. Present in all vascular plants, lignin is an amorphous, hyper-branched, and a cross-linked three-dimensional network polymer with no regular repeating elements (Figure 5). This random structure arises due to the enzymatically initiated free radical polymerisation of three types of phenylpropane units (also known as monolignols) such as *p*-coumaryl, coniferyl, and sinapyl alcohols that form aromatic units of lignin: *p*-hydroxyphenyl, guaiacyl, and syringyl, respectively. The proportion of monomers forming the lignin molecule differs depending on the type of plant, cell, and location in the cell wall structure. For example, hardwood lignin is made up mainly of coniferyl and sinapyl alcohols (guaiacyl-syringyl lignin and GS-lignin), softwood lignin is made of coniferyl alcohol (guaiacyl lignin and G-lignin), while lignin of grass and herbaceous plants contains coniferyl, sinapyl, and *p*-coumaryl alcohol (*p*-hydroxyphenyl-guaiacyl-syringyl lignin and HGS-lignin). Lignin is covalently bonded to hemicelluloses and part of cellulose, serving as a cement between the fibres, supporting the mechanical properties of the cell walls, and protecting the structural polysaccharides from enzymatic microbial degradation [1,42–44].

Each of the main cell wall structural polymers has a different chemical structure and, thus, behaviour. Moreover, they are strongly intertwined and bonded by non-covalent forces (hydrogen bonds and van der Waals interactions) and covalent cross-linkages, forming a highly structured and robust composite matrix. All these, together with a crystalline cellulose nature and low surface area available for enzymes, make lignin complex and highly recalcitrant and resistant to separation and depolymerisation. Therefore, processing lignocellulosic biomass for bioethanol production is challenging, requiring proper knowledge and technologies that employ a combination of chemicals, enzymes, microorganisms, and heat [1,12,34,45].

Among lignocellulose structural polymers, only cellulose and hemicelluloses can be used to produce bioethanol because they are long-chain polysaccharides hydrolysed into a mixture of fermentable pentoses and hexoses that can be further converted to ethanol molecules. However, owing to the recalcitrance of lignocellulosic biomass, obtaining high efficiency and profitability in the process of bioconversion of lignocellulosic substrate into ethanol requires a primary pretreatment process. It is necessary to first release cellulose and hemicelluloses from a complex matrix and make them more accessible towards enzymatic hydrolysis [28,45,46].

2.2. Pretreatment of a Lignocellulosic Biomass

The pretreatment of the lignocellulosic complex is the first and necessary step in its bioconversion to ethanol. During this process, the structure of lignocellulose is disrupted by breaking down cross-linkages between its structural polymers, which helps to separate carbohydrates from lignin, and hydrogen bonds between cellulose chains are broken, thus decreasing cellulose crystallinity and its degree of polymerisation. Pretreatment technologies are intended to improve the accessibility of enzymes to carbohydrates by reducing the size of biomass particles and boosting their surface area and porosity, thus facilitating their hydrolysis and fermentation; they also increase yields of fermentable sugars [45,47,48].

The efficiency of the polysaccharide hydrolysis to monosaccharides, the primary substrates in alcoholic fermentation, is mainly limited by the presence of lignin. On the one hand, the lignin polymer restricts the free access of hydrolytic enzymes to cellulose microfibrils and hemicellulose chains. On the other hand, lignin acts as an adsorbent that binds the enzyme molecules on its surface, thus causing their irreversible inactivation. Therefore, lignin has to be removed in the pretreatment step [1,3,45]. Other limiting factors are inhibitory compounds that can be produced during the pretreatment stage, including furan derivatives (HMF–5-hydroxy-2-methyl-furfural and furfural), phenolic compounds, and weak acids (acetic, formic, and levulinic acid). They adversely affect hydrolysis efficiency by limiting microbial activity and/or disturbing enzymes' efficiency; therefore, their presence is highly undesirable [45].

The pretreatment process should be easy to carry out, cost-effective, and environmentally friendly, producing minimum amounts of inhibitory compounds and allowing a complete utilisation of lignocellulosic biomass, which results in high efficiency of bioethanol production and the proper management of waste lignin. Generally, the existing pretreatment methods can be grouped into four categories, such as physical, chemical, physico-chemical, and biological (Figure 1). Unfortunately, there is no universal pretreatment for all types of biomasses and, usually, a combination of two or more complementary techniques is applied to obtain the most satisfactory results. However, developing the best pretreatment strategies is still a subject of extensive research [45,47–49].

2.2.1. Physical Pretreatment

Physical pretreatment methods employ mechanical forces, irradiation, electric or electromagnetic field, temperature, or pressure to reduce the size of lignocellulosic biomass particles and increase their surface area and pore volume. They usually also decrease the degree of all components' polymerisation and cellulose crystallinity, which facilitates further biomass processing. The physical methods include grinding, milling, chipping, extrusion, freezing, sonication, microwaving, and pulsed electric field treatment [45,47].

Chipping, grinding, and milling are the primary pretreatment techniques to crush lignocellulosic biomass. Depending on the method, the final particle size can be reduced to 10–30 mm or even 0.2–2 mm. Among the mechanical methods, ball milling, colloid milling, hammer milling, two-roll milling, and wet disk milling are commonly used in bioethanol production, with ball milling giving the highest yields of glucose and xylose after enzymatic hydrolysis [45,47,50]. However, milling is relatively expensive due to high energy requirements [45,47].

Extrusion is a thermo-physical method that includes rapid mixing, moderate heating, and high shearing of lignocellulosic biomass, resulting in physical and chemical disruption of its complex structure. The process is highly versatile and efficient, does not produce furfural and HMF, and can be carried out continuously, even for high solids loading. However, due to high energy requirements, it may not be economically the best alternative to conventional pretreatment [45,51,52].

Freeze pretreatment is a relatively new and promising technique. It was shown to significantly increase the enzymatic conversion of rice straw, resulting in enhanced glucose and bioethanol yields. In addition, this method has a low environmental impact and is relatively cost-effective due to the low energy input required and the lack of toxic chemicals involved in the process [53,54].

Sonication employs ultrasound waves to disrupt the lignocellulose complex and make cellulose and hemicelluloses available for enzymatic hydrolysis. As a result, the method enhances the conversion of cellulose to fermentable sugars, increases sugar yields, reduces hydrolysis time, and improves further fermentation. The application of slightly elevated temperatures (about 50 °C) and a change of water into an alkaline medium can additionally ameliorate the pretreatment process [47,55].

Microwave irradiation penetrating through the lignocellulosic feedstock effectively disrupts its recalcitrant structure. This can improve the solubilisation of lignocellulosic biomass, effectively degrade lignin, and alter the structure of the polysaccharides, thus enhancing their susceptibility to hydrolysis. Using higher power and temperatures increases the effectiveness of the process. Microwave-assisted pretreatment can be an interesting alternative to conventional heating due to its uniformity and selectivity, less energy input, and shorter processing time [45,47,56].

Pulsed electric field (PEF) treatment is a novel method that increases biomass porosity and permeability by subjecting it to a series of high voltage (5.0 and 20.0 kV/cm) short-duration (nano to milliseconds) pulses. The technique seems to be cost-efficient due to low energy requirements and the simplicity of instrumentation required that can be easily designed to the biorefinery conditions, and it enhances cellulose hydrolysis resulting in its efficient conversion to bioethanol [47,57].

2.2.2. Chemical Pretreatment

Chemical pretreatment employs various chemicals, including alkalis, acids, gases, salts, ionic liquids, oxidising agents, or organic solvents, to release polysaccharides from the lignocellulosic complex and make them more susceptible to enzymatic hydrolysis [3,45,47].

Acid pretreatment is one of the most frequently used methods to overcome the recalcitrance of lignocellulose in bioethanol production. Biomass is usually treated with mineral acids solutions (HCl and H_2SO_4) at a pressure of 1.5 bar and elevated temperatures ranging from 100 °C to 290 °C for various residence times (up to several hours). The crucial effective parameters for the method include acid concentration, solids loading, temperature, and residence time. When diluted acids are used in the process, their effectiveness is enhanced by increasing the temperature of the process. Acid pretreatment has only a limited effect on lignin while mainly affecting polysaccharides. Hemicelluloses are dissolved and polysaccharide–lignin linkages are broken, thus making cellulose more accessible to enzymes. Its main disadvantages are the high cost of acid recovery and the production of inhibitory by-products. However, the environmental and economic aspects of the method have been improved recently [3,45,47,56].

In alkaline treatment, dilute solutions of NaOH, KOH, $Ca(OH)_2$, or ammonia are usually used to degrade and remove lignin and part of hemicelluloses to make cellulose more available for enzymatic hydrolysis. By breaking crosslinks between hemicellulose and other polymers, the treatment also causes swelling of fibrous cellulose increasing biomass porosity. The process can be performed at elevated temperatures for a short time or at low temperatures for a relatively long period. The advantages of the method are selective lignin removal without a loss of carbohydrates and enhanced porosity of feedstock that improves further enzymatic hydrolysis, as well as biomass disinfection. The main drawback is longer reaction times (several hours up to one day) than other pretreatment methods [3,45,47,56,58].

Solvent pretreatment methods include the application of organic solvents, ionic liquids, and deep eutectic solvents. Several chemicals have been tested in this technique, such as acetone, ethanol, ethylene glycol, glycerol, methanol, n-butanol, phenol, tetrahydrofurfuryl alcohol, and triethylene glycol [3,45,47].

Organic solvent treatment employs a variety of organic solvents, including acetone, amines, alcohols, dioxane, esters, formaldehyde, propionic acid, and phenols with and without a catalyst, for the lignocellulosic biomass pretreatment. The technique is recognised as one of the most prospective pretreatment methods because of its ability to deconstruct lignocellulosic complex and fractionate biomass into lignin, cellulose, and hemicelluloses with high purity. It also allows for easy solvent recovery and reuse. Unfortunately, high energy consumption and the cost of organic solvents make the method not economically viable [47,59,60].

Ionic liquids (mainly salts including a large organic cation and small anion, including ammonium-based, imidazolium-based, phosphonium-based, pyridinium-based, pyrrolidinium-based, and sulfonium-based) have also been extensively studied for their potential to degrade lignin and break down crystalline cellulose structure. The method offers high rates of cellulose recovery and conversion to glucose. However, there are still many challenges to using ionic liquids on a broader scale, including their high price when large amounts are needed for the process, high waste generation with difficult recovery, high energy demands for recycling, and high viscosity of the solution over time that makes them difficult to handle [45,47,56,61,62].

A more "green" approach in bioethanol production involves biomass pretreatment with deep eutectic solvents. They are mixtures of hydrogen bond donors (e.g., amides, alcohols, or carboxylic acids) and acceptors (quaternary ammonium salts) at moderate temperatures of 60–80 °C, which enhance solubilisation of lignocellulosic polymers with higher selectivity towards lignin and without affecting cellulose.

Deep eutectic solvents are considered economical and "green" because they are less toxic than other chemicals used for conventional biomass pretreatment, easily biodegradable and recyclable, and have a great potential for much broader usage in the biorefineries of tomorrow [47,63,64].

The application of various metal salts for biomass pretreatment represents a more novel method that provides high sugar recovery, and its performance can also be further improved by combining it with other pretreatment technologies. The principal advantages of metal salt-based treatments are improved lignin removal, degradation of hemicelluloses, and complete biomass conversion. In addition, these pretreatments also result in enhanced enzymatic hydrolysis, are nontoxic and environmentally safe, and do not require costly non-corrosive reactors [47,65,66].

Another pretreatment method is biomass oxidation, which involves various oxidising agents, with hydrogen peroxide being the most frequently applied chemical. The method results in the degradation of lignin by hydroxyl radicals produced during hydrogen peroxide hydrolysis, which leaves the cell wall polysaccharides more accessible for further enzymatic hydrolysis. Since the method also degrades a part of hemicelluloses, it is not considered one of the most efficient processes for fermentation [47,67].

Ozonolysis is a greener oxidative pretreatment method that employs ozone gas as an oxidant to destruct the lignocellulose complex. Ozone reacts preferably with lignin, which results in effective biomass delignification and a release of sugar during enzymatic hydrolysis. The greatest advantage of this method is that it can be carried out in ambient conditions. Furthermore, the only inhibitory compounds produced are short-chain carboxylic acids, which can be easily removed by washing with water. However, the method is not economically viable due to the high costs of ozone since vast amounts are required in the process [45,68–70].

2.2.3. Physico-Chemical Pretreatment

Physico-chemical methods utilise both physical (high temperature and pressure) and chemical processes to effectively pretreatment of lignocellulosic biomass. Among them, a steam explosion has been the primary technique used for bioethanol production. First, high temperature (160–260 °C) and pressure (0.7–4.8 MPa) are applied to biomass for a few seconds to several minutes; then, a sudden pressure reduction causes explosive decompression in the material. It results in the disruption of the cell wall structure and the solubilisation of hemicellulose and lignin fractions. A steam explosion is effective for all types of biomasses, including that with large particles, without a need for pre-crushing. The method's main advantages are low energy requirements, no additional chemical costs (thus, no recycling), and environmental friendliness, while incomplete lignin removal and the production of some toxic chemicals during the process are the main disadvantages [3,45,47].

Ammonia fibre expansion (AFEX) applies liquid ammonia to lignocellulosic biomass under pressure and elevated temperature, followed by a rapid pressure reduction that expands the fibre structure, increasing its surface area. The treatment also causes the selective delignification of biomass, decrystallisation of cellulose, and partial hemicellulose depolymerisation, which results in high glucose yields in further enzymatic hydrolysis. Other AFEX advantages include that ammonia is a non-polluting and non-corrosive substance that can be easily recovered and reused in the process, and the amount that remains in the biomass serves as a nutrient source for microorganisms used in further bioethanol fermentation. The downside to this method is the small efficiency in the case of feedstock containing significant amounts of lignin [47,71–73].

Supercritical CO_2 explosion is a green pretreatment method that employs supercritical fluid CO_2 as a solvent. During its diffusion through the biomass under high pressure and temperature, carbonic acid is produced, which hydrolyses hemicelluloses. The subsequent explosion releases the gas that penetrated the structure of the lignocellulosic complex, thus weakening the cell wall ultrastructure and increasing the accessible surface area of its polymers for further enzymatic processes. CO_2 necessary for the treatment can be sourced directly from glucose fermentation to ethanol, where it is released as a by-product and continuously recycled in the process without increasing CO_2 emissions into the atmosphere. The method is environmentally friendly and efficient; it enhances glucose yield, facilitates biomass delignification, and allows the extraction of different components from the biomass; it is appropriate for all feedstocks that have retained some moisture. However, due to the costs of reactors suitable for high-pressure conditions, its application is limited [47,74–76].

Liquid hot water (LHW) pretreatment is a simple method that uses water under high pressure, similar to the steam pretreatment technique. Compressed water (at a pressure up to 5 MPa) at a high temperature (170–230 °C) permeates through lignocellulosic biomass, hydrolysing hemicelluloses, removing some lignin, and simultaneously hydrating the cellulose fraction, making it more accessible for enzymes. The method produces minimal inhibitory compounds. It is relatively cost-effective and environmentally friendly because it does not require an energy-demanding preliminary reduction in feedstock size and does not use chemicals or corrosion-resistant hydrolysis reactors [3,47,77].

For lignin-enriched feedstock, wet oxidation is a suitable pretreatment method that produces less inhibitory furan derivatives than a steam explosion or liquid hot water treatments. Oxygen or air is employed as a catalyst, and water or hydrogen peroxide serves as a medium. The process depends mainly on three critical factors, namely temperature, oxygen pressure, and time, and is typically carried out at a high temperature (above 120 °C) and pressure (0.5–2 MPa) for about 30 min. As a result, hemicelluloses undergo solubilisation and hydrolysis to monomers, and some lignin is oxidised, leaving cellulose more available for enzymatic processes. However, high financial expenditures imposed by oxygen and pressure equipment prices prevent the method from becoming the standard industrial application [45,47,78].

2.2.4. Biological Pretreatment

Biological methods use ligninolytic microorganisms (bacterial and fungal strains) or their enzymes to reduce the recalcitrance of lignocellulosic biomass, converting it into compounds more accessible for hydrolysis and subsequent ethanol production. The most effective are white-rot fungi due to their ability to degrade lignin, including the four most frequently industrially used species: *Phanerochaete chrysosporium*, *Trametes versicolor*, *Ceriporiopsis subvermispora*, and *Pleurotus ostreatus*, and also some bacterial strains, including *Clostridium* sp., *Cellulomonas* sp., *Bacillus* sp., *Thermomonospora* sp., and *Streptomyces* sp., are commonly used in biological pretreatment. The crucial parameters affecting the efficiency of biological pretreatments are the type of selected microorganism, the size of biomass particles, and the process conditions, including moisture content, temperature, and time. The main advantages of biological pretreatment methods are their low energy requirements, no chemicals and their recycling costs, low downstream processing costs, the minimal amount of inhibitory compounds produced, relatively simple operating and environmental friendliness. However, the drawbacks, such as ample space requirement, a very slow course of the process, and the necessity of continuous control of microbial growth and activity, preclude more widespread application of these methods in the industry [45,47,56,79–81].

2.2.5. Combined Pretreatment Methods

Apart from the pretreatment methods described above, there are various combinations and several other sophisticated techniques that are being developed to overcome the main drawbacks of the existing techniques to improve the utilisation of the lignocellulosic complex, making the bioethanol production process more economical, efficient and environmentally friendly [47,49,56].

For example, since the most common pretreatment methods require high temperatures (160–290 °C) and pressures (0.69 to 4.9 MPa), and additionally produce inhibitory furan derivatives, there is a need to develop new methods that overcome those disadvantages. One of them is alkaline hydrogen peroxide (AHP) pretreatment that combines the application of NaOH and H_2O_2. The main advantages are high effectiveness for various biomass concentrations providing high efficiency of enzymatic hydrolysis, high lignin and hemicellulose solubilisation values for the liquid fraction, low energy consumption, availability of chemicals needed, no furan derivatives produced, no need for special reactors, compatibility with high solid loadings, and sterility conditions provided by alkaline H_2O_2 without a need to use antibiotics. However, the method is not free from drawbacks, such as the high pH of pretreated biomass, the generation of other inhibitors, such as *p*-coumaric and ferulic acids, the price of chemicals required in high amounts, and the need for an initial grinding of the biomass material. To overcome these shortcomings, some specific approaches are still needed to optimise the whole pretreatment process and keep it efficient while being cost-effective and safe for the environment [82].

In biological pretreatment methods, a long operation time is one of the main disadvantages. To surmount this problem, extensive research has been conducted proposing to combine fungal treatment with various chemical, physical, or physico-chemical methods [83].

- Biological-alkaline pretreatment combination can enhance the delignification of a lignocellulosic complex and help reduce the chemicals' concentration, time, and temperature of alkaline treatment, thus lowering operational expenses [83–85]. However, the treatment may cause a higher loss of carbohydrates from biomass [85].
- Biological-acid combination effectively solubilise the hemicellulose fraction and limit the production of inhibitory compounds while reducing severe acid pretreatment conditions. Moreover, it increases glucose and ethanol yield compared withacid pretreatment alone [83,84,86,87].
- Biological-oxidative pretreatment uses the fact that biomass decay by white-rot fungi involves a Fenton-based oxidation reaction. By mimicking this reaction using other oxidising reagents, e.g., hydrogen peroxide followed a biological pretreatment, it is possible to shorten the residence time and enhance biomass delignification without producing inhibitory by-products, which results in higher sugar yields. This combined pretreatment method seems to be the most effective among biological–chemical treatment combinations [79,83,88,89].
- Biological-organosolv combined pretreatment was studied for woody biomass, resulting in higher saccharification yields and larger amounts of lignin-enriched fractions [83,90,91].
- Biological-LHW treatment allows for the lowering of the temperature of the water during the LHW process while enhancing sugar yield owing to microbial activity [83,92,93].

- Biological-steam explosion combinations significantly increase the net sugar yields compared to the processes applied alone. Using lignin-degrading enzymes also reduces energy consumption, the amount of wastewater, the operational costs of steam explosions, and detoxifies the processed biomass [83,92,94].

However, it should be highlighted that the efficiency of all combined pretreatment methods that involve biological treatment depends strongly on microbial species/strains, culture conditions, biomass type, and the order of pretreatment methods used [83].

Several other combinations of various pretreatments have been studied extensively over the past decade to find the most efficient, economically-viable, and universal solutions, including alkali and metal salt combinations, ultrasound-assisted pretreatment using metal salt with hydrogen peroxide, and a sequential pretreatment comprising of deep eutectic solvents and divalent inorganic salts [92,95–97]. However, an ideal method has not been found yet, and further improvement in the pretreatment step is still necessary to overcome the limitations of optimal utilisation of lignocellulosic biomass and make bioethanol production more common and profitable [30,47,56,83,92,98–101].

2.3. Bioethanol Production

After the pretreatment step, bioethanol production from lignocellulosic biomass requires a series of consecutive processes to obtain a final product, including detoxification, hydrolysis, fermentation, distillation, and dehydration [12,47].

2.3.1. Detoxification

Detoxification aims to remove all the toxic compounds from pretreated biomass or hydrolysates, including fermentation inhibitors (such as furan aldehydes, aliphatic acids, and phenolic compounds) that could minimise the enzymes' efficiency and restrict microbial growth and activity during fermentation. The most common methods to discard inhibitors from biomass and ensure higher bioethanol yield and productivity are, nowadays, various in situ strategies, including membrane extraction, solvent extraction, ion exchange, membrane bioreactors, adsorption, microbial adaptation, using microbial consortium or engineered microorganisms, and several other techniques that are tailored according to pretreatment, hydrolysis, and fermentation methods used in the ethanol production process. Detoxification may be performed separately or integrated into hydrolysis or fermentation [12,102,103].

2.3.2. Hydrolysis

After the pretreatment stage is completed, raw material is subjected to enzymatic hydrolysis. This process is carried out to obtain fermentable sugars, pentoses, and hexoses from polysaccharides present in the pretreated lignocellulosic biomass. Mainly enzymes are employed to catalyse the hydrolysis of cellulose and hemicellulose (xylan), but also acids and alkalis can be used for this purpose (as mentioned in Section 2.2.2) [12,104].

The enzymes capable of hydrolysing cellulose to glucose monomers are known as cellulases. They are multienzyme complexes consisting of mainly three various components, namely endo-1,4-β-D-glucanase (EC 3.2.1.4; breaks intermolecular bonds in cellulose randomly), exo-1,4-β-D-glucanase/exo-cellobiohydrolase (EC 3.2.1.91; removes monomers and dimers from the end of the glucose chain), and β-glucosidase (EC 3.2.1.21; hydrolyses glucose dimers, cellobiose, and other short cellulose oligomers into glucose monomers). Complete hydrolysis of a native cellulose polymer into glucose monomers requires the synergistic action of all three components (Figure 6). Cellulases are sourced from various bacteria and fungi. They are produced by aerobic, anaerobic, mesophilic, and thermophilic microorganisms. Cellulases producing microorganisms include bacterial genera of *Acetovibrio, Clostridium, Cellulomonas, Cellvibrio, Bacillus, Bacteroides, Erwinia, Ruminococcus, Streptomyces,* and *Actinomycetales* genera of *Microbispora* and *Thermomonospora*. Among fungal species, the most common source of cellulase is *Sclerotium rolfsii* and *Phanerochaete chrysosporium* species, as well as some species belonging to the genera of *Aspergillus, Caecomyces,*

Humicola, *Neocallimastix*, *Oprinomyces*, *Penicillium*, *Schizophyllum*, and *Trichoderma* [105–108]. Cellulose hydrolysis is difficult because the cellulose microfibrils are stabilised by internal and external hydrogen bonds and surrounded by hemicellulose polysaccharides (mannans and xylans) joined by covalent and hydrogen bonds; hence, the crucial role of the pretreatment stage emerges [104,109].

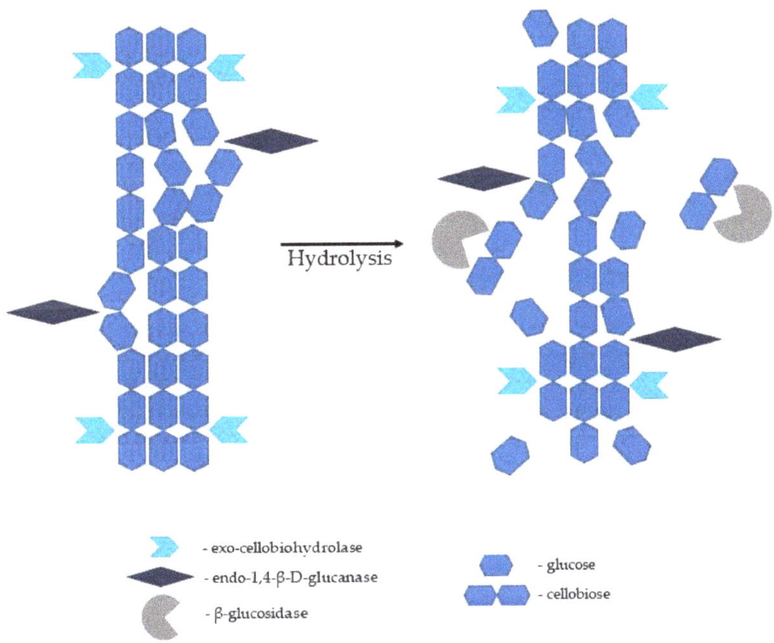

Figure 6. Hydrolysis of native cellulose cellulolytic enzymes (based on [107,110]).

Since hemicelluloses represent 10–30% of lignocellulosic biomass, their conversion to fermentable sugars is also vital for the high yield of bioethanol. Hemicellulose hydrolysis is easier than cellulose due to its more accessible amorphous structure. On the other hand, its more varied composition and structure, with multiple side chains containing various sugar types, requires a complex set of enzymes. Two groups of enzymes are needed for effective hemicellulose hydrolysis: depolymerising core enzymes that can cleave the backbone and de-branching enzymes (so-called ancillary or auxiliary enzymes) that remove side chains posing steric hindrances to core enzymes, thus increasing the total yield of fermentable sugars obtained from lignocellulosic biomass. The core enzymes include β-1-4-mannosidases (EC 3.2.1.25), endo-1,4-β-mannanases (EC 3.2.1.78), endo-β-1,4-xylanases (EC 3.2.1.8), and xylan 1,4-β-xylosidases (EC 3.2.1.37), while de-branching enzymes are acetylxylan esterase (EC 3.1.1.72), α-L-arabinofuranosidase (EC 3.2.1.55), β-glucuronidase (EC 3.2.1.139), ferulic acid esterase (EC 3.1.1.73), and *p*-coumaric acid esterase (EC 3.1.1-). Similar to cellulases, microorganisms are the source of enzymes for hemicellulose hydrolysis. They include fungi, e.g., *Aspergillus niger*, *Aspergillus awamori*, *Trichoderma reesei*, *Penicillium wortmanii*, *Cochliobacillus carbonum*, *Agaricus bisporus*, and other *Aspergillus*, *Agaricus*, *Trichoderma*, and *Sclerotium* genera, and bacteria, e.g., *Thermotoga maritima*, *Clostridium thermocellum*, *C. cellulovorans*, *Thermobacillus xylanilyticus*, *Paenibacillus polymyxa* cel44C-man26A, *Cellvibrio japonicus*, *Caldibacillus cellulovorans*, *Caldicellulosiruptor Rt8b*, *Caldocellum saccharolyticum*, *Bacillus* spp., and *Streptomyces* spp. The synergistic action of various microbial enzymes ensures high sugar yield from lignocellulosic biomass, thus enhancing bioethanol production [108,111].

The most critical parameters during biomass hydrolysis include solid loading, the concentration of sugars, enzyme loading, the shaking speed, hydrolysis time, the concentration of inhibitors, and the effect of various additives [12,112–114].

- Solid loading—High solid loading reduces hydrolysis installation costs and are necessary to obtain syrups with increased sugar concentrations (80–100 g/L), which determines economically viable distillation (i.e., the ethanol concentration in a fermented broth should be above 4% *w*/*w*). It was shown that sugar yield increases with increasing substrate load, but only to some point, after which it decreases. It is mainly because increased cellobiose and glucose concentrations inhibit enzyme activity. Additionally, high solid loading usually translates into a high-viscosity broth, which causes several technical problems due to hampered mixing and impaired mass and heat transfer, affecting the efficiency of enzymes [112,114–117];
- Enzyme loading—Increased doses of enzymes (or enzyme cocktails) enhance saccharification efficiency providing high glucose yield [12].
- Shaking speed—Optimising shaking/mixing speed is necessary to ensure optimal heat and mass transfer that translates into high glucose yield. Lower speed values result in poor mixing and decreased monosugar yields, while too high of a speed produces shearing forces that may destroy enzymes [117–120].
- Hydrolysis time—The long time required for complete hydrolysis limits the commercial production of ethanol from lignocellulosic biomass. Therefore, several approaches have attempted to shorten the process by enhancing hydrolysis efficiency, mainly using engineered enzymes/microorganisms or enzyme cocktails and optimising the parameters of the process [121,122].
- Concentration of inhibitors—Inhibitors produced during biomass pretreatment may slow down or even stop enzymatic hydrolysis. Therefore, the detoxification step (see 2.3.1. Detoxification), performed before or during hydrolysis or selecting pretreatment methods producing only a limited amount of inhibitors, is crucial for the process [12,101–103,123].
- Effect of various additives—Several different substances were successful as additives in the hydrolysis step to improve glucose yield, including polyethylene glycol (PEG)-based polymers (PEG 600, 4000, 6000), non-ionic surfactants (Tween 80 and Triton X-100), non-catalytic protein (bovine serum albumin (BSA)) or novel chemical surfactants, such as Silwet L-77. Their mode of action is based on blocking the interactions between lignin and enzymes, thus intensifying positive substrate-enzyme interactions and recovering cellulose hydrolysability [124–130].

Enzymatic saccharification is the most challenging and relatively expensive stage in bioethanol manufacturing from lignocellulosic biomass, with costs estimated at 20–30% of the total production costs. It has also been recognised as a techno-economical bottleneck in the whole process of biomass-to-ethanol bioconversion. Therefore, all crucial steps impacting the yield of fermentable sugars and total bioethanol require careful optimisation while maintaining minimum operational costs to make the production of lignocellulosic ethanol widespread and profitable [1,12,112].

2.3.3. Ethanol Fermentation

In the bioethanol production from lignocellulosic biomass, both hexoses (glucose, fructose, and sucrose) and pentoses are available for ethanol fermentation (xylose, mannose, galactose, and arabinose), resulting in the production of the respective number of ethanol and carbon dioxide molecules (Figure 7) [12,131,132].

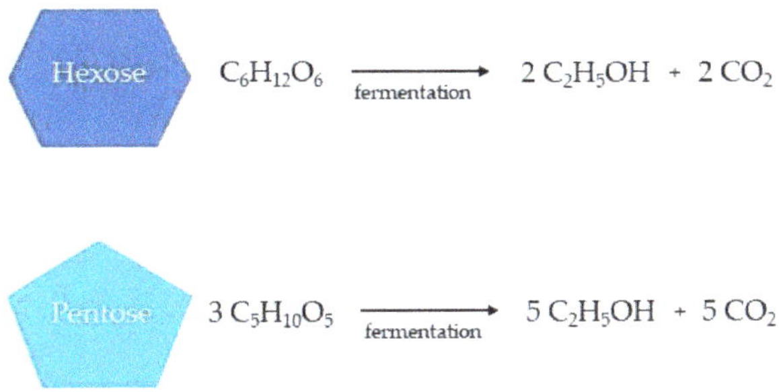

Figure 7. Simplified ethanol production from hexoses and pentoses during the fermentation stage.

For glucose fermentation, industrial strains of *Zymomonas mobilis* and *Saccharomyces cerevisiae* are mainly used, owing to their high ethanol productivity and resistance to high ethanol concentration (up to 120 g/L). However, they are incapable of fermenting pentoses, which limits their use in ethanol production from lignocellulosic raw materials [12,133,134]. Among microorganisms naturally fermenting pentoses are yeasts, such as *Candida shehatae*, *Pachysolen tannophilus*, and *Pichia stipitis* (recently reclassified as *Scheffersomyces stipitis*), and intestinal bacteria; however, the efficiency of the process is minor. Moreover, in the case of pentose-fermenting yeasts, large-scale utilisation is inhibited by their sensitivity to high ethanol concentration (over 40 g/L) and inability to ferment xylose at low pH. In addition, they require microaerophilic conditions and are easily inhibited in the presence of glucose (catabolite repression) and, in a mixed sugar broth, they usually utilise xylose only under glucose-limited conditions [12,135–137].

Due to the lack of natural microorganisms for the efficient simultaneous fermentation of pentoses and hexoses, there is a growing interest in using engineering techniques for metabolic processes to construct organisms with the desired characteristics. Metabolic engineering aims to improve microbial activity due to changing enzymatic, transport, and regulatory functions using recombinant DNA technology. It includes analysing metabolic pathways, designing genetic changes, and creating recombinant cells with enhanced desired properties. The modification goal is to obtain a microorganism able to ferment all sugars in the biomass, tolerating stress conditions, showing high resistance to inhibitors, and producing a mixture of synergistic enzymes necessary for the complete hydrolysis of all lignocellulose carbohydrates [46,134,138–140]. Among the most frequently modified microorganisms are *Saccharomyces cerevisiae*, *Zymomonas mobilis*, and *Escherichia coli* [46,141–144], but also other bacterial and fungal species were tested, including *Fusarium oxysporum* [145], *Thermoanaerobacter mathranii* [146], and *Corynebacterium glutamicum* [147]. Designing perfectly engineered microorganisms with the maximum conversion of monomeric sugars and enhanced tolerance to operational conditions will allow for economically feasible industrial production of bioethanol from lignocellulosic biomass [12,46].

Another way to increase the fermentation efficiency is to use immobilised recombinant microbial cells. Immobilisation is placing intact cells on a suitable carrier using entrapment within a porous matrix, adsorption on the solid carrier surface, fixing to the carrier surface by covalent bonding or cross-linking, or encapsulation without altering their preferred catalytic activity. A carrier should be nontoxic, biodegradable, and cost-effective. For yeasts cells, mainly Ca-alginate, carrageenan, cellulose, chitosan, silica-hydrogel, and pre-polymers are used as carriers [12,136,148–152].

The sugar-to-ethanol conversion process can be conducted as a batch, fed-batch, or continuous fermentation, where the fed-batch mode in a stirred tank is the most frequently used in the industry since it provides the optimum conditions required for the microbial strain applied [152–155].

Industrial biorefineries employ several fermentation technologies to increase ethanol yield and reduce production costs [24,156].

- Separate hydrolysis and fermentation (SHF)—Hydrolysis and fermentation processes are conducted independently in different units. Carbohydrates from pretreated biomass are degraded to monosugars in a hydrolysis reactor and subsequently converted to ethanol in a fermentation unit. It is a time-consuming and cost-intensive process due to the long residence time needed for complete hydrolysis, high enzyme loading, and material costs required for two separate units, and its main drawback is end-product inhibition (Figure 8A) [157–160].
- Simultaneous saccharification and fermentation (SSF)—Hydrolysis and fermentation are carried out in the same unit, which improves hydrolysis rates, yields, and product concentrations compared to SHF due to the continuous removal of the sugars by the yeasts, which reduces the end-product inhibition of the enzyme complex. The main drawback is the difference in optimum temperature between saccharification and fermentation and enzyme inhibition by ethanol, microorganisms, and temperature in the reactor (Figure 8B) [160–162].
- Simultaneous saccharification and co-fermentation (SSCF)—Hydrolysis and fermentation are carried out in the same unit with concurrent co-fermentation of pentoses using pentose-fermenting strains, which allows converting both hexoses and pentoses from lignocellulosic biomass, thus increasing ethanol yield. This process is suitable for xylose-rich biomass, such as hardwood and agricultural residues; however, the ethanol yield is lower compared to SSF (Figure 8C) [163–166].
- Consolidated bioprocessing (CBP)—A single-step process where hydrolysis, fermentation, and enzyme production occur in the same unit. The method employs genetically modified microbes or microbial consortia (e.g., some yeast strains and *Clostridium thermocellum* have already been tested) capable of hydrolysing biomass with enzymes produced on its own and fermenting monosugars to ethanol. The strategy has the potential to revolutionise bioethanol production due to reduced costs for infrastructure and chemicals, making it economically beneficial and environmentally friendly. However, reaching an industrial scale is challenging because of low conversion efficacy, and it still requires further extensive research (Figure 8D) [167–170].

Effective fermentation of monosugars obtained from lignocellulosic biomass is the next bottleneck in bioethanol production. Several factors might affect its efficiency, including temperature, time, pH, inoculum size, sugar concentration, solid-to-liquid ratio, agitation rate, oxygen content, and rotation speed. Additionally, the operating conditions must be adjusted depending on whether the fermentation is conducted simultaneously or separately with saccharification, which is challenging and requires careful optimisation [12,136,150,171,172].

2.3.4. Distillation and Dehydration

Distillation and dehydration are vital steps for obtaining fuel-grade ethanol from lignocellulosic biomass. Distillation allows for the effective separation of a component substance (such as ethanol) from a miscible liquid mixture (such as fermentation broth) through consecutive selective evaporation and condensation processes based on a difference in their volatilities [173–175]. The water content in the post-fermentation mixture is very high, usually exceeding 80% of the dry weight. Therefore, concentrating ethanol up to 96% requires a huge amount of energy, which generates high costs [45]. The first stage of the process is the so-called "drive away the alcohol". The product (about 37% bioethanol) is then concentrated in a rectification column to a concentration of about 95% and finally dehydrated to a high-quality dry product which holds a minimum of 99.5% ethanol by volume [12,46,176].

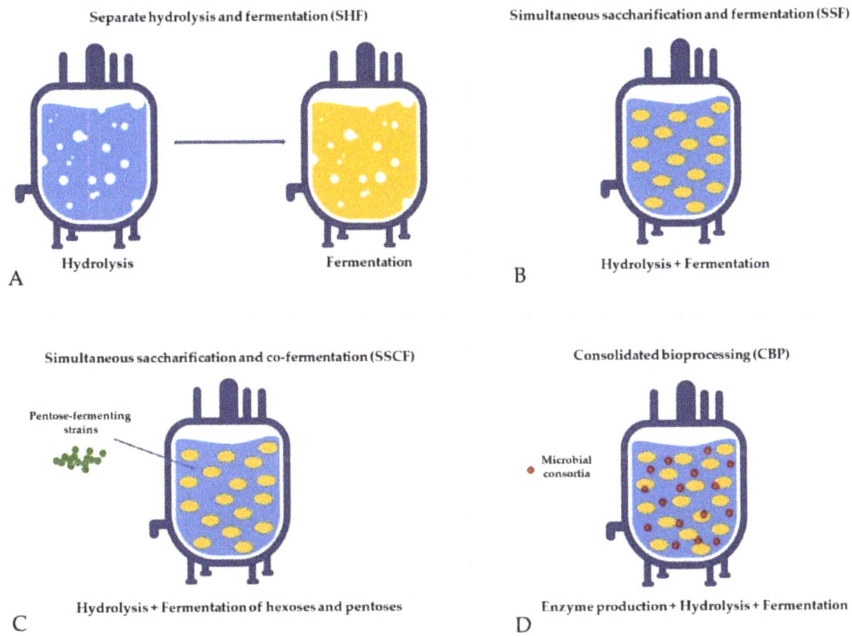

Figure 8. The fermentation strategies used to optimise the process.

Various methods for separating ethanol from a fermentation broth in bioethanol production have been developed, such as adsorption distillation, membrane processes, azeotropic distillation, diffusion distillation, extractive distillation, pervaporation, vacuum distillation, and chemical dehydration, differing in the technique employed, effectiveness and operational costs [17]. Among them, membrane distillation and pervaporation are the most economically viable for bioethanol production.

Membrane distillation is a method that allows for the reduction in the energy expenditure of the process of obtaining ethanol at the stripping stage. During distillation, a membrane separates the fermenting solution from the distillate. Membranes that are used are flat or capillary, porous with gas-filled pores (porosity in the range of 70–85%), hydrophobic (not wetted by liquid), and with high thermal resistance. The process is feasible when there is a pressure difference between molecular components in the gas phase. Different types of membrane distillation have been developed, including contact, air-gap, vacuum, and sweeping gas membrane distillation. The main advantage of using a distillation membrane is the possibility of carrying out the process at a lower temperature. This eliminates the cost of heating the water to the boiling point of ethanol, thus reducing the total costs of bioethanol production. Other advantages of membrane distillation are the possibility of almost 100% retention of non-volatile compounds, lowering the process compared to conventional distillation, obtaining saturated solutions, and implementing durable artificial plastic installations (corrosion-free). Additionally, membrane distillation enables the continuous fermentation process with simultaneous ethanol stripping [17,177–180].

Pervaporation is another type of membrane process that can be employed for obtaining anhydrous bioethanol on an industrial scale. This process uses the difference in ethanol concentrations on both sides of the asymmetric thick polymer membrane. The separation mechanism is based on the differences in the affinity of ethanol and water to the membrane (dissolving and diffusion capacity) and allows the final ethanol dehydration to be 99.8% [17,181–184].

3. Conclusions

The production of second-generation bioethanol has several benefits in offsetting the general use of fossil fuels by increasing global supplies of liquid transport fuels in response to growing energy demand and improving energy security in regions devoid of fossil resource deposits. Thereby, bioethanol contributes to restricting worldwide dependence on fossil supplies and the petroleum industry, thus helping alleviate the energy crisis. Moreover, the transition from petroleum- to biomass-derived fuels reduces net carbon dioxide emissions per unit of energy produced and used, helping tackle anthropogenic climate change and its consequences for people and the environment.

Lignocellulosic biomass used for bioethanol production seems to be a promising renewable energy source. To avoid conflicts of interest, biorefineries should focus on utilising agro- and industry-waste biomass rather than biomass used for animal feed. In this context, lignocellulosic feedstocks employed as an energy source are particularly beneficial. It is abundant, does not threaten food security, and is inexpensive since it can be derived from native vegetation (e.g., invasive species, forest residues and thinnings, and grass), discarded agricultural residues (corn stove and cereal straw), and even industrial urban waste rich in organic matter.

Unfortunately, the production of cellulosic ethanol is highly challenging due to the complexity and recalcitrance of lignocellulose and the diversity of biomass. It requires several steps to release the energy-carrying carbohydrates from the lignocellulosic complex and convert them into ethanol, starting from biomass pretreatment through hydrolysis and fermentation. These three steps contribute their own unique bottlenecks in the entire production process, seriously affecting the final efficiency of the production process and generating high operating costs. Therefore, intensive research has been conducted to develop new technologies that are efficient, economically viable, and universal for various biomass types, while being environmentally friendly.

Although significant progress has been made in this field in the past decade, including the development of advanced engineered microorganisms or attempts to combine pretreatment, hydrolysis, and fermentation, or part of them into a single, more efficient step, there are still several gaps between novel findings and practical applications. Some of the most crucial challenges include the following:

- the selection of a suitable pretreatment strategy that is cost-effective and does not impede the overall efficiency of enzymatic saccharification,
- the improvement of the anaerobic digestibility of biomass,
- limiting carbohydrate degradation and the generation of inhibitors during pretreatment to prevent conversion yield loss,
- downsizing the consumption of toxic chemicals, as well as energy and water,
- the improvement and application of novel biocatalysts that can enhance the efficiency of the saccharification process,
- increasing the efficiency of individual enzymes by designing enzymes with enhanced specific activity, thermal stability, and reduced end-product inhibition, and
- reducing the overall footprint of the process.

Detailed knowledge about the structure and composition of different biomass types is required, as well as the effects of individual pretreatment techniques on various biomass materials at the macro and molecular scales. Additionally, a thorough study of the interactions between biomass, microorganisms, products, and by-products generated during hydrolysis and fermentation at the molecular scale is necessary to establish optimal conditions for those processes. The existing knowledge is broad, but even more comprehensive interdisciplinary research is still needed to bring bioethanol production into a profitable and pervasive light for commercial use. However, it should also be remembered that transitioning from a laboratory to a commercial scale is extremely difficult and requires additional pilot-scale studies with optimisation and high financial expenditure.

As for now, it seems that just using lignocellulosic biomass as a sustainable feedstock for bioethanol production does not guarantee a successful transition from petroleum-based to renewable biomass-derived energy. It seems that the strategy to utilise all components of the lignocellulosic complex by employing cost-competitive manufacturing processes designed with green chemistry is more likely to succeed. The future of this energy sector will be integrated biorefineries that produce both energy and value-added components for the chemical industry based on green chemistry principles with respect to the environment. This is achievable through enhancing the efficiency of all used materials and energy, reducing waste production and toxicity, and reusing resources and by-products. Integrated biorefineries are gaining interest worldwide as they support the circular bioeconomy concept. However, greener processes and technologies are required, such as employing water-based reactions and environmentally friendly oxidants instead of materials and chemicals with high environmental burdens, or those using alternative energy-saving pretreatment methods, such as ultrasound or microwaves, which require time, effort, and financial investment.

Since the 1970s, tremendous progress has been made to alleviate the use of fossil fuels. With the persistent passion of researchers worldwide, there is great optimism for the future of bioethanol from lignocellulosics. This review is meant to not only educate on bioethanol processes and their challenges, but also to illuminate novel and debatable research ideas that will tackle these challenges and build sustainable partnerships in an interdisciplinary fashion to combat the global energy crisis at hand.

Author Contributions: Conceptualisation, M.B., K.S. and D.J.Y.; writing—original draft preparation, K.S., M.B. and D.J.Y.; writing—review and editing, M.B., D.J.Y. and K.S.; visualisation, M.B. All authors have read and agreed to the published version of the manuscript.

Funding: This research received no external funding.

Institutional Review Board Statement: Not applicable.

Informed Consent Statement: Not applicable.

Data Availability Statement: Not applicable.

Conflicts of Interest: The authors declare no conflict of interest.

References

1. Zabed, H.; Sahu, J.N.; Suely, A.; Boyce, A.N.; Faruq, G. Bioethanol Production from Renewable Sources: Current Perspectives and Technological Progress. *Renew. Sustain. Energy Rev.* **2017**, *71*, 475–501. [CrossRef]
2. Abas, N.; Kalair, A.; Khan, N. Review of Fossil Fuels and Future Energy Technologies. *Futures* **2015**, *69*, 31–49. [CrossRef]
3. Safarian, S.; Unnthorsson, R. An Assessment of the Sustainability of Lignocellulosic Bioethanol Production from Wastes in Iceland. *Energies* **2018**, *11*, 1493. [CrossRef]
4. Oh, Y.-K.; Hwang, K.-R.; Kim, C.; Kim, J.R.; Lee, J.-S. Recent Developments and Key Barriers to Advanced Biofuels: A Short Review. *Bioresour. Technol.* **2018**, *257*, 320–333. [CrossRef] [PubMed]
5. Luque, R.; Herrero-Davila, L.; Campelo, J.M.; Clark, J.H.; Hidalgo, J.M.; Luna, D.; Marinas, J.M.; Romero, A.A. Biofuels: A Technological Perspective. *Energy Environ. Sci.* **2008**, *1*, 542–564. [CrossRef]
6. Jeswani, H.K.; Chilvers, A.; Azapagic, A. Environmental Sustainability of Biofuels: A Review. *Proc. R. Soc. A Math. Phys. Eng. Sci.* **2020**, *476*, 20200351. [CrossRef]
7. Liu, Y.; Cruz-Morales, P.; Zargar, A.; Belcher, M.S.; Pang, B.; Englund, E.; Dan, Q.; Yin, K.; Keasling, J.D. Biofuels for a Sustainable Future. *Cell* **2021**, *184*, 1636–1647. [CrossRef]
8. Efemwenkiekie, U.K.; Oyedepo, S.O.; Idiku, U.D.; Uguru-Okorie, D.C.; Kuhe, A. Comparative Analysis of a Four Stroke Spark Ignition Engine Performance Using Local Ethanol and Gasoline Blends. *Procedia Manuf.* **2019**, *35*, 1079–1086. [CrossRef]
9. Gabisa, E.W.; Gheewala, S.H. Can Substitution of Imported Gasoline by Locally Produced Molasses Ethanol in Ethiopia Be Sustainable? An Eco-Efficiency Assessment. *Renew. Sustain. Energy Rev.* **2020**, *123*, 109770. [CrossRef]
10. Iodice, P.; Senatore, A.; Langella, G.; Amoresano, A. Advantages of Ethanol–Gasoline Blends as Fuel Substitute for Last Generation Si Engines. *Environ. Prog. Sustain. Energy* **2017**, *36*, 1173–1179. [CrossRef]
11. Anderson, S.T. The Demand for Ethanol as a Gasoline Substitute. *J. Environ. Econ. Manag.* **2012**, *63*, 151–168. [CrossRef]
12. Robak, K.; Balcerek, M. Current State-of-the-Art in Ethanol Production from Lignocellulosic Feedstocks. *Microbiol. Res.* **2020**, *240*, 126534. [CrossRef] [PubMed]
13. Devi, A.; Bajar, S.; Kour, H.; Kothari, R.; Pant, D.; Singh, A. Lignocellulosic Biomass Valorization for Bioethanol Production: A Circular Bioeconomy Approach. *Bioenerg. Res.* **2022**, *15*, 1820–1841. [CrossRef] [PubMed]

14. Bertrand, E.; Vandenberghe, L.P.S.; Soccol, C.R.; Sigoillot, J.-C.; Faulds, C. First Generation Bioethanol. In *Green Fuels Technology: Biofuels*; Green Energy and Technology; Soccol, C.R., Brar, S.K., Faulds, C., Ramos, L.P., Eds.; Springer International Publishing: Cham, Switzerland, 2016; pp. 175–212. ISBN 978-3-319-30205-8.
15. Murawski de Mello, A.F.; Porto de Souza Vandenberghe, L.; Valladares-Diestra, K.K.; Amaro Bittencourt, G.; Martinez Burgos, W.J.; Soccol, C.R. Corn First-Generation Bioethanol Unities with Energy and Dried Grains with Solubles (DDGS) Production. In *Liquid Biofuels: Bioethanol*; Biofuel and Biorefinery Technologies; Soccol, C.R., Amarante Guimarães Pereira, G., Dussap, C.-G., Porto de Souza Vandenberghe, L., Eds.; Springer International Publishing: Cham, Switzerland, 2022; pp. 109–132. ISBN 978-3-031-01241-9.
16. Mohanty, S.K.; Swain, M.R. Chapter 3—Bioethanol Production From Corn and Wheat: Food, Fuel, and Future. In *Bioethanol Production from Food Crops*; Ray, R.C., Ramachandran, S., Eds.; Academic Press: Cambridge, MA, USA, 2019; pp. 45–59. ISBN 978-0-12-813766-6.
17. Aditiya, H.B.; Mahlia, T.M.I.; Chong, W.T.; Nur, H.; Sebayang, A.H. Second Generation Bioethanol Production: A Critical Review. *Renew. Sustain. Energy Rev.* **2016**, *66*, 631–653. [CrossRef]
18. Baig, K.S.; Wu, J.; Turcotte, G. Future Prospects of Delignification Pretreatments for the Lignocellulosic Materials to Produce Second Generation Bioethanol. *Int. J. Energy Res.* **2019**, *43*, 1411–1427. [CrossRef]
19. Saïdane-Bchir, F.; El Falleh, A.; Ghabbarou, E.; Hamdi, M. 3rd Generation Bioethanol Production from Microalgae Isolated from Slaughterhouse Wastewater. *Waste Biomass Valor.* **2016**, *7*, 1041–1046. [CrossRef]
20. Jambo, S.A.; Abdulla, R.; Azhar, S.H.M.; Marbawi, H.; Gansau, J.A.; Ravindra, P. A Review on Third Generation Bioethanol Feedstock. *Renew. Sustain. Energy Rev.* **2016**, *65*, 756–769. [CrossRef]
21. Veza, I.; Hoang, A.T.; Abbas, M.M.; Tamaldin, N.; Idris, M.; Djamari, D.W.; Sule, A.; Maulana, E.; Putra, N.R.; Opia, A.C. Microalgae and Macroalgae for Third-Generation Bioethanol Production. In *Liquid Biofuels: Bioethanol*; Biofuel and Biorefinery Technologies; Soccol, C.R., Amarante Guimarães Pereira, G., Dussap, C.-G., Porto de Souza Vandenberghe, L., Eds.; Springer International Publishing: Cham, Switzerland, 2022; pp. 301–331. ISBN 978-3-031-01241-9.
22. Association, R.F. Annual Ethanol Production. Available online: https://ethanolrfa.org/markets-and-statistics/annual-ethanol-production (accessed on 28 October 2022).
23. European Union: Biofuels Annual. Available online: https://www.fas.usda.gov/data/european-union-biofuels-annual-1 (accessed on 29 October 2022).
24. Sharma, B.; Larroche, C.; Dussap, C.-G. Comprehensive Assessment of 2G Bioethanol Production. *Bioresour. Technol.* **2020**, *313*, 123630. [CrossRef]
25. Bhatia, L.; Bachheti, R.K.; Garlapati, V.K.; Chandel, A.K. Third-Generation Biorefineries: A Sustainable Platform for Food, Clean Energy, and Nutraceuticals Production. *Biomass Conv. Bioref.* **2022**, *12*, 4215–4230. [CrossRef]
26. Reports. Available online: https://usdabrazil.org.br/en/reports/ (accessed on 29 October 2022).
27. Association, R.F. Ethanol Biorefinery Locations. Available online: https://ethanolrfa.org/resources/ethanol-biorefinery-locations (accessed on 29 October 2022).
28. Tan, K.T.; Lee, K.T.; Mohamed, A.R. Role of Energy Policy in Renewable Energy Accomplishment: The Case of Second-Generation Bioethanol. *Energy Policy* **2008**, *36*, 3360–3365. [CrossRef]
29. Neto, A.C.; Guimarães, M.J.O.C.; Freire, E. Business Models for Commercial Scale Second-Generation Bioethanol Production. *J. Clean. Prod.* **2018**, *184*, 168–178. [CrossRef]
30. Abo, B.O.; Gao, M.; Wang, Y.; Wu, C.; Ma, H.; Wang, Q. Lignocellulosic Biomass for Bioethanol: An Overview on Pretreatment, Hydrolysis and Fermentation Processes. *Rev. Environ. Health* **2019**, *34*, 57–68. [CrossRef]
31. Arefin, M.A.; Rashid, F.; Islam, A. A Review of Biofuel Production from Floating Aquatic Plants: An Emerging Source of Bio-Renewable Energy. *Biofuels Bioprod. Biorefin.* **2021**, *15*, 574–591. [CrossRef]
32. Ge, X.; Burner, D.M.; Xu, J.; Phillips, G.C.; Sivakumar, G. Bioethanol Production from Dedicated Energy Crops and Residues in Arkansas, USA. *Biotechnol. J.* **2011**, *6*, 66–73. [CrossRef] [PubMed]
33. Chen, J.; Zhang, B.; Luo, L.; Zhang, F.; Yi, Y.; Shan, Y.; Liu, B.; Zhou, Y.; Wang, X.; Lü, X. A Review on Recycling Techniques for Bioethanol Production from Lignocellulosic Biomass. *Renew. Sustain. Energy Rev.* **2021**, *149*, 111370. [CrossRef]
34. Zhang, B.; Gao, Y.; Zhang, L.; Zhou, Y. The Plant Cell Wall: Biosynthesis, Construction, and Functions. *J. Integr. Plant Biol.* **2021**, *63*, 251–272. [CrossRef]
35. Baruah, J.; Nath, B.K.; Sharma, R.; Kumar, S.; Deka, R.C.; Baruah, D.C.; Kalita, E. Recent Trends in the Pretreatment of Lignocellulosic Biomass for Value-Added Products. *Front. Energy Res.* **2018**, *6*, 141. [CrossRef]
36. Sakakibara, A. A Structural Model of Softwood Lignin. *Wood Sci.Technol.* **1980**, *14*, 89–100. [CrossRef]
37. Baghaei, B.; Skrifvars, M. All-Cellulose Composites: A Review of Recent Studies on Structure, Properties and Applications. *Molecules* **2020**, *25*, 2836. [CrossRef]
38. Heinze, T. Cellulose: Structure and Properties. In *Cellulose Chemistry and Properties: Fibers, Nanocelluloses and Advanced Materials*; Advances in Polymer Science; Rojas, O.J., Ed.; Springer International Publishing: Cham, Switzerland, 2016; pp. 1–52. ISBN 978-3-319-26015-0.
39. Gu, J.; Catchmark, J.M. The Impact of Cellulose Structure on Binding Interactions with Hemicellulose and Pectin. *Cellulose* **2013**, *20*, 1613–1627. [CrossRef]

40. Abdel-Hamid, A.M.; Solbiati, J.O.; Cann, I.K.O. Chapter One—Insights into Lignin Degradation and Its Potential Industrial Applications. In *Advances in Applied Microbiology*; Sariaslani, S., Gadd, G.M., Eds.; Academic Press: Cambridge, MA, USA, 2013; Volume 82, pp. 1–28.
41. Zhang, N.; Li, S.; Xiong, L.; Hong, Y.; Chen, Y. Cellulose-Hemicellulose Interaction in Wood Secondary Cell-Wall. *Model. Simul. Mat. Sci. Eng.* **2015**, *23*, 085010. [CrossRef]
42. Sheng, Y.; Lam, S.S.; Wu, Y.; Ge, S.; Wu, J.; Cai, L.; Huang, Z.; Le, Q.V.; Sonne, C.; Xia, C. Enzymatic Conversion of Pretreated Lignocellulosic Biomass: A Review on Influence of Structural Changes of Lignin. *Bioresour. Technol.* **2021**, *324*, 124631. [CrossRef] [PubMed]
43. Ralph, J.; Lapierre, C.; Boerjan, W. Lignin Structure and Its Engineering. *Curr. Opin. Biotechnol.* **2019**, *56*, 240–249. [CrossRef] [PubMed]
44. Yelle, D.J.; Ralph, J.; Frihart, C.R. Characterization of Nonderivatized Plant Cell Walls Using High-Resolution Solution-State NMR Spectroscopy. *Magn. Reson. Chem.* **2008**, *46*, 508–517. [CrossRef]
45. Haghighi Mood, S.; Hossein Golfeshan, A.; Tabatabaei, M.; Salehi Jouzani, G.; Najafi, G.H.; Gholami, M.; Ardjmand, M. Lignocellulosic Biomass to Bioethanol, a Comprehensive Review with a Focus on Pretreatment. *Renew. Sustain. Energy Rev.* **2013**, *27*, 77–93. [CrossRef]
46. Zaldivar, J.; Nielsen, J.; Olsson, L. Fuel Ethanol Production from Lignocellulose: A Challenge for Metabolic Engineering and Process Integration. *Appl. Microbiol. Biotechnol.* **2001**, *56*, 17–34. [CrossRef]
47. Das, N.; Jena, P.K.; Padhi, D.; Kumar Mohanty, M.; Sahoo, G. A Comprehensive Review of Characterization, Pretreatment and Its Applications on Different Lignocellulosic Biomass for Bioethanol Production. *Biomass Conv. Bioref.* **2021**, *82*, 1–25. [CrossRef]
48. Awogbemi, O.; Von Kallon, D.V. Pretreatment Techniques for Agricultural Waste. *Case Stud. Chem. Environ. Eng.* **2022**, *6*, 100229. [CrossRef]
49. Hassan, S.S.; Williams, G.A.; Jaiswal, A.K. Emerging Technologies for the Pretreatment of Lignocellulosic Biomass. *Bioresour. Technol.* **2018**, *262*, 310–318. [CrossRef]
50. Shen, F.; Xiong, X.; Fu, J.; Yang, J.; Qiu, M.; Qi, X.; Tsang, D.C.W. Recent Advances in Mechanochemical Production of Chemicals and Carbon Materials from Sustainable Biomass Resources. *Renew. Sustain. Energy Rev.* **2020**, *130*, 109944. [CrossRef]
51. Duque, A.; Manzanares, P.; Ballesteros, M. Extrusion as a Pretreatment for Lignocellulosic Biomass: Fundamentals and Applications. *Renew. Energy* **2017**, *114*, 1427–1441. [CrossRef]
52. Zheng, J.; Rehmann, L. Extrusion Pretreatment of Lignocellulosic Biomass: A Review. *Int. J. Mol. Sci.* **2014**, *15*, 18967–18984. [CrossRef] [PubMed]
53. Rooni, V.; Raud, M.; Kikas, T. The Freezing Pre-Treatment of Lignocellulosic Material: A Cheap Alternative for Nordic Countries. *Energy* **2017**, *139*, 1–7. [CrossRef]
54. Chang, K.-L.; Thitikorn-amorn, J.; Hsieh, J.-F.; Ou, B.-M.; Chen, S.-H.; Ratanakhanokchai, K.; Huang, P.-J.; Chen, S.-T. Enhanced Enzymatic Conversion with Freeze Pretreatment of Rice Straw. *Biomass Bioenergy* **2011**, *35*, 90–95. [CrossRef]
55. Subhedar, P.B.; Ray, P.; Gogate, P.R. Intensification of Delignification and Subsequent Hydrolysis for the Fermentable Sugar Production from Lignocellulosic Biomass Using Ultrasonic Irradiation. *Ultrason. Sonochem.* **2018**, *40*, 140–150. [CrossRef] [PubMed]
56. Rezania, S.; Oryani, B.; Cho, J.; Talaiekhozani, A.; Sabbagh, F.; Hashemi, B.; Rupani, P.F.; Mohammadi, A.A. Different Pretreatment Technologies of Lignocellulosic Biomass for Bioethanol Production: An Overview. *Energy* **2020**, *199*, 117457. [CrossRef]
57. Kumar, P.; Barrett, D.M.; Delwiche, M.J.; Stroeve, P. Pulsed Electric Field Pretreatment of Switchgrass and Wood Chip Species for Biofuel Production. *Ind. Eng. Chem. Res.* **2011**, *50*, 10996–11001. [CrossRef]
58. Broda, M.; Leja, K.; Czaczyk, K.; Grajek, W. The New Methods of Corn Disinfection Used in Bioethanol Production. *J. Biobased Mater. Bioenergy* **2010**, *4*, 430–435. [CrossRef]
59. Zhang, K.; Pei, Z.; Wang, D. Organic Solvent Pretreatment of Lignocellulosic Biomass for Biofuels and Biochemicals: A Review. *Bioresour. Technol.* **2016**, *199*, 21–33. [CrossRef]
60. Joy, S.P.; Krishnan, C. Modified Organosolv Pretreatment for Improved Cellulosic Ethanol Production from Sorghum Biomass. *Ind. Crops Prod.* **2022**, *177*, 114409. [CrossRef]
61. Asim, A.M.; Uroos, M.; Naz, S.; Sultan, M.; Griffin, G.; Muhammad, N.; Khan, A.S. Acidic Ionic Liquids: Promising and Cost-Effective Solvents for Processing of Lignocellulosic Biomass. *J. Mol. Liq.* **2019**, *287*, 110943. [CrossRef]
62. Lin, X.; Jiang, K.; Liu, X.; Han, D.; Zhang, Q. Review on Development of Ionic Liquids in Lignocellulosic Biomass Refining. *J. Mol. Liq.* **2022**, *359*, 119326. [CrossRef]
63. Yoon, L.W.; Rafi, I.S.; Ngoh, G.C. Feasibility of Eliminating Washing Step in Bioethanol Production Using Deep Eutectic Solvent Pretreated Lignocellulosic Substrate. *Chem. Eng. Res. Des.* **2022**, *179*, 257–264. [CrossRef]
64. Hassan, E.-S.R.E.; Mutelet, F. Evaluation of Miscanthus Pretreatment Effect by Choline Chloride Based Deep Eutectic Solvents on Bioethanol Production. *Bioresour. Technol.* **2022**, *345*, 126460. [CrossRef] [PubMed]
65. Loow, Y.-L.; Wu, T.Y.; Tan, K.A.; Lim, Y.S.; Siow, L.F.; Jahim, J.M.; Mohammad, A.W.; Teoh, W.H. Recent Advances in the Application of Inorganic Salt Pretreatment for Transforming Lignocellulosic Biomass into Reducing Sugars. *J. Agric. Food Chem.* **2015**, *63*, 8349–8363. [CrossRef]
66. Banerjee, D.; Mukherjee, S.; Pal, S.; Khowala, S. Enhanced Saccharification Efficiency of Lignocellulosic Biomass of Mustard Stalk and Straw by Salt Pretreatment. *Ind. Crops Prod.* **2016**, *80*, 42–49. [CrossRef]

67. Den, W.; Sharma, V.K.; Lee, M.; Nadadur, G.; Varma, R.S. Lignocellulosic Biomass Transformations via Greener Oxidative Pretreatment Processes: Access to Energy and Value-Added Chemicals. *Front. Chem.* **2018**, *6*, 141. [CrossRef]
68. Ab Rasid, N.S.; Zainol, M.M.; Amin, N.A.S. 14—Pretreatment of Agroindustry Waste by Ozonolysis for Synthesis of Biorefinery Products. In *Refining Biomass Residues for Sustainable Energy and Bioproducts*; Kumar, R.P., Gnansounou, E., Raman, J.K., Baskar, G., Eds.; Academic Press: Cambridge, MA, USA, 2020; pp. 303–336. ISBN 978-0-12-818996-2.
69. Travaini, R.; Martín-Juárez, J.; Lorenzo-Hernando, A.; Bolado-Rodríguez, S. Ozonolysis: An Advantageous Pretreatment for Lignocellulosic Biomass Revisited. *Bioresour. Technol.* **2016**, *199*, 2–12. [CrossRef]
70. Dey, P.; Pal, P.; Kevin, J.D.; Das, D.B. Lignocellulosic Bioethanol Production: Prospects of Emerging Membrane Technologies to Improve the Process—A Critical Review. *Rev. Chem. Eng.* **2020**, *36*, 333–367. [CrossRef]
71. Balan, V.; Bals, B.; Chundawat, S.P.S.; Marshall, D.; Dale, B.E. Lignocellulosic Biomass Pretreatment Using AFEX. In *Biofuels: Methods and Protocols*; Methods in Molecular Biology; Mielenz, J.R., Ed.; Humana Press: Totowa, NJ, USA, 2009; pp. 61–77. ISBN 978-1-60761-214-8.
72. Zhao, C.; Shao, Q.; Chundawat, S.P.S. Recent Advances on Ammonia-Based Pretreatments of Lignocellulosic Biomass. *Bioresour. Technol.* **2020**, *298*, 122446. [CrossRef]
73. Chundawat, S.P.; Pal, R.K.; Zhao, C.; Campbell, T.; Teymouri, F.; Videto, J.; Nielson, C.; Wieferich, B.; Sousa, L.; Dale, B.E. Ammonia Fiber Expansion (AFEX) Pretreatment of Lignocellulosic Biomass. *J. Vis. Exp.* **2020**, *158*, e57488. [CrossRef] [PubMed]
74. Narayanaswamy, N.; Faik, A.; Goetz, D.J.; Gu, T. Supercritical Carbon Dioxide Pretreatment of Corn Stover and Switchgrass for Lignocellulosic Ethanol Production. *Bioresour. Technol.* **2011**, *102*, 6995–7000. [CrossRef] [PubMed]
75. Gu, T. Pretreatment of Lignocellulosic Biomass Using Supercritical Carbon Dioxide as a Green Solvent. In *Green Biomass Pretreatment for Biofuels Production*; SpringerBriefs in Molecular Science; Gu, T., Ed.; Springer: Dordrecht, The Netherlands, 2013; pp. 107–125. ISBN 978-94-007-6052-3.
76. Badgujar, K.C.; Dange, R.; Bhanage, B.M. Recent Advances of Use of the Supercritical Carbon Dioxide for the Biomass Pre-Treatment and Extraction: A Mini-Review. *J. Indian Chem. Soc.* **2021**, *98*, 100018. [CrossRef]
77. Zhuang, X.; Wang, W.; Yu, Q.; Qi, W.; Wang, Q.; Tan, X.; Zhou, G.; Yuan, Z. Liquid Hot Water Pretreatment of Lignocellulosic Biomass for Bioethanol Production Accompanying with High Valuable Products. *Biores. Technol.* **2016**, *199*, 68–75. [CrossRef]
78. Arvaniti, E.; Bjerre, A.B.; Schmidt, J.E. Wet Oxidation Pretreatment of Rape Straw for Ethanol Production. *Biomass Bioenergy* **2012**, *39*, 94–105. [CrossRef]
79. Ummalyma, S.B.; Supriya, R.D.; Sindhu, R.; Binod, P.; Nair, R.B.; Pandey, A.; Gnansounou, E. Chapter 7—Biological Pretreatment of Lignocellulosic Biomass—Current Trends and Future Perspectives. In *Second and Third Generation of Feedstocks*; Basile, A., Dalena, F., Eds.; Elsevier: Amsterdam, The Netherlands, 2019; pp. 197–212. ISBN 978-0-12-815162-4.
80. López-Abelairas, M.; Álvarez Pallín, M.; Salvachúa, D.; Lú-Chau, T.; Martínez, M.J.; Lema, J.M. Optimisation of the Biological Pretreatment of Wheat Straw with White-Rot Fungi for Ethanol Production. *Bioprocess Biosyst. Eng.* **2013**, *36*, 1251–1260. [CrossRef]
81. Vasco-Correa, J.; Ge, X.; Li, Y. Chapter 24—Biological Pretreatment of Lignocellulosic Biomass. In *Biomass Fractionation Technologies for a Lignocellulosic Feedstock Based Biorefinery*; Mussatto, S.I., Ed.; Elsevier: Amsterdam, The Netherlands, 2016; pp. 561–585. ISBN 978-0-12-802323-5.
82. Dutra, E.D.; Santos, F.A.; Alencar, B.R.A.; Reis, A.L.S.; de Fatima Rodrigues de Souza, R.; da Silva Aquino, K.A.; Morais, M.A.M., Jr.; Menezes, R.S.C. Alkaline Hydrogen Peroxide Pretreatment of Lignocellulosic Biomass: Status and Perspectives. *Biomass Conv. Bioref.* **2018**, *8*, 225–234. [CrossRef]
83. Meenakshisundaram, S.; Fayeulle, A.; Leonard, E.; Ceballos, C.; Pauss, A. Fiber Degradation and Carbohydrate Production by Combined Biological and Chemical/Physicochemical Pretreatment Methods of Lignocellulosic Biomass—A Review. *Bioresour. Technol.* **2021**, *331*, 125053. [CrossRef]
84. Si, M.; Liu, D.; Liu, M.; Yan, X.; Gao, C.; Chai, L.; Shi, Y. Complementary Effect of Combined Bacterial-Chemical Pretreatment to Promote Enzymatic Digestibility of Lignocellulose Biomass. *Bioresour. Technol.* **2019**, *272*, 275–280. [CrossRef]
85. Yu, H.; Du, W.; Zhang, J.; Ma, F.; Zhang, X.; Zhong, W. Fungal Treatment of Cornstalks Enhances the Delignification and Xylan Loss during Mild Alkaline Pretreatment and Enzymatic Digestibility of Glucan. *Bioresour. Technol.* **2010**, *101*, 6728–6734. [CrossRef]
86. Martínez-Patiño, J.C.; Lu-Chau, T.A.; Gullón, B.; Ruiz, E.; Romero, I.; Castro, E.; Lema, J.M. Application of a Combined Fungal and Diluted Acid Pretreatment on Olive Tree Biomass. *Ind. Crops Prod.* **2018**, *121*, 10–17. [CrossRef]
87. Yan, X.; Wang, Z.; Zhang, K.; Si, M.; Liu, M.; Chai, L.; Liu, X.; Shi, Y. Bacteria-Enhanced Dilute Acid Pretreatment of Lignocellulosic Biomass. *Bioresour. Technol.* **2017**, *245*, 419–425. [CrossRef] [PubMed]
88. Paudel, S.R.; Banjara, S.P.; Choi, O.K.; Park, K.Y.; Kim, Y.M.; Lee, J.W. Pretreatment of Agricultural Biomass for Anaerobic Digestion: Current State and Challenges. *Bioresour. Technol.* **2017**, *245*, 1194–1205. [CrossRef] [PubMed]
89. Xie, C.; Gong, W.; Yang, Q.; Zhu, Z.; Yan, L.; Hu, Z.; Peng, Y. White-Rot Fungi Pretreatment Combined with Alkaline/Oxidative Pretreatment to Improve Enzymatic Saccharification of Industrial Hemp. *Bioresour. Technol.* **2017**, *243*, 188–195. [CrossRef] [PubMed]
90. Kandhola, G.; Rajan, K.; Labbé, N.; Chmely, S.; Heringer, N.; Kim, J.-W.; Hood, E.E.; Carrier, D.J. Beneficial Effects of Trametes Versicolor Pretreatment on Saccharification and Lignin Enrichment of Organosolv-Pretreated Pinewood. *RSC Adv.* **2017**, *7*, 45652–45661. [CrossRef]

91. Zhuo, S.; Yan, X.; Liu, D.; Si, M.; Zhang, K.; Liu, M.; Peng, B.; Shi, Y. Use of Bacteria for Improving the Lignocellulose Biorefinery Process: Importance of Pre-Erosion. *Biotechnol. Biofuels* **2018**, *11*, 146. [CrossRef]
92. Li, X.; Shi, Y.; Kong, W.; Wei, J.; Song, W.; Wang, S. Improving Enzymatic Hydrolysis of Lignocellulosic Biomass by Bio-Coordinated Physicochemical Pretreatment—A Review. *Energy Rep.* **2022**, *8*, 696–709. [CrossRef]
93. Song, B.; Lin, R.; Lam, C.H.; Wu, H.; Tsui, T.-H.; Yu, Y. Recent Advances and Challenges of Inter-Disciplinary Biomass Valorization by Integrating Hydrothermal and Biological Techniques. *Renew. Sustain. Energy Rev.* **2021**, *135*, 110370. [CrossRef]
94. Moreno, A.D.; Ibarra, D.; Alvira, P.; Tomás-Pejó, E.; Ballesteros, M. Exploring Laccase and Mediators Behavior during Saccharification and Fermentation of Steam-Exploded Wheat Straw for Bioethanol Production. *J. Chem. Technol. Biotechnol.* **2016**, *91*, 1816–1825. [CrossRef]
95. Ramadoss, G.; Muthukumar, K. Mechanistic Study on Ultrasound Assisted Pretreatment of Sugarcane Bagasse Using Metal Salt with Hydrogen Peroxide for Bioethanol Production. *Ultrason. Sonochem.* **2016**, *28*, 207–217. [CrossRef]
96. Moodley, P.; Sewsynker-Sukai, Y.; Gueguim Kana, E.B. Progress in the Development of Alkali and Metal Salt Catalysed Lignocellulosic Pretreatment Regimes: Potential for Bioethanol Production. *Bioresour. Technol.* **2020**, *310*, 123372. [CrossRef]
97. Loow, Y.-L.; Wu, T.Y.; Yang, G.H.; Ang, L.Y.; New, E.K.; Siow, L.F.; Jahim, J.M.; Mohammad, A.W.; Teoh, W.H. Deep Eutectic Solvent and Inorganic Salt Pretreatment of Lignocellulosic Biomass for Improving Xylose Recovery. *Bioresour. Technol.* **2018**, *249*, 818–825. [CrossRef] [PubMed]
98. Ho, M.C.; Ong, V.Z.; Wu, T.Y. Potential Use of Alkaline Hydrogen Peroxide in Lignocellulosic Biomass Pretreatment and Valorization—A Review. *Renew. Sustain. Energy Rev.* **2019**, *112*, 75–86. [CrossRef]
99. Sitotaw, Y.W.; Habtu, N.G.; Gebreyohannes, A.Y.; Nunes, S.P.; Van Gerven, T. Ball Milling as an Important Pretreatment Technique in Lignocellulose Biorefineries: A Review. *Biomass Conv. Bioref.* **2021**, 1–24. [CrossRef]
100. Yu, Y.; Wu, J.; Ren, X.; Lau, A.; Rezaei, H.; Takada, M.; Bi, X.; Sokhansanj, S. Steam Explosion of Lignocellulosic Biomass for Multiple Advanced Bioenergy Processes: A Review. *Renew. Sustain. Energy Rev.* **2022**, *154*, 111871. [CrossRef]
101. Zhao, L.; Sun, Z.-F.; Zhang, C.-C.; Nan, J.; Ren, N.-Q.; Lee, D.-J.; Chen, C. Advances in Pretreatment of Lignocellulosic Biomass for Bioenergy Production: Challenges and Perspectives. *Bioresour. Technol.* **2022**, *343*, 126123. [CrossRef]
102. Moreno, A.D.; Ibarra, D.; Alvira, P.; Tomás-Pejó, E.; Ballesteros, M. A Review of Biological Delignification and Detoxification Methods for Lignocellulosic Bioethanol Production. *Crit. Rev. Biotechnol.* **2015**, *35*, 342–354. [CrossRef] [PubMed]
103. da Nogueira, C.; de Araújo Padilha, C.E.; de Medeiros Dantas, J.M.; de Medeiros, F.G.M.; de Araújo Guilherme, A.; de Santana Souza, D.F.; dos Santos, E.S. In-Situ Detoxification Strategies to Boost Bioalcohol Production from Lignocellulosic Biomass. *Renew. Energy* **2021**, *180*, 914–936. [CrossRef]
104. Tsai, C.-T.; Meyer, A.S. Enzymatic Cellulose Hydrolysis: Enzyme Reusability and Visualization of β-Glucosidase Immobilized in Calcium Alginate. *Molecules* **2014**, *19*, 19390–19406. [CrossRef]
105. Srivastava, N.; Srivastava, M.; Mishra, P.K.; Gupta, V.K.; Molina, G.; Rodriguez-Couto, S.; Manikanta, A.; Ramteke, P.W. Applications of Fungal Cellulases in Biofuel Production: Advances and Limitations. *Renew. Sustain. Energy Rev.* **2018**, *82*, 2379–2386. [CrossRef]
106. Zhang, X.-Z.; Zhang, Y.-H.P. Cellulases: Characteristics, Sources, Production, and Applications. In *Bioprocessing Technologies in Biorefinery for Sustainable Production of Fuels, Chemicals, and Polymers*; John Wiley & Sons, Ltd.: Hoboken, NJ, USA; pp. 131–146. ISBN 978-1-118-64204-7.
107. Singhania, R.R.; Adsul, M.; Pandey, A.; Patel, A.K. 4—Cellulases. In *Current Developments in Biotechnology and Bioengineering*; Pandey, A., Negi, S., Soccol, C.R., Eds.; Elsevier: Amsterdam, The Netherlands, 2017; pp. 73–101. ISBN 978-0-444-63662-1.
108. Mohanram, S.; Amat, D.; Choudhary, J.; Arora, A.; Nain, L. Novel Perspectives for Evolving Enzyme Cocktails for Lignocellulose Hydrolysis in Biorefineries. *Sustain. Chem. Processes* **2013**, *1*, 15. [CrossRef]
109. Kristensen, J.B.; Börjesson, J.; Bruun, M.H.; Tjerneld, F.; Jørgensen, H. Use of Surface Active Additives in Enzymatic Hydrolysis of Wheat Straw Lignocellulose. *Enzym. Microb. Technol.* **2007**, *40*, 888–895. [CrossRef]
110. Malherbe, S.; Cloete, T.E. Lignocellulose Biodegradation: Fundamentals and Applications. *Rev. Environ. Sci. Biotechnol.* **2002**, *1*, 105–114. [CrossRef]
111. Bhattacharya, A.S.; Bhattacharya, A.; Pletschke, B.I. Synergism of Fungal and Bacterial Cellulases and Hemicellulases: A Novel Perspective for Enhanced Bio-Ethanol Production. *Biotechnol. Lett.* **2015**, *37*, 1117–1129. [CrossRef]
112. Abdou Alio, M.; Tugui, O.-C.; Rusu, L.; Pons, A.; Vial, C. Hydrolysis and Fermentation Steps of a Pretreated Sawmill Mixed Feedstock for Bioethanol Production in a Wood Biorefinery. *Bioresour. Technol.* **2020**, *310*, 123412. [CrossRef]
113. Zhang, H.; Zhang, S.; Yuan, H.; Lyu, G.; Xie, J. FeCl3-Catalyzed Ethanol Pretreatment of Sugarcane Bagasse Boosts Sugar Yields with Low Enzyme Loadings and Short Hydrolysis Time. *Bioresour. Technol.* **2018**, *249*, 395–401. [CrossRef]
114. Du, J.; Li, Y.; Zhang, H.; Zheng, H.; Huang, H. Factors to Decrease the Cellulose Conversion of Enzymatic Hydrolysis of Lignocellulose at High Solid Concentrations. *Cellulose* **2014**, *21*, 2409–2417. [CrossRef]
115. da Silva, A.S.; Espinheira, R.P.; Teixeira, R.S.S.; de Souza, M.F.; Ferreira-Leitão, V.; Bon, E.P.S. Constraints and Advances in High-Solids Enzymatic Hydrolysis of Lignocellulosic Biomass: A Critical Review. *Biotechnol. Biofuels* **2020**, *13*, 58. [CrossRef]
116. Qiu, J.; Ma, L.; Shen, F.; Yang, G.; Zhang, Y.; Deng, S.; Zhang, J.; Zeng, Y.; Hu, Y. Pretreating Wheat Straw by Phosphoric Acid plus Hydrogen Peroxide for Enzymatic Saccharification and Ethanol Production at High Solid Loading. *Bioresour. Technol.* **2017**, *238*, 174–181. [CrossRef]

117. Santos, C.A.; Morais, M.A.B.; Terrett, O.M.; Lyczakowski, J.J.; Zanphorlin, L.M.; Ferreira-Filho, J.A.; Tonoli, C.C.C.; Murakami, M.T.; Dupree, P.; Souza, A.P. An Engineered GH1 β-Glucosidase Displays Enhanced Glucose Tolerance and Increased Sugar Release from Lignocellulosic Materials. *Sci. Rep.* **2019**, *9*, 4903. [CrossRef]
118. Chen, X.; Zhai, R.; Shi, K.; Yuan, Y.; Dale, B.E.; Gao, Z.; Jin, M. Mixing Alkali Pretreated and Acid Pretreated Biomass for Cellulosic Ethanol Production Featuring Reduced Chemical Use and Decreased Inhibitory Effect. *Ind. Crops Prod.* **2018**, *124*, 719–725. [CrossRef]
119. Vasić, K.; Knez, Ž.; Leitgeb, M. Bioethanol Production by Enzymatic Hydrolysis from Different Lignocellulosic Sources. *Molecules* **2021**, *26*, 753. [CrossRef] [PubMed]
120. Ingesson, H.; Zacchi, G.; Yang, B.; Esteghlalian, A.R.; Saddler, J.N. The Effect of Shaking Regime on the Rate and Extent of Enzymatic Hydrolysis of Cellulose. *J. Biotechnol.* **2001**, *88*, 177–182. [CrossRef] [PubMed]
121. Fenila, F.; Shastri, Y. Control of Enzymatic Hydrolysis of Lignocellulosic Biomass. *Resour.-Effic. Technol.* **2016**, *2*, S96–S104. [CrossRef]
122. Viikari, L.; Vehmaanperä, J.; Koivula, A. Lignocellulosic Ethanol: From Science to Industry. *Biomass Bioenergy* **2012**, *46*, 13–24. [CrossRef]
123. Devi, A.; Singh, A.; Bajar, S.; Pant, D.; Din, Z.U. Ethanol from Lignocellulosic Biomass: An in-Depth Analysis of Pre-Treatment Methods, Fermentation Approaches and Detoxification Processes. *J. Environ. Chem. Eng.* **2021**, *9*, 105798. [CrossRef]
124. Wei, W.; Jin, Y.; Wu, S.; Yuan, Z. Improving Corn Stover Enzymatic Saccharification via Ferric Chloride Catalyzed Dimethyl Sulfoxide Pretreatment and Various Additives. *Ind. Crops Prod.* **2019**, *140*, 111663. [CrossRef]
125. Tang, S.; Dong, Q.; Fang, Z.; Cong, W.; Miao, Z. High-Concentrated Substrate Enzymatic Hydrolysis of Pretreated Rice Straw with Glycerol and Aluminum Chloride at Low Cellulase Loadings. *Bioresour. Technol.* **2019**, *294*, 122164. [CrossRef]
126. Chen, Y.-A.; Zhou, Y.; Qin, Y.; Liu, D.; Zhao, X. Evaluation of the Action of Tween 20 Non-Ionic Surfactant during Enzymatic Hydrolysis of Lignocellulose: Pretreatment, Hydrolysis Conditions and Lignin Structure. *Bioresour. Technol.* **2018**, *269*, 329–338. [CrossRef]
127. Sun, D.; Yang, Q.; Wang, Y.; Gao, H.; He, M.; Lin, X.; Lu, J.; Wang, Y.; Kang, H.; Alam, A.; et al. Distinct Mechanisms of Enzymatic Saccharification and Bioethanol Conversion Enhancement by Three Surfactants under Steam Explosion and Mild Chemical Pretreatments in Bioenergy Miscanthus. *Ind. Crops Prod.* **2020**, *153*, 112559. [CrossRef]
128. Chaudhary, R.; Kaushal, J.; Singh, G.; Kaur, A.; Arya, S.K. Melioration of Enzymatic Ethanol Production from Alkali Pre-Treated Paddy Straw Promoted by Addition of Surfactant. *Biocatal. Biotransform.* **2022**, 1–10. [CrossRef]
129. Li, Y.; Song, W.; Han, X.; Wang, Y.; Rao, S.; Zhang, Q.; Zhou, J.; Li, J.; Liu, S.; Du, G. Recent Progress in Key Lignocellulosic Enzymes: Enzyme Discovery, Molecular Modifications, Production, and Enzymatic Biomass Saccharification. *Bioresour. Technol.* **2022**, *363*, 127986. [CrossRef] [PubMed]
130. Fan, M.; Zhao, C.; Huang, X.; Zhang, H.; Xie, J. Enhanced Digestibility and Fermentability of Sugarcane Bagasse in Biofuel Production by Surfactant-Assisted Dilute Acid Pretreatment. *Ind. Crops Prod.* **2021**, *172*, 114006. [CrossRef]
131. Dien, B.S.; Hespell, R.B.; Wyckoff, H.A.; Bothast, R.J. Fermentation of Hexose and Pentose Sugars Using a Novel Ethanologenic Escherichia Coli Strain. *Enzym. Microb. Technol.* **1998**, *23*, 366–371. [CrossRef]
132. du Preez, J.C.; Bosch, M.; Prior, B.A. The Fermentation of Hexose and Pentose Sugars by Candida Shehatae and Pichia Stipitis. *Appl. Microbiol. Biotechnol.* **1986**, *23*, 228–233. [CrossRef]
133. Gonçalves, F.A.; Ruiz, H.A.; dos Santos, E.S.; Teixeira, J.A.; de Macedo, G.R. Bioethanol Production by Saccharomyces Cerevisiae, Pichia Stipitis and Zymomonas Mobilis from Delignified Coconut Fibre Mature and Lignin Extraction According to Biorefinery Concept. *Renew. Energy* **2016**, *94*, 353–365. [CrossRef]
134. Ma'As, M.F.; Ghazali, H.M.; Chieng, S. Bioethanol Production from Brewer's Rice by Saccharomyces Cerevisiae and Zymomonas Mobilis: Evaluation of Process Kinetics and Performance. *Energy Sources Part A Recovery Util. Environ. Eff.* **2020**, 1–14. [CrossRef]
135. Mussatto, S.I.; Machado, E.M.S.; Carneiro, L.M.; Teixeira, J.A. Sugars Metabolism and Ethanol Production by Different Yeast Strains from Coffee Industry Wastes Hydrolysates. *Appl. Energy* **2012**, *92*, 763–768. [CrossRef]
136. Wirawan, F.; Cheng, C.-L.; Lo, Y.-C.; Chen, C.-Y.; Chang, J.-S.; Leu, S.-Y.; Lee, D.-J. Continuous Cellulosic Bioethanol Co-Fermentation by Immobilized Zymomonas Mobilis and Suspended Pichia Stipitis in a Two-Stage Process. *Appl. Energy* **2020**, *266*, 114871. [CrossRef]
137. De Bari, I.; De Canio, P.; Cuna, D.; Liuzzi, F.; Capece, A.; Romano, P. Bioethanol Production from Mixed Sugars by Scheffersomyces Stipitis Free and Immobilized Cells, and Co-Cultures with Saccharomyces Cerevisiae. *New Biotechnol.* **2013**, *30*, 591–597. [CrossRef]
138. Bailey, J.E. Toward a Science of Metabolic Engineering. *Science* **1991**, *252*, 1668–1675. [CrossRef] [PubMed]
139. Chandel, A.K.; da Silva, S.S.; Singh, O.V. Detoxification of Lignocellulose Hydrolysates: Biochemical and Metabolic Engineering Toward White Biotechnology. *Bioenerg. Res.* **2013**, *6*, 388–401. [CrossRef]
140. Adegboye, M.F.; Ojuederie, O.B.; Talia, P.M.; Babalola, O.O. Bioprospecting of Microbial Strains for Biofuel Production: Metabolic Engineering, Applications, and Challenges. *Biotechnol. Biofuels* **2021**, *14*, 5. [CrossRef] [PubMed]
141. Romaní, A.; Pereira, F.; Johansson, B.; Domingues, L. Metabolic Engineering of Saccharomyces Cerevisiae Ethanol Strains PE-2 and CAT-1 for Efficient Lignocellulosic Fermentation. *Bioresour. Technol.* **2015**, *179*, 150–158. [CrossRef] [PubMed]
142. Jin, Y.-S.; Cate, J.H. Metabolic Engineering of Yeast for Lignocellulosic Biofuel Production. *Curr. Opin. Chem. Biol.* **2017**, *41*, 99–106. [CrossRef]

143. Kim, J.; Hwang, S.; Lee, S.-M. Metabolic Engineering for the Utilization of Carbohydrate Portions of Lignocellulosic Biomass. *Metab. Eng.* **2022**, *71*, 2–12. [CrossRef]
144. Jeong, D.; Oh, E.J.; Ko, J.K.; Nam, J.-O.; Park, H.-S.; Jin, Y.-S.; Lee, E.J.; Kim, S.R. Metabolic Engineering Considerations for the Heterologous Expression of Xylose-Catabolic Pathways in Saccharomyces Cerevisiae. *PLoS ONE* **2020**, *15*, e0236294. [CrossRef]
145. Anasontzis, G.E.; Kourtoglou, E.; Villas-Boâs, S.G.; Hatzinikolaou, D.G.; Christakopoulos, P. Metabolic Engineering of Fusarium Oxysporum to Improve Its Ethanol-Producing Capability. *Front. Microbiol.* **2016**, *7*, 632. [CrossRef]
146. Yao, S.; Mikkelsen, M.J. Metabolic Engineering to Improve Ethanol Production in Thermoanaerobacter Mathranii. *Appl. Microbiol. Biotechnol.* **2010**, *88*, 199–208. [CrossRef]
147. Jojima, T.; Noburyu, R.; Sasaki, M.; Tajima, T.; Suda, M.; Yukawa, H.; Inui, M. Metabolic Engineering for Improved Production of Ethanol by Corynebacterium Glutamicum. *Appl. Microbiol. Biotechnol.* **2015**, *99*, 1165–1172. [CrossRef]
148. Kourkoutas, Y.; Bekatorou, A.; Banat, I.M.; Marchant, R.; Koutinas, A.A. Immobilization Technologies and Support Materials Suitable in Alcohol Beverages Production: A Review. *Food Microbiol.* **2004**, *21*, 377–397. [CrossRef]
149. Liu, D.-M.; Dong, C. Recent Advances in Nano-Carrier Immobilized Enzymes and Their Applications. *Process Biochem.* **2020**, *92*, 464–475. [CrossRef]
150. Kyriakou, M.; Patsalou, M.; Xiaris, N.; Tsevis, A.; Koutsokeras, L.; Constantinides, G.; Koutinas, M. Enhancing Bioproduction and Thermotolerance in Saccharomyces Cerevisiae via Cell Immobilization on Biochar: Application in a Citrus Peel Waste Biorefinery. *Renew. Energy* **2020**, *155*, 53–64. [CrossRef]
151. Karagoz, P.; Bill, R.M.; Ozkan, M. Lignocellulosic Ethanol Production: Evaluation of New Approaches, Cell Immobilization and Reactor Configurations. *Renew. Energy* **2019**, *143*, 741–752. [CrossRef]
152. Soares, L.B.; da Silveira, J.M.; Biazi, L.E.; Longo, L.; de Oliveira, D.; Furigo Júnior, A.; Ienczak, J.L. An Overview on Fermentation Strategies to Overcome Lignocellulosic Inhibitors in Second-Generation Ethanol Production Using Cell Immobilization. *Crit. Rev. Biotechnol.* **2022**, 1–22. [CrossRef]
153. Palmqvist, E.; Hahn-Hägerdal, B. Fermentation of Lignocellulosic Hydrolysates. I: Inhibition and Detoxification. *Bioresour. Technol.* **2000**, *74*, 17–24. [CrossRef]
154. Flevaris, K.; Chatzidoukas, C. Optimal Fed-Batch Bioreactor Operating Strategies for the Microbial Production of Lignocellulosic Bioethanol and Exploration of Their Economic Implications: A Step Forward towards Sustainability and Commercialization. *J. Clean. Prod.* **2021**, *295*, 126384. [CrossRef]
155. Ghorbanpour Khamseh, A.A.; Miccio, M. Comparison of Batch, Fed-Batch and Continuous Well-Mixed Reactors for Enzymatic Hydrolysis of Orange Peel Wastes. *Process Biochem.* **2012**, *47*, 1588–1594. [CrossRef]
156. Jahnavi, G.; Prashanthi, G.S.; Sravanthi, K.; Rao, L.V. Status of Availability of Lignocellulosic Feed Stocks in India: Biotechnological Strategies Involved in the Production of Bioethanol. *Renew. Sustain. Energy Rev.* **2017**, *73*, 798–820. [CrossRef]
157. Su, T.; Zhao, D.; Khodadadi, M.; Len, C. Lignocellulosic Biomass for Bioethanol: Recent Advances, Technology Trends, and Barriers to Industrial Development. *Curr. Opin. Green Sustain. Chem.* **2020**, *24*, 56–60. [CrossRef]
158. Taherzadeh, M.J.; Karimi, K. Enzymatic-Based Hydrolysis Processes for Ethanol from Lignocellulosic Materials: A Review. *BioResources* **2007**, *2*, 707–738.
159. Srivastava, N.; Rawat, R.; Singh Oberoi, H.; Ramteke, P.W. A Review on Fuel Ethanol Production from Lignocellulosic Biomass. *Int. J. Green Energy* **2015**, *12*, 949–960. [CrossRef]
160. Wright, J.D.; Wyman, C.E., Grohmann, K. Simultaneous Saccharification and Fermentation of Lignocellulose. *Appl. Biochem. Biotechnol.* **1988**, *18*, 75–90. [CrossRef]
161. Hari Krishna, S.; Chowdary, G.V. Optimization of Simultaneous Saccharification and Fermentation for the Production of Ethanol from Lignocellulosic Biomass. *J. Agric. Food Chem.* **2000**, *48*, 1971–1976. [CrossRef] [PubMed]
162. Roberto, I.C.; Castro, R.C.A.; Silva, J.P.A.; Mussatto, S.I. Ethanol Production from High Solid Loading of Rice Straw by Simultaneous Saccharification and Fermentation in a Non-Conventional Reactor. *Energies* **2020**, *13*, 2090. [CrossRef]
163. Bondesson, P.-M.; Galbe, M. Process Design of SSCF for Ethanol Production from Steam-Pretreated, Acetic-Acid-Impregnated Wheat Straw. *Biotechnol. Biofuels* **2016**, *9*, 222. [CrossRef] [PubMed]
164. Koppram, R.; Nielsen, F.; Albers, E.; Lambert, A.; Wännström, S.; Welin, L.; Zacchi, G.; Olsson, L. Simultaneous Saccharification and Co-Fermentation for Bioethanol Production Using Corncobs at Lab, PDU and Demo Scales. *Biotechnol. Biofuels* **2013**, *6*, 2. [CrossRef]
165. Liu, Z.-H.; Chen, H.-Z. Simultaneous Saccharification and Co-Fermentation for Improving the Xylose Utilization of Steam Exploded Corn Stover at High Solid Loading. *Bioresour. Technol.* **2016**, *201*, 15–26. [CrossRef]
166. Liu, L.; Zhang, Z.; Wang, J.; Fan, Y.; Shi, W.; Liu, X.; Shun, Q. Simultaneous Saccharification and Co-Fermentation of Corn Stover Pretreated by H2O2 Oxidative Degradation for Ethanol Production. *Energy* **2019**, *168*, 946–952. [CrossRef]
167. Hasunuma, T.; Kondo, A. Consolidated Bioprocessing and Simultaneous Saccharification and Fermentation of Lignocellulose to Ethanol with Thermotolerant Yeast Strains. *Process Biochem.* **2012**, *47*, 1287–1294. [CrossRef]
168. Singhania, R.R.; Patel, A.K.; Singh, A.; Haldar, D.; Soam, S.; Chen, C.-W.; Tsai, M.-L.; Dong, C.-D. Consolidated Bioprocessing of Lignocellulosic Biomass: Technological Advances and Challenges. *Bioresour. Technol.* **2022**, *354*, 127153. [CrossRef] [PubMed]
169. Periyasamy, S.; Beula Isabel, J.; Kavitha, S.; Karthik, V.; Mohamed, B.A.; Gizaw, D.G.; Sivashanmugam, P.; Aminabhavi, T.M. Recent Advances in Consolidated Bioprocessing for Conversion of Lignocellulosic Biomass into Bioethanol—A Review. *Chem. Eng. J.* **2022**, *453*, 139783. [CrossRef]

170. Liu, Y.; Xie, X.; Liu, W.; Xu, H.; Cao, Y. Consolidated Bioprocess for Bioethanol Production from Lignocellulosic Biomass Using Clostridium Thermocellum DSM 1237. *BioResources* **2020**, *15*, 8355–8368. [CrossRef]
171. Qin, L.; Zhao, X.; Li, W.-C.; Zhu, J.-Q.; Liu, L.; Li, B.-Z.; Yuan, Y.-J. Process Analysis and Optimization of Simultaneous Saccharification and Co-Fermentation of Ethylenediamine-Pretreated Corn Stover for Ethanol Production. *Biotechnol. Biofuels* **2018**, *11*, 118. [CrossRef]
172. Zhao, W.; Zhao, F.; Zhang, S.; Gong, Q.; Chen, G. Ethanol Production by Simultaneous Saccharification and Cofermentation of Pretreated Corn Stalk. *J. Basic Microbiol.* **2019**, *59*, 744–753. [CrossRef] [PubMed]
173. Madson, P.W.; Lococo, D.B. Recovery of Volatile Products from Dilute High-Fouling Process Streams. *Appl. Biochem. Biotechnol.* **2000**, *84*, 1049–1061. [CrossRef]
174. Yang, R.-J.; Liu, C.-C.; Wang, Y.-N.; Hou, H.-H.; Fu, L.-M. A Comprehensive Review of Micro-Distillation Methods. *Chem. Eng. J.* **2017**, *313*, 1509–1520. [CrossRef]
175. Kang, K.E.; Jeong, J.-S.; Kim, Y.; Min, J.; Moon, S.-K. Development and Economic Analysis of Bioethanol Production Facilities Using Lignocellulosic Biomass. *J. Biosci. Bioeng.* **2019**, *128*, 475–479. [CrossRef]
176. Hamelinck, C.N.; Van Hooijdonk, G.; Faaij, A.P. Ethanol from Lignocellulosic Biomass: Techno-Economic Performance in Short-, Middle-and Long-Term. *Biomass Bioenergy* **2005**, *28*, 384–410. [CrossRef]
177. Li, J.; Zhou, W.; Fan, S.; Xiao, Z.; Liu, Y.; Liu, J.; Qiu, B.; Wang, Y. Bioethanol Production in Vacuum Membrane Distillation Bioreactor by Permeate Fractional Condensation and Mechanical Vapor Compression with Polytetrafluoroethylene (PTFE) Membrane. *Bioresour. Technol.* **2018**, *268*, 708–714. [CrossRef]
178. Shirazi, M.A.; Kargari, A. Concentrating of Sugar Syrup in Bioethanol Production Using Sweeping Gas Membrane Distillation. *Membranes* **2019**, *9*, 59. [CrossRef]
179. Khayet, M. Membranes and Theoretical Modeling of Membrane Distillation: A Review. *Adv. Colloid Interface Sci.* **2011**, *164*, 56–88. [CrossRef] [PubMed]
180. Loulergue, P.; Balannec, B.; Fouchard-Le Graët, L.; Cabrol, A.; Sayed, W.; Djelal, H.; Amrane, A.; Szymczyk, A. Air-Gap Membrane Distillation for the Separation of Bioethanol from Algal-Based Fermentation Broth. *Sep. Purif. Technol.* **2019**, *213*, 255–263. [CrossRef]
181. Gaykawad, S.S.; Zha, Y.; Punt, P.J.; van Groenestijn, J.W.; van der Wielen, L.A.M.; Straathof, A.J.J. Pervaporation of Ethanol from Lignocellulosic Fermentation Broth. *Bioresour. Technol.* **2013**, *129*, 469–476. [CrossRef] [PubMed]
182. Peng, P.; Lan, Y.; Liang, L.; Jia, K. Membranes for Bioethanol Production by Pervaporation. *Biotechnol. Biofuels* **2021**, *14*, 10. [CrossRef]
183. Trinh, L.T.P.; Lee, Y.-J.; Park, C.S.; Bae, H.-J. Aqueous Acidified Ionic Liquid Pretreatment for Bioethanol Production and Concentration of Produced Ethanol by Pervaporation. *J. Ind. Eng. Chem.* **2019**, *69*, 57–65. [CrossRef]
184. Jain, A.; Dhabhai, R.; Dalai, A.K.; Chaurasia, S.P. Bioethanol Production in a Pervaporation Membrane Bioreactor. In *Membrane Technology*; CRC Press: Boca Raton, FL, USA, 2018; ISBN 978-1-315-10566-6.

molecules

MDPI

Review

Processing of Biomass Prior to Hydrogen Fermentation and Post-Fermentative Broth Management

Zhila Honarmandrad, Karolina Kucharska * and Jacek Gębicki

Department of Process Engineering and Chemical Technology, Faculty of Chemistry, Gdansk University of Technology, 11/12 Gabriela Narutowicza Street, 80-233 Gdansk, Poland
* Correspondence: karolina.kucharska@pg.edu.pl

Abstract: Using bioconversion and simultaneous value-added product generation requires purification of the gaseous and the liquid streams before, during, and after the bioconversion process. The effect of diversified process parameters on the efficiency of biohydrogen generation via biological processes is a broad object of research. Biomass-based raw materials are often applied in investigations regarding biohydrogen generation using dark fermentation and photo fermentation microorganisms. The literature lacks information regarding model mixtures of lignocellulose and starch-based biomass, while the research is carried out based on a single type of raw material. The utilization of lignocellulosic and starch biomasses as the substrates for bioconversion processes requires the decomposition of lignocellulosic polymers into hexoses and pentoses. Among the components of lignocelluloses, mainly lignin is responsible for biomass recalcitrance. The natural carbohydrate-lignin shields must be disrupted to enable lignin removal before biomass hydrolysis and fermentation. The matrix of chemical compounds resulting from this kind of pretreatment may significantly affect the efficiency of biotransformation processes. Therefore, the actual state of knowledge on the factors affecting the culture of dark fermentation and photo fermentation microorganisms and their adaptation to fermentation of hydrolysates obtained from biomass requires to be monitored and a state of the art regarding this topic shall become a contribution to the field of bioconversion processes and the management of liquid streams after fermentation. The future research direction should be recognized as striving to simplification of the procedure, applying the assumptions of the circular economy and the responsible generation of liquid and gas streams that can be used and purified without large energy expenditure. The optimization of pre-treatment steps is crucial for the latter stages of the procedure.

Keywords: detoxification; biohydrogen; green solvents; biomass; lignocellulose

Citation: Honarmandrad, Z.; Kucharska, K.; Gębicki, J. Processing of Biomass Prior to Hydrogen Fermentation and Post-Fermentative Broth Management. *Molecules* **2022**, 27, 7658. https://doi.org/10.3390/molecules27217658

Academic Editors: Alejandro Rodriguez Pascual, Eduardo Espinosa Víctor and Carlos Martín

Received: 19 October 2022
Accepted: 4 November 2022
Published: 7 November 2022

1. Introduction

The world energy situation is unstable due to environmental, economic, and geopolitical problems. Greenhouse gas emissions from fossil fuel combustion cause global warming, acid rain, climate change, ozone depletion, and biodiversity damage [1]. Declining fossil fuel reserves, increasing pollution, and climate change in the Earth's atmosphere have made the production and use of renewable energy sources that are less polluting an inevitable necessity in the present age. Various fossil resources such as natural gas, coal, gasoline, and oil are used as energy sources to produce electricity (20%) and fuel (80%) [2].

Therefore, fuel systems excluding carbon dioxide need to be developed and applied in the future. Since biomass is the fourth largest source of energy after oil, coal, and gas, it should be considered raw material for further processing [3] Biofuel is a type of solid, liquid, or gaseous fuel that is obtained from a wide range of biomass sources, including a variety of crops, agricultural and forest residues, aquatic plants, animal waste, and municipal waste. Although a wide variety of biofuels is described in the literature, bioethanol, biodiesel, and biogas are best known [4].

Based on the criterion of origin secondary biofuels are distinguished as first-, second-, and third-generation biofuels. The first-generation biofuels are generated on basis of food origin biomass, i.e., wheat, barley, corn, rice, potatoes, canola, soybeans, almonds, sunflower, palm, coconut, ground beet, sugarcane, etc [5,6]. The second-generation biofuels are generated based on non-food biomass, i.e., lignocellulosic material, including waste from agriculture and forestry, sewage, municipal and industrial waste, and trees grown specifically for energy production (such as spruce, poplar, or willow). The third-generation biofuels are generated with the application of algae and microalgae as feed to produce biofuels [7]. Bio-fraction of polysugar-based waste materials (starch-type and lignocellulosic-type biomass) requires the application of pre-treatment steps to improve the saccharification efficiency and post-fermentation purification to remove or recover derivatives or value-added products, i.e., chemical bio compounds from the broth [8].

Lignocellulosic biomass (LB) is an abundant and renewable source of carbohydrates, consisting mainly of polysaccharides including cellulose and hemicelluloses, and an aromatic polymer called lignin. Lignocellulosic biomass has great potential as an alternative to fossil fuels for the production of second-generation biofuels and chemicals and biomaterials without compromising global food security and without addressing the food versus energy debate [9,10]. Most published research is conducted on the homogenous type of biomass. Since waste biomass is usually a mixture, an overview of the mixture should be taken into account while planning the pre-treatment procedure.

Starch is an abundant natural renewable polymer and has been used in biomaterial applications due to its properties such as biodegradability, low toxicity, and stability. Starch is composed of glucose monomers that are linked together by α-1,4-glycoside bonds and branched by α-1,6-glycoside bonds [11,12]. Efficient saccharification of starch before fermentation requires the application of amylolytic enzymes. The pre-treatment of both lignocellulosic and starch-based materials leads to the generation of a carbohydrates cocktail which can be introduced to dark fermentation.

Hydrogen is the most important source of renewable energy, recognized as environmentally friendly, and can be converted into electricity by fuel cells. Hydrogen has the highest energy production of any known fuel. Its production is possible in various ways by using petroleum products, coal, gas, algae, and the fermentation of bacteria [13].

The object of interest for this review paper is hydrolysate generation for biohydrogen production via dark fermentation by microorganisms, especially bacteria and yeasts. There are two types of fermentation to produce biohydrogen by bacteria: one is photo fermentation, which requires a light source, and the other is dark fermentation, which does not require light. Many carbon sources are used in these reactions, all of which are supplied by biomass [14]. The matrix for biohydrogen generation is complexed and therefore, a large group of derivatives may be generated during the pre-treatment step. The effect of the derivatives on the fermentation efficiency must be taken into account [1,5,6]. The unfermentable compounds occurring in the pre-treated biomass hydrolysates must be also considered, as these substances may be further present in the post-fermentation broth and may interact as promoting or inhibiting agents [11,13].

To our best knowledge, dark hydrogen fermentation is the most widespread and promising biological method of hydrogen synthesis [15]. It is characterized by the high synthesis efficiency of the gas desired in the energy industry [7]. Up to date, the published results are focused mainly on pure cultures maintained under mesophilic conditions, sometimes moving towards higher temperatures, i.e., *Clostridiaceae*, *Flexibacteraceae*, *Enterobacter*, and *Klebsiella* [13,15]. Dark fermentation is easy to carry, as it does not require light and therefore issues related to light transmission do not occur [15,16]. The microorganisms able to carry dark fermentation show a temperature optimum ranging from 30 °C to 80 °C. The amount of hydrogen produced during fermentation depends on the value of pH, HRT (hydraulic retention time), and pressure. For optimal hydrogen production, a pH value of 5–6 is recommended, which copes well with the organic acids formulation [16]. Anaerobic bacteria generate biohydrogen via the biotransformation of hexoses, mainly glucose, to pyruvate with simultaneous generation of hydrogen during the regeneration of NADH (nicotinamide adenine dinucleotide). The yield of hydrogen depends on the type of fermentation and activity of ferredoxin oxidoreductase and acetyl-coA. While fermentation leads to acetic and formic acids, the yield of hydrogen may be equal to 4 mol H_2/mol glucose, and when butyric fermentation occurs—up to 2 mol H_2/mol glucose [10,13]. The acids generated in dark fermentation may be applied as carbon sources in photo fermentation with sulfur-free *Rhodospirillaceae*, including the *Rhodopseudomonas, Rhodospirillum, Rhodobacter,* and *Rhodobium*. High purity of hydrogen is an advantage of photo fermentation (gas contains 10–20% carbon dioxide admixture). This eliminates the compulsion of the energetically and time-consuming purification process of the obtained gas. For this reason, photo fermentation with the use of nitrogenase arouses considerable interest among researchers and practitioners dealing with the synthesis of biohydrogen [13–18]. Amongst the disadvantages of the photo fermentation process is the low efficiency, and the economy of biohydrogen generation should be mentioned. Biotechnologists currently use genetic modifications of microorganisms, metabolic engineering, improvement of the reactor structure, or the use of various deposits for cell immobilization, to improve the hydrogen efficiency [18].

Unfortunately, satisfactory results have still not been achieved, and the production of biohydrogen, especially by photo fermentation, of subsequent dark and photo fermentation, remains a crucial problem as energy production on an industrial scale is considered. To make the production of hydrogen by biological methods economically and ecologically feasible, integrated processes need to be developed [15,18,19]. Each planned and optimized set of unit operations in the range from biomass to biohydrogen must take into account the problems that arise in the area of issues related to the processing of the raw materials, conducting fermentation, and management of post-process streams.

This paper presents the current state of knowledge on the relationship between the applied methods of pre-treatment, the derivatives generated during pre-treatment and decomposition products of raw materials, and the yield of hydrogen obtained in the fermentation process. A review of the currently applied techniques enabling the management of post-fermentation broth. The proposed approach is novel, as it considers mixtures of the raw material and the assumptions of circular economy at the early stages of the procedure.

2. State of the Art on the Biomass Recalcitrance

2.1. Biomass Recalcitrance and the Pre-Treatment of Raw Materials

In recent years, due to the intensification of the energy crisis and arising of environmental pollution awareness, special attention has been paid to the production of biofuels and biochemical fuels from biomass. Among the "biomass for biofuels" research direction, the fraction of municipal biowaste, containing starch and lignocellulose-based materials, is often applied due to the scale of their occurrence as raw materials and products in the industry [19]. Annually, 10 to 50 billion tons of dry lignocellulose is obtained in the world, which is about half of the global biomass yield [20,21].

Biomass recalcitrance is related to the chemical and physical properties of the plant cell wall. Lignin, hemicelluloses, pectin, ash, and their spatial bonds create physical barriers to protect cellulose from degradation. Factors affecting the enzymatic hydrolysis of cellulose biomass include lignin, hemicelluloses, the contents of the recalcitrant group, cellulose crystallization, degree of polymerization, specific surface area, pore-volume, and particle size [22,23].

Although these factors limit the enzymatic hydrolysis of biomass and have been extensively studied, the molecular mechanisms of biomass resistance are still unclear. Various methods of pre-treatment have been developed over the past few decades [13,15,24–30]. The biomass recalcitrance occurs due to the presence of diversified monomers in the biomass (Figure 1).

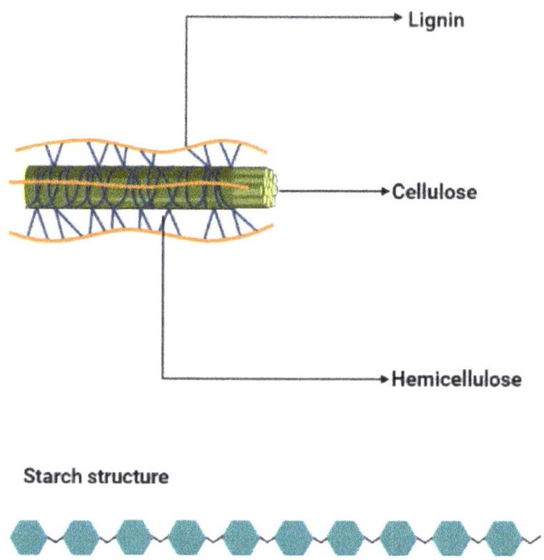

Figure 1. Pictorial diagram of chemical structures of starch and lignocellulose concerning main units in the structure.

The overall purpose of every pre-treatment is to remove recalcitrant barriers to increase the saccharification of cellulose by altering the chemical composition and physical structures of biomass raw materials. Understanding how chemical compounds and physical structures affect biomass recalcitrance and how they affect the saccharification of lignocelluloses can greatly improve existing pretreatment technologies and promote the development of new pretreatment technologies [21]. The main differences between starch and lignocellulose-based materials and their maintenance have been presented in Table 1, concerning selected criteria, i.e., microbial resistance, availability of monosugars, main units, their interactions, and pre-treatment by-products [20–26].

Table 1. Main features affecting the course of pretreatment of starch and lignocellulose raw materials.

Criterion	Starch	Lignocellulose
Microbial resistance	Biodegradable, excluding granular amylase-resistant α-glycans	Resistant to biodegradation due to the strong and compact structure of plant cell walls
Factors affecting access to monosugars	Fiber, physical form	Available surface, pore size, volume, particle size, specific surface area, and degree of polymerization
Chemical compounds or units	Glucose monomers linked with 1,4 and 1,6 linkages; linear polymer—amylose; branched form—amylopectin.	Lignin (amorphous heteropolymer of phenylpropanoid building units, i.e., p-coumaryl, coniferyl, and sinapyl alcohol); hemicellulose (various monosaccharide subunits to form xylans, xyloglucan, mannans, and glucomannans); cellulose (ß-D-glucopyranose units linked via ß-(1,4) glycosidic bonds, with cellobiose as the fundamental repeating unit), extractives
Chemical interactions between polymers	α -1,4 glycosidic bonds;α-1,6 glycosidic bonds	Hydrogen bonds (between cellulose-hemicellulose); Lignin-carbohydrate complex, i.e., the occurrence of phenyl glycosides, γ-esters, benzyl ethers, ferulate esters, coumarate esters; Hemiacetal and acetal linkages at 4-OH and 4-O positions (lignin covalently linked to hemicellulose)
Possible pre-treatment by-products	Monosaccharides conversion by-products and secondary transformation products, i.e., levulinic acid, HMF, furfural	Lignin derivatives, HMF, vanillin, syringole, furfural, p-coumaryl, coniferyl, sinapyl alcohol, oligopeptides, terpenoids, and levulinic acid, monosaccharides conversion by-products, i.e., levulinic acid, HMF, furfural

The need to explore the issues related to biomass pretreatment to achieve a high degradation of polysugars [31] and to minimize the formation of inhibitors for subsequent fermentation steps is a consequence of the objectives presented for biomass in Table 1. Due to chemical properties, it is obvious that the main source of monosugars, i.e., glucose, is strongly correlated with the starch and cellulose content and availability [25,30].

Starch can be almost completely liquefied while an amylolytic enzyme cocktail is applied [8,32]. However, remnant protein, fibers, fats, and their secondary transformation by-products may be present in the starch-based biomass hydrolysates. If the hydrolysis is too short, resistant starch may occur, i.e., a form that is bonded to the fibers, which requires a longer time or more aggressive hydrolysis conditions [8].

Cellulose is a linear polysaccharide in the form of insoluble microfibrils. Amorphous or soluble regions appear in cellulose structure, where the molecules are less compact [32]. However, cellulose fibrils are located in a lignocellulosic matrix which makes them highly resistant to enzymatic hydrolysis. The degree of polymerization (DP) changes due to pretreatment which is associated with changes in structural parameters such as crystallinity and porosity [33]. Lou et al. showed that the DP of cellulose has a negative correlation with cellulose hydrolysis. Long cellulose chains are assumed to contain more hydrogen bonds and are more difficult to hydrolyze, while shorter cellulose chains contain a weaker hydrogen bonding system, which facilitates access to the enzyme or hydrolyzing agent [34]. Additionally, the presence and structure of hemicelluloses affect the process of saccharification, as a result of structural obstacles related to their structure. Hemicelluloses are heterogeneous groups of biopolymers [35,36] and the degree of polymerization of hemicelluloses is in the range of 100–200 units and is easily hydrolyzed by diluted acids or bases as well as enzymes [37]. Hemicellulose is considered a physical barrier restricting cellulose access. Therefore, the removal of hemicellulose could increase the enzymatic digestibility of biomass [25,38]. Additionally, lignin hinders biomass pretreatment. Lignin is a highly complex amorphous heteropolymer of phenylpropanoid monomers (p-coumaryl, coniferyl, and sinapyl alcohol) [39]. Mentioned structures are presented in Figure 2.

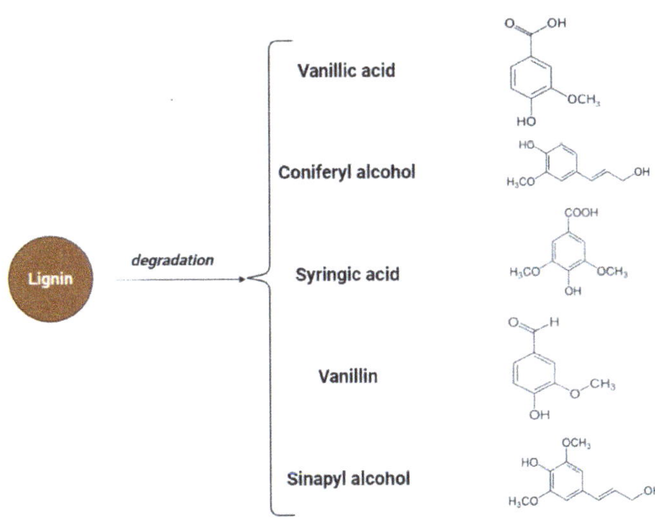

Figure 2. Chemical structures of monomeric compounds in the lignin structure and the main products of lignin degradation.

Lignin is responsible for hydrophobicity and structural stiffness and binds hemicelluloses to cellulose in the cell wall. The presence of lignin must be taken into account, as the secondary derivatives of lignin may affect the dark fermentation step. It is well known that lignin plays a negative role in cellulose conversion which is influenced by several factors such as total lignin content and lignin composition/structure (especially the content of hydroxyl groups). Lignin can block the access of enzymes to cellulose as a physical barrier and thus limit the access of polysaccharides [13]. Lignin permanently absorbs cellulase enzymes, therefore preventing their effect on cellulose. Adsorption of cellulases on the lignin matrix has been observed for pre-treated substrates with dilute acid or vapor explosion. It has been observed that cellulases can be adsorbed on the lignin matrix that pre-treated substrates with dilute acid or steam explosion [21].

2.2. *Issues Related to the Availability and Construction of Biomass Surface*

The accessible surface area is an important limiting factor in the cellulose digestibility process. The available surface is related to biomass particle size, porosity, and pore volume [21]. As the particle size decreases or the pore volume increases, the available surface area increases and, as a result, the enzymatic digestibility of cellulose increases. In terms of microcrystalline cellulose, it has been shown that reducing the size increases the accessible surface area and greatly accelerates the rate of cellulose hydrolysis [40]. For example, by reducing the particle size from 25.52 μm to 0.78 μm, the available surface area increases from 0.24 m^2/g to 25.50 m^2/g, thus increasing the hydrolysis rate [41].

Particle size is a significant parameter that affects cellulose hydrolysis potential. Some studies have shown that particle sizes smaller than 590–350 μm do not significantly improve enzymatic digestibility. However, the available surface depends not only on the particle size but also on the porosity and pore volume [42,43]. The substrate surface is divided into two external (affected by the length and width of the substrate) and internal (pore surface) which is a function of the lumen size and the number of pores and cracks under the substrate. Published studies have shown that there is a direct relationship between the inner surface and the rate of enzymatic hydrolysis and also the most important factor limiting the enzymatic digestibility of biomass is the surface area [44–46].

According to some studies, enzyme access to cellulose is more through the cell wall pores than the outer surface of the substrate. On average, more than 90% of the enzymatic digestibility of the substrate is conducted by available pores and the outer surface plays a lesser role [47,48]. SSA (specific surface area) is the total surface area per unit (volume or mass), and ASA (accessible surface area) represents the area at which cellulases can come into contact with cellulose. In general, ASA is directly related to SSA, and as ASA increases, so does SSA, but the whole surface is not effective for cellulose-associated cellulases, and only pores large enough can allow cellulases to take action [49,50].

Studies have shown that the enzymatic digestibility of biomass decreases after drying due to hornification and reduced pore size, so pore size can be a limiting factor in the enzymatic hydrolysis pretreatment process [51]. Hornification depends on the physical and chemical structure of the cell wall of the undried material, the drying method, and the drying time. Some studies have shown that drying significantly reduces the number of large pores and that the collapse and closure of large pores result in smaller pores that are not accessible to enzymes [52,53]. On the other hand, wet pressing to reduce the moisture content of the material causes an irreversible reduction in the volume of fiber pores and thus reduces the enzymatic digestibility of cellulose. Therefore, the effects of hornification are one of the effective factors in biomass resistance. Although SSA has an important effect on cellulose enzymatic digestibility, some factors such as cellulose crystallinity and degree of polymerase affect enzyme digestibility. As a result, these cellulose-related structures can limit the rate and extent of hydrolysis [54].

2.2.1. Overcoming Cellulose Crystallinity

Cellulose has crystalline and amorphous regions and in these regions of cellulose there is a form of microfibrils in which paracrystalline groups are composed of several dozen (1, 4) β-D-glucan that are longitudinally hydrogen bonded together [7]. Due to this feature (hydrogen bonding), crystalline regions of cellulose are more resistant to enzymatic hydrolysis and microbial attacks than amorphous regions [40,47,48]. Some studies have shown that crystallinity has a negative effect on the enzymatic digestibility of cellulose, especially the initial hydrolysis rate but, on the other hand, the reconstruction of the crystalline hydrogen bonding network can increase the rate of polymerization [53]. Some researchers have shown that the conversion of crystalline allomorph Iβ to IIII by ammonia reduces the number of in-sheet hydrogen bonds of cellulose while increasing the number of inter-sheet hydrogen bonds up to five times [10].

Crystallization is considered one of the most effective and important inhibitory factors of enzymatic hydrolysis. Because the higher the cellulose crystallinity, the lower the availability of biomass for enzymatic hydrolysis. Cellulose crystallinity is caused by intermolecular and intramolecular hydrogen bonds between cellulose chains, which can be modified by biomass pretreatment methods [55,56].

Cellulose consists of two regions, amorphous and crystalline. In order to determine the crystallinity of cellulose in plants, it is very necessary to determine the cellulose content because it is expected that cellulose is the only crystalline compound [57]. Cellulose consists of linear chains of poly [b-1,4-D-anhydroglucopyranose] (C6nH10n + 2O5n + 1 (*n* = degree of polymerization of glucose)) which crystallizes through hydrogen bonding between the chains and has cellobiose as repeating units. The crystal structure of cellulose in higher plants is that of cellulose Iβ, which consists of monoclinic, P21 space groups with cellulose chains oriented along a unique c axis [58,59].

Corn and wheat straws are useful for the production of biofuel after pretreatment and enzymatic hydrolysis of cellulose and hemicellulose to monosaccharides due to their low cellulose content and large cell lumen, which causes low tensile strength. In the enzymatic hydrolysis process, exoglucanase is used for crystalline cellulose and endoglucanase is used for amorphous cellulose to convert cellulose into glucose substrate [60,61].

The effect of crystallinity on hydrolysis is different. Some studies on pretreated wheat straw [56,62], corn [63], switchgrass, and bagasse [64] reported that crystallinity is the most effective inhibitor of enzymatic hydrolysis, so the higher the cellulose crystallinity, the lower the availability of biomass for enzymatic hydrolysis. However, some studies [65–70] showed that crystallinity in limiting hydrolysis is less important than other physical properties such as the DP, pore volume, accessible surface area, and particle size.

Due to the presence of different hydrogen-bonding networks, amorphous celluloses are hydrolyzed three to thirty times faster than high crystalline celluloses [47,48,53]. In the enzymatic hydrolysis process, first, amorphous cellulose is hydrolyzed and then hydrolysis of more solid crystalline compounds takes place. However, in most studies, pure cellulose substrates have been used to investigate the relationship between crystallinity and hydrolysis rate, which does not indicate the heterogeneous lignocellulosic substrate that we encounter during the hydrolysis of pretreated substrates for biotransformation [25–27,71].

Physical pretreatment methods such as ball milling were used to prepare samples with different initial crystallinity degrees to show the effect of crystallinity on hydrolysis [72]. It was found a reduced particle size and an increase in the accessible surface area, which is the most important factor for the enzymatic digestibility of biomass. The most common method for determining the crystallinity of cellulose is X-ray diffraction. The crystallinity index (CrI, Equation (1)) is commonly used to describe the crystalline degree of biomass and pulp, which is defined as follows [72]:

$$\text{Crystallinity index (CRI)\%} - \frac{I_{002} - I_{am}}{I_{002}} \times 100 \quad (1)$$

I_{002} is the diffraction intensity of 002 peaks at $2\theta \approx 22.5^{\circ}$ and I_{am} is the scattering intensity of the amorphous region at $2\theta \approx 18.7^{\circ}$.

CrI measures the relative fraction of crystalline cellulose in total solids and is affected by the presence of lignin and hemicellulose. Removal of lignin and hemicellulose increases the CrI in the pretreated material. Therefore, care should be taken when using CrI to study the effect of pretreatment processes on the change in crystallinity of biomass cellulose.

Drying the sample before analysis is one of the most important limitations of using this method because drying in the air or an oven changes the crystallinity of cellulose [21].

Another method of crystallinity analysis of cellulose is the use of the infrared spectrum. Since the presence of lignin and hemicellulose can interfere with the ratio of amorphous to crystalline cellulose bonds and the ratio of crystalline cellulose polymorphs, the infrared spectrum is used for qualitative rather than quantitative studies [73].

2.2.2. Degree of Polymerization

The number of glucose units in a polymer is called the degree of polymerization (DP) of cellulose. The DP plays an important role in lignocellulose resistance. By changing the DP, other structural parameters, including crystallinity and porosity, also change [9,33]. Some studies have shown that decreasing the DP in cotton by γ-radiation causes very small changes in the rate of saccharification. Long cellulose chains are assumed to consist of more hydrogen bonds that are difficult to hydrolyze, while short cellulose chains are composed of weaker hydrogen bonds that facilitate enzymatic access [34].

The process of enzymatic hydrolysis of cellulose by the synergy of cellulase components is called the process of cellulose depolymerization [15]. The endocellulase breaks down linear cellulose molecules and produces reducing and oxidizing ends that can be attacked by exocellulases or cellobiohydrolase [74]. Exocellulases then remove one of the cellulose molecular strands to create more internal sites for endocellulase binding. Cellobiose is a very strong inhibitor of the activity of endocellulase and exocellulases enzymes, and the conversion of cellobiose to glucose by β-glucosidase reduces its effect and creates the ground for continued cellulolytic activity [75].

Gupta et al. showed that endoglucanase (Endo-G) reacts rapidly with non-crystalline cellulose and reduces the DP by 30 to 60, and then Endo-G is inhibited by non-crystalline cellulose. β-Glucosidase (β-G) can hydrolyze cell-oligosaccharides with a DP less than seven and produce cellulose, while it cannot hydrolyze cell-oligosaccharides with a DP higher than seven. Therefore, due to this mechanism, the hydrolysis rate is faster in shorter cellulose chains [76].

Nahzad et al. showed that beating the pulp speeds up the hydrolysis and also showed that the initial DP of the pulp does not have a significant effect on the final amount of hydrolysis [77]. However, two-thirds of the DP decreased during hydrolysis, which was the same in all hydrolyzed pulp. The DP is similar to cellulose crystallization and is not an independent factor, because a change in the DP is always associated with a change in crystallinity. After beating, the fiber pulp becomes shorter and swells significantly with increasing porosity. As a result, biomass resistance does not arise from a single structural factor. Since the plant cell wall is made of cross-links of chemical compounds and forms a strong and compact spatial structure, there are natural interactions between these factors [77].

The overcoming of biomass recalcitrance more often involves the application of microbial consortia. Enzymes are recognized as expensive agents; their isolation and purification are expensive and complicated. Thus, they are mainly used as mixtures. Therefore, the application of wood-decomposing fungi in consortium with dark fermentation bacteria is common. Rot fungi, i.e., *Phanerochaete chrysosporium, Phlebia radiate, Dichmitus squalene, Rigidosporus lignosus,* and *Jungua separabilima* can produce lignin peroxidase, polyphenol oxidase, and magnesium-dependent peroxidase and cause the hydrolysis of lignocellulose and depolymerization [78].

White rot fungi produce three enzyme fractions [13], i.e., cellulolytic enzymes and hemicellulases (Endo-1,4-β-glucanase, Exo-1,4-β-gluconase, and glucohydrolases, endo-1,4-β-xylanases, β-xylosidases, galactoglucomannazes, or galactosides), hemicellulases; lignosaccharidases (glucose oxidase, pyranose oxidase, oxidoreductase, and cellobiase) and lignin-degrading enzymes (peroxidase, dioxygenases, peroxydismutases, and glyoxal oxidases). Soft rot fungi, i.e., *Trichoderma reesei, Chaetomim sp.* and *Ceratocystis sp., Ascomycota,* and *Deuteromycota,* are effective toward wood with high moisture content since a secondary erosion of cell walls followed by hemicellulose-cellulose complex decomposition occurs and monosaccharides are generated and consumed. The culture of this type of rot should be carried in the presence of fermentative microbes, to avoid losses of monosugars due to rot self-consumption [13].

Future directions and development regarding overcoming biomass recalcitrance by biological methods, should, to our best knowledge, consider the elimination of multistep procedures and therefore the research on the synergistic effect of rot fungi and fermentative bacteria and archaeon should be investigated.

2.3. *Inhibitory Compounds Generation during Pretreatment*

Pretreatment methods have been studied but pretreatment advances are still required since the studies were carried out mainly regarding the mono type of biomass, not biomass mixtures, characterized as more complex matrixes. Technologies of pretreatment of the lignocellulosic biomass are usually classified into physical, physicochemical, chemical, and biological processes. Pretreatment is carried out mainly due to enhancing the fermentation or biorefining processes [79].

Proper selection of a microorganism or a mixture of microorganisms and the control of the process conditions (by affecting the pH during fermentation, temperature, or oxygen content) allows the fermentation to be directed to obtain biocomponents, which are difficult to obtain in the chemical synthesis [15,42,75]. This approach creates a chance for better usage of the raw material and highlights the necessity for carrying out biorafination procedures regarding the fermentation broth [78,79]. Although biogas formed during biological processes contains hydrogen, due to different gaseous ingredients, a gaseous stream purification must be concerned. Additionally, pre- and post-fermentation broths whose composition is based on biomass hydrolysates require purification. Therefore, consideration of inhibitory by-products must be carried out [80].

Starch-based biomass processing generates a very low possibility of inhibitory compound generation. In this case, only secondary transformations of hexoses (glucose) may cause HMF generation, especially in acidic conditions [81]. Additionally, poorly chosen process conditions may lead to incomplete liquefaction of starch [8,74].

Lignocellulosic biomass has a very high potential for the production of biofuels as well as chemicals. In the case of lignocellulose and saccharification, the conversion of complex carbohydrates molecules into simpler sugars is required [79]. Inhibitors may be formed during the hydrolysis process—mainly lignin derivatives—and as secondary transformation products due to saccharification products' transformation under specific conditions. To maximize the fermentation of hexoses (C_6) and pentoses (C_5), and to minimize the presence of inhibitors during the fermentation process, the concentration of possible derivatives must be monitored throughout hydrolysis [82]. Transformation products of pentoses and hexoses include furfural and hydroxymethyl furfural (HMF), considered fermentation inhibitory compounds. For possible lignin derivatives, please refer to Figure 2.

The type of chemical used in the pre-treatment process can have different effects on the structural components of lignocellulose. For example, alkaline pretreatment, ozonolysis, peroxide, and wet oxidation are more effective at removing lignin, while dilute acid pretreatment is more effective at removing hemicellulose [80–82].

2.3.1. Inhibitors Generated during Acidic Pretreatment

Mineral acids such as H_2SO_4 can be used to pre-treat lignocellulosic biomass. Depending on the dose of acid used in the process, it can be divided into concentrated or dilute acid hydrolysis. In the concentrated acid hydrolysis method, lignocellulosic biomass with high concentrations of sulfuric acid is purified at ambient temperature, which results in high sugar yields. The use of this method has the advantage of not using enzymes for saccharification. However, this process also has disadvantages, such as corrosion of equipment, high acid consumption, long reaction time, as well as acid recovery after purification, which to some extent causes limitations in using this method. In the second method, using dilute acid (0.5–1% H_2SO_4) and high temperature can convert cellulose to glucose [83].

The high-temperature conversion of cellulose to glucose is an efficient way to achieve an acceptable rate of hexoses. This method, despite the low concentration of acid, short reaction time, and application of high temperature, accelerates the decomposition of hemicellulose sugars while increasing the corrosion of equipment due to the formation of inhibitory by-products and the need to neutralize the pH for subsequent processes problem [15]. Pre-treatment using the dilute acid method can hydrolyze 100% of the hemicellulose to its constituent sugars depending on the pre-treatment conditions. The main purpose of pretreatment with dilute acid is to increase the sensitivity of cellulose to microbial degradation and enzymatic hydrolysis [84]. A two-step process can be used to prevent the decomposition of sugars. In the first stage, hemicellulose sugars are released under mild conditions, and in the second stage, cellulose-rich solid residues are released under more severe conditions. Depending on the nature of the lignocellulosic material, temperatures of 140 to 190 °C are used for the first stage and 190 to 230 °C for the second stage [84,85].

In the process of acid hydrolysis, temperatures above 110 °C cause the formation of toxic inhibitory compounds such as furfural and 5-hydroxymethyl furfural [79]. These compounds inhibit enzymatic and microbial hydrolysis. Their removal is necessary and possible adsorption on activated carbon or precipitation with calcium hydroxide. Other inhibitors such as chloric, phosphoric, or nitrous acids can be formed with increasing temperature and depend on the hydrolyzing agent and pollution of biomass with inorganic pollutants [13,86]. Acidic pretreatment generates inhibitors that must be removed to minimize downstream processing costs. In Figure 3, possible inhibitory compounds occurring due to cellulose monomers degradation are presented.

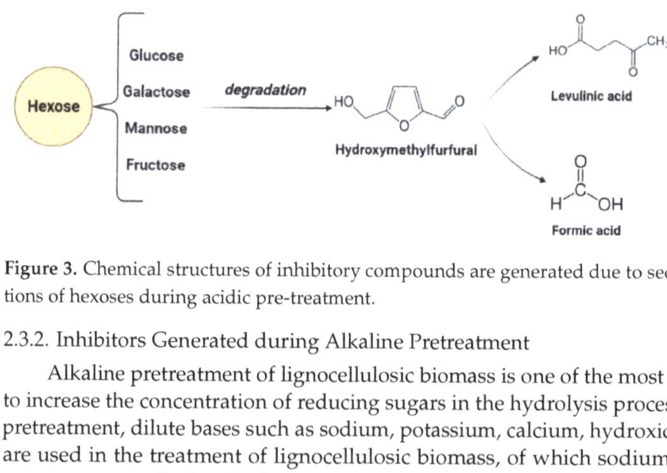

Figure 3. Chemical structures of inhibitory compounds are generated due to secondary transformations of hexoses during acidic pre-treatment.

2.3.2. Inhibitors Generated during Alkaline Pretreatment

Alkaline pretreatment of lignocellulosic biomass is one of the most effective methods to increase the concentration of reducing sugars in the hydrolysis process [29]. In alkaline pretreatment, dilute bases such as sodium, potassium, calcium, hydroxides, and ammonia are used in the treatment of lignocellulosic biomass, of which sodium hydroxide is the most common alkali. Alkaline processes use less temperature and pressure than other methods [87,88].

This process improves the digestibility of cellulose but the decomposition of sugars in this method is less than in acidic pretreatment. However, the main obstacle to this method is the high cost of alkalis. The application of calcium hydroxide, due to its low cost and ability to recover or regenerate ammonia, which is recyclable due to volatility, can be used as a solution to this problem [83,89]. Alkaline pre-treatment also enables hemicellulose degradation. In this case, pentoses may occur [79]. Due to secondary transformations during the hemicellulose structure degradation, secondary derivatives of pentoses may occur (Figure 4).

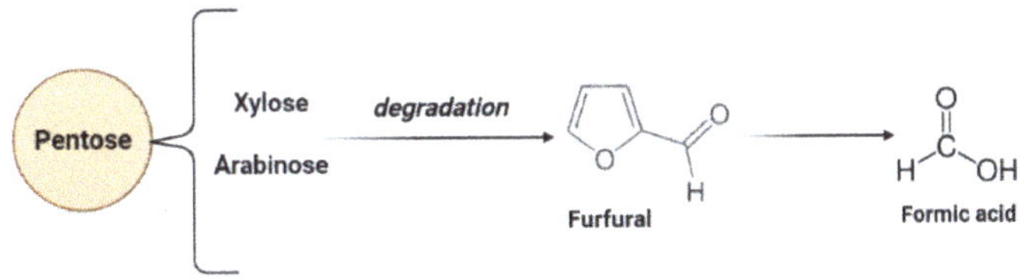

Figure 4. Chemical structures of inhibitory compounds are generated due to secondary transformations of pentoses during alkaline pre-treatment.

2.3.3. Inhibitors Generated during Oxidative Pretreatment

Hydrogen Peroxide

In the oxidation pretreatment process, peroxides such as hydrogen peroxide or alcoholic solutions of acetic acid are used. Oxidizing agents can dissolve amorphous cellulose and lignin, while hemicellulose can be dissolved when separated from the biopolymer. Crystalline cellulose is not dissolved in this method. In this method, processes such as electrophilic substitution, site chain dislocation, and aryl-alkyl bonding cleavage occur [90].

Alkaline hydrogen peroxide is used in the paper industry as bleach, lignin, and xylene remover. This process is very mild and leaves no contamination in the lignocellulosic biomass and also decomposes into water and carbon dioxide [13,79]. In addition, in this process, no by-products and inhibitors of pentose and hexose decomposition are formed. The oxidizing pretreatment reduces the biomass resistance. The hydrogen peroxide only reacts with the aliphatic biopolymer compounds, and in alkaline conditions, due to the presence of the cumene anion, lignin is separated from the lignocellulose structure [91]. Unfortunately, the reagent is unstable under alkaline conditions and decomposes easily in the presence of transition metals such as Mn, Fe, and Cu, so the application of hydrogen peroxide requires special processing conditions, which is hard to accomplish in bio-fraction mixtures. The highest pH value-enabling efficient alkaline pretreatment with H_2O_2 is 11.5. The lowest applicable hydrogen peroxide concentration is 1% with a mass proportion between H_2O_2 and biomass equal to 1: 4 [92].

Wet Oxidation Process

In the wet oxidation process (WO), water and oxygen or air with high pressure and temperature above 120 °C are used for the pre-treatment of lignocellulosic biomass. Combining the alkaline process with wet oxidation, in addition to accelerating the oxidation rate of lignin, prevents the formation of furfural and inhibitory compounds [83]. The WO can be used as an effective pretreatment method to convert lignocellulosic biomass, such as wheat straw, to a soluble hemicellulose fraction and a solid part with high cellulose content with high sensitivity to enzymatic hydrolysis. In this process, the acids produced due to the dissolution of hemicellulose components catalyze subsequent hydrolytic reactions, which decompose hemicellulose into components with low molecular weight and that are soluble in water. At high temperatures, lignin degradation is particularly important because phenolic compounds and carbon–carbon bonds are highly reactive under wet oxidation conditions. In this process, lignin is broken down into CO_2, H_2O, and carboxylic acids, which may be the sole carbon source for photo fermentative bacteria [93].

Ozonolysis

In this method, ozone is used to dissolve lignin and part of hemicellulose. The process of ozonolysis at room temperature can remove lignin without producing any toxic or inhibitory compounds. The main limitation of using this method is the cost [83].

2.4. Summary of Pre-Treatment Methods Advantages and Disadvantages

In Table 2, the advantages and disadvantages of mentioned pre-treatment methods are summarized [13,74,78,86,94–97].

Table 2. Summary of chosen advantages and disadvantages of biomass pre-treatment methods.

Pre-Treatment Method	Advantages	Disadvantages
Enzymatic hydrolysis	The precise method for saccharification, the possibility of process planning, and the selection of an enzymatic liqueur for a specific raw material	The process parameters must be carefully designed and controlled. Enzymes are expensive and not always able to recirculate, loss of activity if the local temperature is unstable, and some saccharification by-products are recognized as enzyme inhibitors
Biological pre-treatment	Ability to design a microbial consortium, reducing the number of pre-treatment steps. Allows the design of a precise liqueur of enzymes at lower costs, a wide range of process parameters, and the possibility to obtain wild rot species for precise raw material	Time-consuming process, the possibility of monosaccharide self-consumption if the consortium is designed inappropriately, risk of infection of the bacterial culture, difficulties in separating the products, and toxicity of the fermentation broth; work in a two-phase system is necessary
Acidic hydrolysis	A method that is cheap, easy to control, widely used, and allows comparison of the results of the pre-treatment. The application of a wide range of acid concentrations allows for controlling the generation of inhibitory compounds	High temperatures, specific requirements of reactor materials, decomposition of parts of the main product and its transformation into inhibitors, and emission of oxides as a result of the fusion of acid particles
Alkaline hydrolysis	An effective method to increase the concentration of reducing sugars in the hydrolysis process. Alkaline processes use less temperature and pressure than other methods	The decomposition of sugars in this method is less than in acidic pretreatment. Possibility of fermentation inhibitors generation and secondary transformation of saccharification products
Oxidative hydrolysis	Dissolves amorphous cellulose and lignin, possibility to remove lignin without derivatives generation, COD lowering effect	Reagents are unstable under alkaline conditions and decompose easily in the presence of transition metals and the application of hydrogen peroxide requires special processing conditions
Ionic liquids	It is an effective method for dissolving the plant cell wall that does not require high temperature to dissolve the cell wall. This method is used in mild processing conditions. It also has low volatility and reusability, selective removal of lignin and hemicellulose as well as cellulose release	Failure to recycle solvents creates toxic substances in the environment and deactivates enzymes
Supercritical fluid CO_2Water	The decomposition of sugars is low and, unlike acid methods, the amount of corrosiveness is significantly reduced. It prevents the degradation of xylose at low temperatures, recovers, and reusesNo need to dry biomass before pretreatment and reduces resistance to mass transfer. Requires a very short reaction time; therefore, the decomposition of glucose, xylose, and arabinose sugars is prevented.	Requires high pressure and temperature, non-change of lignin and hemicellulose, increasing the concentration of xylan and furan for pretreatment of corn
DES	High recovery of sugars during the pre-treatment process, improving the rate of enzymatic saccharification, preventing the degradation of polysaccharides, and preserving carbohydrates. Excellent performance on lignin extraction and biomass saccharification enhancement. Ability to selectively dissolve lignin and hemicellulose	The high viscosity limits their application and the pretreatments are often very complex, with the inhibition effect toward cellulase and acidic DESs destroying polysaccharides

The comparison of the advantages and disadvantages of commonly used pretreatment methods indicates that there is no universal method that will create effective pretreatment options for each known type of biomass. The selection of the appropriate method must be empirical and should take into account the issues related to the processing of the raw material, but also consider the by-products of this treatment. Making it necessary to detoxify the liquefied parts of biomass will significantly affect the technological effectiveness of fermentation processes and their efficiency. In addition, after the fermentation processes, it may be necessary to clean the gas streams from inorganic impurities related to the pre-treatment method used, but also from liquid streams in difficult-to-remove derivatives of hydrolyzed polymers and their decomposition products.

3. Bioconversion of Hydrolysates

3.1. Hydrogen Generation

In nature, anaerobic microorganisms produce hydrogen gas in the absence of oxygen and use the phenomenon of fermentation, but the amount of this gas is low and is not economically justifiable for industrial and domestic use. Therefore, it is necessary to search for methods to increase the efficiency of hydrogen gas production [15]. Anaerobic bacteria generate biohydrogen and organic acids via dark fermentation. The biohydrogen from dark fermentation requires purification and the organic acids may be applied as a sole carbon source in photo fermentation. *Rhodospirillum rubrum* is a frequently studied species, which exhibits unique nitrogenase activity, reducing both molecular nitrogen and protons to molecular hydrogen via photo fermentation [17]. The microorganisms that synthesize biohydrogen in the photo fermentation process with the use of hydrogenase also include purple sulfur bacteria, which are strict anaerobes, e.g., *Allochromatium vinosum, Thiocapsa roseopersicina, Chlorobium vibloroforme, Desulfuromonas acetoxidans,* and *Chloroflexus aurantiacus*. The range of electromagnetic radiation waves absorbed by them ranges from 400 to 950 nm. *Rhodospirillum rubrum,* the commonly used microorganism in photo fermentation is classified as mesophilic, and its temperature optimum is 25–30 °C. It has polar flagella and is a facultative anaerobe [13,17]. Depending on the presence of oxygen, it can carry dark fermentation or oxygen respiration. It is also capable of photosynthesis as it contains carotenoids and bacteriochlorophyll. In addition to the ability to bind carbon dioxide, it can bind nitrogen. It contains both Fe-Mo- (iron-molybdenum) and Fe-nitrogenase. Microorganisms of this type are currently one of the most promising in the field of biohydrogen synthesis by photo fermentation [18].

The main purpose of starch and lignocellulosic pretreatment processes is to reduce the degradation of sugars, minimize the formation of inhibitory compounds, and reduce the consumption of chemical compounds, energy, water, and waste production [98]. Additionally, the digestibility in bioconversion of biomass must be improved as an effect of biomass pre-treatment

The efficiency of bioconversion concerning hydrogen generation for diversified sole carbon sources in the broths is presented in Table 3.

As seen in Table 3, most research carries the discussion on fermentation efficiency on a single substrate. A review of the literature has shown that there is a lack of research based on the fermentation of real mixtures and real wort, in which lignocellulosic and starch polymers can occur simultaneously as carbon sources. Industrial practice in the field of hydrogen fermentation is that in the case of biohydrogen, there must be a departure from the fermentation of only one raw material towards the co-fermentation of many raw materials. This type of approach is more and more often ordered and fits into the assumptions of the circular economy.

Table 3. The efficiency of bioconversion of biomass hydrolysates concerning applied microorganisms and the main sole carbon source present in the fermentation broth.

Substrate	Amount	Organism	Reactor type	pH	Temperature (°C)	HRT	Hydrogen Productivity	Hydrogen Yield	COD Removal (%)	% H_2	Reference
Glucose	10 g/L	*Clostridiaceae Flexibacteraceae* *	Membrane Continuous	5.5	35	3.3	640 mL H_2/(L·h)	4 mol H_2/mol glucose	-	60	[99]
Sucrose	10 g/L	*E. cloacae IIT-BT 08* *	Batch	6	36	-	660 mL H_2/(L·h)	6 mol H_2/mol sucrose	-	92	[100]
Glucose	1%	*E. cloacae* *	Batch	6	36	-	447 mL H_2/(L·h)	2.2 mol H_2/mol glucose	-	-	[100]
D-Xylose	10 g/L	*E. cloacae IIT-BT 08* *	Batch	6	36	-	348 mL H_2/(L·h)	0.95 mol H_2/mol xylose	-	-	[100]
L-Arabinose	10 g/L	*E. cloacae IIT-BT 08* *	Batch	6	36	-	360 mL H_2/(L·h)	1.5 mol H_2/mol arabinose	-	-	[100]
Glucose	10 g/L	Mixed culture from compost	Batch	5.5	60	-	147 mL H_2/(L·h)	2.1 mol H_2/mol glucose	-	-	[101]
Glucose	20 g COD/L	*Clostridia* sp. *	CSTR Continuous	6	28–32	6	7.42 mmol H_2/(gVSS·h)	1.42 mol H_2/mol glucose	-	43	[102]
Glucose	7 g/L	Mixed culture	CSTR Continuous	5.5	36	6	-	2.1 molH_2/mol glucose	-	64	[103]
Glucose	4.85 g COD/L	Mixed culture	UASB Continuous	7.2	70	26.7	11.15 mmol H_2/d	2.46 mol H_2/mol hexose	-	55	[104]
Sucrose	20 g COD/L	Mixed culture	Immobilized bed Continuous	6.7	35	1	1.32 L H_2/(L·h)	-	-	34	[105]
Sucrose	1 g COD/L	Mixed culture	Batch	6	26	-	-	1.8 mol H_2/mol sucrose	-	-	[106]
Sucrose	20 g COD/L	Mixed culture	CSTR Continuous	6.7	35	8	0.105 mol H_2/h	3.47mol H_2/mol sucrose	-	42	[107]
Sucrose	25 g/L	Mixed culture	Fermenter Batch	5.5	35	-	1504 mL H_2/h	2 mol H_2/mol glucose	-	-	[108]
Lactose	29 mmol/L	*C.termolacticum* *	CSTR Continuous	7	58	35.7	2.58 mmol H_2/(L·h)	1.5 mol H_2/mol hexose	-	55	[109]
Glucose	5.5 g/L	*Enterobacter aerogenes* *	Bioreactors		42	48	25.44 mL/g biomass 283.45 _ 1.87 mL/YTRS	-	-	-	[26]

Table 3. *Cont.*

Substrate	Amount	Organism	Reactor type	pH	Temperature (°C)	HRT	Hydrogen Productivity	Hydrogen Yield	COD Removal (%)	% H_2	Reference
Paulownia	-	*photosynthetic consortium HAU-M1* *	-	7	30	26–38	338.41 mL	67.11 mL/g	62	-	[110]
Xylose	20 g/L	*Lactobacillus* and *Sporolactobacillus* spp., *Clostridium* sp *	Dynamic membrane module bioreactor (DMBR)	7.5	37	3–12	30.26 L H_2/L-d	1.40 mol H_2/mol xylose	-	-	[111]
Glucose	10 g/L	*Caldicellulosiruptor* *	Batch	-	70	20	10.55 mmol/L/h	4 mol H_2/mol glucose	-	-	[112]
Glucose and Xylose	50 % mol/mol	*Rhodopseudomonas palustris* *	Bioreactor	7	30	-	30.6 mL h^{-1} L^{-1}	1.63 (mol H_2/mol carbon)	-	-	[113]
Glucose, xylose	41.17 g/L	*Rhodospirillum rubrum* *	Batch	4.5	60	-	819 mL H_2/L medium m/7 d	-	82	-	[114]
Rice husk	5 g dw	*Clostridium termitidis ATCC-21846* * and *Clostridium intestinale ATCC-BAA 1027*	Batch	7.5	37	-	0.023 mL H_2 g^{-1} dw rice husk h	5.9 mL g^{-1} dw Rice husk	-	29.26 mL	[115]
Corncob	10 g	*HAU-M1* * photosynthetic bacteria	Batch	7	50	48	-	27.34 mL/g TS	-	80.94	[116]
Duckweed and corn straw	5:1	photosynthetic strain *HAU-M1*	Batch	8	30	18.57	-	85.6 mL/g TS	-		[117]

* The kind of bacteria.

Co-fermentation uses a mixture of biowaste from different sources. Joint fermentation allows to obtain proper hydration of the fermentation mass, improvement of the balance of biogenic elements, or an increase in the load of easily biodegradable matter, which contributes to a more stable course of the process, and also allows for the synergy effect, which increases the efficiency of organic mass decomposition and biohydrogen generation efficiency.

The most frequently tested dark fermentation feeds are hydrolysates based on glucose and sucrose. A review of the literature [26,99–117] showed that for most model studies, substrate concentrations of about 10 g/L were neutral towards the slightly acidic pH of the fermentation broths. Only a few studies [104] were carried out for a pH > 7; however, as it appears from the content of the study, this is only the starting pH, which drops during fermentation due to the formation of organic acids during fermentation in a continuous UASB reactor. Researchers of processes mentioned in Table 3 have not decided to conduct continuous processes, but there are exiles with two-stage fermentation or in a continuous rotary system. Based on experiments, it can be concluded that in most cases when pure cultures [26,100,111] are used, slightly higher yields of hydrogen can be obtained than when mixed cultures are used [104–108]. Unfortunately, the nature of these pure cultures often requires the use of relatively high temperatures, since they are thermophilic strains. Based on the published data, it is impossible to balance the productivity of hydrogen and the energy sense of the processes performed. Therefore, tests for real broth conditions are required, to compare the effectiveness of the fermentation data or to adjust a universal model for fermentation conditions. This kind of model must raise issues related not only to gaseous products but also to the composition of the post-fermentative liquid. When increasing the profitability of the process is possible, more attention is paid not only to detoxification but also to potential methods of post-fermentation broth management.

3.2. Genetic Modifications of Microorganisms

Genetic modification of microorganisms is an interesting aspect of enhancing hydrogen productivity with the simultaneous adaptation of microorganisms against the resulting fermentation by-products. Industrial conversion of biomass to fuel currently involves heat and chemical treatment to overcome the biomass recalcitrance of starch, cellulose, hemicellulose, and lignin, followed by enzymatic hydrolysis to dissolve the plant cell wall to produce a fermentable substrate for fuel-producing microorganisms. All of these methods add costs and, on the other hand, produce hydrolysates that are toxic to microorganisms and harmful to the sugars in biomass [118]. An approach involving genetic engineering techniques to manipulate the metabolism of microorganisms may also be applied. The efficiency of biohydrogen production can be improved and, consequently, the total processing costs can be reduced. Expression of the genes responsible for the production of organic acids may be turned off, while the strains with multiple enzyme systems, including cellulase and xylanase, which are responsible for the breakdown of cellulose and hemicellulose, may be introduced and developed [15]. Genetic engineering provides the basis for increasing biohydrogen production. There are several methods, including deletion of a competitive gene, overexpression of a homologous or heterologous gene, creation of artificial pathways, culture, and identification of indirect hydrogen-producing organisms to improve hydrogen metabolic function through genetic engineering. A genetic modification of enzyme activity can be effective when the specific amount of that enzyme is limited [119]. The possible paths and the effects of genetic engineering of different species of microorganisms on biohydrogen production efficiency are presented in Table 4.

Table 4. Effect of genetic modification in various species of bacteria on hydrogen efficiency in bioconversion.

Microorganism	Strain	Genetic Modification	mol H_2/mol Glucose	Reference
Caldicellulosiruptor bescii *	-	deletion of L-lactate dehydrogenase gene (ldh)	2.5	[118]
Escherichia coli *	SR15	modifying ΔldhA, ΔfrdBC	1.82	[120]
Escherichia coli *	-	production BW25113 hyaB hybC hycA fdoG frdC ldhA aceE	2	[121]
Escherichia coli *	MC4100, wild-type FTD89, mutant	deletion of Hyd-1 + Hyd-2; hyaB + hybC	1.043	[122]
Escherichia coli *	FTD67, mutant	deletion of Hyd-2; hybC	1.024	[122]
Escherichia coli *	W3110, wild-type SR15, mutant	deletion of ldhA + frdBC	1.82	[123]
Escherichia coli *	W3110, wild-type SR14, mutant	deletion of ldhA + frdBC overexpression of fhlA	1.87	[123]
Escherichia coli *	Bl-21 recombinant, mutant	deletion of hydA	3.12	[124]
Escherichia coli *	BL21(DE3] _iscR pYdbK pAF, mutant	deletion of iscR + MCS2 overexpression of YdbK + CpFdx + hydA + hydF + hydG + hydE	1.46	[125]
Clostridium paraputrificum *	M-21 pJIR751, mutant	overexpression of hydA	2.4	[126]
Clostridium acetobutylicum *	DSM 792 [pSOS], mutant	overexpression of thl promoter	1.77	[127]
Clostridium acetobutylicum *	DSM 792 [pSOShydACa], mutant	overexpression of hydA	1.81	[127]
Clostridium acetobutylicum *	DSM 792 (pSOShydACb), mutant	overexpression of hydA	1.80	[127]
Clostridium tyrobutyricum *	PAK-Em, mutant	deletion of ack	2.16	[128]
Clostridium tyrobutyricum *	PAK-Em, mutant	deletion of ack	2.61	[129]
Clostridium tyrobutyricum *	ATCC 25,755 PPTA-Em, mutant	deletion of pta	1.08	[129]
Enterobacter aerogenes *	IAM1183 A, mutant	deletion of hycA	1.20	[130]
Enterobacter aerogenes *	IAM1183 O, mutant	deletion of hybO	1.27	[130]
Enterobacter aerogenes *	IAM1183 AO, mutant	deletion of hycA + hybO	1.36	[130]
Enterobacter aerogenes *	ATCC 13048/hydA, mutant	overexpression of hydA	2.31	[131]
Enterobacter aerogenes *	IAM1183 Ea (pMCL-fdhF), mutant	overexpression of fdhF	1.16	[132]
Enterobacter aerogenes *	IAM1183 A (pMCL-fdhF), mutant	deletion of hycA overexpression of fdhF	1.19	[133]
Enterobacter aerogenes *	IAM1183 (pCOM 10-fdh1), mutant	deletion of ldh overexpression of Fdh1	1.70	[134]

* The kind of bacteria.

To the best knowledge of the authors, research in the field of genetic modification of microorganisms for adaptation to the composition of hydrolysates is not currently carried out. This type of approach could, however, allow the implementation of large-scale processes and independence from difficult ingredients in fermentation broths, which may occur as a result of pre-treatment. However, the designed genetically modified microorganisms have to be tested on real fermentation broths since research has shown that the adaptation of model conditions to real broths may require time-consuming optimization.

4. Post-Fermentative Broth Detoxification and Management Methods

The inhibitory compounds generated during pre-treatment may affect the efficiency of bioconversion; the broth composed mainly from the hydrolysates must be processed for detoxification, if necessary [79].

The degrees of inhibition of lignocellulosic hydrolysates and also the degree of inhibition tolerance of various microorganisms are different. The choice of detoxification methods depends on the source of lignocellulosic hydrolysate and the microorganisms used. Therefore, detoxification methods can be divided into three groups: physical, physicochemical, and biological [135].

As chemical structures are a criterion for classification, inhibitory compounds may be divided into four groups [79]: substances produced by hemicelluloses (acetic acid which is the source of deacetylation of xylan); substances that are produced from the degradation of lignin (phenolic compounds and other aromatic compounds); materials obtained from the destruction of pentoses (furan derivatives, furfural) and hexoses (5-hydroxymethylfurfural); And metals leached from inorganic pollutants and/or equipment (copper, chromium, nickel, and iron) [101,114]. All of these compounds can individually or synergistically affect the physiology of fermenting microorganisms during bioconversion [134]. Therefore, the removal or reduction of the amount of these compounds is necessary to increase the efficiency during the microbial fermentation process of biomass hydrolysates [135].

The processing of post-fermentation broths may be crucial for their management, especially since the broth may contain not only the products of microbial metabolism but also remnants of the microbes present in the broth during bioconversion. The methods of purification of the remnant broths before and after fermentation would therefore be the same, and the applicable ideas are presented in Figure 5.

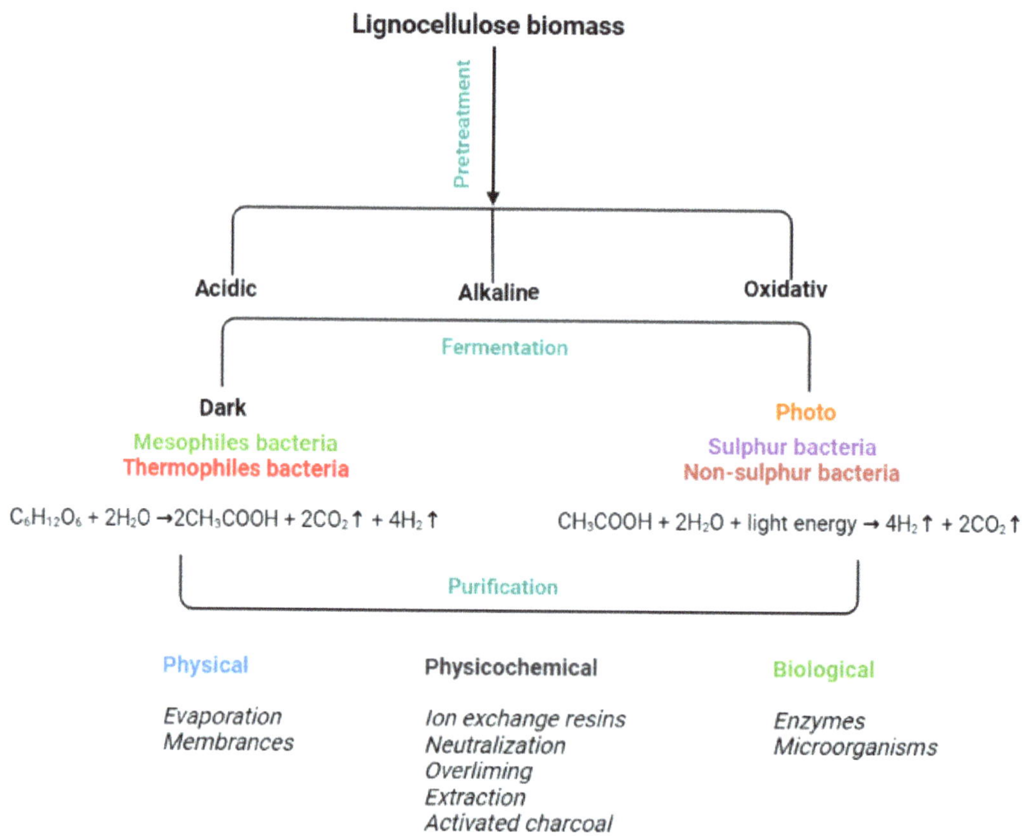

Figure 5. The procedure of biomass pre-treatment and the applied methods for broths purification.

4.1. Physical Methods

4.1.1. Evaporation

The evaporation process can be used to remove toxic inhibitory compounds, such as acetic acid, furfural, and vanillin. One of the disadvantages of this method is the increase of non-volatile toxic compounds as extractives [134]. Evaporation may be used especially for post-fermentation broth purification, since an increase in the temperature may affect the structures of microbes in the broth. Increased temperature causes the denaturation of proteins and, therefore, allows for the removal of the proteins and microbial remnants from the post-fermentation broth, especially when coupled with centrifugation and sediment separation.

4.1.2. Membranes

The membrane process can be used as one of the detoxification methods. This method prevents the aqueous phase (hydrolyzate) from mixing with the organic phase (solvents) which is toxic to microorganisms [136]. Each membrane has surface functional groups that can remove inhibitory compounds such as acetic acid, 5-hydroxymethyl furfural, furfural, formic, levulinic, and sulfuric acids from hydrolysate solutions [137].

4.2. Physicochemical

4.2.1. Ion Exchange Resins

The process of ion exchange resins is one of the most effective detoxification methods. In this process, inhibitory compounds derived from lignocellulose hydrolysis, including lignin, acetic acid, and furfural, are removed, thus improving the efficiency of the fermentation process [138]. The main advantage of this method is that they are recoverable and can be reused without affecting the detoxification efficiency. On the other hand, this method also has disadvantages, such as increased high-pressure drop across the bed during work, long processing time due to slow pore diffusion, and the possibility of degradation of fragile biological product molecules, In this method, a significant amount of fermentable sugars are lost after the process [115].

4.2.2. Neutralization

Because the pH is low after acid hydrolysis pretreatment, the pH neutralization process approaches the fermentation conditions. Additionally, the inhibitory compounds (phenol and furfural) are removed by precipitation. This method uses chemical compounds such as calcium hydroxide and sodium hydroxide. In the calcium hydroxide neutralization method, $CaSO_4$ precipitates are produced which must be removed from the environment by centrifugation in the next step, so the production of precipitates can cause problems in the fermentation process [139].

4.2.3. Overliming

Among detoxification methods, the $CaSO_4$ process has been reported as one of the most widely used methods [140]. In this process, first, the pH of acidic hydrolysis increases and then decreases to the desired pH for fermentation. During this pH increase, toxic, inhibitory, and unstable compounds precipitate. This method has high efficiency in removing these compounds, especially furan compounds, and is an economically desirable method [141,142].

4.2.4. Activated Charcoal

Activated charcoal is another widely used method for detoxification. This method is very low-cost and efficient. In this method, most phenolic compounds are removed and also do not cause many changes in the level of fermentable sugars. Important factors in improving this process are the ratio of activated carbon to hydrolyses, pH, temperature, and contact time [135,143].

4.2.5. Extraction

Solvent extraction is an efficient method for the removal of highly available toxic inhibitory compounds such as acetic acid, furfural, vanillin, hydroxy-benzoic acid, and low molecular weight phenolics. Ethyl acetate, chloroform, and trichloroethylene are among the most common solvents used in this process [144].

4.3. Biological Methods

In the biological process, special enzymes and microorganisms are used to remove or induce changes in the composition of the inhibitory compound [134]. The advantages of this method include the following: less waste production, the possibility of detoxification in the fermentation vessel, being environmentally friendly, fewer side reactions, and lower energy requirement [145].

This group of methods requires a long process time. Enzymes such as laccase and peroxidase derived from white-rot fungi are used to remove phenolic compounds from lignocellulose hydrolyzers. The main mechanism of detoxification of these enzymes may include oxidative polymerization of low molecular weight phenolic compounds. White rot fungi may be applied both in pre-treatment and broth management steps. They also catalyze the oxidation of alternative phenols, anilines, and aromatic thiols [146,147]. Another disadvantage of enzymatic detoxification is the long incubation time and high cost. While the advantage of this method is that it takes place in mild environmental conditions (neutral pH and mesophilic temperature) [148].

Microorganisms such as yeasts, fungi, and bacteria can be used to absorb inhibitory and toxic compounds. Some microorganisms can release cellulose and hemicellulose during incubation and only decompose lignin, so this method creates a lignocellulosic substrate that can decompose fermentable sugars in a mild and short time [146,148].

4.4. Perspectives of Broth Detoxification and Management Methods

Known methods of post-fermentation broth management focus primarily on lowering those parameters of the liquid, which may affect the possibility of discharging the broth to wastewater. Most often this includes metal ion content, total organic carbon level, chemical and biological oxygen demand, solids content, and phenolic compounds content. The authors, however, have some experience in model post-fermentation broth management [149] which, to our best knowledge, may be adopted under real conditions.

Interest in the topic of broth management methods has shifted to the possibility of extracting added-value products from fermentation broths [134–139]. If for some reason, on the side of microorganisms, the complete utilization of monosaccharides by fermentation is not achieved, it is possible to control the detoxification process in such a way as to convert monosaccharides into substances with potential use in the synthesis of green solvents [149]. One must remember that the post-fermentation broth may contain some inhibitors present due to pre-treatment residual products, but also other chemicals that have been recognized as fermentation inhibitors that were generated during the fermentation process. Some of these chemical compounds could become precursors of deep eutectic solvents that can be obtained in situ, directly from post-fermentation broths.

The pilot research carried out by the team [149] indicates that this type of action allows us to successfully obtain DES based on LA, HMF, and furfural. At the same time, since this process requires the conversion of monosaccharide residues, which is carried by increasing the temperature and reducing the pH, the microorganisms die and decay, and the proteins suspended in the broth are denatured. It is therefore possible to generate DES in situ and to precipitate morphotic cell debris. The treatment of the broth by precipitation of solid particles and their subsequent centrifugation allows at the same time to reduce the parameter related to the content of solid particles, reducing the values of COD and BOD. Only after the separation of the liquid and solid streams, the pre-processed fermentation broth, often rich in organic acids being a product of the metabolism of dark fermentation bacteria, can be directed to photo fermentation, where the organic acids will be processed by microorganisms, i.e., the *Rodospirillum rubrum* sp [114].

A promising direction of broth management is biorefining, which allows, under specific conditions, to obtain a wide spectrum of end products. Biorefining mainly concerns raw materials of plant origin and the process requires the separation of botanical and chemical components [18,31,46,99]. After the processing of biomass to obtain hydrogen, the remaining debris, solids, and liquids can be converted from one form to another. Such transformations require the use of enzymes, microorganisms, and other biological agents. As a result of the biorefining process, plant-specific ingredients, such as proteins and lipids, are recovered from plants [46].

The disposal of post-fermentative broth during a co-fermentation in a biohydrogen plant allows carrying sufficient management of the broth. After dark and photo fermentation, several nutrients and microelements still occur in the broth. Therefore, an addition of post-fermentation broth to a co-digester may have a promoting effect on methane generation. Finally, the leftovers from the biohydrogen plant may be applied as fertilizers after maturation.

Future perspectives should also consider problems related to scale-up, for each discusses the pre-treatment method. These types of problems arise in the very first part of the technology when pre-treatment is considered. Acidic pretreatment with the application of high acid concentrations causes corrosion of the equipment, which can involve a lot of maintenance and repair costs. Due to the low pH, it requires the neutralization of hydrolyzed biomass before fermentation. Due to the production of inhibitory compounds, it needs to add a detoxification step to remove these compounds, thus the costs will be increased. Optimization of the process parameters before scale-up is required. Alkaline pretreatment with the use of sodium, potassium, calcium, and ammonia hydroxide in high concentrations has a very high cost and, on the other hand, their entry into the environment causes environmental problems, so there is a need for recycling, wastewater treatment, and waste handling that will increase the costs. Oxidative pre-treatment is cost-consuming and requires special processing conditions and safety regulations. Biological pretreatment requires specific control for the growth conditions of microorganisms and a skilled operator. It needs a lot of space and time since it has a low efficiency compared to other methods. Enzymatic pretreatment requires optimal conditions and constant monitoring of hydrolysis temperature, time, pH, enzyme loading, and substrate concentration, so it has a difficult design and operation. During ionic liquids pretreatment, the recovery of the hydrolyzing agent and the reuse of ionic liquids are major problems for industrial applications of biomass pretreatment. As supercritical CO_2 pretreatment is considered, the use of high temperature and pressure is an important economic problem on a large scale. DES pretreatment is very complicated due to the responsibility of many variables, so skilled operators are needed and, on the other hand, the high viscosity of DES severely limits their use.

5. Summary

Investigations on a multistep bioconversion process of biohydrogen generation should be carried out concerning the sustainable development and minimization of energy expenditure for the purification of post-production streams. Most published papers focus on the purification of gaseous streams after fermentation, since the generation of biofuels is the main direction of the conducted studies. Since several intermediate liquid streams occur during the multistep process of bioconversion, their disposal, detoxification, and management must be taken into account. Hydrolysates of starch and lignocellulose may become a source of vanillin, syringole, cumarol, furfural, HMF, levulinic acid, and many other value-added products. The literature lacks models that would allow us to unambiguously balance the productivity of biohydrogen, and at the same time assess the environmental effect of the fermentation carried out with the potential liquid pollutants generated as a result of the conducted fermentation.

Several paths of pre-treatment and derivatives occurring in the broth due to this process were discussed. According to the literature, the purification of intermediate streams does not differ from the purification after the bioconversion and, therefore, if the applied temperature does not increase above the temperature allowing for protein denaturation, it may be carried to obtain liquid streams allowing to improve the bioconversion to gaseous biofuels, i.e., biohydrogen. On the other hand, the post-fermentation broth after dark fermentation contains organic acids and therefore may be purified by subjecting it to photo fermentation. This approach allows us to improve the overall efficiency of hydrogen. The insights of the process based on broths composed of organic acids should become a significant direction of research.

In the opinion of the authors, the used biomass pre-treatment method influences the possible paths of managing the fermentation broth. If the possibility of using diversified raw materials, different cultures, and process parameters is considered, it turns out that despite the multitude of research in the field of fermentation, there is no universal method that would allow the management of all types of broth. This justifies the need for further research, an important aspect of which is a complete approach to the topic, which will enable the description of each of the streams and an unequivocal way of its management. Many investigations are focused on finding a novel bio-refining path that would allow for improving the economic issues in biofuel generation. Methods allowing the generation of green solvents, i.e., DES, or allowing the disposal of post-fermentation broth as an addition in a biohydrogen plant and as a fertilizer precursor seem promising.

Author Contributions: Conceptualization, Z.H., J.G., K.K.; methodology, Z.H.; formal analysis, K.K.; investigation, Z.H.; resources, J.G.; data curation, Z.H.; writing—original draft preparation, Z.H.; writing—review and editing, K.K.; visualization, Z.H.; supervision, J.G.; project administration, K.K, J.G.; funding acquisition, J.G. All authors have read and agreed to the published version of the manuscript.

Funding: This research was funded by the National Science Centre, Poland, under research project No. UMO-2021/41/B/ST8/02395.

Institutional Review Board Statement: Not applicable.

Informed Consent Statement: Not applicable.

Data Availability Statement: Not applicable.

Conflicts of Interest: The authors declare no conflict of interest. The funders had no role in the design of the study; in the collection, analyses, or interpretation of data; in the writing of the manuscript; or in the decision to publish the results.

Sample Availability: Samples of the compounds are not available from the authors.

References

1. Peng, L.; Fu, D.; Chu, H.; Wang, Z.; Qi, H. Biofuel production from microalgae: A review. *Environ. Chem. Lett.* **2020**, *18*, 285–297. [CrossRef]
2. Sharma, R.B.; Parey, A. Modelling of acoustic emission generated due to pitting on spur gear. *Eng. Fail. Anal.* **2018**, *86*, 1–20. [CrossRef]
3. Lay, C.; Dharmaraja, J.; Shobana, S.; Arvindnarayan, S.; Priya, R.K.; Jeyakumar, R.B.; Saratale, R.G.; Park, Y.; Kumar, V.; Kumar, G. Lignocellulose biohydrogen towards net zero emission: A review on recent developments. *Bioresour. Technol.* **2022**, *364*, 128084. [CrossRef] [PubMed]
4. Jayakody, L.N.; Ferdouse, J.; Hayashi, N.; Kitagaki, H. Identification and detoxification of glycolaldehyde, an unattended bioethanol fermentation inhibitor. *Crit. Rev. Biotechnol.* **2017**, *37*, 177–189. [CrossRef] [PubMed]
5. Soltanian, S.; Aghbashlo, M.; Almasi, F.; Hosseinzadeh-Bandbafha, H.; Nizami, A.-S.; Ok, Y.S.; Lam, S.S.; Tabatabaei, M. A critical review of the effects of pretreatment methods on the exergetic aspects of lignocellulosic biofuels. *Energy Convers. Manag.* **2020**, *212*, 112792. [CrossRef]
6. Bhatia, S.K.; Jagtap, S.S.; Bedekar, A.A.; Bhatia, R.K.; Rajendran, K.; Pugazhendhi, A.; Rao, C.V.; Atabani, A.; Kumar, G.; Yang, Y.-H. Renewable biohydrogen production from lignocellulosic biomass using fermentation and integration of systems with other energy generation technologies. *Sci. Total Environ.* **2021**, *765*, 144429. [CrossRef]

7. Lv, Y.; Liu, X.; Zhou, S.; Yu, Q.; Xu, Y. Microbial saccharification—Biorefinery platform for lignocellulose. *Ind. Crops Prod.* **2022**, *189*, 115761. [CrossRef]
8. Mahmoodi, P.; Karimi, K.; Taherzadeh, M.J. Efficient conversion of municipal solid waste to biofuel by simultaneous dilute-acid hydrolysis of starch and pretreatment of lignocelluloses. *Energy Convers. Manag.* **2018**, *166*, 569–578. [CrossRef]
9. Zoghlami, A.; Paës, G. Lignocellulosic biomass: Understanding recalcitrance and predicting hydrolysis. *Front. Chem.* **2019**, *7*, 874. [CrossRef]
10. Sharma, R.; Joshi, R.; Kumar, D. Present status and future prospect of genetic and metabolic engineering for biofuels production from lignocellulosic biomass. In *Genetic and Metabolic Engineering for Improved Biofuel Production from Lignocellulosic Biomass*; Elsevier: Amsterdam, The Netherlands, 2020; pp. 171–192.
11. Vu, H.P.; Nguyen, L.N.; Vu, M.T.; Johir, M.A.H.; McLaughlan, R.; Nghiem, L.D. A comprehensive review on the framework to valorise lignocellulosic biomass as biorefinery feedstocks. *Sci. Total Environ.* **2020**, *743*, 140630. [CrossRef]
12. Li, H.; Liu, Y.; Gao, X.; Li, X. Preparation and characterization of cassava starch-based adsorbents for separating of azeotropic ethanol-water in biofuels ethanol production. *J. Chem. Technol. Biotechnol.* **2016**, *91*, 977–984. [CrossRef]
13. Kucharska, K.; Rybarczyk, P.; Hołowacz, I.; Łukajtis, R.; Glinka, M.; Kamiński, M. Pretreatment of lignocellulosic materials as substrates for fermentation processes. *Molecules* **2018**, *23*, 2937. [CrossRef]
14. Wang, Y.; Huang, W.; Sathitsuksanoh, N.; Zhu, Z.; Zhang, Y.-H.P. Biohydrogenation from biomass sugar mediated by in vitro synthetic enzymatic pathways. *Chem. Biol.* **2011**, *18*, 372–380. [CrossRef]
15. Łukajtis, R.; Hołowacz, I.; Kucharska, K.; Glinka, M.; Rybarczyk, P.; Przyjazny, A.; Kamiński, M. Hydrogen production from biomass using dark fermentation. *Renew. Sustain. Energy Rev.* **2018**, *91*, 665–694. [CrossRef]
16. Hawkes, F.R.; Hussy, I.; Kyazze, G.; Dinsdale, R.; Hawkes, D.L. Continuous dark fermentative hydrogen production by mesophilic microflora: Principles and progress. *Int. J. Hydrogen Energy* **2007**, *32*, 172–184. [CrossRef]
17. Adessi, A.; Philippis, R.D. Hydrogen production: Photofermentation. In *Microbial Technologies in Advanced Biofuels Production*; Springer: Berlin/Heidelberg, Germany, 2012; pp. 53–75.
18. Ashokkumar, V.; Venkatkarthick, R.; Jayashree, S.; Chuetor, S.; Dharmaraj, S.; Kumar, G.; Chen, W.; Ngamcharussrivichai, C. Recent advances in lignocellulosic biomass for biofuels and value-added bioproducts—A critical review. *Bioresour. Technol.* **2022**, *344*, 126195. [CrossRef]
19. Kamm, B.; Kamm, M. Biorefineries–multi product processes. *White Biotechnol.* **2007**, *105*, 175–204.
20. Claassen, P.; Van Lier, J.; Lopez Contreras, A.; Van Niel, E.; Sijtsma, L.; Stams, A.; de Vries, S.S.; Weusthuis, R.A. Utilisation of biomass for the supply of energy carriers. *Appl. Microbiol. Biotechnol.* **1999**, *52*, 741–755. [CrossRef]
21. Zhao, X.; Zhang, L.; Liu, D. Biomass recalcitrance. Part I: The chemical compositions and physical structures affecting the enzymatic hydrolysis of lignocellulose. *Biofuels Bioprod. Biorefin.* **2012**, *6*, 465–482. [CrossRef]
22. Zheng, Y.; Pan, Z.; Zhang, R. Overview of biomass pretreatment for cellulosic ethanol production. *Int. J. Agric. Biol. Eng.* **2009**, *2*, 51–68.
23. Pauly, M.; Keegstra, K. Plant cell wall polymers as precursors for biofuels. *Curr. Opin. Plant Biol.* **2010**, *13*, 304–311. [CrossRef] [PubMed]
24. Alvira, P.; Tomás-Pejó, E.; Ballesteros, M.; Negro, M. Pretreatment technologies for an efficient bioethanol production process based on enzymatic hydrolysis: A review. *Bioresour. Technol.* **2010**, *101*, 4851–4861. [CrossRef] [PubMed]
25. Kucharska, K.; Hołowacz, I.; Konopacka-Łyskawa, D.; Rybarczyk, P.; Kamiński, M. Key issues in modeling and optimization of lignocellulosic biomass fermentative conversion to gaseous biofuels. *Renew. Energy* **2018**, *129*, 384–408. [CrossRef]
26. Kucharska, K.; Cieśliński, H.; Rybarczyk, P.; Słupek, E.; Łukajtis, R.; Wychodnik, K.; Kamiński, M. Fermentative conversion of two-step pre-treated lignocellulosic biomass to hydrogen. *Catalysts* **2019**, *9*, 858. [CrossRef]
27. Łukajtis, R.; Kucharska, K.; Hołowacz, I.; Rybarczyk, P.; Wychodnik, K.; Słupek, E.; Nowak, P.; Kamiński, M. Comparison and optimization of saccharification conditions of alkaline pre-treated triticale straw for acid and enzymatic hydrolysis followed by ethanol fermentation. *Energies* **2018**, *11*, 639. [CrossRef]
28. Kucharska, K.; Słupek, E.; Cieśliński, H.; Kamiński, M. Advantageous conditions of saccharification of lignocellulosic biomass for biofuels generation via fermentation processes. *Chem. Pap.* **2020**, *74*, 1199–1209. [CrossRef]
29. Kucharska, K.; Łukajtis, R.; Słupek, E.; Cieśliński, H.; Rybarczyk, P.; Kamiński, M. Hydrogen production from energy poplar preceded by MEA pre-treatment and enzymatic hydrolysis. *Molecules* **2018**, *23*, 3029. [CrossRef]
30. Yu, G.; Yano, S.; Inoue, H.; Inoue, S.; Endo, T.; Sawayama, S. Pretreatment of Rice Straw by a Hot-Compressed Water Process for Enzymatic Hydrolysis. *Appl. Biochem. Biotechnol.* **2010**, *160*, 539–551. [CrossRef]
31. Sharma, H.K.; Xu, C.; Qin, W. Biological pretreatment of lignocellulosic biomass for biofuels and bioproducts: An overview. *Waste Biomass Valorization* **2019**, *10*, 235–251. [CrossRef]
32. Agbor, V.B.; Cicek, N.; Sparling, R.; Berlin, A.; Levin, D.B. Biomass pretreatment: Fundamentals toward application. *Biotechnol. Adv.* **2011**, *29*, 675–685. [CrossRef]
33. Yoo, C.G.; Yang, Y.; Pu, Y.; Meng, X.; Muchero, W.; Yee, K.L.; Thompson, O.A.; Rodriguez, M.; Bali, G.; Engle, N.L.; et al. Insights of biomass recalcitrance in natural *Populus trichocarpa* variants for biomass conversion. *Green Chem.* **2017**, *19*, 5467–5478. [CrossRef]
34. Lu, H.; Wang, X.; Zang, M.; Zhou, J.; Wang, J.; Guo, W. Degradation pathways and kinetics of anthraquinone compounds along with nitrate removal by a newly isolated *Rhodococcus pyridinivorans* GF3 under aerobic conditions. *Bioresour. Technol.* **2019**, *285*, 121336. [CrossRef]

35. Chandel, A.K.; Garlapati, V.K.; Singh, A.K.; Antunes, F.A.F.; da Silva, S.S. The path forward for lignocellulose biorefineries: Bottlenecks, solutions, and perspective on commercialization. *Bioresour. Technol.* **2018**, *264*, 370–381. [CrossRef]
36. Mota, T.R.; Oliveira, D.; Marchiosi, R.; Ferrarese-Filho, O.; Santos, W. Plant cell wall composition and enzymatic deconstruction. *AIMS Bioeng.* **2018**, *5*, 63–77. [CrossRef]
37. Kruyeniski, J.; Ferreira, P.J.; Carvalho, M.d.G.V.S.; Vallejos, M.E.; Felissia, F.E.; Area, M.C. Physical and chemical characteristics of pretreated slash pine sawdust influence its enzymatic hydrolysis. *Ind. Crops Prod.* **2019**, *130*, 528–536. [CrossRef]
38. Ragauskas, A.J.; Beckham, G.T.; Biddy, M.J.; Chandra, R.; Chen, F.; Davis, M.F.; Davison, B.H.; Dixon, R.A.; Gilna, P.; Keller, M.; et al. Lignin valorization: Improving lignin processing in the biorefinery. *Science* **2014**, *344*, 1246843. [CrossRef]
39. Whitham, J.M.; Moon, J.-W.; Rodriguez, M.; Engle, N.L.; Klingeman, D.M.; Rydzak, T.; Abel, M.M.; Tschaplinski, T.J.; Guss, A.M.; Brown, S.D. Clostridium thermocellum LL1210 pH homeostasis mechanisms informed by transcriptomics and metabolomics. *Biotechnol. Biofuels* **2018**, *11*, 1–13. [CrossRef]
40. Zhu, J.; Wang, G.; Pan, X.; Gleisner, R. Specific surface to evaluate the efficiencies of milling and pretreatment of wood for enzymatic saccharification. *Chem. Eng. Sci.* **2009**, *64*, 474–485. [CrossRef]
41. Yeh, A.-I.; Huang, Y.-C.; Chen, S.H. Effect of particle size on the rate of enzymatic hydrolysis of cellulose. *Carbohydr. Polym.* **2010**, *79*, 192–199. [CrossRef]
42. Wen, Z.; Liao, W.; Chen, S. Hydrolysis of animal manure lignocellulosics for reducing sugar production. *Bioresour. Technol.* **2004**, *91*, 31–39. [CrossRef]
43. Moniruzzaman, M.; Dale, B.; Hespell, R.; Bothast, R. Enzymatic hydrolysis of high-moisture corn fiber pretreated by AFEX and recovery and recycling of the enzyme complex. *Appl. Biochem. Biotechnol.* **1997**, *67*, 113–126. [CrossRef]
44. Cosgrove, D.J. Growth of the plant cell wall. *Nat. Rev. Mol. Cell Biol.* **2005**, *6*, 850–861. [CrossRef] [PubMed]
45. Grous, W.R.; Converse, A.O.; Grethlein, H.E. Effect of steam explosion pretreatment on pore size and enzymatic hydrolysis of poplar. *Enzym. Microb. Technol.* **1986**, *8*, 274–280. [CrossRef]
46. Shevchenko, S.; Chang, K.; Robinson, J.; Saddler, J. Optimization of monosaccharide recovery by post-hydrolysis of the water-soluble hemicellulose component after steam explosion of softwood chips. *Bioresour. Technol.* **2000**, *72*, 207–211. [CrossRef]
47. Zhao, X.; Peng, F.; Cheng, K.; Liu, D. Enhancement of the enzymatic digestibility of sugarcane bagasse by alkali–peracetic acid pretreatment. *Enzym. Microb. Technol.* **2009**, *44*, 17–23. [CrossRef]
48. Zhao, X.; Song, Y.; Liu, D. Enzymatic hydrolysis and simultaneous saccharification and fermentation of alkali/peracetic acid-pretreated sugarcane bagasse for ethanol and 2, 3-butanediol production. *Enzym. Microb. Technol.* **2011**, *49*, 413–419. [CrossRef]
49. Ximenes, E.; Kim, Y.; Mosier, N.; Dien, B.; Ladisch, M. Deactivation of cellulases by phenols. *Enzym. Microb. Technol.* **2011**, *48*, 54–60. [CrossRef]
50. Nakagame, S.; Chandra, R.P.; Saddler, J.N. The influence of lignin on the enzymatic hydrolysis of pretreated biomass substrates. *Sustain. Prod. Fuels Chem. Fibers For. Biomass* **2011**, *1067*, 145–167.
51. Wang, Q.; He, Z.; Zhu, Z.; Zhang, Y.H.; Ni, Y.; Luo, X.; Zhu, J. Evaluations of cellulose accessibilities of lignocelluloses by solute exclusion and protein adsorption techniques. *Biotechnol. Bioeng.* **2012**, *109*, 381–389. [CrossRef]
52. Luo, X.; Zhu, J. Effects of drying-induced fiber hornification on enzymatic saccharification of lignocelluloses. *Enzym. Microb. Technol.* **2011**, *48*, 92–99. [CrossRef]
53. Huang, R.; Su, R.; Qi, W.; He, Z. Understanding the key factors for enzymatic conversion of pretreated lignocellulose by partial least square analysis. *Biotechnol. Prog.* **2010**, *26*, 384–392. [CrossRef]
54. Luo, X.; Zhu, J.; Gleisner, R.; Zhan, H. Effects of wet-pressing-induced fiber hornification on enzymatic saccharification of lignocelluloses. *Cellulose* **2011**, *18*, 1055–1062. [CrossRef]
55. Bian, J.; Peng, F.; Peng, X.-P.; Xiao, X.; Peng, P.; Xu, F.; Sun, R.-C. Effect of [Emim] Ac pretreatment on the structure and enzymatic hydrolysis of sugarcane bagasse cellulose. *Carbohydr. Polym.* **2014**, *100*, 211–217. [CrossRef]
56. Morais, A.R.C.; Pinto, J.V.; Nunes, D.; Roseiro, L.B.; Oliveira, M.C.; Fortunato, E.; Bogel-Łukasik, R. Imidazole: Prospect Solvent for Lignocellulosic Biomass Fractionation and Delignification. *ACS Sustain. Chem. Eng.* **2016**, *4*, 1643–1652. [CrossRef]
57. Thygesen, A.; Oddershede, J.; Lilholt, H.; Thomsen, A.B.; Ståhl, K. On the determination of crystallinity and cellulose content in plant fibres. *Cellulose* **2005**, *12*, 563–576. [CrossRef]
58. Sarko, A.; Muggli, R. Packing analysis of carbohydrates and polysaccharides. III. *Valonia* cellulose and cellulose II. *Macromolecules* **1974**, *7*, 486–494. [CrossRef]
59. Nishiyama, Y.; Langan, P.; Chanzy, H. Crystal structure and hydrogen-bonding system in cellulose Iβ from synchrotron X-ray and neutron fiber diffraction. *J. Am. Chem. Soc.* **2002**, *124*, 9074–9082. [CrossRef]
60. Felby, C.; Klinke, H.; Olsen, H.; Thomsen, A. *Ethanol from Wheat Straw Cellulose by Wet Oxidation Pretreatment and Simultaneous Saccharifleation and Fermentation*; ACS Publications: Columbus, OH, USA, 2003.
61. Varga, E.; Réczey, K.; Zacchi, G. (Eds.) Optimization of steam pretreatment of corn stover to enhance enzymatic digestibility. In Proceedings of the Twenty-Fifth Symposium on Biotechnology for Fuels and Chemicals, Breckenridge, CO, USA, 4–7 May 2003; Springer: Berlin/Heidelberg, Germany, 2004.
62. Pihlajaniemi, V.; Sipponen, M.H.; Liimatainen, H.; Sirviö, J.A.; Nyyssölä, A.; Laakso, S. Weighing the factors behind enzymatic hydrolyzability of pretreated lignocellulose. *Green Chem.* **2016**, *18*, 1295–1305. [CrossRef]
63. Liu, Z.-H.; Qin, L.; Li, B.-Z.; Yuan, Y.-J. Physical and Chemical Characterizations of Corn Stover from Leading Pretreatment Methods and Effects on Enzymatic Hydrolysis. *ACS Sustain. Chem. Eng.* **2015**, *3*, 140–146. [CrossRef]

64. Chang, V.S.; Holtzapple, M.T. (Eds.) Fundamental factors affecting biomass enzymatic reactivity. In *Twenty-First Symposium on Biotechnology for Fuels and Chemicals*; Springer: Berlin/Heidelberg, Germany, 2000.
65. Mansfield, S.D.; Mooney, C.; Saddler, J.N. Substrate and enzyme characteristics that limit cellulose hydrolysis. *Biotechnol. Prog.* **1999**, *15*, 804–816. [CrossRef]
66. Ioelovich, M.; Morag, E. Effect of cellulose structure on enzymatic hydrolysis. *BioResources* **2011**, *6*, 2818–2835. [CrossRef]
67. Aldaeus, F.; Larsson, K.; Srndovic, J.S.; Kubat, M.; Karlström, K.; Peciulyte, A.; Olsson, L.; Larsson, P.T. The supramolecular structure of cellulose-rich wood pulps can be a determinative factor for enzymatic hydrolysability. *Cellulose* **2015**, *22*, 3991–4002. [CrossRef]
68. Auxenfans, T.; Crônier, D.; Chabbert, B.; Paës, G. Understanding the structural and chemical changes of plant biomass following steam explosion pretreatment. *Biotechnol. Biofuels* **2017**, *10*, 36. [CrossRef] [PubMed]
69. Meng, X.; Pu, Y.; Yoo, C.G.; Li, M.; Bali, G.; Park, D.Y.; Gjersing, E.; Davis, M.F.; Muchero, W.; Tuskan, G.A.; et al. An in-depth understanding of biomass recalcitrance using natural poplar variants as the feedstock. *ChemSusChem* **2017**, *10*, 139–150. [CrossRef] [PubMed]
70. Zhang, H.; Li, J.; Huang, G.; Yang, Z.; Han, L. Understanding the synergistic effect and the main factors influencing the enzymatic hydrolyzability of corn stover at low enzyme loading by hydrothermal and/or ultrafine grinding pretreatment. *Bioresour. Technol.* **2018**, *264*, 327–334. [CrossRef] [PubMed]
71. Liu, D.; Yu, Y.; Wu, H. Differences in water-soluble intermediates from slow pyrolysis of amorphous and crystalline cellulose. *Energy Fuels* **2013**, *27*, 1371–1380. [CrossRef]
72. Ahvenainen, P.; Kontro, I.; Svedström, K. Comparison of sample crystallinity determination methods by X-ray diffraction for challenging cellulose I materials. *Cellulose* **2016**, *23*, 1073–1086. [CrossRef]
73. Sathitsuksanoh, N.; Zhu, Z.; Wi, S.; Percival Zhang, Y.H. Cellulose solvent-based biomass pretreatment breaks highly ordered hydrogen bonds in cellulose fibers of switchgrass. *Biotechnol. Bioeng.* **2011**, *108*, 521–529. [CrossRef]
74. Yao, L.; Yoo, C.G.; Meng, X.; Li, M.; Pu, Y.; Ragauskas, A.J.; Yang, H. A structured understanding of cellobiohydrolase I binding to poplar lignin fractions after dilute acid pretreatment. *Biotechnol. Biofuels* **2018**, *11*, 96. [CrossRef]
75. Sánchez, C. Lignocellulosic residues: Biodegradation and bioconversion by fungi. *Biotechnol. Adv.* **2009**, *27*, 185–194. [CrossRef]
76. Gupta, R.; Lee, Y. Mechanism of cellulase reaction on pure cellulosic substrates. *Biotechnol. Bioeng.* **2009**, *102*, 1570–1581. [CrossRef]
77. Nazhad, M.; Ramos, L.; Paszner, L.; Saddler, J. Structural constraints affecting the initial enzymatic hydrolysis of recycled paper. *Enzym. Microb. Technol.* **1995**, *17*, 68–74. [CrossRef]
78. Maki, M.; Leung, K.T.; Qin, W. The prospects of cellulase-producing bacteria for the bioconversion of lignocellulosic biomass. *Int. J. Biol. Sci.* **2009**, *5*, 500. [CrossRef]
79. Jönsson, L.J.; Martín, C. Pretreatment of lignocellulose: Formation of inhibitory by-products and strategies for minimizing their effects. *Bioresour. Technol.* **2016**, *199*, 103–112. [CrossRef]
80. Galbe, M.; Zacchi, G. A review of the production of ethanol from softwood. *Appl. Microbiol. Biotechnol.* **2002**, *59*, 618–628. [CrossRef]
81. Sanchez, O.J.; Cardona, C.A. Trends in biotechnological production of fuel ethanol from different feedstocks. *Bioresour. Technol.* **2008**, *99*, 5270–5295. [CrossRef]
82. Tomas-Pejo, E.; Oliva, J.; Ballesteros, M. Realistic approach for full-scale bioethanol production from lignocelluloses: A review. *J. Sci. Ind. Res.* **2008**, *67*, 874–884.
83. Talebnia, F.; Karakashev, D.; Angelidaki, I. Production of bioethanol from wheat straw: An overview on pretreatment, hydrolysis and fermentation. *Bioresour. Technol.* **2010**, *101*, 4744–4753. [CrossRef]
84. Taherzadeh, M.J.; Karimi, K. Acid-based hydrolysis processes for ethanol from lignocellulosic materials: A review. *BioResources* **2007**, *2*, 472–499.
85. Talebnia, F.; Taherzadeh, M.J. In situ detoxification and continuous cultivation of dilute-acid hydrolyzate to ethanol by encapsulated *S. cerevisiae*. *J. Biotechnol.* **2006**, *125*, 377–384. [CrossRef]
86. Singh, D.P.; Trivedi, R.K. Acid and Alkaline pretreatment of lignocellulosic biomass to produce ethanol as biofuel. *Int. J. ChemTech Res.* **2013**, *5*, 727–734.
87. Kumar, P.; Barrett, D.M.; Delwiche, M.J.; Stroeve, P. Methods for pretreatment of lignocellulosic biomass for efficient hydrolysis and biofuel production. *Ind. Eng. Chem. Res.* **2009**, *48*, 3713–3729. [CrossRef]
88. Mosier, N.; Hendrickson, R.; Ho, N.; Sedlak, M.; Ladisch, M.R. Optimization of pH controlled liquid hot water pretreatment of corn stover. *Bioresour. Technol.* **2005**, *96*, 1986–1993. [CrossRef] [PubMed]
89. Kim, S.; Holtzapple, M.T. Effect of structural features on enzyme digestibility of corn stover. *Bioresour. Technol.* **2006**, *97*, 583–591. [CrossRef] [PubMed]
90. Ozmihci, S.; Kargi, F. Effects of feed sugar concentration on continuous ethanol fermentation of cheese whey powder solution (CWP). *Enzym. Microb. Technol.* **2007**, *41*, 876–880. [CrossRef]
91. Rabelo, S.C.; Amezquita Fonseca, N.A.; Andrade, R.R.; Maciel Filho, R.; Costa, A.C. Ethanol production from enzymatic hydrolysis of sugarcane bagasse pretreated with lime and alkaline hydrogen peroxide. *Biomass Bioenergy* **2011**, *35*, 2600–2607. [CrossRef]
92. Hendriks, A.T.W.M.; Zeeman, G. Pretreatments to enhance the digestibility of lignocellulosic biomass. *Bioresour. Technol.* **2009**, *100*, 10–18. [CrossRef]

93. Klinke, H.B.; Ahring, B.K.; Schmidt, A.S.; Thomsen, A.B. Characterization of degradation products from alkaline wet oxidation of wheat straw. *Bioresour. Technol.* **2002**, *82*, 15–26. [CrossRef]
94. Da Costa Lopes, A.M.; João, K.G.; Morais, A.R.C.; Bogel-Łukasik, E.; Bogel-Łukasik, R. Ionic liquids as a tool for lignocellulosic biomass fractionation. *Sustain. Chem. Process.* **2013**, *1*, 3. [CrossRef]
95. Escobar, E.L.N.; Da Silva, T.A.; Pirich, C.L.; Corazza, M.L.; Pereira Ramos, L. Supercritical fluids: A promising technique for biomass pretreatment and fractionation. *Front. Bioeng. Biotechnol.* **2020**, *8*, 252. [CrossRef]
96. Toscan, A.; Morais, A.R.C.; Paixão, S.M.; Alves, L.; Andreaus, J.; Camassola, M.; Dillon, A.J.P.; Lukasik, R.M. High-pressure carbon dioxide/water pre-treatment of sugarcane bagasse and elephant grass: Assessment of the effect of biomass composition on process efficiency. *Bioresour. Technol.* **2017**, *224*, 639–647. [CrossRef]
97. Wang, W.; Lee, D.-J. Lignocellulosic biomass pretreatment by deep eutectic solvents on lignin extraction and saccharification enhancement: A review. *Bioresour. Technol.* **2021**, *339*, 125587. [CrossRef]
98. Margeot, A.; Hahn-Hagerdal, B.; Edlund, M.; Slade, R.; Monot, F. New improvements for lignocellulosic ethanol. *Curr. Opin. Biotechnol.* **2009**, *20*, 372–380. [CrossRef]
99. Oh, S.E.; Iyer, P.; Bruns, M.A.; Logan, B.E. Biological hydrogen production using a membrane bioreactor. *Biotechnol. Bioeng.* **2004**, *87*, 119–127. [CrossRef]
100. Kumar, N.; Das, D. Enhancement of hydrogen production by Enterobacter cloacae IIT-BT 08. *Process Biochem.* **2000**, *35*, 589–593. [CrossRef]
101. Morimoto, M.; Atsuko, M.; Atif, A.; Ngan, M.; Fakhru'l-Razi, A.; Iyuke, S.; Bakir, A.M. Biological production of hydrogen from glucose by natural anaerobic microflora. *Int. J. Hydrogen Energy* **2004**, *29*, 709–713. [CrossRef]
102. Lin, C.-Y.; Chang, R.-C. Fermentative hydrogen production at ambient temperature. *Int. J. Hydrogen Energy* **2004**, *29*, 715–720. [CrossRef]
103. Fang, H.H.; Liu, H. Effect of pH on hydrogen production from glucose by a mixed culture. *Bioresour. Technol.* **2002**, *82*, 87–93. [CrossRef]
104. Kotsopoulos, T.A.; Zeng, R.J.; Angelidaki, I. Biohydrogen production in granular up-flow anaerobic sludge blanket (UASB) reactors with mixed cultures under hyper-thermophilic temperature (70 °C). *Biotechnol. Bioeng.* **2006**, *94*, 296–302. [CrossRef]
105. Chang, J.-S.; Lee, K.-S.; Lin, P.-J. Biohydrogen production with fixed-bed bioreactors. *Int. J. Hydrogen Energy* **2002**, *27*, 1167–1174. [CrossRef]
106. Logan, B.E.; Oh, S.-E.; Kim, I.S.; Van Ginkel, S. Biological hydrogen production measured in batch anaerobic respirometers. *Environ. Sci. Technol.* **2002**, *36*, 2530–2535. [CrossRef] [PubMed]
107. Chen, C.; Lin, C.; Chang, J. Kinetics of hydrogen production with continuous anaerobic cultures utilizing sucrose as the limiting substrate. *Appl. Microbiol. Biotechnol.* **2001**, *57*, 56–64. [PubMed]
108. Mu, Y.; Yu, H.-Q.; Wang, G. Evaluation of three methods for enriching H2-producing cultures from anaerobic sludge. *Enzym. Microb. Technol.* **2007**, *40*, 947–953. [CrossRef]
109. Collet, C.; Adler, N.; Schwitzguébel, J.-P.; Péringer, P. Hydrogen production by *Clostridium thermolacticum* during continuous fermentation of lactose. *Int. J. Hydrogen Energy* **2004**, *29*, 1479–1485. [CrossRef]
110. Zhang, Q.; Jin, P.; Li, Y.; Zhang, Z.; Zhang, H.; Ru, G.; Jiang, D.; Jing, Y.; Zhang, X. Analysis of the characteristics of paulownia lignocellulose and hydrogen production potential via photo fermentation. *Bioresour. Technol.* **2022**, *344*, 126361. [CrossRef]
111. Baik, J.-H.; Jung, J.-H.; Sim, Y.-B.; Park, J.-H.; Yang, J.; Kim, S.-H. High-rate biohydrogen production from xylose using a dynamic membrane bioreactor. *Bioresour. Technol.* **2022**, *344*, 126205. [CrossRef]
112. Byrne, E.; Björkmalm, J.; Bostick, J.P.; Sreenivas, K.; Willquist, K.; van Niel, E.W. Characterization and adaptation of *Caldicellulosiruptor* strains to higher sugar concentrations, targeting enhanced hydrogen production from lignocellulosic hydrolysates. *Biotechnol. Biofuels* **2021**, *14*, 199–212. [CrossRef]
113. Zhang, C.; Wang, G.; Ma, S.; Huang, H.; Ma, Y.; Li, Z. Enhancing Hydrogen Productivity of Photosynthetic Bacteria from the Formulated Carbon Source by Mixing Xylose with Glucose. *Appl. Biochem. Biotechnol.* **2021**, *193*, 3996–4017. [CrossRef]
114. Kucharska, K.; Makoś-Chełstowska, P.; Słupek, E.; Gębicki, J. Management of Dark Fermentation Broth via Bio Refining and Photo Fermentation. *Energies* **2021**, *14*, 6268. [CrossRef]
115. Tosuner, Z.V.; Taylan, G.; Özmıhçı, S. Effects of rice husk particle size on biohydrogen production under solid state fermentation. *Int. J. Hydrogen Energy* **2019**, *44*, 18785–18791. [CrossRef]
116. Zhang, Z.; Fan, X.; Li, Y.; Jin, P.; Jiao, Y.; Ai, F.; Zhang, H.; Zhang, Q. Photo-fermentative biohydrogen production from corncob treated by microwave irradiation. *Bioresour. Technol.* **2021**, *340*, 125460. [CrossRef] [PubMed]
117. Zhang, X.; Jiang, D.; Zhang, H.; Wang, Y.; Zhang, Z.; Lu, C.; Zhang, Q. Enhancement of the biohydrogen production performance from the mixed substrate by photo-fermentation: Effects of initial pH and inoculation volume ratio. *Bioresour. Technol.* **2021**, *319*, 124153. [CrossRef] [PubMed]
118. Cha, M.; Chung, D.; Elkins, J.G.; Guss, A.M.; Westpheling, J. Metabolic engineering of *Caldicellulosiruptor bescii* yields increased hydrogen production from lignocellulosic biomass. *Biotechnol. Biofuels* **2013**, *6*, 85. [CrossRef] [PubMed]
119. Colletti, P.F.; Goyal, Y.; Varman, A.M.; Feng, X.; Wu, B.; Tang, Y.J. Evaluating factors that influence microbial synthesis yields by linear regression with numerical and ordinal variables. *Biotechnol. Bioeng.* **2011**, *108*, 893–901. [CrossRef] [PubMed]
120. Yoshida, A.; Nishimura, T.; Kawaguchi, H.; Inui, M.; Yukawa, H. Enhanced hydrogen production from glucose using ldh- and frd-inactivated *Escherichia coli* strains. *Appl. Microbiol. Biotechnol.* **2006**, *73*, 67–72. [CrossRef] [PubMed]

121. Maeda, T.; Sanchez-Torres, V.; Wood, T.K. Enhanced hydrogen production from glucose by metabolically engineered *Escherichia coli*. *Appl. Microbiol. Biotechnol.* **2007**, *77*, 879–890. [CrossRef]
122. Redwood, M.D.; Mikheenko, I.P.; Sargent, F.; Macaskie, L.E. Dissecting the roles of *Escherichia coli* hydrogenases in biohydrogen production. *FEMS Microbiol. Lett.* **2008**, *278*, 48–55. [CrossRef]
123. Yoshida, A.; Nishimura, T.; Kawaguchi, H.; Inui, M.; Yukawa, H. Enhanced hydrogen production from formic acid by formate hydrogen lyase-overexpressing *Escherichia coli* strains. *Appl. Environ. Microbiol.* **2005**, *71*, 6762–6768. [CrossRef]
124. Chittibabu, G.; Nath, K.; Das, D. Feasibility studies on the fermentative hydrogen production by recombinant *Escherichia coli* BL-21. *Process Biochem.* **2006**, *41*, 682–688. [CrossRef]
125. Akhtar, M.K.; Jones, P.R. Construction of a synthetic YdbK-dependent pyruvate: H2 pathway in *Escherichia coli* BL21 (DE3). *Metab. Eng.* **2009**, *11*, 139–147. [CrossRef]
126. Morimoto, K.; Kimura, T.; Sakka, K.; Ohmiya, K. Overexpression of a hydrogenase gene in *Clostridium paraputrificum* to enhance hydrogen gas production. *FEMS Microbiol. Lett.* **2005**, *246*, 229–234. [CrossRef] [PubMed]
127. Klein, M.; Ansorge-Schumacher, M.B.; Fritsch, M.; Hartmeier, W. Influence of hydrogenase overexpression on hydrogen production of *Clostridium acetobutylicum* DSM 792. *Enzym. Microb. Technol.* **2010**, *46*, 384–390. [CrossRef]
128. Liu, X.; Zhu, Y.; Yang, S.T. Construction and characterization of ack deleted mutant of *Clostridium tyrobutyricum* for enhanced butyric acid and hydrogen production. *Biotechnol. Prog.* **2006**, *22*, 1265–1275. [CrossRef] [PubMed]
129. Liu, X.; Zhu, Y.; Yang, S.-T. Butyric acid and hydrogen production by *Clostridium tyrobutyricum* ATCC 25755 and mutants. *Enzym. Microb. Technol.* **2006**, *38*, 521–528. [CrossRef]
130. Zhao, H.; Ma, K.; Lu, Y.; Zhang, C.; Wang, L.; Xing, X.-H. Cloning and knockout of formate hydrogen lyase and H2-uptake hydrogenase genes in *Enterobacter aerogenes* for enhanced hydrogen production. *Int. J. Hydrogen Energy* **2009**, *34*, 186–194. [CrossRef]
131. Zhao, J.-F.; Song, W.-L.; Cheng, J.; Zhang, C.-X. Heterologous expression of a hydrogenase gene in *Enterobacter aerogenes* to enhance hydrogen gas production. *World J. Microbiol. Biotechnol.* **2010**, *26*, 177–181. [CrossRef]
132. Lu, Y.; Zhao, H.; Zhang, C.; Lai, Q.; Xing, X.-H. Perturbation of formate pathway for hydrogen production by expressions of formate hydrogen lyase and its transcriptional activator in wild *Enterobacter aerogenes* and its mutants. *Int. J. Hydrogen Energy* **2009**, *34*, 5072–5079. [CrossRef]
133. Lu, Y.; Zhao, H.; Zhang, C.; Lai, Q.; Wu, X.; Xing, X.-H. Alteration of hydrogen metabolism of ldh-deleted *Enterobacter aerogenes* by overexpression of NAD (+)-dependent formate dehydrogenase. *Appl. Microbiol. Biotechnol.* **2010**, *86*, 255–262. [CrossRef]
134. Anish, R.; Rao, M. Bioethanol from lignocellulosic biomass part III hydrolysis and fermentation. In *Handbook of Plant-based Biofuels;* CRC Press: Boca Raton, FL, USA, 2009; pp. 159–173.
135. Canilha, L.; Chandel, A.K.; Suzane dos Santos Milessi, T.; Antunes, F.A.F.; Luiz da Costa Freitas, W.; das Graças Almeida Felipe, M.; da Silva, S.S. Bioconversion of sugarcane biomass into ethanol: An overview about composition, pretreatment methods, detoxification of hydrolysates, enzymatic saccharification, and ethanol fermentation. *J. Biomed. Biotechnol.* **2012**, *2012*, 989572. [CrossRef]
136. Grzenia, D.L.; Schell, D.J.; Wickramasinghe, S.R. Membrane extraction for detoxification of biomass hydrolysates. *Bioresour. Technol.* **2012**, *111*, 248–254. [CrossRef]
137. Chandel, A.K.; da Silva, S.S.; Singh, O.V. Detoxification of lignocellulosic hydrolysates for improved bioethanol production. *Biofuel Prod.-Recent Dev. Prospect.* **2011**, *10*, 225.
138. Canilha, L.; Carvalho, W.; Giulietti, M.; Felipe, M.D.G.A.; Almeida, E.; Silva, J.B. Clarification of a wheat straw-derived medium with ion-exchange resins for xylitol crystallization. *J. Chem. Technol. Biotechnol.* **2008**, *83*, 715–721. [CrossRef]
139. Gírio, F.M.; Fonseca, C.; Carvalheiro, F.; Duarte, L.C.; Marques, S.; Bogel-Łukasik, R. Hemicelluloses for fuel ethanol: A review. *Bioresour. Technol.* **2010**, *101*, 4775–4800. [CrossRef]
140. Zhu, J.J.; Yong, Q.; Xu, Y.; Yu, S.Y. Comparative detoxification of vacuum evaporation/steam stripping combined with overliming on corn stover prehydrolyzate. In Proceedings of the 2009 International Conference on Energy and Environment Technology, Guilin, China, 16–18 October 2009.
141. Antunes, F.; Milessi, T.; Oliveira, I.; Chandel, A.; Silva, S. Characterization of sugarcane bagasse hemicellulosic hydrolysate after detoxification with overliming and activated charcoal. In Proceedings of the 20th European Biomass Conference and Exhibition, Milan, Italy, 18–22 June 2012.
142. Carvalho, W.; Canilha, L.; Silva, S.S. Semi-continuous xylose-to-xylitol bioconversion by Ca-alginate entrapped yeast cells in a stirred tank reactor. *Bioprocess Biosyst. Eng.* **2008**, *31*, 493–498. [CrossRef]
143. Converti, A.; Domínguez, J.M.; Perego, P.; Da Silva, S.; Zilli, M. Wood hydrolysis and hydrolyzate detoxification for subsequent xylitol production. *Chem. Eng. Technol.* **2000**, *23*, 1013–1020. [CrossRef]
144. Cantarella, M.; Cantarella, L.; Gallifuoco, A.; Spera, A.; Alfani, F. Comparison of different detoxification methods for steam-exploded poplar wood as a substrate for the bioproduction of ethanol in SHF and SSF. *Process Biochem.* **2004**, *39*, 1533–1542. [CrossRef]
145. Zhang, H.-R.; Qin, X.-X.; Silva, S.S.; Sarrouh, B.F.; Cai, A.-H.; Zhou, Y.-H.; Jin, K.; Xiang, Q. Novel isolates for biological detoxification of lignocellulosic hydrolysate. *Appl. Biochem. Biotechnol.* **2009**, *152*, 199–212. [CrossRef]
146. Ch, A.K.; Ch, G.; Radhika, K.; Ravinder, R.; Ravindra, P. Bioconversion of pentose sugars into ethanol: A review and future directions. *Biotechnol. Mol. Biol. Rev.* **2011**, *6*, 8–20.

147. Moreno, A.D.; Ibarra, D.; Fernández, J.L.; Ballesteros, M. Different laccase detoxification strategies for ethanol production from lignocellulosic biomass by the thermotolerant yeast *Kluyveromyces marxianus* CECT 10875. *Bioresour. Technol.* **2012**, *106*, 101–109. [CrossRef]
148. Parawira, W.; Tekere, M. Biotechnological strategies to overcome inhibitors in lignocellulose hydrolysates for ethanol production. *Crit. Rev. Biotechnol.* **2011**, *31*, 20–31. [CrossRef]
149. Makoś-Chełstowska, P.; Słupek, E.; Kucharska, K.; Kramarz, A.; Gębicki, J. Efficient Extraction of Fermentation Inhibitors by Means of Green Hydrophobic Deep Eutectic Solvents. *Molecules* **2022**, *27*, 157. [CrossRef] [PubMed]

molecules

Article

Characterisation of a Novel Acetyl Xylan Esterase (BaAXE) Screened from the Gut Microbiota of the Common Black Slug (*Arion ater*)

Henry Madubuike * and Natalie Ferry *

School of Science, Engineering and Environment, University of Salford, Manchester M5 4WT, UK
* Correspondence: h.k.madubuike1@salford.ac.uk (H.M.); n.ferry@salford.ac.uk (N.F.);
Tel.: +44-(0)161-295-3886 (N.F.)

Abstract: Acetyl xylan esterases (AXEs) are enzymes capable of hydrolysing the acetyl bonds in acetylated xylan, allowing for enhanced activity of backbone-depolymerizing enzymes. Bioprospecting novel AXE is essential in designing enzyme cocktails with desired characteristics targeting the complete breakdown of lignocellulose. In this article, we report the characterisation of a novel AXE identified as Gene_id_40363 in the metagenomic library analysed from the gut microbiota of the common black slug. The conserved domain description was identified with an NCBI BLASTp search using the translated nucleotide sequence as a query. The activity of the recombinant enzyme was tested on various synthetic substrates and acetylated substrates. The protein sequence matched the conserved domain described as putative hydrolase and aligned closely to an uncharacterized esterase from *Buttiauxella agrestis*, hence the designation as BaAXE. BaAXE showed low sequence similarity among characterized CE family proteins with an available 3D structure. BaAXE was active on 4-nitrophenyl acetate, reporting a specific activity of 78.12 U/mg and a Km value of 0.43 mM. The enzyme showed optimal activity at 40 °C and pH 8 and showed high thermal stability, retaining over 40% activity after 2 h of incubation from 40 °C to 100 °C. BaAXE hydrolysed acetyl bonds, releasing acetic acid from acetylated xylan and β-D-glucose pentaacetate. BaAXE has great potential for biotechnological applications harnessing its unique characteristics. In addition, this proves the possibility of bioprospecting novel enzymes from understudied environments.

Keywords: acetyl xylan esterase; lignocellulose; hemicellulose; novel enzyme bioprospecting; carboxylesterase; hydrolase

check for updates

Citation: Madubuike, H.; Ferry, N. Characterisation of a Novel Acetyl Xylan Esterase (BaAXE) Screened from the Gut Microbiota of the Common Black Slug (*Arion ater*). *Molecules* **2022**, *27*, 2999. https://doi.org/10.3390/molecules27092999

Academic Editors: Alejandro Rodriguez Pascual and Maksymilian Chruszcz

Received: 31 March 2022
Accepted: 3 May 2022
Published: 7 May 2022

1. Introduction

The utilisation of fossil fuels has led to several concerns, such as greenhouse gas emissions, environmental pollution, and resource depletion, prompting the development of renewable energy alternatives, such as lignocellulose biomass, to mitigate the disadvantages associated with fossil fuel use [1–3]. Lignocellulose biomass (LCB) is the most abundant and widely distributed organic chemical on earth [4]. LCB includes hardwood, such as poplar and oak; softwood, such as pine and spruce; agricultural wastes, such as wheat straw and sugarcane bagasse; and dedicated energy crops, such as jatropha and *Miscanthus* sp. [5,6]. LCB is a potential and promising feedstock source for the production of fuels and platform chemicals in the concept of biorefinery [7]. Refining LCB from various sources would produce cellulose and hemicellulose, which can be processed into fuels and chemicals, and lignin, which can be harnessed to produce energy and chemicals [8]. However, the full utilisation of LCB is challenged by its recalcitrant nature, and the need for expensive pretreatment to release the components of LCB makes the technology unsustainable [9,10]. Pretreatment methods, such as physical and chemical pretreatments, are energy-consuming, require special equipment, and produce compounds that can inhibit hydrolytic enzymes and fermentation [11]. For instance, acid treatment of LCB leads to the formation of inhibitory compounds, such as furfural and 5-hydroxyl furfurals, and a large amount of basic solution is required to restore the pH needed for downstream processes [8]. Biological pretreatment with isolated enzymes or enzyme cocktails offers a sustainable pretreatment alternative characterized by a low energy requirement and selective isolation of the different components of LCB and devoid of the formation of inhibitory compounds. Hence, effective biological pretreatment presents a promising approach to achieving LCB degradation with positive economic and environmental impacts [12].

Hemicellulose is the second-most abundant polysaccharide type, composed of pentoses, such as xylose, and hexoses, such as glucose [13]. It is an important plant-derived polysaccharide with applications in the development of bio-based chemicals, biofuels, pharmaceuticals, and prebiotics [14,15]. Hemicellulose is heterogeneous and comprises different subunits, such as glucomannans, arabinoglucuronoxylans, xyloglucans, mannans, xylan, and arabinogalactans. The composition of hemicellulose subunits is varied among the various biomass sources [16–18]. For instance, hardwoods are essentially composed of xylans in the form of acetylated xylan and glucomannans [19]. The enzymatic hydrolysis of hemicellulose involves various hemicellulases, such as xylanases (endo- and exo-xylanases); β-xylosidases; and accessory enzymes, such as arabinofuranosidases and acetyl xylan esterases (Figure 1) [20]. Xylanases act on the xylan chain to release shorter chains, such as xylooligosaccharide and xylobiose, which are then hydrolysed to xylose units by β-xylosidase [21]. The xylan backbones are often substituted with different side chains, such as acetyl, 4-O-methyl glucuronic acid, feruloyl and ρ-coumaroyl, and arabinosyl residues [14,22]. Acetylation, like other substitutions, restricts the access of xylanases, resulting in reduced release of sugars available for fermentation [23]. Hence, the removal of xylan substitutions is crucial for improved hydrolysis of hemicellulose. The treatment of LCB with accessory enzymes has been shown to increase hydrolysis efficiency and improve the activity of backbone depolymerizing enzymes [23]. For example, the synergetic action of acetyl xylan esterase from *Neocallimastix patriciarum* and xylanase (XynA) released higher amounts of reducing sugars compared with hydrolysis with xylanase alone [24].

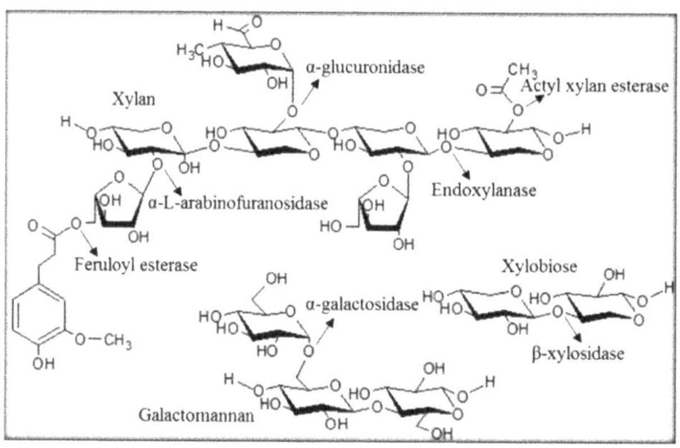

Figure 1. Hemicellulose subunits and breakdown. Structure of major xylan types and enzymes involved in subunit breakdown are identified with arrows indicating enzyme specificity.

Acetyl xylan esterases (AXEs; EC 3.1.1.72) are accessory enzymes able to hydrolyse ester linkages, liberating acetic acid from acetylated hemicellulose [25]. Hardwoods are highly acetylated, with approximately 50–70% of their xylose units acetylated at the C-2 and/or C-3 hydroxyl positions [26]. Hence, acetyl residues are considered the most abundant substitution in hemicellulose [27]. AXEs belong to the α/β hydrolase superfamily, characterised by the α/β- hydrolase fold and the canonical catalytic triad Ser-His-Asp [28]. The sequence of AXEs has the consensus motif Gly-X-Ser-X-Gly around the active site serine [28]. According to the CAZy classification, AXEs are classified into nine carbohydrate esterase (CE) families—CE 1–7, 12, and 16, with the CE 10 family reported as esterases acting on non-carbohydrate substrates [28]. Substrate characterisation of AXEs shows varied substrate activity. Most of the characterised AXEs hydrolysed acetylated xylooligosaccharides and were active on acetylated oligosaccharides/monosaccharides, such as acetylated glucose [29]. For instance, a novel CE (BD-FAE) showed bifunctional attributes, with acetyl xylan esterase activity on acetylated glucuronoxylan from birchwood and feruloyl esterase activity on feruloylated xylooligosaccharides from corn fibre [16]. AXEs have been discovered from a wide range of microbial sources (fungi and bacteria), such as *Thermotoga maritima, Bacillus pumilus, Trichoderma reesei, Clostridium thermocellum, Coriolus versicolor, Schizophyllum commune, Aspergillus versicolor, Streptomyces albicans* [12,30], and plant and animal sources [31].

Novel AXEs have been mined from different environmental sources such as compost, termite gut, and bovine rumen returning libraries of putative enzymes using a metagenomic approach [32,33]. Although several AXEs have been identified and characterised, the importance of AXEs in lignocellulose degradation supports the need for continuous bioprospecting of novel AXEs with improved properties. The identification of novel AXEs with improved catalytic properties promotes a robust enzyme reservoir for plant biomass degradation under operating conditions [34]. For example, a novel acetyl esterase targeting 4-O-methylglucopyranosyluronic acid substitution in glucuronoxylan was identified from a polysaccharide utilization locus (PUL), having been identified as a protein of unknown function [35]. The common black slug (*Arion ater*) is a plant-feeding organism and one of the most widespread plant pest species in Western Europe and North America [36]. The gut microbiota metagenomics of this organism showed a community of bacteria species known for lignocellulose degradation and well-characterised bacteria plant pathogens [37]. Over 3383 genes with putative activity in lignocellulose degradation were identified from the functional metagenomics screening of this organism [37]. In this work, we report the biochemical and functional properties of a novel AXE identified from the metagenomic library of the common black slug.

2. Results

2.1. Bioinformatics and Homology Modelling

Gene_id_40363 was annotated as a Carbohydrate Esterase 1 (CE1) family protein in the metagenomic library of the slug's gut microbiota. This library comprises proteins with putative functions classified under different CAZyme families (Table 1). The translated nucleotide sequence of Gene_id_40363 returned a conserved domain function description for putative esterase and showed 92.7% similarity with an uncharacterized esterase from *Buttiauxella agrestis* (Max score: 420 and E-value: 7×10^{-148}). The sequence of Gene_id_40363 contained the predicted SGNH hydrolase-type esterase domain and belonged to the GFSQ-like lipase abhydrolase family (abhydrolase_2; Pfam domain PF02230; Figure 2).

The CE1 family contains over 5000 protein entries and covers activities including acetyl xylan esterase, cinnamoyl esterase, feruloyl esterase, carboxylesterase, S-formylglutathione hydrolase, diacylglycerol *O*-acyltransferase, and trehalose 6-*O*-mycolyltransferase and several other esterases, such as poly (β-hydroxybutyrate) depolymerase. However, only 11 proteins (10 from eukaryote sources and 1 from bacteria sources) have been characterised so far in this family. Homology models developed in I-Tasser using the closest structural similarity based on the threading program returned the α/β fold model of the abhydrolase protein family. The homology model features 5 α-helices and 6 β-sheets, and the catalytic triad was observed at positions S111, D159, and H191. The homology model of Gene_id_40363 showed the absence of lid structure, but three loop structures were observed around the active site region (Figure 2). We accessed relatedness with members of other CE families. Gene_id_40363 shares low sequence identity with characterised proteins in the CE family that have a 3D structure model available in PDB (Figure 3).

Table 1. Hemicellulases screened from the metagenomic library of the slug's gut microbiota.

CAZy Family	Activity *	No. of Genes
CE 1	Acetyl xylan esterase	120
CE 10	Esterase	38
GH 43	β-Xylosidase, α-L-arabinofuranosidase, xylanase and arabinanase	55
GH 5	Endo-β-1,4-xylanase, β-mannosidase	15
GH 12	Endoglucanase, xyloglucan hydrolase	13
GH 16	Xyloglucanse, endo-1,3-β-glucanase	117
GH 10	Endo-1,4-β-xylanase, endo-1,3-β-xylanase	16
GH 67	α-Glucuronidase, xylan α-1,2-glucuronidase	1
GH 51	Endoglucanase, endo-β-1,4-xylanase, β-xylosidase, α-L-arabinofuranosidase	3
CE 6 and CE 7	Acetyl xylan esterase	5
GH 8	Endo-1,4-β-xylanase	11
GH 38	α-Mannosidase	39
GH 39	β-Xylosidase	279

* Protein family function as described on CAZy.org.

A.

B.

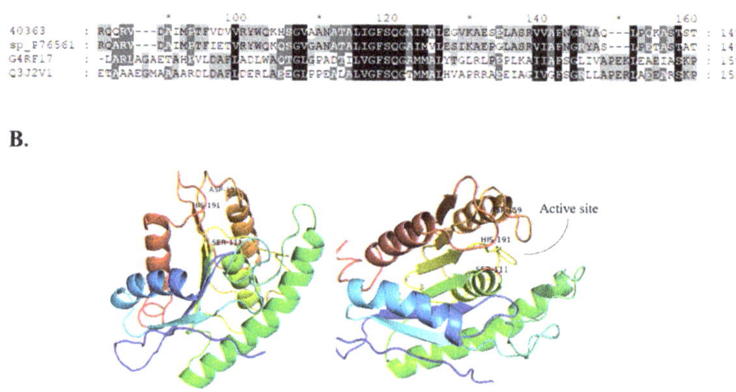

Figure 2. MSA and homology model. (**A**) The sequence of Gene_id_40363 was aligned with selected members of the abhydrolase_2 protein family. The function and source of aligned sequences (from top to bottom) are 40363 (Gene_id_40363, metagenomic sequence), sp_P76561 (esterase YpfH from *Escherichia coli*), Q3J2V1 (phospholipase/carboxylesterase from *Rhodobacter sphaeroides*), and G4RF17 (phospholipase/carboxylesterase from *Pelagibacterium halotolerans*). (**B**) The homology model of Gene_id_40363 was developed in I-Tasser and displayed in PyMOL. Catalytic residues are annotated as S111, D159, and H191.

2.2. Recombinant Protein and Activity Assay

Recombinant protein of Gene_id_40363 (BaAXE) was expressed in *E. coli* BL21 (DE3) with a C-terminal 6x His-tag to enhance detection and purification. The recombinant protein showed a molecular weight of 27 kDa estimated by SDS-PAGE and confirmed with Western blot detection. The observed size corresponds to the theoretical value estimated by ProtParam hosted on Expasy. The purified protein was eluted with buffers containing 105 mM and 120 mM imidazole (Supplementary Data Figure S1). BaAXE predicted as an esterase was active on acetate substrates—4-nitrophenyl acetate (specific activity: 78.12 U/mg) and 4-methylumbelliferyl acetate (specific activity: 2.49 U/mg). Furthermore, BaAXE was active on 4-nitrophenyl butyrate (specific activity: 99.73 U/mg) but showed no clear activity on longer-chain acyl chain esters (C8–C16; Figure 4).

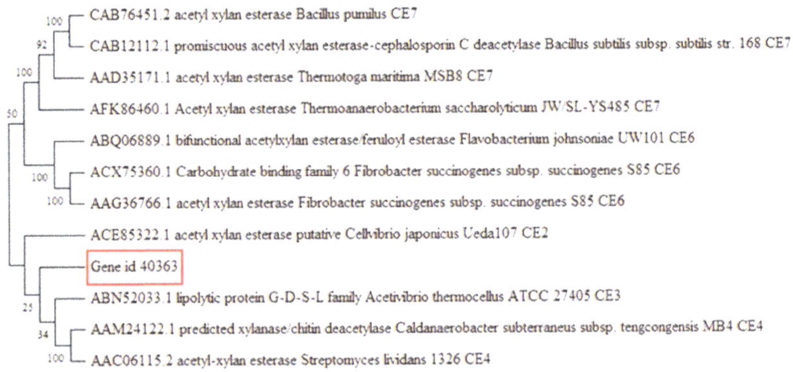

Figure 3. Phylogenetic tree. The nucleotide sequence of Gene_id_40363 was used to construct a phylogenetic tree with characterised proteins of CE family 1–7 with 3D models available in PDB. A Neighbour-Joining phylogenetic tree was constructed in MEGAX with 1000 bootstrap replicates. Proteins are identified with their GenBank number and described with their function(s) and CAZy classification.

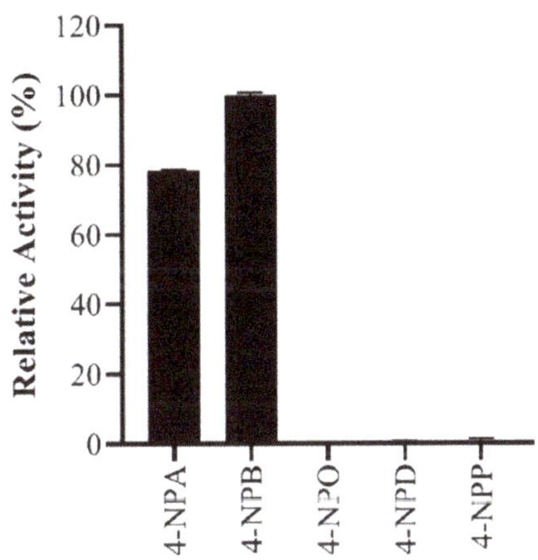

Figure 4. Activity assay. BaAXE was assayed using various acyl chain esters (C2–C16). 4-NPA—4-nitrophenyl acetate; 4-NPB—4-nitrophenyl butyrate; 4-NPO—4-nitrophenyl octanoate; 4-NPD—4-nitrophenyl decanoate; 4-NPP—4-nitrophenyl palmitate. The assay reaction was carried out with a 1 mM concentration of each substrate, and the reaction proceeded for 20 min at 40 °C. Absorbance readings were taken at 410 nm at 1 min intervals. Each bar represents activity on each substrate, $n = 3$ and the error bar corresponds to the standard error of the mean (SEM).

2.3. Biochemical Properties

The biochemical properties of BaAXE were investigated in assays with 4-NPA as the substrate. BaAXE showed optimal activity at pH 8 and 40 °C. The km, kcat, and kcat/km (catalytic efficiency) values were calculated as 0.43 mM, 122.4 s^{-1}, and 282 mM^{-1} s^{-1}, respectively. A thermostability assay showed that the BaAXE retained around 40% activity after incubating the enzyme for 2 h at 40–100 °C but showed no clear activity at acidic pHs. At pH 7 and 9, BaAXE retained over 80% activity after incubating the enzyme for 4 h (Figure 5).

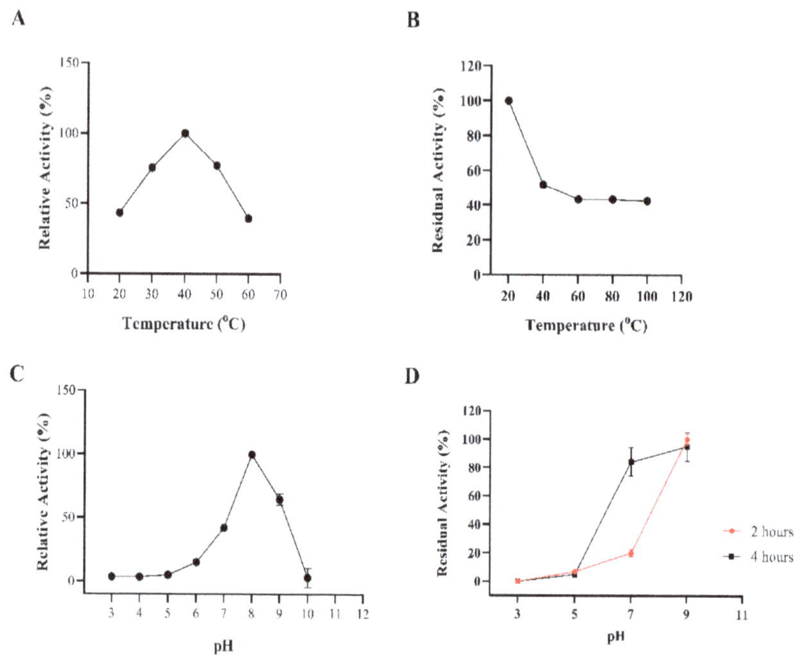

Figure 5. Biochemical properties. The biochemical properties of BaAXE were determined in assays with 4-NPA. (**A**) The optimal temperature was determined in assays performed at different temperatures: 20 °C, 30 °C, 40 °C, 50 °C, and 60 °C. (**B**) Thermostability assay determined after incubating the enzyme for 2 h at different temperatures. (**C**) The optimal pH was determined in assays performed at different pHs at the optimal temperature. (**D**) pH stability was determined after incubation of the enzyme for 4 h at 20 °C. After 2 h of incubation, residual activity was determined, and after 4 h, the residual activity was determined. Absorbance readings were measured at 410 nm. Assay without the enzyme was used as blank. Assays were performed in triplicate, and error bars correspond to SEM.

We investigated the effect of organic solvents and additives on enzyme activity. BaAXE was not severely impacted by organic solvents at 20% (*v*/*v*) retaining over 60% activity in most of the solvents tested after a 17 h incubation. None of the metal ions tested severely impacted its activity, as the enzyme retained over 40% activity after incubation in a 1 mM concentration of the metal salt. The activity of the enzyme was severely impacted when incubated with 1% SDS, retaining only 5% activity after incubation (Figure 6).

2.4. Functional Characterisation

BaAXE was assayed on various substrates to investigate the functional characterization of the enzyme. The enzyme showed no activity on the α-L-arabinofuranosidase substrate—4-nitrophenyl arabinofuranoside, and no activity was detected on ferulic esterase substrates—chromogenic acid and benzyl cinnamate. Furthermore, the enzyme was tested for methylesterase activity and showed no activity on pectin substrate (poly-D-galacturonic acid methyl ester). However, the enzyme released acetic acid from acetylated monosaccharaides and acetylated xylooligosaccharides (Figure 7).

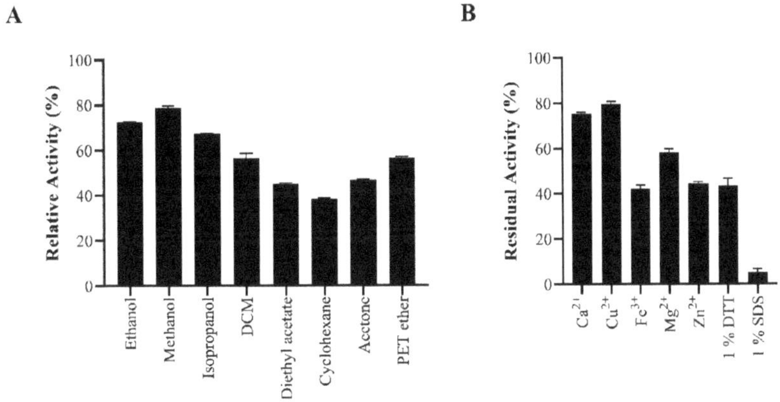

Figure 6. Inhibition by organic solvents and additives. The enzyme was assayed in the presence of (**A**) 20% selected organic solvent and (**B**) 1 mM metal ion, 1% DTT, and 1% SDS. Reactions proceeded for 10 min, and absorbance reading was measured at 410 nm. Assays were performed in triplicate, and error bars correspond to SEM. Relative activity was determined relative to a control reaction lacking any additive or organic solvent. DCM—dichloromethane, PET—petroleum ether.

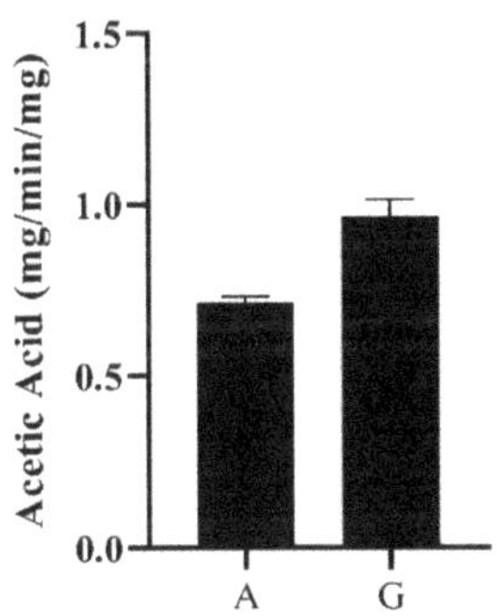

Figure 7. Acetic acid release. The activity of the enzyme on acetylated monosaccharides and xylooligosaccharides was determined by the hydrolytic release of acetic acid from these substrates. Acetic acid release was measured with GC–MS after incubating the enzyme with birchwood xylan (partially acetylated) (A) and β-D-glucosepentaacetate (G). The acetic acid release was estimated from an acetic acid standard plot.

3. Discussion

The recalcitrance of biomass is a major obstacle to harnessing its full potential. Reducing acetylation is a promising approach to reducing recalcitrancy in LCB, allowing for the sustainable production of chemicals and fuels [38]. Here, we reported the characterisation of a novel acetyl xylan esterase (designated as BaAXE), which was previously annotated as Gene_id_40363 in the metagenomic library of the gut microbiota of the common black slug. Carboxylesterases are enzymes distinguished by the α/β-hydrolase fold containing the Ser-His-Asp catalytic triad with serine residue as the nucleophile. Acetyl xylan esterases are carboxylesterases known to hydrolyse ester linkages in acetylated xylan to liberate acetic acid. Hence, they are important auxiliary enzyme components for the degradation of LCB [39].

The translated nucleotide sequence matched functions described as predicted hydrolase and shared 92% sequence similarity with an uncharacterised esterase from *Buttiauxella agrestis*. BaAXE was shown through multiple sequence alignment to contain the GFSQG motif around the active site serine. Acetyl xylan esterases are reported as serine-type esterases but might differ in the residues surrounding the active site serine [40]. However, BaAXE shares low sequence similarity among characterised carbohydrate esterase (CE) with 3D structures submitted in PDB but was assigned to the abhydrolase_2 (pfam02230) family, consisting of both phospholipases and carboxylesterases that have broad substrate specificity and are structurally related to the α/beta hydrolases (pfam00561). Furthermore, in the metagenomic library in which it was identified, it was designated as a member of the CE1 family, which consists of over 5000 entries and 11 characterised proteins. Characterized proteins in this family are AXEs from fungal sources, and no AXE from bacteria has been reported in this family. Hence, BaAXE is likely the first characterized AXE from the CE1 family. Enzymes from bacterial sources have several advantages over enzymes from fungal sources; for example, bacteria exist almost everywhere, are highly adaptable, and are easier to genetically modify [8]. Hydrolases of utmost importance are classed as lipases (triacylglycerol hydrolases) and esterases (carboxyl ester hydrolases). Carboxylester hydrolases, such as pectin methylesterases, acetyl xylan esterase, and feruloyl esterase, are known to act on plant cell wall polysaccharides [41]. The homology three-dimensional (3D) model of BaAXE confirms the presence of the canonical α/β hydrolase fold of the abhydrolase family and the presence of the active site residues S111, D159, and H191, also referred to as the catalytic triad [42]. The crystal structure of a carboxylesterase from *Pseudomonas fluorescens* showed the α/β hydrolase fold containing the Ser-His-Asp catalytic triad, and the active site clefts were relatively open—solvent-exposed—similar to our deduction from the 3D homology model of BaAXE [43]. Lipases are distinguished from esterase by the occurrence of interfacial activation, which is due to the lid domain covering the active sites of lipases. This domain is lacking in BaAXE, as observed in the modelled structure; hence, BaAXE is likely to be identified as an esterase [44,45].

The recombinant enzyme BaAXE showed clear activity toward C2 and C4 acyl chain substrates and was inactive on longer chain acyl chains (C > 6), confirming its classification as an esterase and not a lipase. Carboxylesterases are known to catalyse the hydrolysis of short-chain aliphatic and aromatic esters with broad specificity [46]. Furthermore, BaAXE did not exhibit FAE activity or methyl esterase activity but released acetic acid from acetylated monosaccharides and xylooligosaccharides, confirming its acetyl xylan esterase classification. An acetyl xylan esterase from *Aspergillus oryzae*, designated as AoAXEC, released acetic acid from wheat arabinoxylan but was inactive on methyl esters of ferulic, ρ-coumaric, caffeic, or sinapic acids [28]. To investigate the biochemical activity of BaAXE, we assayed the pH and temperature profile of the enzyme using 4-nitrophenyl acetate as the substrate. The recombinant enzyme showed optimum activity at pH 8.0 and 40 °C which is similar to previously characterised acetyl xylan esterases (Table 2). Most notably, the enzyme showed interesting thermostability, retaining more than 40% residual activity after incubation at 40 °C–100 °C for 2 h. Most characterised AXEs did not show activity at acidic pHs (2–5) and showed an optimal pH in an alkaline range (7.5–8.5). BaAXE retained over 80% residual activity after incubation at pH 7–9 for 4 h. This is an interesting characteristic, as stability in alkaline environments is of interest in several bio-industrial applications, especially in direct applications in alkaline-pretreated biomass [47]. A novel acetyl xylan esterase from *Flavobacterium johnsoniae* (FjoAcXE) showed similar biochemical properties to BaAXE; however, FjoAcXE had no residual activity after 5 min of incubation at 60 °C [35]. Most of the characterised AXEs are thermolabile, especially when incubated at over 60 °C (Table 2). BaAXE obeyed the classical Michaelis–Menten kinetics, reporting a Km value of 0.4 mM and Kcat value of 122.4 s^{-1}. This property of BaAXE further supports its classification as an esterase, as lipases do not obey the classical Michaelis–Menten kinetics and require only a minimum substrate concentration before activity is observed [42].

We reported the biochemical and functional characterisation of a novel acetyl xylan esterase that was not previously classified in the six families of AXEs in the ESTHER database [28]. However, the functional characterisation of BaAXE reported here strongly recommends its assignment into one of the nine AXE families in the CAZy database [48]. Harnessing the unique properties of BaAXE, most notably its stability in a wide range of temperatures and stability at an alkaline pH, would benefit industrial applications. BaAXE liberated acetic acid from xylan (birchwood, partially acetylated) and β-D-glucopyranose pentaacetate, indicating that BaAXE is capable of deacetylation at various acetyl substitutions. Cell wall polysaccharides are either mono- or di-acetylated, and the positions of these acetylations vary among cell wall polysaccharides. Acetylation in xylan usually occurs at position O-2 or O-3, and a xylopyranose may be acetylated at position O-3 [49]. It would be interesting to investigate the synergetic activity of BaAXE with xylanases. The simultaneous treatment of AXE from *Lactobacillus antri* with xylanase showed a 1.44-fold increase in the degradation of beechwood xylan compared with xylanase treatment alone [40]. Furthermore, carboxylesterases have been implicated in the biosynthesis of compounds and the resolution of racemic mixtures [41]. It would be interesting to investigate the stereospecific synthetic ability of BaAXE, notably the synthesis of platform chemicals (such as muconic acid) from lignocellulose substrates [50], as well as the deacetylation prospects of BaAXE on non-xylooligosaccharide substrates, such as polyvinyl acetate [51]. CE1 enzymes have been described with functions relating to the degradation of natural polyesters, such as poly 3-hydroxybutyrate (PHB) depolymerase, and esterases have been implicated as promising candidates for the degradation of polyester-based plastics [52]. It would be informative to explore other applications of this enzyme, as BaAXE promises to be a promising candidate in developing an enzyme consortium targeting the complete degradation of lignocellulose biomass and the development of novel chemicals.

Table 2. Characteristics of acetyl xylan esterases.

Protein Designation	Source Organism	Temp. Opt.	pH Opt.	Temp. Stability	Reference
axeA	*Aspergillus awamori*	-	7	Enzyme activity decreased at temperatures higher than 40 °C.	[53]
axeA	*Aspergillus ficuum*	40 °C	7	Thermal stability decreased at temperatures above 40 °C.	[54]
rAoAXE	*Aspergillus oryzae*	45 °C	6	Unstable at 40 degrees Celsius, with a half-life of less than 60 min at 40 °C and 10 min at 50 °C	[55]
AXE	*Bacillus pumilus* PS213	55 °C	8	Stable at 50 °C but rapidly inactivated at a temperature higher than 60 °C, with a half-life of about 1 h at this temperature	[56]
TM0077	*Thermotoga maritima*	100 °C	7.5	Unstable at 100 °C with a half-life of <5 min.	[57]
AxeA	*Thermotoga maritima*	90 °C	6.5	Retained over 60% residual activity after 8 h incubation at 98 °C	[58]
AXE	*Ochrovirga pacifica*	50 °C	8	Maintained over 70% residual activity after incubation at 55 °C for 120 min	[25]
rAoAXEC	*Aspergillus oryzae*	50 °C	7	Stable up to 50 °C, with a half-life of approximately 2 h at 50 °C and 40 min at 60 °C	[28]
LaAXE	*Lactobacillus antri*	50 °C	7	At 70 °C, the residual activity decreased to 48.4%	[40]
AXE1	*Thermoanaerobacterium* sp.	80 °C	7	At 75 °C, the enzyme showed a half-life of 1 h.	[59]

4. Materials and Methods

4.1. Materials

The slug gut metagenome DNA was provided by Dr Natalie Ferry, University of Salford, Manchester, United Kingdom. The substrates—4-nitrophenyl acetate, 4-nitrophenyl butyrate, 4-nitrophenyl octanoate, 4-nitrophenyl decanoate, and 4-nitrophenyl palmitate, β-D-glucose pentaacetate, pectin, chlorogenic acid, benzyl cinnamate, and 4-methylumbelliferyl acetate—were purchased from Sigma-Aldrich (Gillingham, United Kingdom). Xylan (birchwood, partially acetylated) was purchased from Megazyme (Wicklow, Ireland). Other chemicals were of molecular grade and purchased from Sigma-Aldrich (Gillingham, United Kingdom). Primers for PCR were synthesised and sequenced by Eurofins (Konstanz, Germany). BL21 and TOP10 competent cells were purchased from ThermoFisher Scientific (Leicester, United Kingdom).

4.2. Methods

4.2.1. Bioinformatics

The output of the functional metagenomic screening of the slug gut microbiota was in the format of a translated nucleotide sequence in FASTA format (http://www.ebi.ac.uk/ena/data/view/PRJEB21599, accessed on 25 January 2022). The amino acid sequence identified in the metagenomic library as Gene_id_40363 was selected for characterisation. To verify the putative protein function, the amino acid sequence was used to query the non-redundant protein sequence database using the default algorithm for the BLASTp program hosted on the NCBI website (https://www.ncbi.nlm.nih.gov/, accessed on 10 March 2022). Amino acid sequences selected for phylogenetic tree construction were aligned with MUSCLE (Codons) available in MEGA software (version X) [60] using the default algorithm, and the UPGMA (unweighted pair group method) was selected as the cluster method. The amino acid sequences of proteins that had been previously functionally characterised with available 3D models in PDB (as reported in the CAZy database) were selected for constructing the phylogenetic tree. Tree construction and phylogenetic analysis were performed in MEGA software (version X) [60] with the neighbour-joining method and the Poisson model. Trees were validated with the bootstrapping method using 1000 Bootstrap replications [61].

The annotation of protein features was achieved with Multiple Sequence Alignment (MSA). Multiple sequence alignment was performed with ClustalO hosted on the EMBL-EBI website using the default algorithm (https://www.ebi.ac.uk/Tools/, accessed on 10 March 2022). The amino acid sequence of Gene_id_40363 was aligned with the amino acid sequence of the top hit protein when queried against the swissprot database and closely related proteins identified from the constructed phylogenetic tree [25]. The annotation and display of the sequence alignment were performed in GeneDoc (version 2.7). The molecular weight and isoelectric point (pI) of proteins were determined using the ProtParam tool available on ExPASy (https://web.expasy.org/protparam/, accessed on 15 March 2022). The three-dimensional (3D) model of Gene_id_40363 was generated with homology modelling performed in Iterative Threading ASSEmbly Refinement (I-TASSER; https://zhanglab.ccmb.med.umich.edu/I-TASSER, accessed on 15 March 2022) using multiple sequences of crystalised proteins as a template [61,62]. The modelled structure was validated based on the C-score (confidence score), TM-score (template modelling score), and RMSD (root-mean-square deviation), as previously described [62]. The final protein model was visualised and annotated in PyMOL version 2.0 (Schrodinger New York, NY, USA).

4.2.2. Amplification, Cloning, and Bacterial Transformation

Gene_id_40363 was amplified from the slug's gut metagenomic DNA using gene-specific primers—FP: CACCATGAAACATGACCAC, RP: GCCTCATCGAAATAGTGC—and the following PCR conditions: initial denaturation at 98 °C for 1 min; 35 cycles of denaturation at 98 °C for 40 s, an annealing temperature of 61 °C for 30 s, and extension for 19 s at 72 °C; and a final extension time of 10 min. PCR amplification was conducted with a high-fidelity thermostable DNA polymerase Q5 (NEB, Hitchin, UK) in a 25 μL reaction containing 12.5 μL of Q5 Hi-Fidelity 2X Master Mix, 1.25 μL of (10 μM forward and 10 μM reverse primers), and ~60 ng of template DNA.

The PCR-amplified product (654 bp) was cloned into an entry vector pENTR/SD/D/TOPO (ThermoFischer Scientific Leicester, UK) following the manufacturer's protocol. To generate the recombinant expression vector (Figure 8), pENTR: Gene_id_40363 was recombined with the destination vector pDEST42 (ThermoFischer Scientific, Leicester, UK) following the manufacturer's protocol. The recombinant expression vector was used to transform *Escherichia coli* (*E. coli*) DE3 cells. The desired transformant was screened with restriction digest and sequencing.

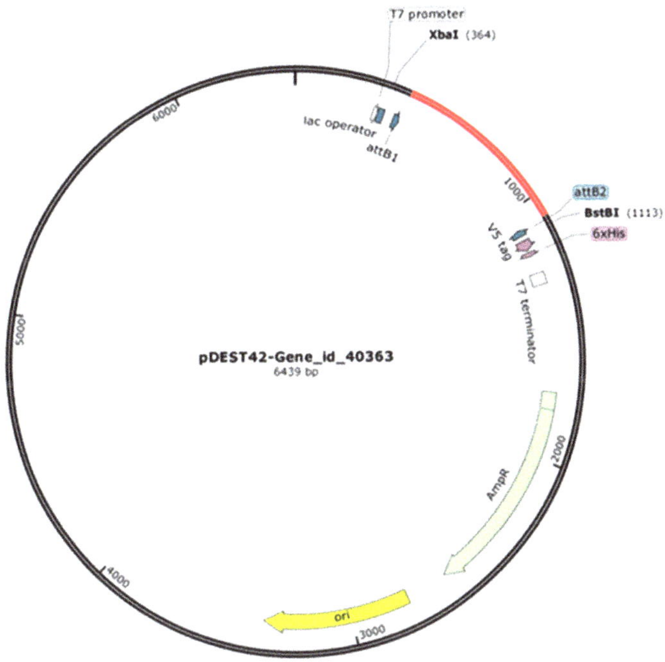

Figure 8. Expression vector construct. The expression plasmid with Gene_id_40363 ligated was modelled in SnapGene (Dotmatics San Diego, USA). Position of gene insert is indicated by the red segment, and selected plasmid features are also annotated.

4.2.3. Protein Expression and Purification

Recombinant plasmid (pDEST42-Gene_id_40363) was used to transform *E. coli* BL21 (DE3) cells for recombinant protein expression with IPTG induction. Transformed cells were grown to an OD600 of 0.5 before induction with 0.6 mM IPTG. Protein expression was allowed for 6 h at 30 °C with shaking at 225× *g*. Cells pellets were harvested by centrifugation at 100× *g* for 10 min at 4 °C. Cell pellets were lysed in a lysis buffer (50 mM sodium phosphate, 150 mM NaCl, 0.1% triton x-100, and 10 mM imidazole) for 30 min at 4 °C. After cell lysis, the soluble fraction was recovered by centrifugation at 4600 rpm for 15 min at 4 °C. The recombinant His-tagged protein was purified with a His GraviTrap TALON gravity flow column charged with cobalt ions, following the manufacturer's protocol (Cytiva, Sheffield, United Kingdom). The elution step was performed with a gradient imidazole concentration (30 mM–150 mM imidazole). Imidazole and excess salt were dialysed out using SnakeSkin dialysis tubing following the manufacturer's protocol (ThermoFisher Scientific Leicester, UK). The purified protein was concentrated using the Amicon ultra centrifugal filter unit (10 kDa cut-off, Millipore, Watford, UK).

4.2.4. Enzyme Assay and Biochemical Properties

The purified and concentrated recombinant protein was assayed for activity using synthetic substrates. Esterase activity was determined spectrophotometrically with 4-nitrophenyl acetate (4-NPA) as the substrate. The assay reaction contained 50 mM sodium phosphate, pH 7.5; 150 mM NaCl; and 0.01% triton x-100. A stock solution (250 mM) of the substrate was prepared in dichloromethane, and a final concentration of 1 mM substrate in 100 μL was used in the assay. The enzyme reaction was initiated by adding a properly diluted enzyme solution (3–5 μg). The enzyme reaction was performed at 40 °C, and the absorbance measurement was monitored every minute for 15 min at 410 nm using a Varioskan LUX Multimode Microplate Reader (ThermoFischer Scientific, Leicester, UK). The background hydrolysis of the substrate was subtracted by performing a blank reaction without the enzyme. One unit of enzymatic activity was defined as the amount of the protein releasing 1 μmol of 4-nitrophenol per minute at the assay conditions described.

4.2.5. Biochemical Properties

The enzyme's temperature profile was determined in an assay with a 1 mM concentration of 4-NPA as described above at varying temperatures. The pH profile was determined at the optimum temperature under various experimental pHs. pH 3 and 4 were prepared in sodium citrate buffer, pH 5 and 6 were prepared in Hepes buffer, pH 7 and 8 were prepared in sodium phosphate buffer, and pH 9 and 10 were prepared in glycine NaOH buffer. The assay was performed for 10 min, and activity was expressed as the percentage relative activity with respect to the maximum activity. The enzyme kinetics assay was carried out at pH 8 using 4-NPA at concentrations from 0.1 mM to 2 mM. The kinetic parameters—Km and Vmax—were calculated by a nonlinear regression fit based on the Michaelis–Menten model using GraphPad Prism 9 [62,63].

4.2.6. Thermostability

The thermal stability and pH stability were determined as described by Razeq et al. (2018). For temperature stability, 4 μg of enzyme in 50 mM sodium phosphate (pH 7.0)—final volume of 40 μL—was incubated at 20 °C, 30 °C, 40 °C, and 50 °C for 2 h. After incubation, the enzyme was ice-cooled for 5 min before the residual activity was determined in the standard assay with 4-NPA.

4.2.7. Effect of Organic Solvents and Additives

The effect of organic solvents on enzyme activity was assayed by incubating 4 μg of the enzyme in 20% concentrations of various organic solvents for 17 h at 4 °C. The residual activity was determined in a standard assay with 1 mM 4-NPA. Residual activity is expressed as a percentage relative to a control (assay without organic solvent addition) set at 100%.

The effect of additives, such as detergents and divalent ions, was determined. The effect of metal ions was determined in reactions with 1 mM 4-NPA as the substrate. The reaction mixture contained a 1mM concentration of the following metal ions: Ca^{2+}, Co^{2+}, Cu^{2+}, Fe^{3+}, Mg^{2+}, and Zn^{2+} (all as chloride salts). The effect of other additives was determined by the addition of 20 mM EDTA, 1% Triton x-100, 1% tween-20, 1% (w/v) dithiothreitol (DTT), and 1% (w/v) SDS. The residual activity was determined in a standard assay with 1 mM 4-NPA.

4.2.8. Acyl Chain Substrate

The substrate specificity of Gene_id_40363 was determined by assaying its activity on various 4-NP alkyl esters—4-nitrophenyl acetate, 4-nitrophenyl butyrate (4-NPB), 4-nitrophenyl octanoate (4-NPO), 4-nitrophenyl decanoate, and 4-nitrophenyl palmitate (4-NPP). The final concentration of the substrate in each assay was 1 mM. Reactions were performed in 50 mM sodium phosphate, as described in Section 4.2.4, and initiated by adding 4 μg of protein. The reaction was allowed for 15 min, and the reaction without the enzyme was used as a blank.

4.2.9. 4-Methylumberiferyl Acetate (4-MUA)

The activity of Gene_id_40363 was assayed on 4-methylumberiferyl acetate (4-MUA). The reaction mixture contained 50 mM sodium phosphate (pH 8.0) and 1 mM 4-MUA dissolved in DMSO. The assay was initiated by adding 4 μg of protein in a total reaction volume of 200 μL. The release of 4-umberiferinone was monitored by reading the fluorescence (excitation/emission 340/520) using a Varioskan LUX Multimode Microplate Reader (ThermoFischer Scientific, Leicester, UK). The assay reaction was incubated at 40 °C, and readings were taken every 1 min for 30 min. Reaction without the enzyme was used as blank. One unit of enzymatic activity was defined as the amount of the protein releasing 1 μmol of 4-methylumbelliferone per minute at the assay conditions described.

4.2.10. Hydrocinnamate Substrates

The activity of Gene_id_40363 towards hydrocinnamate substrates (chlorogenic acid and Benzyl cinnamate) was assayed as previously described [47]. The substrates were solved in DMSO, and the assay was initiated with 4 μg of purified protein. The decrease in substrate concentration was spectrophotometrically quantified following absorbance at 340 nm. The assay was quantified with reference to a standard curve. One unit of enzyme is defined as the amount of enzyme releasing 1 μg of substrate in 1 min under the assay conditions.

4.2.11. Acetic Acid Release from Acetylated Substrates

The activity of Gene_id_40363 was assayed on acetylated xylan (Figure 9) and β-D-glucose pentaacetate. Acetylated xylan (1%) or β-D-glucose pentaacetate (0.1%) was incubated with 4 μg of enzyme at 40 °C for 10 min. After incubation, the reaction was equilibrated 1:1 (*v*/*v*) with acetonitrile. Acetic acid release was analysed with GC–MS (59778B GC/MSD, 8860 GC system, Agilent technologies, Cheadle, UK). The samples were run for 20 min using an Agilent 19091s-433UI HP-5ms Ultra Inert (30 m × 250 μm × 0.25 μm) column. The MS parameters were set as follows: ion source—EI, source temperature—230 °C, quad temperature–150 °C, fixed electron energy—70 eV, solvent delay—2 min. The acquisition was set to scan a mass range of 40–200. The method report for the MS run is reported in the Supplementary Materials section (Figure S2). Reaction mixtures without the enzymes were used as a negative control. Acetic acid content was quantified using the abundance data from various concentrations of acetic acid standards.

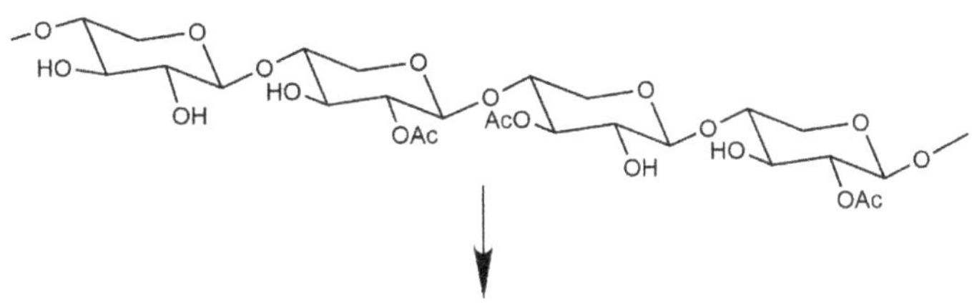

Figure 9. Acetylated xylan. The release of acetic acid is achieved by the action of acetyl xylan esterase on acetylated xylan.

Supplementary Materials: The following supporting information can be downloaded at: https://www.mdpi.com/article/10.3390/molecules27092999/s1, Figure S1: Purification of Gene_id_40363; Figure S2: MS parameter report.

Author Contributions: Conceptualisation, H.M. and N.F.; data curation, H.M.; formal analysis, H.M.; funding acquisition, H.M.; investigation, H.M. and N.F.; methodology, H.M.; supervision, N.F.; writing—original draft, H.M.; writing—review and editing, H.M. and N.F. All authors have read and agreed to the published version of the manuscript.

Funding: The authors declare that no funds, grants, or other support were received during the preparation of this manuscript. The APC was funded by the University of Salford Institutional Fund.

Institutional Review Board Statement: Not applicable.

Data Availability Statement: Not applicable.

Acknowledgments: Many thanks to Lee Harman for supporting this study with the GC–MS instrumentation.

Conflicts of Interest: The authors declare no conflict of interest.

Sample Availability: The plasmid construct used in this study is available from H.M.

Abbreviations

LCB—lignocellulose biomass; AXE—acetyl xylan esterase—GC–MS—gas chromatography–mass spectrometry; 4-NPA—4-nitrophenyl acetate; 4-NPB—4-nitrophenyl butyrate; 4-NPO—4-nitrophenyl octanoate; 4-NPD—4-nitrophenyl decanoate; 4-NPP—4-nitrophenyl palmitate; CE—carbohydrate esterase; GH—glycosyl hydrolase.

References

1. Kumar, R.; Strezov, V.; Weldekidan, H.; He, J.; Singh, S.; Kan, T.; Dastjerdi, B. Lignocellulose Biomass Pyrolysis for Bio-Oil Production: A Review of Biomass Pre-Treatment Methods for Production of Drop-in Fuels. *Renew. Sustain. Energy Rev.* **2020**, *123*, 109763. [CrossRef]
2. Aghbashlo, M.; Khounani, Z.; Hosseinzadeh-Bandbafha, H.; Gupta, V.K.; Amiri, H.; Lam, S.S.; Morosuk, T.; Tabatabaei, M. Exergoenvironmental Analysis of Bioenergy Systems: A Comprehensive Review. *Renew. Sustain. Energy Rev.* **2021**, *149*, 111399. [CrossRef]
3. Rittmann, B.E. Opportunities for Renewable Bioenergy Using Microorganisms. *Biotechnol. Bioeng.* **2008**, *100*, 203–212. [CrossRef] [PubMed]
4. Lorenci Woiciechowski, A.; Dalmas Neto, C.J.; Porto de Souza Vandenberghe, L.; de Carvalho Neto, D.P.; Novak Sydney, A.C.; Letti, L.A.J.; Karp, S.G.; Zevallos Torres, L.A.; Soccol, C.R. Lignocellulosic Biomass: Acid and Alkaline Pretreatments and Their Effects on Biomass Recalcitrance—Conventional Processing and Recent Advances. *Bioresour. Technol.* **2020**, *304*, 122848. [CrossRef]
5. Abraham, A.; Mathew, A.K.; Park, H.; Choi, O.; Sindhu, R.; Parameswaran, B.; Pandey, A.; Park, J.H.; Sang, B.I. Pretreatment Strategies for Enhanced Biogas Production from Lignocellulosic Biomass. *Bioresour. Technol.* **2020**, *301*, 122725. [CrossRef]
6. Sindhu, R.; Binod, P.; Pandey, A. Biological Pretreatment of Lignocellulosic Biomass—An Overview. *Bioresour. Technol.* **2016**, *199*, 76–82. [CrossRef]
7. Hazeena, S.H.; Sindhu, R.; Pandey, A.; Binod, P. Lignocellulosic Bio-Refinery Approach for Microbial 2,3-Butanediol Production. *Bioresour. Technol.* **2020**, *302*, 122873. [CrossRef]
8. Kumar, B.; Bhardwaj, N.; Agrawal, K.; Chaturvedi, V.; Verma, P. Current Perspective on Pretreatment Technologies Using Lignocellulosic Biomass: An Emerging Biorefinery Concept. *Fuel Process. Technol.* **2020**, *199*, 106244. [CrossRef]
9. Rebello, S.; Anoopkumar, A.N.; Aneesh, E.M.; Sindhu, R.; Binod, P.; Pandey, A. Sustainability and Life Cycle Assessments of Lignocellulosic and Algal Pretreatments. *Bioresour. Technol.* **2020**, *301*, 122678. [CrossRef]
10. Raghavi, S.; Sindhu, R.; Binod, P.; Gnansounou, E.; Pandey, A. Development of a Novel Sequential Pretreatment Strategy for the Production of Bioethanol from Sugarcane Trash. *Bioresour. Technol.* **2016**, *199*, 202–210. [CrossRef]
11. Sindhu, R.; Kuttiraja, M.; Prabisha, T.P.; Binod, P.; Sukumaran, R.K.; Pandey, A. Development of a Combined Pretreatment and Hydrolysis Strategy of Rice Straw for the Production of Bioethanol and Biopolymer. *Bioresour. Technol.* **2016**, *215*, 110–116. [CrossRef] [PubMed]
12. Xu, J.; Zhao, X.; Yao, Q.; Zong, W.; Dai, S.; Deng, Z.; Liu, S.; Yun, J.; Yang, X.; Li, H. Cloning, Characterization of a Novel Acetyl Xylan Esterase, and Its Potential Application on Wheat Straw Utilization. *All Life* **2021**, *14*, 622–635. [CrossRef]
13. Farhat, W.; Venditti, R.; Ayoub, A.; Prochazka, F.; Fernández-de-Alba, C.; Mignard, N.; Taha, M.; Becquart, F. Towards Thermoplastic Hemicellulose: Chemistry and Characteristics of Poly-(ε-Caprolactone) Grafting onto Hemicellulose Backbones. *Mater. Des.* **2018**, *153*, 298–307. [CrossRef]
14. Huang, L.Z.; Ma, M.G.; Ji, X.X.; Choi, S.E.; Si, C. Recent Developments and Applications of Hemicellulose from Wheat Straw: A Review. *Front. Bioeng. Biotechnol.* **2021**, *9*, 690773. [CrossRef]
15. Ahmad, N.; Tayyeb, D.; Ali, I.; Alruwaili, N.K.; Ahmad, W.; ur Rehman, A.; Khan, A.H.; Iqbal, M.S. Development and Characterization of Hemicellulose-Based Films for Antibacterial Wound-Dressing Application. *Polymers* **2020**, *12*, 548. [CrossRef] [PubMed]
16. Hameleers, L.; Penttinen, L.; Ikonen, M.; Jaillot, L.; Fauré, R.; Terrapon, N.; Deuss, P.J.; Hakulinen, N.; Master, E.R.; Jurak, E. Polysaccharide Utilization Loci-Driven Enzyme Discovery Reveals BD-FAE: A Bifunctional Feruloyl and Acetyl Xylan Esterase Active on Complex Natural Xylans. *Biotechnol. Biofuels* **2021**, *14*, 127. [CrossRef]
17. Marasinghe, S.D.; Jo, E.; Hettiarachchi, S.A.; Lee, Y.; Eom, T.Y.; Gang, Y.; Kang, Y.H.; Oh, C. Characterization of Glycoside Hydrolase Family 11 Xylanase from *Streptomyces* Sp. Strain J103; Its Synergetic Effect with Acetyl Xylan Esterase and Enhancement of Enzymatic Hydrolysis of Lignocellulosic Biomass. *Microb. Cell Factories* **2021**, *20*, 129. [CrossRef]
18. Binod, P.; Gnansounou, E.; Sindhu, R.; Pandey, A. Enzymes for Second Generation Biofuels: Recent Developments and Future Perspectives. *Bioresour. Technol. Rep.* **2019**, *5*, 317–325. [CrossRef]
19. Shen, D.K.; Gu, S.; Bridgwater, A.V. The Thermal Performance of the Polysaccharides Extracted from Hardwood: Cellulose and Hemicellulose. *Carbohydr. Polym.* **2010**, *82*, 39–45. [CrossRef]

20. Park, S.H.; Yoo, W.; Lee, C.W.; Jeong, C.S.; Shin, S.C.; Kim, H.W.; Park, H.; Kim, K.K.; Kim, T.D.; Lee, J.H. Crystal Structure and Functional Characterization of a Cold-Active Acetyl Xylan Esterase (PbAcE) from Psychrophilic Soil Microbe *Paenibacillus* sp. *PLoS ONE* **2018**, *13*, e0206260. [CrossRef]
21. Joshi, N.; Sharma, M.; Singh, S.P. Characterization of a Novel Xylanase from an Extreme Temperature Hot Spring Metagenome for Xylooligosaccharide Production. *Appl. Microbiol. Biotechnol.* **2020**, *104*, 4889–4901. [CrossRef] [PubMed]
22. Houfani, A.A.; Anders, N.; Spiess, A.C.; Baldrian, P.; Benallaoua, S. Insights from Enzymatic Degradation of Cellulose and Hemicellulose to Fermentable Sugars—A Review. *Biomass Bioenergy* **2020**, *134*, 105481. [CrossRef]
23. Ohta, K.; Fujii, S.; Higashida, C. Characterization of a Glycoside Hydrolase Family-51 α-l-Arabinofuranosidase Gene from Aureobasidium Pullulans ATCC 20524 and Its Encoded Product. *J. Biosci. Bioeng.* **2013**, *116*, 287–292. [CrossRef] [PubMed]
24. Wang, L.; Han, X.; Wang, Y.; Wei, X.; Liu, S.; Shao, S.; Yang, S.; Sun, L.; Xin, F. Rational Design for Broadened Substrate Specificity and Enhanced Activity of a Novel Acetyl Xylan Esterase from Bacteroides Thetaiotaomicron. *J. Agric. Food Chem.* **2021**, *69*, 6665–6675. [CrossRef]
25. Hettiarachchi, S.A.; Kwon, Y.K.; Lee, Y.; Jo, E.; Eom, T.Y.; Kang, Y.H.; Kang, D.H.; de Zoysa, M.; Marasinghe, S.D.; Oh, C. Characterization of an Acetyl Xylan Esterase from the Marine Bacterium Ochrovirga Pacifica and Its Synergism with Xylanase on Beechwood Xylan. *Microb. Cell Fact.* **2019**, *18*, 122. [CrossRef]
26. Puchart, V.; Gjermansen, M.; Mastihubová, M.; Mørkeberg Krogh, K.B.R.; Biely, P. Positional Specificity of Flavobacterium Johnsoniae Acetylxylan Esterase and Acetyl Group Migration on Xylan Main Chain. *Carbohydr. Polym.* **2020**, *232*, 115783. [CrossRef]
27. Zhang, Y.; Yang, H.; Yu, X.; Kong, H.; Chen, J.; Luo, H.; Bai, Y.; Yao, B. Synergistic Effect of Acetyl Xylan Esterase from Talaromyces Leycettanus JCM12802 and Xylanase from Neocallimastix Patriciarum Achieved by Introducing Carbohydrate-Binding Module-1. *AMB Express* **2019**, *9*, 13. [CrossRef]
28. Kato, T.; Shiono, Y.; Koseki, T. Identification and Characterization of an Acetyl Xylan Esterase from Aspergillus Oryzae. *J. Biosci. Bioeng.* **2021**, *132*, 337–342. [CrossRef]
29. Zhang, Y.; Ding, H.-T.; Jiang, W.-X.; Zhang, X.; Cao, H.-Y.; Wang, J.-P.; Li, C.-Y.; Huang, F.; Zhang, X.-Y.; Chen, X.-L. Active Site Architecture of an Acetyl Xylan Esterase Indicates a Novel Cold Adaptation Strategy. *J. Biol. Chem.* **2021**, *297*, 100841. [CrossRef]
30. Sista Kameshwar, A.K.; Qin, W. Understanding the Structural and Functional Properties of Carbohydrate Esterases with a Special Focus on Hemicellulose Deacetylating Acetyl Xylan Esterases. *Mycology* **2018**, *9*, 273–295. [CrossRef]
31. Burlacu, A.; Israel-Roming, F.; Cornea, C.P. Screening of microorganisms displaying acetyl xylan esterase activity. *Sci. Pap. Ser. B Hortic.* **2018**, *62*, 715–720.
32. Rashamuse, K.; Ronneburg, T.; Sanyika, W.; Mathiba, K.; Mmutlane, E.; Brady, D. Metagenomic Mining of Feruloyl Esterases from Termite Enteric Flora. *Appl. Microbiol. Biotechnol.* **2014**, *98*, 727–737. [CrossRef] [PubMed]
33. Ferrer, M.; Golyshina, O.V.; Chernikova, T.N.; Khachane, A.N.; Reyes-Duarte, D.; Martins Dos Santos, V.A.P.; Strompl, C.; Elborough, K.; Jarvis, G.; Neef, A.; et al. Novel Hydrolase Diversity Retrieved from a Metagenome Library of Bovine Rumen Microflora. *Environ. Microbiol.* **2005**, *7*, 1996–2010. [CrossRef]
34. Adesioye, F.A.; Makhalanyane, T.P.; Biely, P.; Cowan, D.A. Phylogeny, Classification and Metagenomic Bioprospecting of Microbial Acetyl Xylan Esterases. *Enzym. Microb. Technol.* **2016**, *93–94*, 79–91. [CrossRef] [PubMed]
35. Razeq, F.M.; Jurak, E.; Stogios, P.J.; Yan, R.; Tenkanen, M.; Kabel, M.A.; Wang, W.; Master, E.R. A Novel Acetyl Xylan Esterase Enabling Complete Deacetylation of Substituted Xylans. *Biotechnol. Biofuels* **2018**, *11*, 74. [CrossRef] [PubMed]
36. Peláez, M.L.; Valdecasas, A.G.; Martinez, D.; Horreo, J.L. Towards the Unravelling of the Slug A. Ater-A. Rufus Complex (Gastropoda Arionidae): New Genetic Approaches. *Web Ecol.* **2018**, *18*, 115–119. [CrossRef]
37. Joynson, R.; Pritchard, L.; Osemwekha, E.; Ferry, N. Metagenomic Analysis of the Gut Microbiome of the Common Black Slug Arion Ater in Search of Novel Lignocellulose Degrading Enzymes. *Front. Microbiol.* **2017**, *8*, 2181. [CrossRef]
38. Wang, Z.; Pawar, P.M.A.; Derba-Maceluch, M.; Hedenström, M.; Chong, S.L.; Tenkanen, M.; Jönsson, L.J.; Mellerowicz, E.J. Hybrid Aspen Expressing a Carbohydrate Esterase Family 5 Acetyl Xylan Esterase Under Control of a Wood-Specific Promoter Shows Improved Saccharification. *Front. Plant Sci.* **2020**, *11*, 380. [CrossRef]
39. Yang, Y.; Yang, J.; Liu, J.; Wang, R.; Liu, L.; Wang, F.; Yuan, H. The Composition of Accessory Enzymes of Penicillium Chrysogenum P33 Revealed by Secretome and Synergistic Effects with Commercial Cellulase on Lignocellulose Hydrolysis. *Bioresour. Technol.* **2018**, *257*, 54–61. [CrossRef]
40. Kim, M.J.; Jang, M.U.; Nam, G.H.; Shin, H.; Song, J.R.; Kim, T.J. Functional Expression and Characterization of Acetyl Xylan Esterases CE Family 7 from Lactobacillus Antri and Bacillus Halodurans. *J. Microbiol. Biotechnol.* **2020**, *30*, 155–162. [CrossRef]
41. McKay, A.M. Microbial Carboxylic Ester Hydrolases (EC 3.1.1) in Food Biotechnology. *Lett. Appl. Microbiol.* **1993**, *16*, 1–6. [CrossRef]
42. Bornscheuer, U.T. Microbial Carboxyl Esterases: Classification, Properties and Application in Biocatalysis. *FEMS Microbiol. Rev.* **2002**, *26*, 73–81. [CrossRef] [PubMed]
43. Kyu Kim, K.; Kyu Song, H.; Hae Shin, D.; Yeon Hwang, K.; Choe, S.; Joon Yoo, O.; Won Suh, S. Crystal Structure of Carboxylesterase from Pseudomonas Fluorescens, an a/b Hydrolase with Broad Substrate Specificity. *Structure* **1997**, *5*, 1571–1584. [CrossRef]
44. Khan, F.I.; Lan, D.; Durrani, R.; Huan, W.; Zhao, Z.; Wang, Y. The Lid Domain in Lipases: Structural and Functional Determinant of Enzymatic Properties. *Front. Bioeng. Biotechnol.* **2017**, *5*, 16. [CrossRef] [PubMed]

45. Verger, R. 'Interfacial Activation' of Lipases: Facts and Artifacts. *Trends Biotechnol.* **1997**, *15*, 32–38. [CrossRef]
46. Sood, S.; Sharma, A.; Sharma, N.; Kanwar, S.S. Carboxylesterases: Sources, Characterization and Broader Applications. *Insights Enzym. Res.* **2018**, *1*, 1. [CrossRef]
47. Li, X.; Griffin, K.; Langeveld, S.; Frommhagen, M.; Underlin, E.N.; Kabel, M.A.; de Vries, R.P.; Dilokpimol, A. Functional Validation of Two Fungal Subfamilies in Carbohydrate Esterase Family 1 by Biochemical Characterization of Esterases From Uncharacterized Branches. *Front. Bioeng. Biotechnol.* **2020**, *8*, 694. [CrossRef]
48. Ali, S.; Mahmood, S. Mutagenesis of a Thermophilic Alkalibacillus Flavidus for Enhanced Production of an Extracellular Acetyl Xylan Esterase in Semi-Solid Culture of Linseed Meal. *Waste Biomass Valorization* **2020**, *11*, 3327–3335. [CrossRef]
49. Qaseem, M.F.; Wu, A.M. Balanced Xylan Acetylation Is the Key Regulator of Plant Growth and Development, and Cell Wall Structure and for Industrial Utilization. *Int. J. Mol. Sci.* **2020**, *21*, 7875. [CrossRef]
50. Dilokpimol, A.; Verkerk, B.; Bellemare, A.; Lavallee, M.; Frommhagen, M.; Underlin, E.N.; Kabel, M.A.; Powlowski, J.; Tsang, A.; de Vries, R.P. Characterization of New Fungal Carbohydrate Esterase Family 1 Proteins Leads to the Discovery of Two Novel Dual Feruloyl/Acetyl Xylan Esterases. 2020. Available online: https://www.researchsquare.com/article/rs-17222/latest.pdf (accessed on 1 March 2022).
51. Yin, C.-F.; Xu, Y.; Deng, S.-K.; Yue, W.-L.; Zhou, N.-Y. A Novel Esterase, DacA Pva, from *Comamonas* Sp. Strain NyZ500 with Deacetylation Activity for the Acetylated Polymer Polyvinyl Alcohol. *Appl Environ Microbiol.* **2021**, *8*, e03016-20. [CrossRef]
52. Urbanek, A.K.; Mirończuk, A.M.; García-Martín, A.; Saborido, A.; de la Mata, I.; Arroyo, M. Biochemical Properties and Biotechnological Applications of Microbial Enzymes Involved in the Degradation of Polyester-Type Plastics. *Biochim. Et Biophys. Acta-Proteins Proteom.* **2020**, *1868*, 140315. [CrossRef] [PubMed]
53. Koseki, T.; Furuse, S.; Iwano, K.; Sakai, H.; Matsuzawa, H. An Aspergillus Awamori Acetylesterase: Purification of the Enzyme, and Cloning and Sequencing of the Gene. *Biochem. J.* **1997**, *326*, 485–490. [CrossRef] [PubMed]
54. Chung, H.-J.; Park, S.-M.; Kim, H.-R.; Yang, M.-S.; Kim, D.-H. Cloning the Gene Encoding Acetyl Xylan Esterase from Aspergillus Ficuum and Its Expression in Pichia Pastoris. *Enzym. Microb. Technol.* **2002**, *31*, 384–391. [CrossRef]
55. Koseki, T.; Miwa, Y.; Akao, T.; Akita, O.; Hashizume, K. An Aspergillus Oryzae Acetyl Xylan Esterase: Molecular Cloning and Characteristics of Recombinant Enzyme Expressed in Pichia Pastoris. *J. Biotechnol.* **2006**, *121*, 381–389. [CrossRef]
56. Degrassi, G.; Okeke, B.C.; Bruschi, C.V.; Venturi, V. Purification and Characterization of an Acetyl Xylan Esterase from Bacillus Pumilus. *Appl. Environ. Microbiol.* **1998**, *64*, 789–792. [CrossRef] [PubMed]
57. Levisson, M.; Han, G.W.; Deller, M.C.; Xu, Q.; Biely, P.; Hendriks, S.; ten Eyck, L.F.; Flensburg, C.; Roversi, P.; Miller, M.D.; et al. Functional and Structural Characterization of a Thermostable Acetyl Esterase from Thermotoga Maritima. *Proteins: Struct. Funct. Bioinform.* **2012**, *80*, 1545–1559. [CrossRef] [PubMed]
58. Drzewiecki, K.; Angelov, A.; Ballschmiter, M.; Tiefenbach, K.J.; Sterner, R.; Liebl, W. Hyperthermostable Acetyl Xylan Esterase. *Microb. Biotechnol.* **2010**, *3*, 84–92. [CrossRef]
59. Shao, W.; Wiegel, J. Purification and Characterization of Two Thermostable Acetyl Xylan Esterases from *Thermoanaerobacterium* sp. *Strain JW/SL-YS485. Appl. Environ. Microbiol.* **1995**, *61*, 729–733. [CrossRef]
60. Kumar, S.; Stecher, G.; Li, M.; Knyaz, C.; Tamura, K. MEGA X: Molecular Evolutionary Genetics Analysis across Computing Platforms. *Mol. Biol. Evol.* **2018**, *6*, 1547–1549. [CrossRef]
61. Ontañon, O.M.; Ghio, S.; Marrero Díaz de Villegas, R.; Piccinni, F.E.; Talia, P.M.; Cerutti, M.L.; Campos, E. EcXyl43 β-Xylosidase: Molecular Modeling, Activity on Natural and Artificial Substrates, and Synergism with Endoxylanases for Lignocellulose Deconstruction. *Appl. Microbiol. Biotechnol.* **2018**, *102*, 6959–6971. [CrossRef]
62. Yang, J.; Zhang, Y. Protein Structure and Function Prediction Using I-TASSER. *Curr. Protoc. Bioinform.* **2015**, *52*, 5.8.1–5.8.15. [CrossRef] [PubMed]
63. Wierzbicka-Woś, A.; Henneberger, R.; Batista-García, R.A.; Martínez-Ávila, L.; Jackson, S.A.; Kennedy, J.; Dobson, A.D.W. Biochemical Characterization of a Novel Monospecific Endo-β-1,4-Glucanase Belonging to GH Family 5 from a Rhizosphere Metagenomic Library. *Front. Microbiol.* **2019**, *10*, 1342. [CrossRef] [PubMed]

molecules

MDPI

Review

Durability of Cellulosic-Fiber-Reinforced Geopolymers: A Review

Jie Liu [1] and Chun Lv [2,*]

[1] College of Light-Industry and Textile Engineering, Qiqihar University, Qiqihar 161006, China; 01250@qqhru.edu.cn
[2] College of Architecture and Civil Engineering, Qiqihar University, Qiqihar 161006, China
* Correspondence: lvc@qqhru.edu.cn; Tel.: +86-452-2725797

Abstract: Geopolymers have high early strength, fast hardening speed and wide sources of raw materials, and have good durability properties such as high temperature resistance and corrosion resistance. On the other hand, there are abundant sources of plant or cellulose fibers, and it has the advantages of having a low cost, a light weight, strong adhesion and biodegradability. In this context, the geopolymer sector is considering cellulose fibers as a sustainable reinforcement for developing composites. Cellulosic-fiber-reinforced geopolymer composites have broad development prospects. This paper presents a review of the literature research on the durability of cellulosic-fiber-reinforced geopolymer composites in recent years. In this paper, the typical properties of cellulose fibers are summarized, and the polymerization mechanism of geopolymers is briefly discussed. The factors influencing the durability of cellulosic-fiber-reinforced geopolymer composites were summarized and analyzed, including the degradation of fibers in a geopolymer matrix, the toughness of fiber against matrix cracking, the acid resistance, and resistance to chloride ion penetration, high temperature resistance, etc. Finally, the influence of nanomaterials on the properties of geopolymer composites and the chemical modification of fibers are analyzed, and the research on cellulosic-fiber-reinforced geopolymer composites is summarized.

Keywords: cellulose fiber; plant fiber; geopolymer composites; durability; alkaline degradation; acid resistance

Citation: Liu, J.; Lv, C. Durability of Cellulosic-Fiber-Reinforced Geopolymers: A Review. *Molecules* **2022**, *27*, 796. https://doi.org/10.3390/molecules27030796

Academic Editors: Alejandro Rodriguez Pascual, Eduardo Espinosa Víctor and Carlos Martín

Received: 17 December 2021
Accepted: 24 January 2022
Published: 25 January 2022

1. Introduction

Geopolymer is a kind of inorganic silico-aluminum cementitious material with a spatial structure prepared by the reaction of active low-calcium silico-alumina material with an alkaline activator, with a three-dimensional network composed of SiO_4 and AlO_4 tetrahedral unit structure. The synthesis of geopolymers requires active solid aluminosilicates and alkaline solutions containing alkali metals and silicates. Among them, the alkaline solution acts as a binder, alkali activator and dispersant [1]. Compared to cement-based composites, geopolymers have the advantages of a high early strength, fast hardening speed and wide range of raw materials [2]. Geopolymers have lower energy consumption and less pollutants in the production process, and they are considered to be the material with the highest potential to replace cement [3–5]. The concept of geopolymers was originally proposed to describe the inorganic aluminosilicate polymers synthesized with natural materials by French scientist Davidovits in 1978 [6]. His team used alkali metal silicate solutions to stimulate geological minerals to form polymeric aluminum silicate materials under strong alkaline conditions [7]. Subsequently, other solid silicate raw materials including fly ash [8,9], pozzolan [10], ground blast furnace slag [11] and other wastes [12–14] successfully prepared geopolymers.

Traditional cement-based composites have poor durability such as high temperature resistance and corrosion resistance. Geopolymer composites overcome this shortcoming well [15]. However, geopolymers are similar to ceramics, their flexural strength and tensile strength are poor, and they are very sensitive to microcracks. In order to solve the problem of the brittleness of geopolymers, the toughness of composites can be improved by incorporating fibers. Adding fibers to the geopolymer can limit the growth of cracks, and at the same time can enhance the ductility, toughness and tensile strength of the geopolymer [16,17]. In recent years, many scholars have conducted research on the durability of geopolymers. This type of research mostly focuses on the durability of geopolymers such as sulfate resistance, freeze-thaw, weathering, water absorption, abrasion resistance, alternating wet and dry effects, and chloride ion resistance [18,19]. By adjusting the ratio of silicon to aluminum, alkaline solution, curing conditions, and adding fibers, rice husk ash, etc., the mechanical properties and durability of the composites are improved. The presence of fibers improves the bending strength and fracture behavior of the material and promotes the toughening mechanism of the material. Fiber-reinforced geopolymers have better durability than cement-based materials of the same grade [20].

At present, the fibers used in composites mainly include natural fibers, metal fibers [21,22], inorganic fibers [23–25] and synthetic fibers [26–28]. Among them, there are many studies on synthetic-fiber-reinforced geopolymers, such as polyvinyl alcohol (PVA), polypropylene (PP), etc., but their production process pollutes the environment and faces difficulties in meeting the requirements of sustainable development [29,30]. Among natural fibers, plant or cellulose fibers (CFs) are most commonly used. Plant fiber is also called natural cellulosic fiber. It has the advantages of having a low cost, a light weight, strong adhesion, simple manufacturing process and biodegradability, which attract more and more scholars' attentions [31–33].

In recent years, fiber-reinforced geopolymers have been studied from different perspectives. The effects of different types of fibers on the enhanced performance of geopolymer, and of cellulose fiber fabrics on the properties of cementitious composites and geopolymers were studied [34,35]. At present, there is no special review on the durability of plant- or cellulosic-fiber-reinforced geopolymer composites (CFGCs). The paper presents a review of the literature research on the durability of CFGCs for the past few years, and briefly summarizes the polymerization mechanism of geopolymers. Then, according to the performance characteristics of CFs, the factors that affect the durability of CFGCs are summarized and analyzed, including the degradation of CFs in the geopolymer matrix, the toughness of CF against matrix cracking, and the performance of acid resistance, anti-chloride ion penetration, high temperature resistance and so on. Finally, the addition of nanomaterials and the chemical modification of CFs affecting geopolymer composites are analyzed. However, there are few research cases about the freezing-thawing resistance and carbonization resistance of CFGCs, so this paper does not carry out in-depth discussion on relevant issues, and further relevant studies are needed in the future.

2. Typical Properties of CFs

CF is one of the most abundant natural resources in the world, and it is widely found in agricultural residues, such as rice straw, rice husk, maize straw, bagasse, wood shavings, wood chips, bamboo chips, etc. These agricultural residues are mainly composed of cellulose, hemicellulose, lignin, pectin, wax and some water-soluble materials. Cellulose is the most important component of CF, and its chemical formula is $(C_6H_{10}O_5)_n$. Cellulose is a macromolecular polysaccharide composed of glucose, which is a straight-chain polymer formed by linking countless D-glucopyranose anhydrides with β(1–4) glycosides, and its structure is regular and unbranched. Cellulose has a large number of hydroxyl groups on the molecular chain, which promote the formation of intramolecular and intermolecular hydrogen bonds [31]. CFs commonly used for geopolymer reinforcement include bast fibers, leaf fibers, stem fibers, etc. [35,36], as shown in Figure 1.

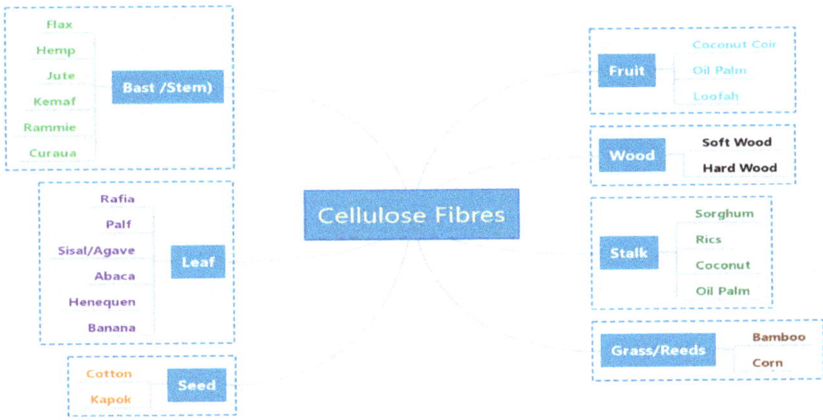

Figure 1. Classification of CFs used for reinforcement the geopolymers. Reprinted with permission from ref. [36]. Copyright 2020 Copyright MDPI.

It can be seen from Table 1 that the density of CFs is roughly similar, with little difference, between 1.1 and 1.6 g·cm^{-3}. The density of bast fiber is basically about 1.5 g·cm^{-3}, its tensile strength is relatively large, and the tensile strength of fruit coconut husk fiber is relatively small.

Table 1. Mechanical properties of typical fibers.

Fiber Type	Fiber Name	Density/ (g cm^{-3})	Tensile Strength/MPa	Specific Strength/(S ρ^{-1})	Tensile Modulus/GPa	Specific Modulus/(E ρ^{-1})	Elongation at Break/%	Ref.
Bast	Flax	1.5	800–1500	535–1000	27.6–80	18.4–53	1.2–3.2	[37]
	Hemp	1.48	550–900	372–608	70	47.3	2–4	[38]
	Jute	1.46	393–800	269–548	10–30	6.85–20.6	1.5–1.8	[39]
	Kenaf	1.45	930	641	53	36.55	1. 6	[40]
	Ramie	1.5	220–938	147–625	44–128	29.3–85	2–3.8	[41]
Leaf	Abaca	1.5	400	267	12	8	3–10	[42]
	Sisal	1.45	530–640	366–441	9.4–22	6.5–15.2	3–7	[11]
	Banana Leaf	1.35	600	444	17.85	13.2	3.36	[41]
	Coconut leaf	1.15	500	435	2. 5	2.17	20	[43]
Seed	cotton	1.6	287–597	179–373	5.5–12.6	3.44–7.9	7–8	[43]
Grass	bamboo	1.1	500	454	35.91	32.6	1.4	[43]
Fruit	Coconut shell	1.2	175	146	4–6	3.3–5	30	[41]
Wood	Soft wood	1.5	1000	667	40	26.67	4.4	[43]

3. Fiber-Reinforced Geopolymer Composites

Geopolymers can be prepared in two ways: alkali excitation and acid excitation. According to the different active raw materials, alkali excitation methods are mainly divided into alkali-silicate glass body cementing materials and alkali-silicate mineral cementing materials. Alkali-silicate glass body cementing materials mainly use amorphous silicate glass bodies as raw materials, such as slag, fly ash, various metallurgical slags, coal gangue, etc., and the main raw alkali-silicate mineral cementing material is a crystalline mineral, such as clay, feldspar and other tailings.

3.1. The Polymerization Mechanism of Geopolymer

Under the condition of strong alkali, the silicon-oxygen bonds and aluminum-oxygen bonds of active materials such as kaolin are broken to form oligomers of polymer monomers, namely oligomeric silicon-oxygen tetrahedra and aluminum-oxygen tetrahedra. Under the same conditions, oligomeric silicon-oxygen tetrahedrons and aluminum-oxygen tetrahedrons are dehydrated and polymerized to form geopolymers with a three-dimensional network structure in space [44]. It is generally believed that the reaction of geopolymers can be divided into four processes: dissolution, diffusion, polymerization and solidification. Using metakaolin as the active material and (NaOH) or (KOH) as the alkali activator, the reaction mechanism of the resulting geopolymer is shown in Figure 2 [45,46].

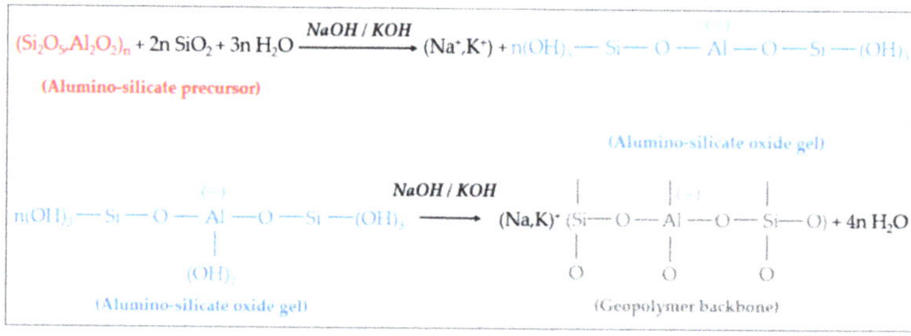

Figure 2. Hydrolysis and polycondensation of aluminum-silicate precursors and formation of geopolymers.

It can be seen from Figure 2 that the aluminosilicate raw materials (precursors) gradually dissolve in (NaOH) or (KOH) alkali activator, producing a large amount of silicon and aluminum monomers. These monomers gradually diffuse in the solution from the surface to the inside, and quickly undergo a polycondensation reaction to form silico−alumina oligomers. The oligomer gel phase solidifies and hardens to form geopolymers.

3.2. Fiber Matrix Interface Bonding Mechanisms

A geopolymer composite is composed of fiber and a matrix with different properties, and the interface between the fiber and matrix is formed. The interface of the composite includes the geometric surface of the matrix and the fiber in contact with each other and the transition area, which is an extremely complex microstructure. Adjusting the bonding state of the fiber and the matrix interface, and optimizing the characteristics of the interface layer between the fiber and the matrix can make the geopolymer composites achieve the best performance. Improving the interfacial adhesion between the fiber reinforcement and the matrix is the most critical factor in the interface control technology of composites. The bonding forms of the fiber and matrix interface generally include interdiffusion, electrostatic adhesion, chemical bonding and mechanical interlocking [46,47]. According to the microscopic morphology of the bonding of fibers and geopolymers, the interface bonding is usually mainly in the form of mechanical interlocking.

4. Research Status of the Durability of CFGCs

Durability refers to the ability of a material to resist the long-term destructive effects of both itself and the natural environment. Generally, the better the durability of a material, the longer its service life will be [48,49]. At present, scholars have conducted a lot of research on the properties of CFGCs, such as crack resistance, acid corrosion resistance, chloride ion penetration resistance, dry and wet cycle, and high temperature resistance. Microscopic analysis shows [50] that geopolymers can form an impermeable layer under the action of fibers, making the geopolymer matrix structure more compact. The good adhesion between the fiber and the matrix enables the geopolymer composite to prevent crack propagation, resist freeze-thaw and penetration erosion, and enhance its durability.

4.1. The Alkaline Degradation Mechanism of CFs

The durability of CFGCs involves the durability of the matrix and the durability of the CFs in the matrix. As we all know, the amorphous components of CFs in cement concrete will be degraded to varying degrees in an alkaline environment [36]. Similarly, on the one hand, the fiber has good compatibility with the polymer matrix, on the other hand, the fiber in the matrix also degrades [51,52]. First, lignin and part of the hemicellulose were degraded, and then the hemicellulose was completely degraded, destroying the integrity and stability of the CF cell wall. This results in the peeling of the cellulosic fiber from the CF cell wall and the failure of the cellulose fiber, which leads to the complete degradation of the CF. Due to the alkaline hydrolysis of cellulose fiber chains, hemicellulose and lignin, the integrity of the fiber and geopolymer matrix interface area is lost, thereby damaging the mechanical properties and durability of CFGCs.

Geopolymer-based materials composed of cellulose hemicellulose and lignin have unique microstructures. A total of 5% mass content of cellulose hemicellulose and lignin could effectively improve the flexural and compressive strength of geopolymer [53]. In fact, the increase in the content of lignin and hemicellulose leads to the porosity, low density and brittleness of the composite, which reduces the flexural and compressive strength of the composite; the alkaline degradation of hemicellulose reduces the degree of polymerization of the composite. However, as the cellulose content increases, the matrix structure becomes denser, with fewer pores, and the toughness of the composite increases. The geopolymer matrix and the cellulose fibers also showed good bonding without significant degradation.

Different types of CFs and different external environmental conditions make the degree of degradation of cellulosic fibers different, and also have great influence on the durability of composite materials. The flax-fabric-reinforced composite was immersed in water, seawater and 5% sodium hydroxide alkaline solution. After aging for one year, the tensile and bending properties of the composite were tested. The results showed that the degradation of the composites was most serious in 5% sodium hydroxide alkaline solution [54]. After impregnating bamboo pulp and nanocellulose fibers with cement concrete and geopolymer, lignin was removed from the fiber surface, and hemicellulose and cellulose were degraded to a certain extent. The tensile strength of pulp sheet decreased by 70% and 34%, respectively [55]. Due to the inherent properties of CFs, although geopolymers did not contain calcium hydroxide, the high alkalinity of the slurry also sped up the degradation process. The mineralization and partial degradation of hemicellulose of black locust and longleaf acacia grains and bagasse were found in the geopolymers, indicating that the durability of black locust and longleaf acacia grains in alkaline substrates deteriorated [56].

Although the degradation degree of CF in the geopolymer matrix is relatively weaker than that in cement, the degradation of the alkaline matrix of the geopolymer also affects the durability of CFGC.

4.2. *Crack Resistance and Toughness of CFGCs*

In recent years, there have been many reports on the physical, thermal and mechanical properties of CFGCs, but few studies on their durability. The current research on the durability of geopolymers mainly focuses on the toughening and cracking resistance of materials, the resistance to sulfate erosion, resistant to high temperatures, chloride ion corrosion resistance, frost resistance, etc. CFs can inhibit and stabilize the development of micro-cracks in the geopolymer, which is an effective way to alleviate the performance degradation of geopolymer composites. As we all know, the durability of geopolymer composites is closely related to its compactness and crack resistance, and good toughness helps to improve the durability of geopolymers.

4.2.1. The Effect of Bast Fiber on the Toughness of CFGCs

Most bast fibers have good strength and are widely used in the manufacture of ropes, twine, packaging materials and industrial thick cloth. Bast fibers mainly include hemp, flax, jute, ramie and kenaf. Tests show that most bast fibers have a good strengthening and toughening effect on CFGCs.

Hemp fiber has a positive effect on the microstructure and fracture structure of the geopolymers and can effectively improve the mechanical properties of the composites. The matrix with relatively brittle initial strength shows a higher increase in relative toughness. Eyerusalem et al. [57] used randomly oriented hemp as a reinforcement fiber and observed the fiber pulling out after loading, which showed the toughening ability of fiber. The tensile strength of the composite reinforced by 9% hemp fiber by volume was especially improved, and its tensile strength was about 5.5 MPa. At the same time, the addition of hemp fiber also slightly affected the density water absorption, compressive strength and flexural strength of the composite, but significantly improved the energy absorption capacity of the composite [58]. With the increase of fiber content, the compressive strength of the composite decreases continuously, indicating that the fiber distribution in the matrix becomes more and more uneven.

Bast fibers can change the initial brittle behavior of the geopolymer matrix to make it a ductile material, and their arrangement direction and treatment mode in the matrix also affect the toughening effect of composites. Trindade et al. [59] adopted a one-way and bidirectional arrangement of sisal and jute as reinforcement materials and showed that the composite material exhibits strain and flexural hardening behavior under tensile and bending action and produces multiple cracks. Sáez-Pérez et al. [60] found that under the two experimental conditions of fresh and wet storage for 6 months, similar to other pretreatment methods to improve the properties of hemp fiber, wet storage caused an increase in cellulose content and improved the mechanical properties of the geopolymer. Na et al. [61] soaked alkali-treated Kenaf fiber in 1 mol/L $CaCl_2$ solution, which improved the compatibility between fiber and matrix, increased the flexural strength by 69.1%, and increased the toughness by 473%. After loading, the fiber had a typical failure mode of toughness.

The effects of different content of bast fiber on the properties of the geopolymer composites are different. Assaedi et al. [62] used flax fiber as reinforcement material to significantly improve the flexural strength, compressive strength hardness and fracture toughness of geopolymers.

Figure 3 shows typical stress–strain curves of pure geopolymer and composites with different fiber content. It can be seen that the composite with a fiber content of 4.1 wt% has the highest flexural strength among all composites. The flexural strength of the composite has been increased from 4.5 MPa for pure geopolymer to 23 MPa. It shows that increasing the content of flax fiber can significantly improve the flexural strength of the composite.

However, the fiber content should not be too high. Korniejenko et al. [63] also showed that the addition of flax fiber led to the decline of mechanical properties of composite materials. When flax fiber was added at 8%, the compressive strength of the material decreased by about 50%. Similarly, the bending strength of the pure geopolymer matrix was 3.45 MPa, and that of 8% flax-fiber-reinforced composite was 2.13 MPa. The results show that the fracture toughness of the composite increased the most when the flax fiber content was 4.1 wt%.

Generally speaking, fiber debonding and bridging slow down the crack propagation of composites and increase the fracture energy. The fracture toughness of the composite containing flax fiber is significantly higher than that of pure polymer, and the higher the fiber content, the higher the fracture toughness. This enhancement is due to the unique anti-breaking ability of flax fiber, which leads to increased energy dissipation of fiber matrix interface crack deflection, fiber debonding, fiber bridging, fiber pull-out and fracture, as shown in Figure 4a–d. It can be seen that the geopolymer adheres to the surface of the fiber, showing good adhesion between the fiber and the matrix. SEM images show various toughening mechanisms including crack bridging fiber pulling out and fiber fracture. Because of the degradation of flax fiber, flax-fiber-reinforced geopolymer exhibits higher net weight loss than pure geopolymer.

It can be seen that bast fibers have a good toughening effect on CFGCs, which further reduces the possibility of CFGCs cracking, thus effectively improving their durability.

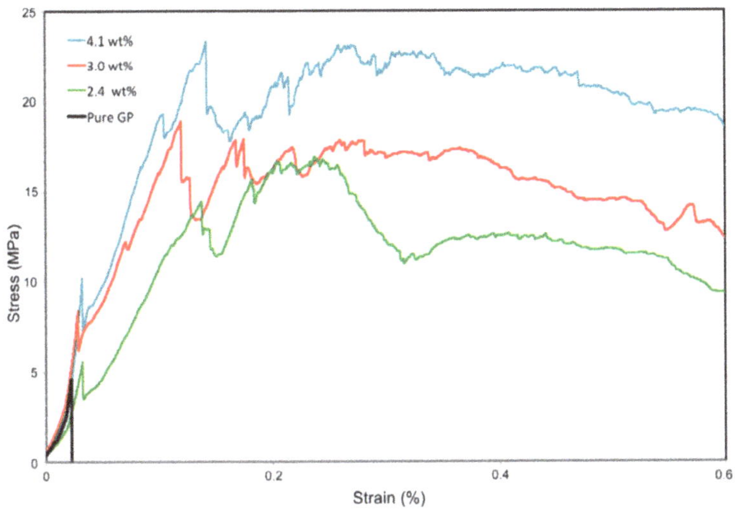

Figure 3. Typical stress–strain curves of pure geopolymer and composites with various fiber contents. Reprinted with permission from ref. [62]. Copyright 2015 Copyright Springer Nature.

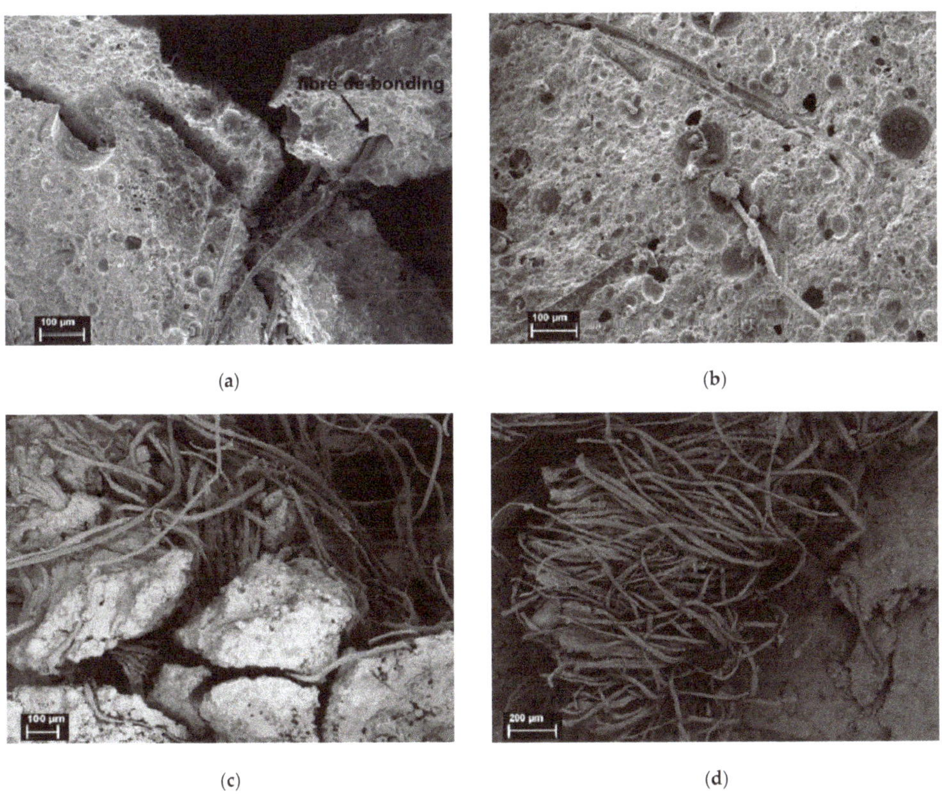

Figure 4. Fracture of flax-fiber-reinforced geopolymer, (**a**) fiber debonding, (**b**) fiber creasing and pulling out, (**c**) fiber bridging cracks, (**d**) fiber breaking. Reprinted with permission from ref. [62]. Copyright 2015 Copyright Springer Nature.

4.2.2. The Effect of Leaf Fiber on the Toughness of CFGCs

Leaf fibers are vascular bundle fibers obtained from the leaves of monocotyledonous plants. There are many varieties of leaf fiber, including raffia fiber, pineapple leaf fiber, sisal fiber, abaca fiber, agave fiber and so on. Similar to bast fibers, leaf fibers are often used for CFGCs, among which sisal-fiber-reinforced geopolymers are more studied.

The properties of the composites can be affected by the different content and length of leaf fiber. Ampol et al. [64] blended sisal fiber and coconut fiber into geopolymers with volume fractions of 0%, 0.5%, 0.75% and 1.0%, tested the mechanical properties of the geopolymers, and combined them with glass fiber. The results showed that, compared with glass fiber, adding sisal fiber and coconut fiber as reinforcement materials significantly improves the tensile and flexural strength properties. At the same time, the processing performance, dry density, ultrasonic pulse speed and compressive strength values all have a tendency to decrease. Zulfiat et al. [65] conducted compressive strength tests on pineapple fibers with lengths of 10, 20 and 30 mm, and geopolymers with fiber weight percentages of 0, 0.25 and 0.50%, respectively. The composite with a fiber content of 0.50% and a fiber length of 30 mm had a compressive strength of 41.468 MPa and a maximum bending strength of 9.209 MPa. Studies have shown that the compressive strength and flexural strength of the geopolymer mortar reinforced by 0.5% by mass pineapple fiber is higher than that of the geopolymer reinforced by 0.25% by mass pineapple fiber.

However, whether it is leaf fiber or bast fiber, the excessive fiber content leads to the disharmony between the fiber volume and the matrix volume, which reduces the mechanical properties of the geopolymer, indicating that the appropriate amount of fiber can improve the mechanical properties of the geopolymer.

4.2.3. The Effect of Seed Fiber on the Toughness of CFGCs

Seed fibers are single-cell fibers grown from epidermal cells of plant seeds. Mainly include cotton fiber, kapok fiber and so on. Cotton fiber is amongst the most well-known of CFs. Cotton fiber has unique properties such as high cellulose content, good moisture absorption, excellent heat resistance, light resistance, alkali resistance and higher breaking strength, so it can play an important role in geopolymers. Compared to the pure geopolymer, the addition of cotton fiber gradually improves the fracture toughness of the fiber-reinforced geopolymer composite. Cotton fiber has the characteristics of energy absorption through fiber fracture, fiber matrix interface debonding, fiber pull-out and fiber bridging, etc., which slow down the propagation of cracks and increase the fracture energy, thus playing an important role in enhancing the toughness of the matrix [66]. Tests have shown that the addition of cotton, sisal or coir fiber composites can improve its bending properties [67].

However, it should be noted that the increase in the volume of the hydrophilic natural fibers in the geopolymer matrix will adversely affect the strength of the composite. The fly-ash-based geopolymer composite reinforced with cotton fabric can prevent the cotton fabric from degrading at high temperatures. When the fabric is arranged in a horizontal direction with respect to the applied load, it achieves a higher load and greater deformation resistance than a vertically arranged fabric [68,69]. The results show that cotton fabric orientation affects the bending strength, compressive strength, hardness and fracture toughness of geopolymer composites.

4.2.4. The Effect of Fruit Fiber on the Toughness of CFGCs

Fruit fiber refers to fiber obtained from the fruit of a plant. It is mainly composed of cellulose and associated biomass and intercellular substance, such as coconut fiber. Kroehong et al. [70] found that the addition of oil palm fiber had a significant impact on the physical and mechanical properties and microstructure of geopolymer with high calcium fly ash. The increase of oil palm fiber content reduced the compressive strength of geopolymer but improved the bending strength and toughness of the material and changed the failure behavior of the composite. In addition, with the increase of fiber content, the pore size and total porosity of the material increased, while the thermal conductivity decreased.

In fact, the strength and deformation curve of the fruit fiber is similar to that of the bast fiber and generally the strength value of the fruit fiber is lower than that of the bast fiber. Mazen [71] used loofah fiber as a geopolymer reinforcement material. Compared with pure geopolymer, the compressive strength of the composite was increased from 13 to 31 MPa, and the bending strength was increased from 3.4 to 14.2 MPa. After 20 months of aging time, the flexural yield strength of the composite with 10% loofah fiber content increased from 8.6 to 9.8 MPa, as shown in Figure 5. The increase in yield strength is due to the continuous polymerization of geopolymers as the aging progresses. During the 20-month aging period, the ultimate flexural strength, strain hardening and flexural modulus all changed slightly. The aging study showed that the mechanical properties of the composite material did not decrease significantly within 20 months.

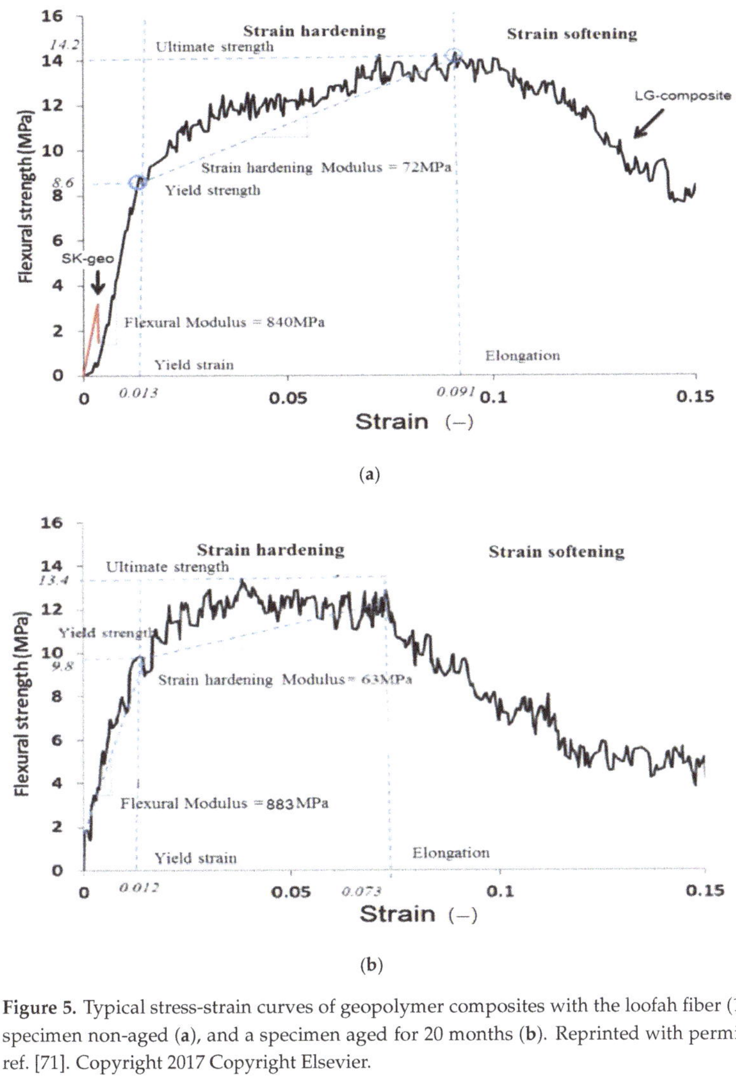

(a)

(b)

Figure 5. Typical stress-strain curves of geopolymer composites with the loofah fiber (10% v/v), a specimen non-aged (**a**), and a specimen aged for 20 months (**b**). Reprinted with permission from ref. [71]. Copyright 2017 Copyright Elsevier.

In another study, Gabriel et al. [72] synthesized wood fiber-reinforced geopolymer composites with fly ash, sand and wood fibers, and added 5, 10, 15, 20, 25, 30 and 35 wt% variable wood fibers. As the amount of wood fiber added increased, the mechanical properties decreased [73]. Su [74] used fly ash and slag as raw materials to prepare geopolymers and improved its crack resistance and strength by adding lignin fibers, polypropylene fibers and alkali-resistant glass fibers. When the fiber addition amount was 0.75%, the strength of fiber-reinforced geopolymer was the best. The coherence between natural fiber and geopolymer matrix is lower than that of artificial fiber. The improvement effects of fiber reinforcement and shrinkage resistance are, in order, PP fiber, alkali-resistant glass fiber and lignin fiber. The fiber not only prevents the separation from the geopolymer matrix, but also inhibits the generation and expansion of cracks, and ultimately improves the strength of the fiber.

4.2.5. The Effect of Stem Fiber on the Toughness of CFGCs

Agricultural waste straws are mostly stem fibers, such as rice, wheat, sorghum, bagasse, and so on. Chen et al. [75] reported that when the fiber content was less than 2.0%, the increase in the content of sweet sorghum fiber in the geopolymer caused the density and unconfined compressive strength of the composite to continue to decrease, while the bending strength, tensile strength and peak toughness had been significantly improved. It shows that the main function of fiber is not to improve the compressive strength of composites, but to improve its flexural performance and control the further development of matrix concrete cracks.

4.2.6. The Effect of Grass/Reeds Fiber on the Toughness of CFGCs

Grass/reeds fiber includes reed, bamboo fiber, corn fiber, and so on. Kaushik et al. [76] used 5 wt% of untreated bamboo fiber to reinforce potassium-based metakaolin geopolymer to obtain a four-point flexural strength of 7.5 MPa.

As mentioned above, cellulosic fiber is the main source of toughness of CF reinforced geopolymers, whether it is bast fiber, fruit fiber, leaf fiber, seed fiber, wood fiber or grass fiber. The strength and fracture resilience of CFGCs with different fiber contents are quite different. Relevant experiments show that, compared with pure geopolymer, the fracture toughness of geopolymer containing 2% cellulosic fiber can be increased 4-fold [77,78].

The microstructure analysis shows that there may be chemical interaction between organic fiber and inorganic polymer chain, and the failure dynamics of geopolymer matrix composites include crack bridging, fiber pulling out and fiber tearing mechanism [62]. Compared with other fibers, CFs have higher specific modulus and elongation at break, and are more evenly distributed in the geopolymer matrix. The presence of CFs usually increases the tensile strength, flexural strength and toughness of the geopolymer composite, thereby improving the durability of the geopolymer composite. On the other hand, the use rate of CF should not be too high because too much CF will increase the porosity of the geopolymer composite, make the fiber and the matrix poorly bonded, and fail to achieve the desired strengthening and toughening effect.

4.3. Resistance to Sulfate Attack of CFGCs

Sulfate resistance is one of the important indexes of cement-based material durability. As a new cementing material which can replace the traditional Portland cement, geopolymer has better sulfate resistance than cement. After sulfate solution erosion, the mechanical strength, microstructure and surface morphology of CFGCs will change to some extent.

Fan et al. [79] used sisal fiber and polyvinyl alcohol fiber (PVA) to strengthen metakaolin-based geopolymer, indicating that fiber incorporation can greatly improve the physical properties of the geopolymer, while fiber doping can improve the sulfate erosion resistance of the geopolymer. Compared with the low concentration of sulfate, the high concentration of sulfate erodes the geopolymer with more cracks and pores, and the compressive strength decreases more obviously. The results showed that the fiber-reinforced geopolymer prepared by 0.5 wt% PVA and 0.75 wt% sisal fiber had the highest compressive and flexural strength and was the most stable.

The microstructure of the fiber-reinforced metakaolin-based polymer samples after curing for 28 d and after being eroded by sulfate at a concentration of 5 wt% and 15 wt% for 28 d is shown in Figure 6 [79]. With the increase of sulfate concentration, it can be seen that there are certain cracks and holes. This is because the sulfate gradually infiltrates into the geopolymer sample during the erosion process, occupying some voids in the geopolymer. The accumulation continues, causing the development of cracks.

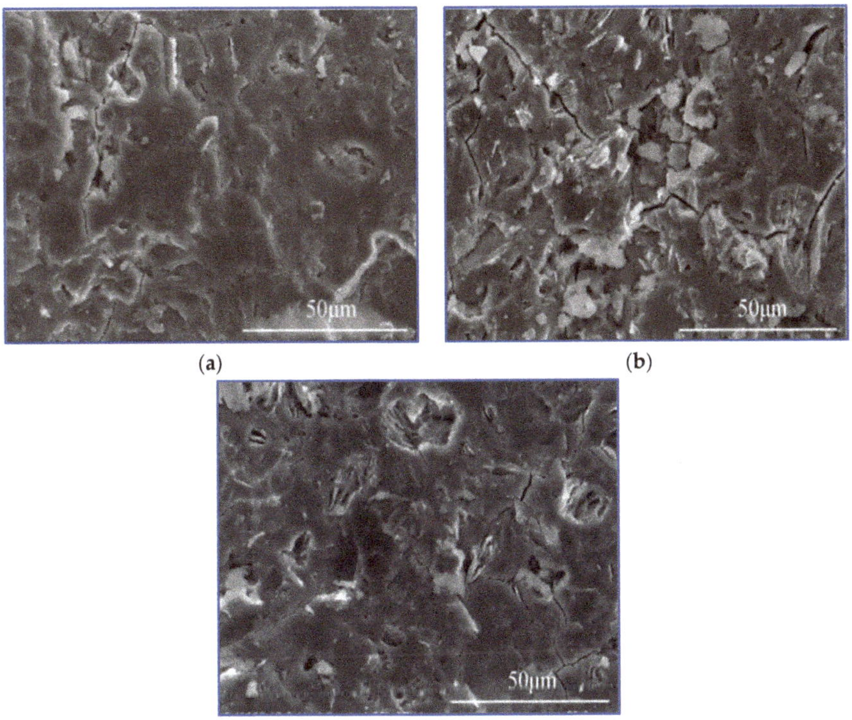

(c)

Figure 6. SEM images of fiber-reinforced metakaolin-based geopolymer, (**a**) before sulfate attack; (**b**) after sulfate attack at a concentration of 5 wt% and (**c**) 15 wt%. Reprinted with permission from ref. [79]. Copyright 2020 Copyright The Chinese Ceramic Society.

Similar to the above, cellulosic geopolymer composites can be stabilized by acid rain leaching over long periods of time. Jin et al. [80] mixed straw with a mass fraction of 4% into the geopolymer, and the compressive strength of the composite was greater than 30 MPa. Under acid rain leaching with a pH value above 3, the compressive strength was maintained at about 36 MPa, and the acid rain leaching resistance was good. When the composite was immersed in thiobacillus thioxide, the compressive strength of the composite was 26.3 MPa for 21 days and 18.4 MPa for 28 days. The results showed that although thiobacillus thioxide had a certain effect on the compressive strength of the straw geopolymer, its amorphous three-dimensional network silicoaluminate structure still existed.

However, in a strong acid solution, the properties of CFGCs also change greatly. Jin [81] synthesized a metakaolin-based geopolymer from rice straw fiber. The straw geopolymer was soaked in sulfuric acid solution at pH 1.0, 3.0 and 5.0 and sodium hydroxide solution at pH 9.0, 11.0 and 13.0. After being immersed in an acid solution with a pH of 1.0 and an alkali solution with a pH of 13.0 for 28 days, the compressive strength of the samples decreased from 50.28 to 28.90 MPa and 40.00 MPa, respectively. Soaking the sample in a sodium hydroxide solution with a pH of 9.0 had the best effect.

The durability of CFGCs can be judged by soaking in sulfuric acid solution to test its cross-section reduction, weight loss, compressive strength loss and other changes. Maan et al. [82] prepared several concrete mixtures by using fly ash and crushed palm oil clinker (POC) as lightweight aggregate and oil palm trunk fiber (OPTF) as natural fiber reinforcement. In the concrete mixture, POC was added to the mixture in proportions of 25%, 50%, 75% and 100% by mass, and OPTF was added to the mixture in proportions of 1%, 2% and 3% by volume, respectively. The results showed that when the compressive strength was above 30 MPa, the optimal replacement rate of POC content was 75%, and OPTF content was 1%. The addition of POC and OPTF reduced the acid resistance of concrete.

Table 2 shows the strength and water absorption test results of the composites in 28 days. The addition of POC significantly improved the water absorption of the mixture. Due to the porosity of POC, the water absorption value increased by one-third when POC content was 25%, and 4.3-fold when it was completely replaced, as shown in Table 2. The water absorption value of geopolymer increased exponentially when added to OPTF. When the OPTF content was 1%, the increase was about two-thirds. When the content of OPTF was 3%, the increase was 3.1-fold. Through the durability test of immersion in sulfuric acid solution, the section reduction, weight loss and compressive strength loss have no obvious change, indicating that the corrosion resistance of sulfuric acid is good.

Table 2. Strength value and water absorption test results of CFGCs.

POC/(%)	OPTF/(%)	Water Reducing Agent/(%)	Tensile/(MPa)	Shear/(MPa)	Flexural/(MPa)	Water Absorption/(%)	Cross-Section Reduction/(%)	Weight Loss/(%)
0	0	0	4.55	9.41	6.31	0.6	−1.15	−2.2
25	0	0	4.31	9.04	6.02	0.8	−1.2	−2.2
50	0	0	3.94	7.94	5.56	1.4	−1.2	−2.6
75	0	0	3.62	7.09	5.10	1.8	−1.5	−2.8
100	0	0	2.91	6.42	4.78	3.2	−2	−3.1
100	0	0.5	3.05	6.48	4.83	3	−1.9	−3
100	1	0.5	4.41	7.19	6.86	4.9	−2	−3.4
100	2	0.5	3.44	6.93	5.34	7.8	−2.6	−3.9
100	3	0.5	3.21	6.54	5.03	12.5	−3.9	−5.7

4.4. Resistance to Chloride Ion Penetration of CFGCs

The durability of CFGCs is an important property in engineering application. In fact, through the study of the long-term durability of fly-ash-based geopolymer exposed for 10 years under deep burial and complete saturation conditions in a severe salt lake environment, it was found that [83], compared with cement concrete, geopolymer had an adverse effect on chloride ion transport, with higher chloride ion diffusion coefficient and lower bond ability.

Zhou et al. [84] conducted a hydrochloric acid erosion test on biogeopolymer of cotton stalk powder and found that the addition of untreated cotton stalk fiber reduced the density and compressive strength of geopolymer, while the flexural strength slightly increased. The compressive strength and flexural strength of cotton stalk fiber after alkali treatment were 4.8% and 11.5% higher than those without alkali treatment, respectively. The treated cotton stalk powder could effectively improve the compressive strength of geopolymer but reduce the acid corrosion resistance of geopolymer. Through analysis, the effect of cotton stalk powder on geopolymer is mainly filling and cementation. The sugar precipitated from cotton stems in an alkaline environment reduces the compactness of the geopolymer gel.

The effects of different concentrations of hydrochloric acid on the resistance of CFGCs are very different. Ribeiro et al. [85] immersed bamboo fiber geopolymers in sulfuric acid and hydrochloric acid of 0, 5 and 15 wt% for 7, 28 and 112 days to study their appearance, quality changes and compressive strength behavior. No mass loss was observed in 0% acid

(100% water), indicating durability in water. The mass loss of geopolymer increased from 5, 10 and 15 wt% to 2.7%, 3.5% and 4.4%, respectively, with the increase of acid concentration.

Figure 7 shows the microscopic morphology of bamboo-fiber-reinforced geopolymer before and after chloride ion erosion. (a) shows SEM images of 4.1 wt% bamboo-fiber-reinforced geopolymer soaked in 15 wt% hydrochloric acid for 28 days. The phenomenon of micropores formed by acid leaching is obvious. (b) shows the parallel periodic micro-cracking in the geopolymer matrix. (c) shows the undissolved chopped bamboo fiber in 15 wt% sulfuric acid treatment for 28 days. (d) shows the formation of periodic parallel microcrack complexes in bamboo-fiber-reinforced geopolymer.

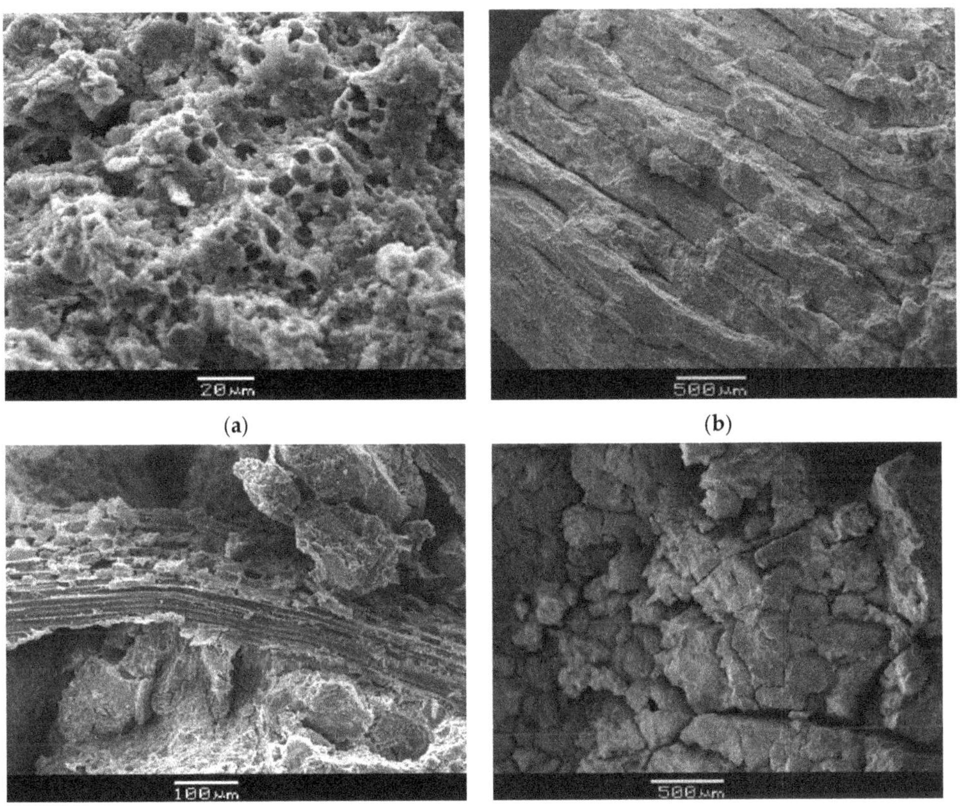

(a) (b) (c) (d)

Figure 7. SEM micrograph of bamboo fiber-reinforced geopolymer, (**a**) acid leaching to form micropores; (**b**) parallel periodic micro-cracking in the geopolymer matrix; (**c**) undissolved chopped Bamboo fiber (**d**) a compound with periodic parallel microcracks. Reprinted with permission from ref. [85]. Copyright 2021 Copyright Elsevier.

4.5. Performance of CFGCs against Wetting/Drying Cycles

The performance of composites against wetting/drying cycles is an important part of durability. Trindade et al. [86] formulated two geopolymers of 100% metakaolin (MK) and 60% (MK) + 40% blast furnace slag (BFS). The mechanical behavior of the two matrices is changed by the jute reinforcing fiber to make it ductile and change its crack mode. Jute fiber promotes the formation of (C-A-S-H) gel and significantly improves the compressive strength of the material. The five-layer jute-fabric-reinforced composite exhibits strain and flexural hardening behavior under tension and bending, and produces multiple cracks.

It can be seen in Figure 8 that after 15 wetting/drying cycles, the first crack strength values of the two composites decreased significantly, but the ultimate strength did not change significantly. This behavior shows that after 15 wetting/drying cycles, the ultimate mechanical capacity of the composite does not change significantly. The composites all exhibit deflection hardening behavior, which leads to smaller crack openings, and various cracks are formed after accelerated aging. The fibers did not degrade significantly after 15 wetting/drying cycles, indicating that the jute fabric-reinforced geopolymer has superior durability compared with Portland cement matrix composites.

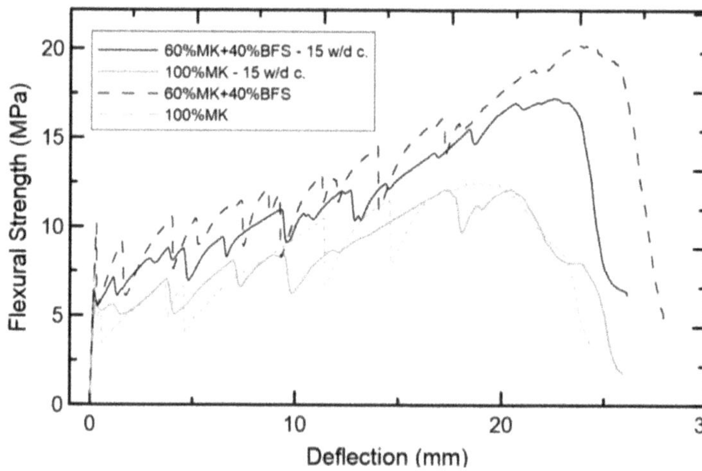

Figure 8. Flexural Strength–Deflection curves. Reprinted with permission from ref. [86]. Copyright 2017 Copyright The American Ceramic Society.

A similar report was also found in ref. [87], Nkwaju et al. used iron-rich red clay and bagasse fibers as raw materials to prepare geopolymer composites, and found that the addition of fibers facilitated the transition of the fracture behavior of geopolymers from brittleness to toughness. With the increase of fiber content, when the fiber mass fraction was 3%, the elastic modulus increased by 50%. After 20 wetting/drying cycles, the performance of the geopolymer composite material had been improved, and the ductility had been improved. It shows that the wetting/drying cycles improves the fiber matrix bond, thereby increasing the ductility of the composite. Santos et al. [88] evaluated durability by accelerating aging through 10 wetting and drying cycles. The composite (0 cycles) was about 15 MPa in the bending test, and the aged composite reached 11 MPa, indicating that the wetting and drying cycles had good durability. The fiber of the composite that had been naturally aged for 3 years has almost no degradation, and the composite had good durability.

The mechanical behavior of the composites at the inelastic stage, such as cracking mechanism, strength and ductility, was tested by a bending test. Canpolat et al. [89] studied the influence of wetting-drying curing system on the performance of fiber reinforced metakaolin-based geopolymer composites. Similarly, Asante et al. [90] also found that the specific strength of the pine and eucalyptus particle geopolymer composite material was reduced by 15.32% after multiple soaking and drying.

4.6. High Temperature Tolerance of CFGCs

Compared with cement-based materials, geopolymer materials have better durability than cement-based materials. Alomayri et al. [91] tested a geopolymer composite material containing 0.83% wt% cotton fabric by exposing it to high temperatures of 200 °C, 400 °C, 600 °C, 800 °C and 1000 °C. As the temperature increased, the compressive strength, flexural strength and fracture toughness of geopolymers all decreased. The high temperature severely degrades the cotton fiber, resulting in holes and small channels in the composite material, which makes the composite material exhibit brittle behavior. When the temperature reaches above 600 °C, the mechanical properties of the composite material are significantly reduced. Alomayri et al. [92] used cotton-fabric-reinforced polymer composites to characterize their thermal properties through thermogravimetric analysis, and evaluated their mechanical properties, such as flexural strength, fracture toughness, flexural modulus and impact strength. When the fiber content was 2.1 wt%, the mechanical properties of the fiber were improved. The thermal analysis results showed that the fly-ash-based polymer can prevent the degradation of cotton fabrics at high temperatures. Amalia et al. [93] studied the high temperature resistance of fly-ash-based hybrid geopolymers with pineapple leaf fibers as aggregates, and the results showed that pineapple leaf fibers had great potential in geopolymer reinforcement or lightweight aggregates.

Another experiment [86] found that there were fine crack networks on pure geopolymer heated at 400 and 600 °C. After over 600 °C, severe cracks appear on the surface. When cotton fiber was added, no cracks were found on the surface of geopolymer at the same temperature. It shows that cotton fiber is very effective in preventing the matrix from forming cracks at high temperature. The structure of the composite becomes more porous, and the expanded water vapor escapes without causing major damage to the microstructure. This degradation of cotton fibers will facilitate the behavior of geopolymers under heat exposure. The porosity and small channels produced by the degradation of cotton fibers can reduce the internal vapor pressure, thereby reducing the possibility of cracking.

Compared to ordinary Portland cement, geopolymer has better acid resistance and sulfate resistance, but its resistance to carbonization is slightly worse. Frost resistance is also an important performance for the durability of plant-fiber-reinforced geopolymers, but there are few studies in this area. Xuan et al. [94] prepared geopolymer-based CF composites using industrial waste slag and agricultural and forestry residue bagasse as raw materials. The bending strength was analyzed, and its appearance and microstructure were analyzed. It shows that CFGCs have better frost resistance.

5. Other Factors Affecting the Durability of CFGCs

In general, there are basically two ways for improving the durability of CFGCs. One way is to add nanomaterials into the matrix; the other way is to modify CFs.

5.1. The Effect of Nanomaterial Addition on the Durability of CFGCs

Pore structure is the basis of the theory of geopolymer mix ratio design and its relationship with mechanical properties and is closely related to the macro-mechanical properties and durability of geopolymers. Nanomaterials have unique nano-effects such as volume effect, surface effect, quantum size, quantum tunnel, etc., which lead to unique physical and chemical properties of nanomaterials and nanostructures. Incorporating nano-SiO_2 can not only speed up the polymer polymerization reaction process, in which the unreacted nano-SiO_2 particles are wrapped by the geopolymer, but also play the role of particle filling, so that the overall structure of the composite material is more compact, and the polymerization is improved, as is the durability of objects [95].

Generally, CFs will deteriorate to varying degrees after being exposed to an alkaline environment. The attack of alkali ions leads to the weakening of cellulose and hemicellulose, and the mineralization of fiber cell walls in the geopolymer pulp leads to fiber brittleness. Assaedi [96] evaluated that after 32 weeks of aging, the flexural strength of geopolymer composites decreased. Figure 9a,b show the load deflection behavior of fiber-reinforced geopolymer (GP/FF) composites and nano-SiO_2 fiber-reinforced geopolymer (GPNS-1/FF) composites at 4 and 32 weeks, respectively. Among the two composites, the composite containing nano-SiO_2 has a higher load capacity.

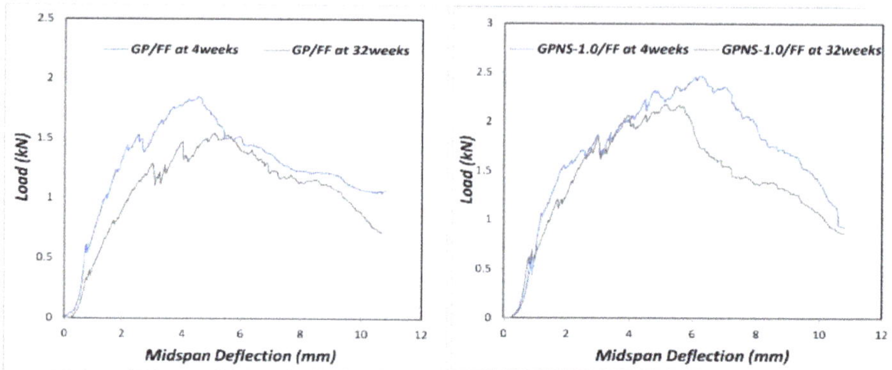

(a) (b)

Figure 9. Load–deflection diagrams of (**a**) GP/FF composite and (**b**) GPNS-1/FF nanocomposite at 4 and 32 weeks. Reprinted with permission from ref. [96]. Copyright 2019 Copyright MDPI.

The flexural strength of the flax-fiber-reinforced geopolymer composite material was reduced by 22.4%, while the flexural strength of GPNS-1/FF was reduced by approximately 10.3%. It shows that after 32 weeks of aging, the flexural strength of the nanocomposite decreases less. The analysis shows that nano-silica consumes the alkaline solution and reduces the alkalinity of the system, thereby reducing the degradation of flax fibers. In addition, nano-silica accelerates the geopolymer reaction and increases the geopolymer gel in the matrix. The amount of it improves the density of the matrix and actually improves the adhesion between the fiber and the geopolymer matrix, thereby enhancing the durability of the geopolymer.

In a similar way, Saulo et al. [97] improved the compressive strength and stiffness of geopolymers for 7 days by adding microcrystalline cellulose fibers. After 28 days, due to the degradation of microcrystalline cellulose, the geopolymer reduced its mechanical properties. The group of Cut [98] studied the effect of different cellulose nanocrystal concentrations on the mechanical properties of geopolymers. Lower concentrations of cellulose nanocrystals (<0.5%) could produce higher strength geopolymers. The higher concentration of cellulose nanocrystals prevents the pyrolysis of the geopolymer in an unstable solidification environment and enhances the corrosion resistance of the composite. Rahman et al. [99] studied the synergistic effect of silicon dioxide and silicon carbide whiskers derived from rice husk ash. It showed that the spherical silica nanoparticles prepared from rice husk ash reduced the nanoporosity of the geopolymer by 20% and doubled the compressive strength. When rice husk ash and silicon carbide whiskers were added at the same time, the flexural strength increased by 27% and 97%, respectively. The increased compressive strength of silica nanoparticles is related to the decrease of porosity, and silicon carbide whiskers can effectively improve the bridge network and crack resistance.

Assaedi et al. [100] synthesized a geopolymer composite material reinforced by flax fabric and nano-clay flakes, and added nano-clay flakes to strengthen the geopolymer matrix at 1.0, 2.0 and 3.0% by mass. The 2.0 wt% nanoclay had the best effect, increasing the density and reducing the porosity, thereby increasing the flexural strength and toughness. Rahmawati et al. [101] used Typha as a new raw material for separating cellulose nanocrystals and extracted cellulose from stem fibers by alkaline and bleaching methods. Then, cellulose nanocrystals were separated from the extracted cellulose by acid hydrolysis. The acid hydrolyzed cellulose nanocrystal had good thermal stability at 240 °C, which was higher than that of raw meal. Cellulose nanocrystals have the potential as a geopolymer cement reinforcement agent.

Based on the above studies, it can be concluded that nanomaterials can be used not only as a filler to improve the microstructure of the binder, but also as an activator to support the geopolymer reaction and produce a higher content of geopolymer gel. This enhances the adhesion between the geopolymer matrix and CF, thereby improving the properties of CFGCs.

5.2. The Effect of Fiber Modification on the Durability of CFGCs

Compared with other types of fibers, some limitations of CFs, such as biodegradation, UV degradation or weak bonding, affect their mechanical properties and durability. In order to improve the performance of CFs in the geopolymer matrix, a modification treatment method is usually used. Alkalization is one of the chemical modification techniques of bio-based materials. The purpose of the treatment is to have less impurities in the fiber and increase the adhesion of its contact surface.

Scholars have many research cases on the modification of bast fibers. Kumar et al. [41] used 10% (NaOH) solution for alkali treatment of ramie with a maximum treatment temperature of 160 °C and obtained strong fibers with low lignin content and good fiber separation. Lazorenko [102] et al. treated alkaline media with 5% (NaOH) solution mercerizing and ultrasonic (22 kHz, 500 W). The combined treatment of alkali and high-intensity ultrasound is an effective way to treat and modify fiber-reinforced polymer composites, which has the best technical and economic effects. Maichin et al. [103] studied the influence of sodium hydroxide concentration on the pretreatment performance of hemp fiber and the self-treatment behavior of hemp fiber in geopolymer composites. The self-treatment behavior of hemp fiber in the geopolymer can improve the final performance of the hardened product. The self-treatment process is to add fibers to the geopolymer mixture without any pre-alkaline treatment. Similarly, Maichin et al. [104] also discussed the influence of geopolymer alkalinity on fiber self-treatment. The results showed that the self-treatment process is controlled by the alkaline environment in the geopolymer system. After the fiber is alkalized, the surface of the fiber is modified, so that the fiber has stronger cohesiveness and better compatibility with cement paste. Pickering et al. [38] discussed different methods of chemical modification of hemp fiber to make it have good water repellency, chemical resistance and good mechanical properties. Georg et al. [105] used pretreatment and surface modification to remove short hemicellulose, lignin, pectin and wax, increase the interface adhesion between the matrix and flax fibers, and optimize the rheology of the geopolymer slurry performance. Roy et al. [106] applied different chemical treatments to the abaca fiber to change its surface characteristics and improve the adhesion to the fly-ash-based polymer matrix. It showed that the tensile strength of abaca fiber without alkali pretreatment and soaked in ($Al_2(SO_4)_3$) solution with pH6 for 12 h had the highest tensile strength. Chemical treatment and deposition of aluminum compounds make the surface rougher. This improves the interfacial bonding between the geopolymer matrix and the fibers, while the geopolymer protects the treated fibers from thermal degradation.

Sisal is typical of leaf fiber. Figure 10 shows the microscopic morphology of sisal fiber-reinforced metakaolin-based polymer. It can be observed that the fiber is broken or torn at the section, as shown in Figure 10a,b. The phenomenon of the fibrillation of sisal fiber indicates that the sisal fiber plays a role in bearing force during the crack propagation and fracture process of the sample, which causes the damage and tear of the fiber. As shown in Figure 10c,d, the interfacial bonding ability between the alkali-treated sisal fiber and the metakaolin-based polymer material has been significantly improved. It can be seen that the sisal fiber breaks under external force. The fiber is not drawn out due to external force, but breaks during the stress process, indicating that the interface bonding force between the sisal fiber and the geopolymer is greater than the maximum stress that the fiber can withstand, which shows that the alkali treatment can greatly improve the interface bonding between sisal fiber and metakaolin-based polymer material.

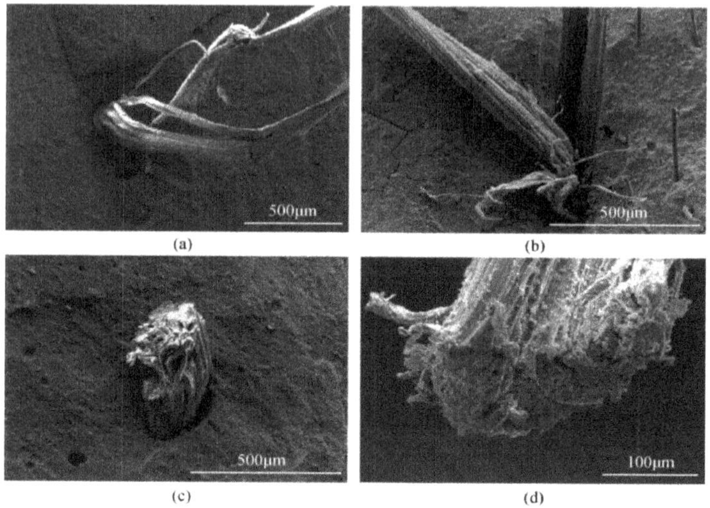

Figure 10. SEM images of sisal-fiber-reinforced metakaolin-based geopolymer, (**a**,**b**) the fiber is broken or torn at the section; (**c**,**d**) the damage and tear of the fiber. Reprinted with permission from ref. [77]. Copyright 2018 Copyright MDPI.

The modification treatment of stem fiber is mainly alkali treatment. Huang et al. [107] found that both untreated and alkali-treated rice straw can significantly increase the flexural strength of geopolymers. The bonding effect of the geopolymer matrix and straw after alkali treatment was better than that of untreated rice straw. During the curing time of 28 days, the flexural strength of the alkali-treated straw-reinforced geopolymer composite with a fiber content of 10% reached 13.6 MPa. Workiye et al. [108] chemically treated corn stalks with 98% pure sodium hydroxide for 30 min, and prepared 0, 0.1, 0.2, 0.6 and 1% by mass corn stalk monocellulose-reinforced geopolymer composites. The material indicates that the proper addition of single fiber of corn stover can improve the compressive strength of the calcined kaolinite base polymer.

Ribeiro et al. [109] found that bamboo fibers and strips enhance the compressive strength of geopolymers. The alkali-treated micro-bamboo fiber had a compressive strength of 23–38 MPa, which was lower than the 56 MPa of pure geopolymers, but still had good structural applications. Alkali treatment and water treatment have no significant difference in the bending strength of bamboo fiber and bamboo strips, and both achieve the effect of toughening and cracking resistance. Similarly, water treatment methods are also applicable to wood fibers. Asante et al. [109] studied the effect of hot water treatment of wood particles on the physical properties and specific compressive strength of geopolymers before and after immersion and drying. Hot water washing resulted in a reduction of 47% and 67% in the extract content of pine and eucalyptus particles, respectively, and the specific strength values of pine and eucalyptus particle geopolymers increased by 27% and 3%, respectively. Hot water pretreatment significantly increased the specific compressive strength of the pine base polymer, while the specific compressive strength of the eucalyptus base polymer did not increase. It showed that the hot water washed away the unique extracts of pine trees, so that there was better compatibility between geopolymer and wood.

6. Conclusions

In this paper, the larger part is the durability of CFGCs such as crack resistance and toughness, sulfate corrosion resistance, chloride ion penetration resistance, dry and wet cycle resistance, and high temperature resistance. Compared with cement-based materials, the weak alkalinity of CFGCs slows down the degradation of CFs. CFs have been widely used in CFGCs due to its excellent properties. Meanwhile, the mechanical properties of the interface between CFs and matrix should be further improved to enhance the durability of composites and lay a foundation for their engineering application. The main conclusions are as follows:

- All types of natural cellulose fibers can be used to reinforce geopolymers. Among the bast fibers, hemp, flax and jute, and leaf fiber sisal are the most widely used, and there is also more related research;
- An appropriate amount of plant fiber has a beneficial effect on the mechanical properties of the geopolymer, toughening and cracking resistance, and other types of durability. Too much mixing will have a negative effect. In CFGCs, the CF content range is mostly 0.1–10%, and the best content is usually 2–4% volume content;
- The alkaline degradation of CF in the geopolymer matrix has an adverse effect on the mechanical properties of the composites. Chemical modification and self-modification can be used to adjust the adhesion state of the fiber and matrix interface and optimize the properties of the interface layer between the fiber and matrix to achieve the best properties of the geopolymer;
- Nanomaterials can improve the microstructure of CFGCs, make the material matrix more compact, reduce the degradation rate of CF and improve the durability of CFGCs;
- CFGCs have good properties of resistance to sulfate and chloride ion erosion and can prevent degradation of fibers at high temperatures. However, the sugar precipitated from CFs in alkaline environment reduces the compactness of geopolymer gel and has a negative effect on its durability.

Author Contributions: Conceptualization, J.L. and C.L.; methodology, J.L.; validation, J.L. and C.L.; formal analysis, J.L.; investigation, J.L.; resources, J.L.; data curation, J.L.; writing—original draft preparation, J.L.; writing—review and editing, J.L.; supervision, C.L.; project administration, C.L.; funding acquisition, C.L. All authors have read and agreed to the published version of the manuscript.

Funding: This research was funded by the general items of basic scientific research and business expenses of provincial universities in Heilongjiang Province, grant number 145109126; 135409313.

Institutional Review Board Statement: Not applicable.

Informed Consent Statement: Not applicable.

Data Availability Statement: Not applicable.

Conflicts of Interest: The authors declare no conflict of interest.

References

1. Parathi, S.; Nagarajan, P.; Pallikkara, S.A. Ecofriendly geopolymer concrete: A comprehensive review. *Clean Technol. Environ. Policy* **2021**, *23*, 1701–1713. [CrossRef]
2. Ren, B.; Zhao, Y.; Bai, H.; Kang, S.; Zhang, T.; Song, S. Eco-friendly geopolymer prepared from solid wastes: A critical review. *Chemosphere* **2021**, *267*, 128900. [CrossRef] [PubMed]
3. Francesco, C.; Ilenia, F.; Marta, T.; Cinzia, S.; Raffaele, C.; Antonella, P. Eco-efficient industrial waste recycling for the manufacturing of fibre reinforced innovative geopolymer mortars: Integrated waste management and green product development through LCA. *J. Clean. Prod.* **2021**, *312*, 127777.
4. Buchwald, A. What are geopolymers? Current state of research and technology, the opportunities they offer and their significance for the precast industry. *Betonw. Und Fert.-Tech.* **2006**, *72*, 42–49.
5. Favier, A.; Hot, J.; Habert, G.; Roussela, N.; Lacaillerie, J. Flow properties of mk-based geopolymer pastes. a comparative study with standard portland cement pastes. *Soft Matter* **2013**, *10*, 1134–1141. [CrossRef] [PubMed]
6. Huang, Y.; Gong, L.; Pan, Y.; Li, C.; Zhou, T.; Cheng, X. Facile construction of the aerogel/geopolymer composite with ultra-low thermal conductivity and high mechanical performance. *RSC Adv.* **2018**, *8*, 2350–2356. [CrossRef]
7. Davidovits, J. Geopolymers. *J. Therm. Anal. Calorim.* **1991**, *37*, 1633–1656. [CrossRef]
8. Kozub, B.; Bazan, P.; Mierzwiński, D.; Korniejenko, K. Fly-ash-based geopolymers reinforced by melamine fibers. *Materials* **2021**, *14*, 400. [CrossRef] [PubMed]
9. Zhang, P.; Wang, K.; Wang, J.; Guo, J.; Ling, Y. Macroscopic and microscopic analyses on mechanical performance of metakaolin/fly ash based geopolymer mortar. *J. Clean. Prod.* **2021**, *294*, 126193. [CrossRef]

10. Lemougna, P.N.; Mackenzie, K.J.; Melo, U.C. Synthesis and thermal properties of inorganic polymers (geopolymers) for structural and refractory applications from volcanic ash. *Ceram. Int.* **2011**, *37*, 3011–3018. [CrossRef]
11. Mahesh, K.; Komal, K.; Manjunath, G.S. Workability and strength study on fiber reinforced geopolymer concrete. *IUP J. Struct. Eng.* **2018**, *11*, 41–57.
12. Yan, S.; Sagoe-Crentsil, K. Properties of wastepaper sludge in geopolymer mortars for masonry applications. *J. Environ. Manag.* **2012**, *112*, 27–32. [CrossRef] [PubMed]
13. Gábor, M.; Ágnes, S.; Sándor, N. Fiber reinforced geopolymer from synergetic utilization of fly ash and waste tire. *J. Clean. Prod.* **2018**, *178*, 429–440.
14. Saloni; Parveen; Pham, T.M. Enhanced properties of high-silica rice husk ash-based geopolymer paste by incorporating basalt fibers. *Constr. Build. Mater.* **2020**, *245*, 118422. [CrossRef]
15. Fu, Q.; Xu, W.; Zhao, X.; Bu, M.; Niu, D. The microstructure and durability of fly ash-based geopolymer concrete: A review. *Ceram. Int.* **2021**, *47*, 29550–29566. [CrossRef]
16. Rashad, A.M. The effect of polypropylene, polyvinyl-alcohol, carbon and glass fibres on geopolymers properties. *Mater. Sci. Technol.* **2019**, *35*, 127–146. [CrossRef]
17. Manfaluthy, M.L.; Ekaputri, J.J. The application of PVA fiber to improve the mechanical properties of geopolymer concrete. *MATEC Web Conf.* **2017**, *138*, 1020. [CrossRef]
18. Zhuang, X.; Chen, L.; Komarneni, S.; Zhou, C.; Tong, D.; Yang, H.; Yu, W.; Wang, H. Fly ash-based geopolymer: Clean production, properties and applications. *J. Clean. Prod.* **2016**, *125*, 253–267. [CrossRef]
19. Ganesan, N.; Abraham, R.; Raj, S.D. Durability characteristics of steel fibre reinforced geopolymer concrete. *Constr. Build. Mater.* **2015**, *93*, 471–476. [CrossRef]
20. Annalisa, N.; Valentina, M.; Elena, L. Production and thermomechanical characterization of wool-geopolymer composites. *J. Am. Ceram. Soc.* **2017**, *100*, 2822–2831.
21. Rabiaa, E.; Mohamed, R.A.S.; Sofi, W.H.; Taher, A.T. Developing geopolymer concrete properties by using nanomaterials and steel fibers. *Adv. Mater. Sci. Eng.* **2020**, *21*, 1–12. [CrossRef]
22. Ibraheem, M.; Butt, F.; Waqas, R.M.; Hussain, K.; Tufail, R.F.; Ahmad, N.; Usanova, K.; Musarat, M.A. Mechanical and Microstructural Characterization of Quarry Rock Dust Incorporated Steel Fiber Reinforced Geopolymer Concrete and Residual Properties after Exposure to Elevated Temperatures. *Materials* **2021**, *14*, 6890. [CrossRef] [PubMed]
23. Ma, P.; Xin, M.; Zhang, Y.; Ge, S.; Wang, D.; Jiang, C.; Zhang, L.; Cheng, X. Facile synthesis of novel dopamine-modified glass fibers for improving alkali resistance of fibers and flexural strength of fiber-reinforced cement. *RSC Adv.* **2021**, *11*, 18818–18826. [CrossRef]
24. Al-mashhadani Mukhallad, M.; Orhan, C.; Yurdakul, A.; Mucteba, U.; Savaş, E. Mechanical and microstructural characterization of fiber reinforced fly ash based geopolymer composites. *Constr. Build. Mater.* **2018**, *167*, 505–513. [CrossRef]
25. Guo, X.; Xiong, G. Resistance of fiber-reinforced fly ash-steel slag based geopolymer mortar to sulfate attack and drying-wetting cycles. *Constr. Build. Mater.* **2021**, *269*, 121326. [CrossRef]
26. Aygrmez, Y.; Canpolat, O.; Al-Mashhadani, M.M. Assessment of geopolymer composites durability at one year age. *J. Build. Eng.* **2020**, *32*, 101453. [CrossRef]
27. Kheradmand, M.; Mastali, M.; Abdollahnejad, Z.; Pacheco-Torgal, F. Experimental and numerical investigations on the flexural performance of geopolymers reinforced with short hybrid polymeric fibres. *Compos. Part B Eng.* **2017**, *126*, 108–118. [CrossRef]
28. Al-Majidi, M.H.; Lampropoulos, A.; Cundy, A.B. Tensile properties of a novel fibre reinforced geopolymer composite with enhanced strain hardening characteristics. *Compos. Struct.* **2017**, *168*, 402–427. [CrossRef]
29. Mohammed, F.; Aamer, B.; Nemkumar, B. Tensile performance of eco-friendly ductile geopolymer composites (EDGC) incorporating different micro-fibers. *Cem. Concr. Compos.* **2019**, *103*, 183–192.
30. Shaikh, A. Review of mechanical properties of short fibre reinforced geopolymer composites. *Constr. Build. Mater.* **2013**, *43*, 37–49. [CrossRef]
31. Liu, J.; Lv, C. Research progress on durability of cellulose fiber-reinforced cement-based composites. *Int. J. Polym. Sci.* **2021**, *2021*, 1014531. [CrossRef]
32. Janne, P.S.N.; Michael, A.B.P. Development of abaca fiber-reinforced foamed fly ash geopolymer. *MATEC Web Conf.* **2018**, *156*, 5018. [CrossRef]
33. Sankar, K.; Kriven, W.M. Sodium geopolymer reinforced with jute weave. *Ceram. Eng. Sci. Proc.* **2014**, *35*, 5–8.
34. Ribeiro, M.; Kriven, W.M. A Review of Particle-and Fiber-Reinforced Metakaolin-Based Geopolymer Composites. *J. Ceram. Sci. Technol.* **2017**, *8*, 307–322.
35. Yan, L.; Kasal, B.; Huang, L. A review of recent research on the use of cellulosic fibres, their fibre fabric reinforced cementitious, geo-polymer and polymer composites in civil engineering. *Compos. Part B Eng.* **2016**, *92*, 94–132. [CrossRef]
36. Camargo, M.M.; Taye, E.A.; Roether, J.A.; Redda, D.T.; Boccaccini, A.R. A review on natural fiber-reinforced geopolymer and cement-based composites. *Materials* **2020**, *13*, 4603. [CrossRef]
37. Bos, H.L.; Oever, M.; Peters, O. Tensile and compressive properties of flax fibres for natural fibre reinforced composites. *J. Mater. Sci.* **2002**, *37*, 1683–1692. [CrossRef]
38. Pickering, K.L.; Beckermann, G.W.; Alam, S.N.; Foreman, N.J. Optimising industrial hemp fibre for composites. *Compos. Part A Appl. Sci. Manuf.* **2007**, *38*, 461. [CrossRef]

39. Summerscales, J.; Dissanayake, N.; Virk, A.S.; Hall, W. A review of bast fibres and their composites. Part1-fibres as reinforcements. *Compos. Part A Appl. Sci. Manuf.* **2010**, *41*, 1329–1335. [CrossRef]
40. Mustafa, A.; Abdollah, M.; Shuhimi, F.F.; Ismail, N.; Amiruddin, H.; Umehara, N. Selection and verification of kenaf fibres as an alternative friction material using weighted decision matrix method. *Mater. Des.* **2015**, *67*, 577–582. [CrossRef]
41. Kumar, R.; Obrai, S.; Sharma, A. Chemical modifications of natural fiber for composite material. *Chem. Sin.* **2011**, *2*, 219–228.
42. Haque, M.; Rahman, R.; Islam, N.; Huque, M.; Hasan, M. Mechanical properties of polypropylene composites reinforced with chemically treated coir and abaca fiber. *J. Reinf. Plast. Compos.* **2010**, *29*, 2253–2261. [CrossRef]
43. Huda, M.S.; Drzal, L.T.; Mohanty, A.K.; Misra, M. Chopped glass and recycled newspaper as reinforcement fibers in injection molded poly (lactic acid) (pla) composites: A comparative study. *Compos. Sci. Technol.* **2015**, *66*, 1813–1824. [CrossRef]
44. Arioza, E.; Ariozb, O.; Mete Kockar, O. An Experimental Study on the Mechanical and Microstructural Properties of Geopolymers. *Procedia Eng.* **2012**, *42*, 100–105. [CrossRef]
45. Sambucci, M.; Sibai, A.; Valente, M. Recent advances in geopolymer technology. A potential eco-friendly solution in the construction materials industry: A review. *J. Compos. Sci.* **2021**, *5*, 109. [CrossRef]
46. Duxson, P.; Lukey, G.C.; Separovic, F.; van Deventer, J.S.J. Effect of alkali cations on aluminum incorporation in geopolymeric gels. *Ind. Eng. Chem. Res.* **2005**, *44*, 832–839. [CrossRef]
47. Rao, J.; Zhou, Y.; Fan, M. Revealing the interface structure and bonding mechanism of coupling agent treated WPC. *Polymers* **2018**, *10*, 266. [CrossRef]
48. Ren, D.; Yan, C.; Duan, P.; Zhang, Z.; Li, L.; Yan, Z. Durability performances of wollastonite, tremolite and basalt fiber-reinforced metakaolin geopolymer composites under sulfate and chloride attack. *Constr. Build. Mater.* **2017**, *134*, 56–66. [CrossRef]
49. Ranjbar, N.; Zhang, M. Fiber-reinforced geopolymer composites: A review. *Cem. Concr. Compos.* **2020**, *107*, 103498. [CrossRef]
50. Ali, A.A.; Mucteba, U.; Arın, Y.; Al-mashhadani Mukhallad, M.; Orhan, C.; Furkan, Ş.; Yurdakul, A. Influence of wetting-drying curing system on the performance of fiber reinforced metakaolin-based geopolymer composites. *Constr. Build. Mater.* **2019**, *225*, 909–926.
51. Tan, J.; Lu, W.; Huang, Y.; Wei, S.; Xuan, X.; Liu, L. Preliminary study on compatibility of metakaolin-based geopolymer paste with plant fibers. *Constr. Build. Mater.* **2019**, *225*, 772–775. [CrossRef]
52. Wei, J.; Meyer, C. Degradation mechanisms of natural fiber in the matrix of cement composites. *Cem. Concr. Res.* **2015**, *73*, 1–16. [CrossRef]
53. Ye, H.; Zhang, Y.; Yu, Z. Effects of cellulose, hemicellulose, and lignin on the morphology and mechanical properties of metakaolin-based geopolymer. *Constr. Build. Mater.* **2018**, *173*, 10–16. [CrossRef]
54. Yan, L.; Chouw, N.; Jayaraman, K. Effect of UV and water spraying on the mechanical properties of flax fabric reinforced polymer composites used for civil engineering applications. *Mater. Des.* **2015**, *71*, 17–25. [CrossRef]
55. Correia, V.D.C.; Ardanuy, M.; Claramunt, J.; Savastano, H. Assessment of chemical and mechanical behavior of bamboo pulp and nanofibrillated cellulose exposed to alkaline environments. *Cellulose* **2019**, *26*, 9269–9285. [CrossRef]
56. Olayiwola, H.O.; Amiandamhen, S.O.; Meincken, M. Investigating the suitability of fly ash/metakaolin-based geopolymers reinforced with South African alien invasive wood and sugarcane bagasse residues for use in outdoor conditions. *Eur. J. Wood Prod.* **2021**, *79*, 611–627. [CrossRef]
57. Eyerusalem, A.; Judith, A.; Dirk, W.; Daniel, T.; Aldo, R. Hemp fiber reinforced red mud/fly ash geopolymer composite materials: Effect of fiber content on mechanical strength. *Materials* **2021**, *14*, 14030511. [CrossRef]
58. Poletanovic, B.; Dragas, J.; Ignjatovic, I.; Komljenovic, M.; Merta, I. Physical and mechanical properties of hemp fibre reinforced alkali-activated fly ash and fly ash/slag mortars. *Constr. Build. Mater.* **2020**, *259*, 119677. [CrossRef]
59. Trindade, A.C.; Arêas, I.O.; Almeida, D.C.; Alcamand, H.A.; Borges, P.H.; Silva, F.A. Mechanical behavior of geopolymeric composites reinforced with natural fibers. In *International Conference on Strain-Hardening Cement-Based Composites*; Mechtcherine, V., Slowik, V., Kabele, P., Eds.; RILEM Bookseries; Springer: Berlin/Heidelberg, Germany, 2017; Volume 15.
60. Sáez-Pérez, M.; Brümmer, M.; Durán-Suárez, A. Effect of the state of conservation of the hemp used in geopolymer and hydraulic lime concretes. *Constr. Build. Mater.* **2021**, *285*, 122853. [CrossRef]
61. Na, Z.; Hya, B.; Dpa, B.; Yang, Z. Effects of alkali-treated kenaf fiber on environmentally friendly geopolymer-kenaf composites: Black liquid as the regenerated activator of the geopolymer. *Constr. Build. Mater.* **2021**, *297*, 123787.
62. Assaedi, H.; Alomayri, T.; Shaikh, F.U.A.; Low, I. Characterisation of mechanical and thermal properties in flax fabric reinforced geopolymer composites. *J. Adv. Ceram.* **2015**, *4*, 272–281. [CrossRef]
63. Korniejenko, K.; Łach, M.; Hebdowska-Krupa, M.; Mikuła, E.J. The mechanical properties of flax and hemp fibres reinforced geopolymer composites. In *IOP Conference Series: Materials Science and Engineering, International Conference Building Materials, Products and Technologies*; IOP: Bristol, UK, 2018; Volume 379.
64. Ampol, W.; Ronnakrit, K.; Sakchai, N.; Vanchai, S.; Prinya, C. Natural fiber reinforced high calcium fly ash geopolymer mortar. *Constr. Build. Mater.* **2020**, *241*, 118143.
65. Zulfiati, R.; Saloma; Idris, Y. Mechanical properties of fly ash-based geopolymer with natural fiber. *J. Phys. Conf. Ser.* **2019**, *1198*, 082021. [CrossRef]
66. Low, I.; Alomayri, T.; Hasan, A. *Cotton and Flax Fibre-Reinforced Geopolymer Composites: Synthesis, Properties and Applications*; Springer: Berlin/Heidelberg, Germany, 2021. [CrossRef]

67. Korniejenko, K.; Frczek, E.; Pytlak, E.; Adamski, M. Mechanical properties of geopolymer composites reinforced with natural fibers. *Procedia Eng.* **2016**, *151*, 388–393. [CrossRef]
68. Kornejenko, K.; Ach, M.; Salamtmur, N.D.; Furtos, G.; Mkua, J. The overview of mechanical properties of short natural fiber reinforced geopolymer composites. *Environ. Res. Technol.* **2020**, *3*, 21–32. [CrossRef]
69. Alomayri, T.; Shaikh, F.U.A.; Low, I.M. Effect of fabric orientation on mechanical properties of cotton fabric reinforced geopolymer composites. *Mater. Des.* **2014**, *57*, 360–365. [CrossRef]
70. Kroehong, W.; Chai, J.; Pothisiri, T.; Chindaprasirt, P. Effect of oil palm fiber content on the physical and mechanical properties and microstructure of high-calcium fly ash geopolymer paste. *Arab. J. Sci. Eng.* **2018**, *11*, 1–10. [CrossRef]
71. Mazen, A.; Abu, M.S.; Juma'a, A.; Yasair, A.; Tarek, F.; Abderrazek, K.; Fernando, R. Fabrication, microstructural and mechanical characterization of Luffa Cylindrical Fibre-Reinforced geopolymer composite. *Appl. Clay Sci.* **2017**, *143*, 125–133.
72. Gabriel, F.; Laura, S.; Petru, P.; Codruta, S.; Kinga, K. Mechanical properties of wood fiber reinforced geopolymer composites with sand addition. *J. Nat. Fibers* **2021**, *18*, 285–296.
73. Mourak, A.; Hajjaji, M.; Alagui, A. Cured alkali-activated heated clay-cellulose composites: Microstructure, effect of glass addition and performances. *Boletín Soc. Española. Cerámica. Vidr.* **2021**, *20*, 62–72. [CrossRef]
74. Su, Z.; Guo, L.; Zhang, Z.; Duan, P. Influence of different fibers on properties of thermal insulation composites based on geopolymer blended with glazed hollow bead. *Constr. Build. Mater.* **2019**, *203*, 525–540. [CrossRef]
75. Chen, R.; Ahmari, S.; Zhang, L. Utilization of sweet sorghum fiber to reinforce fly ash-based geopolymer. *J. Mater. Sci.* **2014**, *49*, 2548–2558. [CrossRef]
76. Kaushik, S.; Sá Ribeiro Ruy, A.; Sá Ribeiro Marilene, G.; Kriven Waltraud, M.; Colombo, P. Potassium-based geopolymer composites reinforced with chopped bamboo fibers. *J. Am. Ceram. Soc.* **2017**, *100*, 49–55.
77. Gianmarco, T.; Enrico, B.; Ivo, D. Mechanical performance of glass-based geopolymer matrix composites reinforced with cellulose fibers. *Materials* **2018**, *11*, 2395. [CrossRef]
78. Amalia, F.; Akifah, N.; Nurfadilla; Subaer. Development of coconut trunk fiber geopolymer hybrid composite for structural engineering materials. In *IOP Conference Series: Materials Science and Engineering*; IOP Publishing: Bristol, UK, 2017; Volume 180, p. 12014.
79. Fan, J.; Jiang, Y.; Wang, L.; Ding, F.; Guo, L.; Duan, P. Sulfate attack resistance of Sisal-PVA hybrid fiber reinforced geopolymer. *Bull. Chin. Ceram. Soc.* **2020**, *39*, 1430–1437, 1443.
80. Jin, M.; Xia, R.; Sun, Y. Environmental stability of straw-reinforced geopolymer composites. *J. Zhejiang Univ. Technol.* **2017**, *45*, 43–46.
81. Jin, M.; Xia, R.; Zhu, C.; Zheng, Y.; Jin, Z. Resistance of straw-geopolymer materials to acid and alkali attack. *Mater. Sci. Forum* **2016**, *852*, 1409–1412. [CrossRef]
82. Malkawi Ahmad, B.; Maan, H.; Jamal, A.; Yazan, A. Engineering properties of fibre reinforced lightweight geopolymer concrete using palm oil biowastes. *Aust. J. Civ. Eng.* **2020**, *18*, 82–92. [CrossRef]
83. Pasupathy, K.; Cheema, D.S.; Sanjayan, J. Durability performance of fly ash-based geopolymer concrete buried in saline environment for 10 years. *Constr. Build. Mater.* **2021**, *281*, 122596. [CrossRef]
84. Zhou, B.; Wang, L.; Ma, G.; Zhao, X.; Zhao, X. Preparation and properties of bio-geopolymer composites with waste cotton stalk materials. *J. Clean. Prod.* **2019**, *245*, 118842. [CrossRef]
85. Ribeiro, M.G.S.; Ribeiro, M.G.S.; Keane, P.F.; Sardela, M.R.; Kriven, W.M.; Ribeiro, R.A.S. Acid resistance of metakaolin-based, bamboo fiber geopolymer composites. *Constr. Build. Mater.* **2021**, *302*, 124194. [CrossRef]
86. Trindade, A.C.C.; Silva, F.A.; Alcamand, H.A.; Borges, P.H.R. On the durability behavior of natural fiber reinforced geopolymer. In Proceedings of the 41st International Conference on Advanced Ceramics and Composites: Ceramic Engineering and Science Proceedings, Daytona Beach, FL, USA, 22–27 January 2017; The American Ceramic Society: Columbus, OH, USA, 2017; Volume 38.
87. Nkwaju, R.Y.; Djobo, J.N.Y.; Nouping, J.N.F.; Huisken, P.W.M.; Deutou, J.G.N.; Courard, L. Iron-rich laterite-bagasse fibers based geopolymer composite: Mechanical, durability and insulating properties. *Appl. Clay Sci.* **2019**, *183*, 105333. [CrossRef]
88. Santos, G.Z.B.D.; Oliveira, D.P.D.; Filho, J.D.A.M.; Silva, N.M.D. Sustainable geopolymer composite reinforced with sisal fiber: Durability to wetting and drying cycles. *J. Build. Eng.* **2021**, *43*, 102568. [CrossRef]
89. Canpolat, O.; Arslan, A.A.; Uysal, M.; Ylmaz, A.; Aygrmez, Y. Influence of wetting-drying curing system on the performance of fiberreinforced metakaolin-based geopolymer composites. *Constr. Build. Mater.* **2019**, *225*, 909–926.
90. Asante, B.; Schmidt, G.; Teixeira, R.; Krause, A.; Junior, H.S. Influence of wood pretreatment and fly ash particle size on the performance of geopolymer wood composite. *Eur. J. Wood Wood Prod.* **2021**, *79*, 597–609. [CrossRef]
91. Alomayri, T.; Vickers, L.; Shaikh, F.U.A.; Low, I.M. Mechanical properties of cotton fabric reinforced geopolymer composites at 200–1000 °C. *J. Adv. Ceram.* **2014**, *3*, 184–193. [CrossRef]
92. Alomayri, T.; Shaikh, F.U.A.; Low, I.M. Thermal and mechanical properties of cotton fabric-reinforced geopolymer composites. *J. Mater. Sci.* **2013**, *48*, 6746–6752. [CrossRef]
93. Amalia, N.S.; Haris, A.; Subaer. Physico-mechanics properties of hybrid composite geopolymers-pineapple leaf fiber (PLF). In Proceedings of the AIP Conference Proceedings, Ho Chi Minh, Vietnam, 29 April 2018.
94. Xuan, X.; Tan, J.; Huang, Y.; Xie, M.; Zheng, G. Preparation and freeze resistance of geopolymer-bagasse fiber composites. *Bull. Chin. Ceram. Soc.* **2020**, *39*, 532–536.

95. Sumesh, M.; Alengaram, U.J.; Jumaat, M.Z.; Mo, K.H.; Alnahhal, M.F. Incorporation of nano-materials in cement composite and geopolymer based paste and mortar-A review. *Constr. Build. Mater.* **2017**, *148*, 62–84. [CrossRef]
96. Assaedi, H.; Alomayri, T.; Shaikh, F.; Low, I.-M. Influence of nano silica particles on durability of flax fabric reinforced geopolymer composites. *Materials* **2019**, *12*, 1459. [CrossRef]
97. Saulo, R.F.; Neven, U.; Keoma, D.C.S.; Silva, L.E. Effect of microcrystalline cellulose on geopolymer and Portland cement pastes mechanical performance. *Constr. Build. Mater.* **2021**, *288*, 123053.
98. Cut, R.; Sri, A.; Taufiq, S. Current development of geopolymer cement with nanosilica and cellulose nanocrystals. *J. Phys. Conf. Ser.* **2021**, *1783*, 012056.
99. Rahman, A.S.; Shah, C.; Gupta, N. Simultaneous effects of rice husk silica and silicon carbide whiskers on the mechanical properties and morphology of sodium geopolymer. *J. Compos. Mater.* **2020**, *54*, 4611–4620. [CrossRef]
100. Assaedi, H.; Shaikh, F.U.A.; Low, I.M. Characterizations of flax fabric reinforced nanoclay-geopolymer composites. *Compos. Part B Eng.* **2016**, *95*, 412–422. [CrossRef]
101. Rahmawati, C.; Aprilia, S.; Saidi, T.; Aulia, T.B.; Ahmad, I. Preparation and characterization of cellulose nanocrystals from Typha sp. as a reinforcing agent. *J. Nat. Fibers* **2021**, *18*, 1–14. [CrossRef]
102. Lazorenko, G.; Kasprzhitskii, A.; Yavna, V. Effect of pre-treatment of flax tows on mechanical properties and microstructure of natural fiber reinforced geopolymer composites. *Environ. Technol. Innov.* **2020**, *20*, 101105. [CrossRef]
103. Maichin, P.; Suwan, T.; Jitsangiam, P. Hemp fiber reinforced geopolymer composites: Effects of NaOH concentration on fiber pre-treatment process. *Key Eng. Mater.* **2020**, *841*, 166–170. [CrossRef]
104. Maichin, P.; Suwan, T.; Jitsangiam, P. Effect of self-treatment process on properties of natural fiber-reinforced geopolymer composites. *Mater. Manuf. Processes* **2020**, *35*, 1120–1128. [CrossRef]
105. Georgy, L.; Anton, K.; Alexander, K.; Vasilii, M.; Victor, Y. Sustainable geopolymer composites reinforced with flax tows. *Ceram. Int.* **2020**, *46*, 12870–12875.
106. Roy, M.; Janne, N.; Michael, P. Chemical treatment of waste abaca for natural fiber-reinforced geopolymer composite. *Materials* **2017**, *10*, 579.
107. Huang, Y.; Tan, J.; Xuan, X.; Liu, L.; Xie, M.; Liu, H.; Yu, S.; Zheng, G. Study on untreated and alkali treated rice straw reinforced geopolymer composites. *Mater. Chem. Phys.* **2021**, *262*, 124304. [CrossRef]
108. Workiye, A.; Woldsenbet, E. Development of maize stalk cellulose fiber reinforced calcined kaolinite clay geopolymer composite. *Proc. Eng. Technol. Innov.* **2020**, *16*, 30–38. [CrossRef]
109. Ribeiro, R.A.S.; Ribeiro, M.G.S.; Sankar, K.; Kriven, W.M. Geopolymer-bamboo composite–A novel sustainable construction material. *Constr. Build. Mater.* **2016**, *123*, 501–507. [CrossRef]

molecules

MDPI

Article

Pressurized Solvent Extraction of *Paulownia* Bark Phenolics

Paula Rodríguez-Seoane [1], Beatriz Díaz-Reinoso [2] and Herminia Domínguez [1,*]

[1] Chemical Engineering Department, Universidade de Vigo, Edificio Politecnico, As Lagoas, 32004 Ourense, Spain; paurodriguez@uvigo.es

[2] CITI (Center of Research, Transfer and Innovation), Universidade de Vigo, Parque Tecnoloxico de Galicia, Rua Galicia N° 2, 32900 Ourense, Spain; bdreinoso@uvigo.es

* Correspondence: herminia@uvigo.es

Abstract: *Paulownia* bark is mostly utilized jointly with wood, but the possibility of a separate valorization through the pressurized extraction of bark bioactives has been assessed. Subcritical water extraction and supercritical CO_2 extraction are green technologies allowing shorter times than conventional solvent extraction under atmospheric shaken conditions. Subcritical water extraction was carried out at temperatures ranging from 140 to 240 °C and supercritical CO_2 extraction was performed at different pressures (10, 20 and 30 MPa), temperatures (35, 45 and 55 °C) and ethanol concentrations (0, 10 and 15% (*w*/*w*)). Subcritical water extraction under a non-isothermal operation during heating up to 160 °C (19 min) provided extraction yields up to 30%, and the extracts contained up to 7% total phenolics with an ABTS (2,2′-azino-bis(3-ethylbenzothiazoline-6-sulfonic acid)) radical scavenging capacity equivalent to 35% the activity of Trolox, whereas at 240 °C, the yield decreased to 20%, but the phenolic content reached 21%, and the antiradical activity was equivalent to 85% of Trolox. Supercritical CO_2 extraction at 30 MPa, 45 °C and 30 min reached a global yield of 2% after 180 min of extraction, but the product showed very low antiradical capacity. Gallic acid, vanillic acid, vanillin and apigenin were the major phenolic compounds found in the extracts.

Keywords: *Paulownia*; subcritical water; supercritical CO_2; antioxidants; valorization

Citation: Rodríguez-Seoane, P.; Díaz-Reinoso, B.; Domínguez, H. Pressurized Solvent Extraction of *Paulownia* Bark Phenolics. *Molecules* **2022**, *27*, 254. https://doi.org/10.3390/molecules27010254

Academic Editors: Alejandro Rodriguez Pascual, Eduardo Espinosa Víctor and Carlos Martín

Received: 16 December 2021
Accepted: 29 December 2021
Published: 31 December 2021

1. Introduction

The genus *Paulownia* has been used as a folk remedy, especially in traditional Chinese medicine, and more recently has been considered an interesting source of secondary metabolites, such as flavonoids, lignans, phenolic acids and terpenoids [1]. Different parts of the tree (bark, fruit, xylem, and leaves) have shown therapeutic properties, mainly due to antioxidant, anti-inflammatory, antimicrobial and anticancer activities [2–7].

It has been suggested that the phenolic components with antioxidant properties may be responsible for the observed activity in traditionally used bark extracts [5,6,8–10], and they can also be used as chemotaxonomic markers [6]. Antioxidants can protect from the damage caused by free radicals and can protect lipid-containing food and cosmetic stuffs from oxidation. Among other mechanisms, antioxidants can exert their effect by scavenging free radicals. Therefore, the search for natural compounds with antiradical properties can provide alternative efficient antioxidants, which could offer additional biological properties and could be used as therapeutics.

Different extraction and purification strategies have been proposed to selectively recover *P. tomentosa* bark bioactives. Solvent fractionation of a crude ethanolic extract using a successive liquid–liquid extraction in *n*-hexane, dichloromethane, ethyl acetate and *n*-butanol has been proposed to obtain compounds with antioxidant properties [10]. Ethyl acetate provided the most active radical scavenging fraction and contained glucodistylin, luteolin, ellagic acid, cistanoside F, campneoside II, isocampneoside II, verbascoside and isoverbascoside [10]. Flavonoids (such as naringenin and quercetin), phenolic acids, (such as cinnamic acid and gallic acid) and phenylpropanoid glycosides (such as cistanoside F, acteoside, isoacteoside, campneoside II and isocampneoside II) were chromatographically isolated from the *n*-butanol soluble fraction [6]. Apigenin derivatives have been identified in acetone extracts [3]. A methanol extract of *P. tomentosa* stem bark, with phenylethanoid glycosides verbascoside and isoverbascoside as the predominant compounds, showed anti-inflammatory effects in RAW 264.7 macrophages and in an acute lung injury murine model [7]. A methanolic *P. tomentosa* bark extract and the fractions obtained by sequential liquid–liquid extraction with *n*-hexane, chloroform and water and chromatographic separation showed antiviral activity [4].

However, the use of greener solvents is preferred to obtain natural bioactives. In addition, obtaining wood and bark extractives is highly interesting to propose the valorization of important chemicals for future biorefinery valorization schemes [11]. In this context, the application of more selective and environmentally friendly extraction techniques would offer additional value of the final products [12]. Pressurized solvent extraction has advantages in relation to the enhanced efficiency in terms of extraction time [13]. Subcritical water extraction (SWE) is an eco-friendly technology performed at temperatures in the range of 100–374 °C, corresponding with the boiling point and critical point of water, respectively. Pressure is switched in the range of 2–15 MPa to maintain the water in its liquid state [14,15]. Under these conditions, water presents unique properties (low dielectric constant and high ionic product) that favor the extraction of apolar compounds and the partial breakage of the polysaccharidic structures [16]. Supercritical fluid extraction (SCFE) is a green and clean (GLEAN) extractive technique that operates with a fluid at temperatures and pressures over the critical point, a state without distinct liquid and gaseous phases, while simultaneously possessing properties of both phases [17,18]. The most frequent solvent used in this extraction method is carbon dioxide (CO_2), which presents low critical conditions (7.4 MPa, 32 °C). CO_2 is relatively highly available at relatively low cost, it is non-toxic, non-flammable and a good solvent for compounds with low volatility and polarity. It can be highly selective by adequately selecting operational conditions. Recent studies, using wood or bark as the starting material, employed the sc-CO_2 extraction to the enrichment of different compounds. In this context, Barbini et al. [19] extracted pine bark using subcritical and supercritical CO_2. The first step allowed a selective enrichment of the extract with unsaturated fatty acids despite the low extraction yields obtained. After removing most of the lipophilic compounds, sc-CO_2 using ethanol as a polar cosolvent facilitated the recovery of more polar compounds, such as phenolics.

The aim of this work is to explore the potential of pressurized solvent extraction with subcritical water and with supercritical carbon dioxide to obtain extracts from *Paulownia elongata x fortune* with antiradical properties.

2. Results and Discussion

2.1. Composition

The proximal composition of *Paulownia bark* is shown in Table 1. The saccharidic fraction, accounting for almost 50% of the dry weight, was the most abundant, with glucose being the major component. Ethanol extractives accounted for 21%, whereas protein and ash represented a lower fraction. Qi et al. [20] reported a similar ash content (2.8%) for *Paulownia tomentosa* bark, with 80% volatile matter content and 17% fixed carbon content.

2.2. Subcritical Water Extraction

The influence of the final extraction temperature on the extraction yields is shown in Figure 1. A maximum of 33% was observed as a plateau region at 160–200 °C, with a steady decrease at higher values. However, the phenolic content showed a continuous increase in the range studied, more marked at the highest tested temperature, accounting for up to 21% of the dried extract. Lower solubilization can be attained with organic solvents, i.e., in three stages with methanol during 24 h [7].

Table 1. Composition of *Paulownia elongata x fortunei* bark.

Component	Content (%, d.w.)
Ash	3.00 ± 0.17
Ethanol extractives	21.32 ± 0.28
Protein	4.95 ± 0.51
Acid insoluble residue	25.62 ± 0.79
Carbohydrates	
Galacturonic acid	5.12 ± 0.13
Glucose	20.63 ± 0.15
Xylose	5.34 ± 0.30
Mannose	0.43 ± 0.11
Galactose	3.81 ± 0.09
Rhamnose	1.93 ± 0.16
Arabinose	1.83 ± 0.02

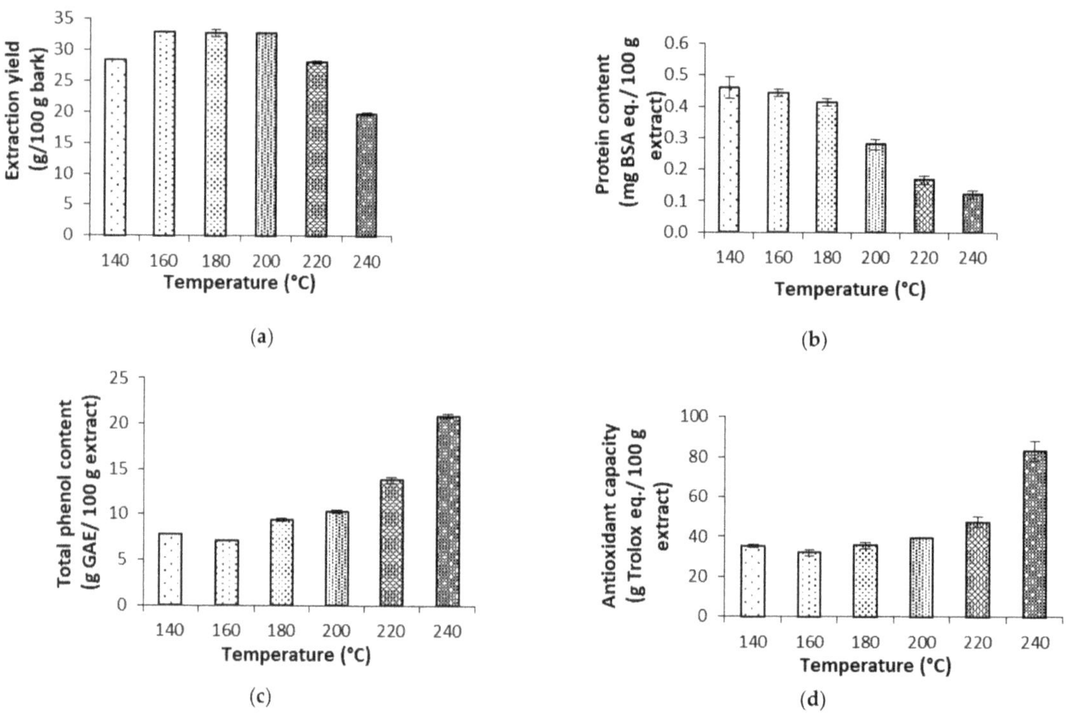

Figure 1. Influence of the final subcritical water extraction temperature on total extraction yield (**a**), protein content (**b**), the total phenolic content (**c**) and the antiradical capacity against ABTS (2,2′-azino-bis-(3-ethylbenzothiazoline-6-sulfonic acid)), (**d**), for the aqueous extracts from bark.

According to data from Table 2, the saccharidic fraction was mostly found as oligomers, with glucose, arabinose and galactose as the most abundant. The operation temperature affected the different compounds differently; glucose in oligomeric units decreased with temperature, and the monomers only decreased at the highest values; galactose remained in oligomers up to 200 °C, but the monomer concentration decreased with temperature; and both xylose and arabinose content in oligomeric form increased with temperature, but the monomers were less affected. As expected, the organic acid concentration increased with operation temperature. Maximal extraction yields of 30% in the range of 160–200 °C were attained in subcritical water extraction, and at higher treatment temperatures, decreased values were observed until 20%. However, the phenolic content of the extracts increased more markedly with increasing temperatures, reaching more than 20%, and the antiradical capacity showed a similar trend. The ABTS radical scavenging capacity increased from 35% to 85% of the activity of Trolox. The protein content of the extract was under 0.5% and lowered with increasing extraction temperature.

Table 2. Influence of the final extraction temperature on monomers (**a**) and oligomers (**b**) from saccharidic fraction of subcritical water extraction of *Paulownia bark*.

(a)

Monomers (g/100 g Extract)	Temperature (°C)					
	140	160	180	200	220	240
Trehalose	7.5	6.07	5.29	2.93	-	-
Glucuronic acid	-	-	-	-	-	-
Galacturonic acid	-	-	-	-	-	-
Glucose	11.63	10.1	12.72	11.36	3.08	2
Xylose	0.7	0.51	0.21	0.25	1.29	0.87
Galactose	8.36	1.91	1.57	1.9	1.57	0.6
Rhamnose	0.22	0.15	0.65	2.55	1.29	0.5
Arabinose	-	0.08	0.84	3.2	1.42	0.92
Mannose	2.34	2.32	4.12	2.95	0.92	0.86
Mannitol	0.14	0.12	0.12	0.14	0.16	0.59
Formic acid	2.72	-	0.8	3.61	1.99	15.16
Acetic acid	-	0.27	0.61	2.36	8.2	20.57

(b)

Oligomers (g/100 g Extract)	Temperature (°C)					
	140	160	180	200	220	240
Trehalose	-	-	-	-	-	-
Glucuronic acid	0.88	1.17	1.39	1	1.06	1.47
Galacturonic acid	8.41	11.57	5.65	1.48	0.05	0.07
Glucose	20.91	18.25	15.9	14.73	15.19	11.37
Xylose	0.7	0.37	1.26	9.07	16.15	1.25
Galactose	8.36	8.04	9.13	9.2	5.56	0.83
Rhamnose	6.72	6.69	6.83	5.13	1.76	-
Arabinose	3.24	13.01	17.74	13.28	1.8	-
Mannose	-	-	-	-	2.73	2.03
Formic acid	2.72	2.33	3.35	2.29	7.34	-
Acetic acid	0.94	0.88	1.34	2.97	3.88	-

2.3. Supercritical Fluid Extraction

Figure 2 shows the influence of pressure and temperature on the extraction yield and the antiradical properties of the sc-CO_2 extracts from *P. elongata x fortunei* bark. As can be seen in Figure 2a, the use of higher pressures led to an improvement of extraction yield as a consequence of the increase in solubility caused by the supercritical CO_2 density enhacement with pressure. Maximum extraction yields up to 0.8% were obtained when operating at the highest pressure. On the other hand, the effect of temperature on the extraction yield was the result of the dual effect that an increase in temperature may cause on the parameters, affecting solubility, i.e., the decrease in CO_2 density and the increase in solute vapor pressure. In this work, the use of a lower temperature had a positive effect on the extraction yield during operation at the lower pressure, whereas at the higher pressure, a positive effect of temperature on the extraction yield was observed since the decrease in solvent density was compensated by the increase in the solute vapor pressure. This behavior, despite the important deviations observed in the experimental data at 30 MPa, suggests that the crossover point of the different isotherms could be in the range of 15–25 MPa. The ABTS radical scavenging test was selected for its simplicity, and since it is widely used, this assay can be valid for a rapid comparison of the bioactive potential of *Paulownia bark* extractives. The activity against ABTS radical was very low, under 2.5% Trolox equivalents, and no clear effect of temperature at the studied pressure conditions was found (Figure 2b).

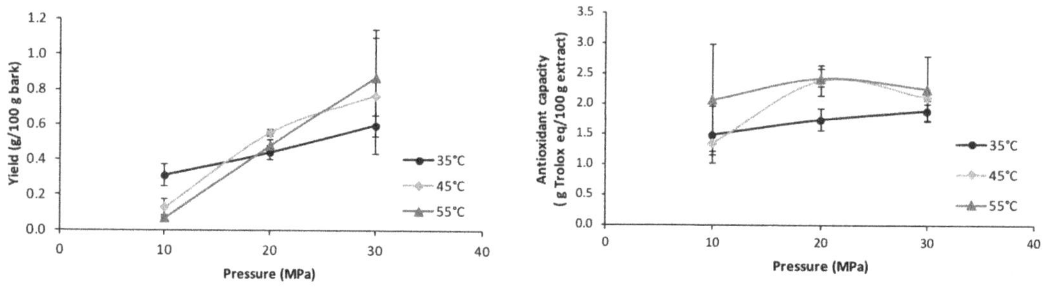

(**a**) (**b**)

Figure 2. Influence of the extraction pressure and temperature on the extraction yield (**a**), and the antiradical capacity determined as ABTS (**b**) of *Paulownia elongata x fortunei* bark samples extracted with pure sc-CO_2 for 30 min at the temperatures of 35 °C (dark gray), 45 °C (light gray) and 55 °C (intermediate gray).

The effect of the addition of ethanol as a polar modifier on the extraction yield and antioxidant activity of the extracts was evaluated (Figure 3). The use of ethanol as a solvent is highly frequent for its greener and renewable character, and in some cases offered higher yield than others, such as ethylacetate or water [12]. This influence was studied at 10 MPa and 35 °C and at 30 MPa and 55 °C, selected as the highest and lowest values of pressure operation, at temperatures providing higher yields and an extraction time of 30 min. At 10 MPa and 35 °C, the extraction yield increased significantly up to 0.8% with 10% ethanol as a modifier, whereas at 30 MPa and 55 °C, no significant effect of ethanol content on yield was observed after 30 min of extraction. Regarding the antioxidant capacity against the ABTS radical, a gradual increase with ethanol content was observed. The extraction yields were in the range of those reported for solvent fractions from a 95% ethanolic extract obtained at room temperature during 5 days, yielding 0.79% in the ethyl acetate fraction, which was also the most active radical scavenger. Other fractions represented lower yields—0.12% for the *n*-hexane and 0.10% for CH_2Cl_2, 0.01% for *n*-BuOH, whereas the water fraction represented 6.28% of the sample [10].

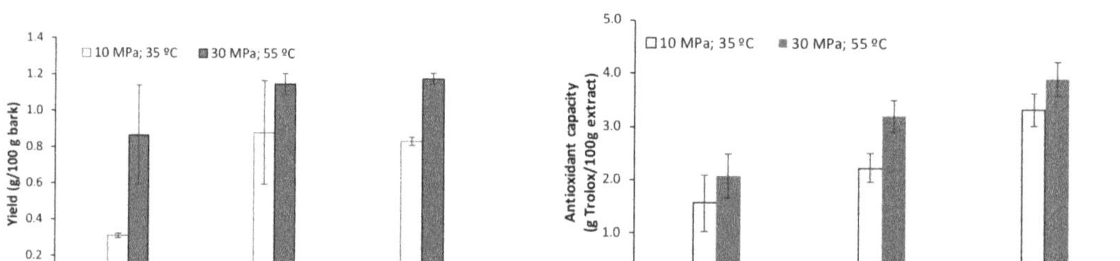

(**a**) (**b**)

Figure 3. Influence of the ethanol concentration as a modifier on the extraction yield (**a**) and the antiradical properties, determined and TEAC value for ABTS scavenging capacity (**b**) of *Paulownia elongata x fortunei* bark at 10 MPa, 35 °C (white); and 30 MPa, 55 °C (gray) for 30 min.

The extraction yield was higher than that of the fractions in dichloromethane, ethyl acetate or *n*-butanol fractions from a 95% ethanolic extract obtained at room temperature during 5 days from *P. tomentosa* [10]. The major compounds present in extracts were gallic acid, vanillic acid, vanillin and apigenin.

Supercritical fluid extraction of phenolic compounds from tree barks has been frequently reported and performed similarly to the present study. Bukhanko et al. [11] reported that supercritical carbon dioxide provided 2% yield from Norway spruce bark, a value slightly lower than from needles; the values were more than double using Soxhlet extraction. This technique proved suitable for the extraction of *Eucalyptus globulus* bark triterpenic acids [21] and *Pinus brutia* bark extractives 20 MPa, 60 °C and 3% ethanol [22]. Supercritical solvent extracts from *Eucalyptus globulus* bark obtained at 30 MPa, 70 °C, 20% ethanol and 10 g of CO_2/min, with a yield of 0.48%, contained 6% gallic acid equivalents. Higher phenolic content could be found in ethanol:water extracts (15.9%) and with methanol:water (40.7%) [12].

Kinetic experiments were carried out with CO_2 at the pressure conditions that provided the highest extraction yields and at an intermediate temperature (30 MPa and 45 °C). The total extraction yield achieved after 330 min of extraction was 2%. The kinetic curves and their modeling offer information on the mechanism of the extraction process and on further scaling up. Among all the models proposed in the literature for fitting kinetic data, the BIC model [23] is the most widely used. Both experimental data and the fitted curve obtained with the broken-intact cell (BIC) model are shown in Figure 4. The parameters necessary for this model are presented in Table 3. The solubility of the solute (Ys) was initially estimated as the slope of the first part of the curve and then fitted to the experimental data along with the adjustable parameters of the model. The initial extractable solutes mass ratio in the raw material (x_0) was fixed as the asymptotic value at infinite time.

The adjustable parameters of the model are also presented in Table 3. The experimental kinetic curve obtained can be divided in three periods controlled by different mass transfer mechanisms: (1) a constant extraction rate (CER), characterized by a fast extraction rate where the solutes present on the particle surface are transferred by convection to the solvent; (2) a falling extraction rate (FER), or transition period where the easily accessible solutes on the particle surface begin to be depleted and resistance to mass transfer at the fluid–solid interface starts to become significant; and (3) a diffusion-controlled period (DC), a slow extraction rate period where the mass transfer of the less accessible solutes is governed exclusively by diffusion. According to the results obtained by the model, the duration of the CER period was of 53 min, whereas the diffusion-controlled period (DC period) was not achieved until 120 min. The mass transfer coefficient in the solid phase, k_xa, was three orders of magnitude lower than the mass transfer coefficient in the fluid phase, k_ya, evidencing a strong limitation of solute mass transfer in the solid phase.

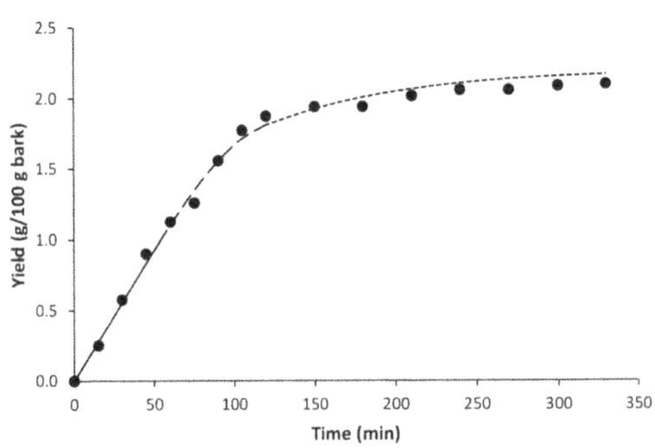

Figure 4. Kinetic extraction curve for the supercritical extraction of *Paulownia bark* at 30 MPa and 45 °C. Lines represents the BIC model: (—) CER period; (– – –) FER period; (---) DC period.

Table 3. BIC Model Parameters Obtained for the sc-CO_2 Extraction of *Paulownia bark* at 30 MPa and 45 °C.

Parameter	Value
k_xa (min^{-1})	6.22×10^{-3}
k_ya (min^{-1})	1.95
r	0.74
Ys (kg/kg)	1.09×10^{-4}
xo	0.022
t_{CER} (min)	53.3
t_{FER} (min)	117.7
AARD (%)	7.09

k_xa: Mass transfer coefficient in the solid phase; k_ya: mass transfer coefficient in the fluid phase; r: fraction of broken cells. Ys: solubility of the extract in the solvent; xo; initial solute mass ratio in the raw material; t_{CER}; extration time at the end of the CER period; t_{FER}: extraction time at the end of the FER period; AARD: absolute average relative deviation.

3. Materials and Methods

3.1. Raw Material

The hybrid *Paulownia elongata x fortunei* was harvested by Maderas Álvarez Oroza in Nois (Foz, Lugo, Galicia, Spain). Bark was collected in June 2017 and transported to the laboratory, where it was air dried at room temperature for 15 days, ground and stored in a dark and fresh place. Before drying, the average moisture content of bark was 72.13%.

3.2. Extraction

3.2.1. Subcritical Water Extraction

Subcritical water extraction under non-isothermal conditions was performed. Milled bark was mixed with distilled water at a liquid–solid mass ratio 7:1 (*w*/*w*). The reaction was carried out in a pressurized reactor (Parr Instrument Company, Moline, IL, USA) with 600 mL vessel capacity under constant stirring (150 rpm). Final heating temperatures in the range of 140–240 °C were studied, based on previous preliminary experiments. Once the selected temperature was reached, the system was cooled with water through a stainless steel coil located inside the vessel. Then, solid and liquid phases were separated by filtration. Operating with the highest heating rate, the temperature profiles at the operation temperatures, manually monitored, are shown in Figure 5.

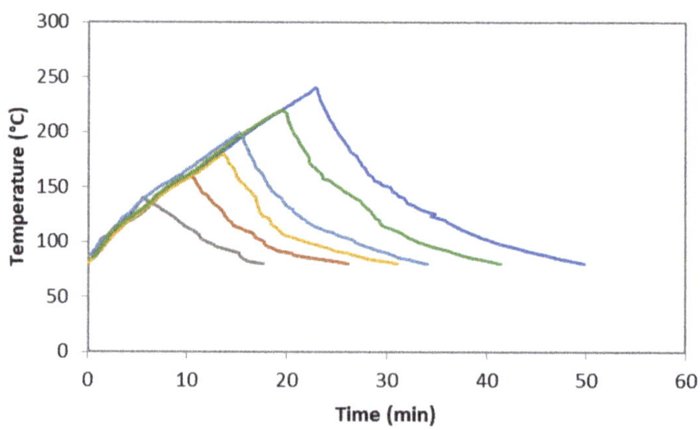

Figure 5. Heating profiles during non-isothermal treatments of *Paulownia bark*.

3.2.2. Supercritical Carbon Dioxide Extraction

Supercritical CO_2 was performed in a supercritical fluid equipment (Thar Process, Inc., Pittsburg, PA, USA) with a 1000 mL extraction cell operating with a solvent mass flow of CO_2 (99.99%, purity) fixed at 25 g/min. The amount of bark introduced into the reactor cell was 10 g with a moisture content of about 9–10%. The extractor was filled with glass balls until occupying the existing space. The study of influence of pressure and temperature was proposed. Experiments were performed at 10, 20 and 30 MPa and 35, 45 and 55 °C during 30 min. These conditions have been selected based on literature information [12,22] and preliminary experiments. Furthermore, the influence of absolute ethanol in concentrations of 10% and 15% (*w*/*w*) as a modifier was studied. Dynamic extractions were performed when the experimental conditions in the extractor were achieved. Extracts were recovered in a separator vessel with absolute ethanol.

Additionally, a kinetic curve was constructed during 330 min at 30 MPa and 45 °C, collecting the extract from the separator at intervals of 15–30 min. The modelling of the experimental data was carried out using the broken-intact cell model (BIC model) proposed by Sovová [23].

3.3. Analytical Methods

3.3.1. Raw Material Characterization

The characterization of raw material was performed. For moisture determination (ISO 638 method), bark was dried in an oven at 105 °C for 24–48 h to reach a constant weight. Ash content was determined (ISO 776 method) after calcination at 575 °C for 6 h. The ethanol extractive content was gravimetrically determined after Soxhlet extraction. Total nitrogen was measured by elemental analysis (FlashEA 1112 Elemental analyzer, Thermo, Waltham, MA, USA) using 130 mL/min of He as a carrier gas and 100 mL/min as a reference gas. The oxygen flow was 250 mL/min, and the temperatures of the oxidation and reduction ovens were 900 °C and 680 °C, respectively. Protein content was determined by converting total nitrogen content using the factor 6.25.

Acid hydrolysis in two stages was performed before the chromatographic quantification of carbohydrates. In the first stage, the sample was digested with 72% sulfuric acid in a bath water at 30 °C for 1 h to break the polysaccharides. Then, the reaction was stopped by adding water. The second stage was performed with 4% sulfuric acid in an autoclave for 40 min in order to obtain monosaccharides. The liquid phase obtained was filtered through 0.45 μm cellulose acetate membranes and analyzed by HPLC using a refractive index detector (Model 1200, Agilent Technologies, Santa Clara, CA, USA). An Aminex HPX-87H (430150) column at 50 °C was used with 0.003 M H_2SO_4 at 0.6 mL/min as the mobile phase and an Aminex HPX-87P (424351) column at 80 °C operating with ultra-pure water at 0.4 mL/min as the mobile phase were used to determine different sugars. External standards for all the compounds were used.

3.3.2. Extraction Yield

Extraction yield was gravimetrically determined. Aliquots of extract of known weight were dried in an oven at 105 °C until reaching a constant weight. Assays were carried out by triplicate.

3.3.3. Total Phenolic Content and Antioxidant Profile

Total phenolic content (TPC) was determined by the Folin–Ciocalteu method proposed by Singleton and Rossi [24]. The results were expressed as grams of gallic acid equivalents (GAE). In this method, 0.5 mL extract or standard (gallic acid) and 3.75 mL distilled water were mixed. Then, 0.25 mL Folin–Ciocalteu reagent diluted 1:1 (*v*/*v*) and 0.50 mL sodium carbonate solution (10%, *w*/*v*) were also incorporated to the mixture. After one hour in the darkness, the absorbance of samples was measured at 765 nm. All assays were carried out by triplicate.

The phenolic compounds found in the extracts were analyzed using an Agilent HPLC 1100 instrument equipped with a Waters Spherisorb ODS-2 column (5 mm, 250 mm × 4.6 mm) and diode-array detector, operating at 30 °C, using 20 mL injection volume at 1 mL/min. Solvent A (acetonitrile/5% (*v*/*v*) formic acid in water, 10:90) and solvent B (acetonitrile/5% (*v*/*v*) formic acid in water, 90:10) were used in a non-linear gradient: 0 min, 100% A; 40 min, 85% A, 15% B; 45 min, 100% B; 55 min, 100% B; 60 min, 100% A; 65 min, 100% A. Identification was carried out by comparing the retention time and UV-visible spectral data with those of authentic compounds.

ABTS radical scavenging assay or a Trolox equivalent antioxidant capacity (TEAC) method was proposed by Re et al. [25] to obtain the ABTS (2,2′-azino-bis(3-ethylbenzothiazo line-6-sulfonic acid)) diammonium salt radical cation scavenging capacity of the above samples. To make 10 µL of extract or 6-hydroxy-2,5,7,8-tetramethylchroman-2-carboxylic acid standard (Trolox), 1.0 mL of diluted $ABTS^{\cdot+}$ solution was added. The $ABTS^{\bullet+}$ solution was obtained by the addition of phosphate buffer saline (PBS) (pH 7.4) to a 7 mM ABTS stock solution until reaching an absorbance of 0.70 at 734 nm. The mixture of the extracts with the $ABTS^{\bullet+}$ solution was measured at 734 nm after 6 min of incubation at 30 °C. A calibration curve with Trolox solutions was prepared following the same procedure described above, and the results were expressed as milligrams of Trolox equivalents. All assays were carried out by triplicate.

3.3.4. Saccharidic Fraction of Hydrothermal Extracts

The saccharidic fraction of hydrothermal aqueous extracts were chromatographically analyzed. Monomers were obtained from direct analysis of the extracts, and before the oligosaccharide determination, a posthydrolysis process was necessary. Aliquots of the liquid extract were mixed with sulfuric acid to a final concentration of 4% and then were autoclaved for 20 min at 121 °C. The content of oligomers was determined by the difference between the content of monomers in the subcritical water extracts and in the posthydrolysis liquid phase. All samples were filtered through a cellulose acetate filter and analyzed by HPLC, following the same method used for carbohydrate determination described in Section 3.3.1.

3.4. Statistical Analysis

Significant differences between results were calculated by an analysis of variance (ANOVA) using the MINITAB 19 (Minitab Inc., State College, PA, USA) software. The significant differences ($p < 0.05$) were evaluated by Tukey's test. Mean values and their standard deviations were calculated and presented on the figure as error bars.

4. Conclusions

Two pressurized extraction methods were applied to *Paulownia bark*. Subcritical water extraction exhibited better results in terms of extraction yield and antiradical properties than sc-CO_2 extraction. The maximum extraction yield was attained at intermediate temperatures of hydrothermal extraction (33%), whereas the total phenolic content and the antioxidant capacity rose with temperature and reached the highest values at 240 °C, exhibiting 20 g GAE/100 g extract and 85 g Trolox equivalents/100 g extract, respectively. Attending to the saccharidic fraction, the monomers of glucose and the oligomers of glucose, galactose and arabinose were the most abundant. On the other hand, supercritical fluid extraction yield (0.8%) was improved with pressure, being maximum at 30 MPa and 55 °C. The addition of ethanol as a polar modifier enhanced the extraction yield at 10 MPa and 35 °C. This positive influence of ethanol was also observed in the antioxidant capacity against the ABTS radical under both studied conditions. Further studies on the detailed characterization and potential applications are ongoing.

Author Contributions: Conceptualization, P.R.-S., B.D.-R. and H.D.; methodology, P.R.-S. and B.D.-R.; validation, P.R.-S. and B.D.-R.; investigation, P.R.-S. and B.D.-R.; resources, H.D.; data curation, P.R.-S., B.D.-R. and H.D.; writing—original draft preparation, P.R.-S., B.D.-R. and H.D.; writing—review and editing, P.R.-S., B.D.-R. and H.D.; supervision, B.D.-R. and H.D.; project administration, H.D.; funding acquisition, H.D. All authors have read and agreed to the published version of the manuscript.

Funding: This work was funded by the Ministry of Economy, Industry and Competitiveness of Spain through the project CTM2015-68503-R. P.R.-S. thanks the Ministry of Economy, Industry and Competitiveness of Spain for her predoctoral grant (BES-2016-076840).

Institutional Review Board Statement: Not applicable.

Informed Consent Statement: Not applicable.

Data Availability Statement: Data are contained within the article.

Conflicts of Interest: The authors declare no conflict of interest.

Sample Availability: Samples of the raw material and products obtained are available from the authors.

References

1. Schneiderová, K.; Šmejkal, K. Phytochemical profile of Paulownia tomentosa (Thunb). Steud. *Phytochem. Rev.* **2015**, *14*, 799–833. [CrossRef]
2. He, T.; Vaidya, B.; Perry, Z.; Parajuli, P.; Joshee, N. Paulownia as a Medicinal Tree: Traditional Uses and Current Advances. *Eur. J. Med. Plants* **2016**, *14*, 1–15. [CrossRef]
3. Si, C.-L.; Wu, L.; Zhu, Z.-Y.; Kim, J.-K.; Kwon, D.-J.; Bae, Y.-S. Apigenin derivatives from *Paulownia tomentosa* Steud. var. tomentosa stem barks. *Holzforschung* **2009**, *63*, 440–442. [CrossRef]
4. Kang, K.H.; Huh, H.; Kim, K.; Lee, L. An antiviral furanoquinone from *Paulownia tomentosa* Steud. *Phyther. Res.* **1999**, *13*, 624–626. [CrossRef]
5. Cheng, C.-L.; Jia, X.-H.; Xiao, C.-M.; Tang, W.-Z. Paulownia C-geranylated flavonoids: Their structural variety, biological activity and application prospects. *Phytochem. Rev.* **2019**, *18*, 549–570. [CrossRef] [PubMed]
6. Si, C.L.; Lu, Y.Y.; Qin, P.P.; Sun, R.C.; Ni, Y.H. Phenolic extractives with chemotaxonomic significance from the bark of Paulownia tomentosa var. Tomentosa. *BioResources* **2011**, *6*, 5086–5098. [CrossRef]
7. Lee, J.-W.; Seo, K.-H.; Ryu, H.W.; Yuk, H.J.; Park, H.A.; Lim, Y.; Ahn, K.-S.; Oh, S.-R. Anti-inflammatory effect of stem bark of *Paulownia tomentosa* Steud. in lipopolysaccharide (LPS)-stimulated RAW264.7 macrophages and LPS-induced murine model of acute lung injury. *J. Ethnopharmacol.* **2018**, *210*, 23–30. [CrossRef] [PubMed]
8. Sticher, O.; Lahloub, M.F. Phenolic glycosides of *Paulownia tomentosa* bark. *Planta Med.* **1982**, *46*, 145–148. [CrossRef] [PubMed]
9. Kim, J.-K.; Si, C.-L.; Bae, Y.-S. Epimeric phenylpropanoid glycosides from inner bark of *Paulownia coreana* Uyeki. *Holzforschung* **2007**, *61*, 161–164. [CrossRef]
10. Si, C.-L.; Liu, S.-C.; Hu, H.-Y.; Jiang, J.-Z.; Yu, G.-J.; Ren, X.-D.; Xu, G.-H. Activity-guided screening of the antioxidants from *Paulownia tomentosa* var. Tomentosa bark. *BioResources* **2013**, *8*, 628–637. [CrossRef]
11. Bukhanko, N.; Attard, T.; Arshadi, M.; Eriksson, D.; Budarin, V.; Hunt, A.J.; Geladi, P.; Bergsten, U.; Clark, J. Extraction of cones, branches, needles and bark from Norway spruce (*Picea abies*) by supercritical carbon dioxide and soxhlet extractions techniques. *Ind. Crops Prod.* **2020**, *145*, 112096. [CrossRef]
12. Santos, S.A.O.; Villaverde, J.J.; Silva, C.M.; Neto, C.P.; Silvestre, A.J.D. Supercritical fluid extraction of phenolic compounds from *Eucalyptus globulus* Labill bark. *J. Supercrit. Fluids* **2012**, *71*, 71–79. [CrossRef]
13. Lachos-Perez, D.; Brown, A.B.; Mudhoo, A.; Martinez, J.; Timko, M.T.; Rostagno, M.A.; Forster-Carneiro, T. Applications of subcritical and supercritical water conditions for extraction, hydrolysis, gasification, and carbonization of biomass: A critical review. *Biofuel Res. J.* **2017**, *14*, 611–626. [CrossRef]
14. Munir, M.T.; Kheirkhah, H.; Baroutian, S.; Quek, S.Y.; Young, B.R. Subcritical water extraction of bioactive compounds from waste onion skin. *J. Clean. Prod.* **2018**, *183*, 487–494. [CrossRef]
15. Yan, Z.; Zhang, H.; Dzah, C.S.; Zhang, J.; Diao, C.; Ma, H.; Duan, Y. Subcritical water extraction, identification, antioxidant and antiproliferative activity of polyphenols from lotus seedpod. *Sep. Purif. Technol.* **2020**, *236*, R713–R715. [CrossRef]
16. Wiboonsirikul, J.; Adachi, S. Extraction of functional substances from agricultural products or by-products by subcritical water treatment. *Food Sci. Technol. Res.* **2008**, *14*, 319–328. [CrossRef]
17. Díaz-Reinoso, B.; Moure, A.; Domínguez, H.; Parajó, J.C. Supercritical CO_2 extraction and purification of compounds with antioxidant activity. *J. Agric. Food Chem.* **2006**, *54*, 2441–2469. [CrossRef]
18. Chai, Y.H.; Yusup, S.; Ruslan, M.S.H.; Chin, B.L.F. Supercritical fluid extraction and solubilization of *Carica papaya* Linn. leaves in ternary system with CO2 + ethanol solvents. *Chem. Eng. Res. Des.* **2020**, *156*, 31–42. [CrossRef]
19. Barbini, S.; Jaxel, J.; Karlström, K.; Rosenau, T.; Potthast, A. Multistage fractionation of pine bark by liquid and supercritical carbon dioxide. *Bioresour. Technol.* **2021**, *341*, 125862. [CrossRef] [PubMed]
20. Qi, Y.; Yang, C.; Hidayat, W.; Jang, J.H.; Kim, N.H. Solid bioenergy properties of *Paulownia tomentosa* grown in Korea. *J. Korean Wood Sci. Technol.* **2016**, *44*, 890–896. [CrossRef]

21. Domingues, R.M.A.; De Melo, M.M.R.; Neto, C.P.; Silvestre, A.J.D.; Silva, C.M. Measurement and modeling of supercritical fluid extraction curves of *Eucalyptus globulus* bark: Influence of the operating conditions upon yields and extract composition. *J. Supercrit. Fluids* **2012**, *72*, 176–185. [CrossRef]
22. Yesil-Celiktas, O.; Otto, F.; Gruener, S.; Parlar, H. Determination of Extractability of Pine Bark Using Supercritical CO_2 Extraction and Different Solvents: Optimization and Prediction. *J. Agric. Food Chem.* **2008**, *57*, 341–347. [CrossRef] [PubMed]
23. Sovová, H. Rate of the vegetable oil extraction with supercritical CO_2—I. Modelling of extraction curves. *Chem. Eng. Sci.* **1994**, *49*, 409–414. [CrossRef]
24. Singleton, V.L.; Rossi, J.A. Colorimetry of total phenolics with phosphomolybdic-phosphotungstic acid reagents. *Am. J. Enol. Vitic.* **1965**, *16*, 144–158.
25. Re, R.; Pellegrini, N.; Proteggente, A.; Pannala, A.; Yang, M.; Rice-Evans, C. Antioxidant activity applying an improved ABTS radical cation decolorization assay. *Free Radic. Biol. Med.* **1999**, *26*, 1231–1237. [CrossRef]

molecules

Article

Polish Varieties of Industrial Hemp and Their Utilisation in the Efficient Production of Lignocellulosic Ethanol

Aleksandra Wawro *, Jolanta Batog and Weronika Gieparda

Institute of Natural Fibres and Medicinal Plants, National Research Institute, Wojska Polskiego 71B, 60-630 Poznan, Poland; jolanta.batog@iwnirz.pl (J.B.); weronika.gieparda@iwnirz.pl (W.G.)
* Correspondence: aleksandra.wawro@iwnirz.pl; Tel.: +48-61-84-55-814

Abstract: Nowadays, more and more attention is paid to the development and the intensification of the use of renewable energy sources. Hemp might be an alternative plant for bioenergy production. In this paper, four varieties of Polish industrial hemp (Białobrzeskie, Tygra, Henola, and Rajan) were investigated in order to determine which of them are the most advantageous raw materials for the effective production of bioethanol. At the beginning, physical and chemical pretreatment of hemp biomass was carried out. It was found that the most effective is the alkaline treatment with 2% NaOH, and the biomasses of the two varieties were selected for next stages of research: Tygra and Rajan. Hemp biomass before and after pretreatment was analyzed by FTIR and SEM, which confirmed the effectiveness of the pretreatment. Next, an enzymatic hydrolysis process was carried out on the previously selected parameters using the response surface methodology. Subsequently, the two approaches were analyzed: separated hydrolysis and fermentation (SHF) and a simultaneous saccharification and fermentation (SSF) process. For Tygra biomass in the SHF process, the ethanol concentration was 10.5 $g \cdot L^{-1}$ (3.04 $m^3 \cdot ha^{-1}$), and for Rajan biomass at the SSF process, the ethanol concentration was 7.5 $g \cdot L^{-1}$ (2.23 $m^3 \cdot ha^{-1}$). In conclusion, the biomass of Polish varieties of hemp, i.e., Tygra and Rajan, was found to be an interesting and promising raw material for bioethanol production.

Keywords: hemp biomass; alkaline pretreatment; SEM; FTIR; response surface methodology; SHF; SSF; bioethanol

Citation: Wawro, A.; Batog, J.; Gieparda, W. Polish Varieties of Industrial Hemp and Their Utilisation in the Efficient Production of Lignocellulosic Ethanol. *Molecules* **2021**, *26*, 6467. https://doi.org/10.3390/molecules26216467

Academic Editors: Alejandro Rodriguez Pascual, Eduardo Espinosa Víctor and Carlos Martín

Received: 10 June 2021
Accepted: 21 October 2021
Published: 26 October 2021

1. Introduction

The European Union countries have been obliged to achieve a certain share of biofuels in transport and to take measures to reduce greenhouse gas emissions. It is, therefore, necessary to replace diesel and gasoline with biofuels which are produced from lignocellulosic raw materials and represent an advantageous option for the fuels currently in use due to their renewable nature and the emission of an acceptable quality exhaust gases. Currently, mainly three biofuels are produced: bioethanol, biodiesel, and biogas. According to the EU RED II directive, the contributions of advanced biofuels and biogas produced from raw materials listed in Annex IX, part A to this directive, including lignocellulosic feedstocks as a share of final energy consumption in the transport sector are expected to be at least: 0.2% in 2022, 1% in 2025, and 3.5% in 2030 [1]. The production of biofuels from plant biomass is innovative and contributes to the solution of the key issue in the production of biofuels for transport fuels.

In Poland, high expectations are associated with plant biomass due to a significant amount of waste, including that from agri-food sector and the available acreage of agricultural land that can be used for the cultivation of energy crops. In recent years, there has been an increase in the acreage of cultivated industrial hemp (*Cannabis sativa* L.) in Poland (over 1000 ha). The cultivation of hemp for seed purposes is intensively developed, and there is unused hemp biomass in the field, which can be a suitable raw material for the production of lignocellulosic ethanol. Hemp is an environmentally friendly plant characterized by a short vegetation period (3–4 months), a rapid growth up to 4 m in height, and a dry matter yield up to 15 Mg·ha^{-1}. These plants improve soil quality and are useful for the remediation of degraded land (e.g., in the region of lignite mine). Hemp is also extremely resistant, perfectly adapts to various climatic conditions, is resistant to various pests, requires a slight number of pesticide treatments, and the cultivation of 1 ha of hemp in one season absorbs approximately 11 Mg of CO_2 from the atmosphere [2–4].

The dry matter yields of these hemp varieties described in this study were as follows: Białobrzeskie 8–10 Mg·ha^{-1}, Tygra 8–11 Mg·ha^{-1}, Henola 7–8 Mg·ha^{-1}, and Rajan 9–12 Mg·ha^{-1}. Based on the data from 2016 (the own research of the Institute of Natural Fibres and Medicinal Plants—National Research Institute), it was estimated that, in Poland, the acreage of devastated and degraded land requiring reclamation and constituting a potential area for hemp cultivation for energy purposes amounts to approximately 65,000 ha. In the INF&MP-NRI, the Hemp Program is carried out, the aim of which is to develop hemp cultivation for seed reproduction. As part of this program, cultivation acreage is significantly increasing every year: 420 ha in 2018, 1000 ha in 2019, and 2000 ha in 2020.

The use of hemp waste straw for the production of lignocellulosic ethanol is beneficial for the environment, as it results in a rational management of bio-waste. The straw remaining after the ginning of the hemp panicles is not suitable for textile purposes, but it can be used, e.g., as a raw material for the production of bioethanol. Additionally, the dynamic growth of the hemp cultivation acreage in Poland may significantly contribute to increasing the efficiency process of obtaining bioethanol.

The process of plant biomass conversion to lignocellulosic ethanol includes several stages, from the preparation of plant material (effective pretreatment), through enzymatic hydrolysis, i.e., the decomposition of polysaccharides into fermentable sugars (the selection of effective enzymatic preparations) and to ethanol fermentation (the selection of appropriate microorganisms). The production of bioethanol from lignocellulosic raw material consists of the deconstruction of cell walls into individual polymers and the hydrolysis of carbohydrates into simple sugars. Currently, one of the main challenges is to increase the efficiency of the fermentation of organic substrates, and alternative solutions that directly affect the quantity and the quality composition of the final product are still being researched.

Hemp biomass as a lignocellulosic raw material contains a polymer complex, i.e., lignocellulose, which is relatively resistant to biodegradation. It is found in cell walls and consists of cellulose, hemicelluloses, and lignin. Cellulose and hemicelluloses are potential substrates in the fermentation process, while lignin adversely affects the conversion of hemp biomass. This necessitates the use of the pretreatment of biomass, the purpose of which is to fragmentate the solid phase and loosen the compact structure of lignocellulose. Recent advances in biomass pretreatment can be found in the review article by Bing et al. [5]. The second important stage in the process of obtaining bioethanol from hemp biomass is enzymatic hydrolysis, which determines the amount of simple sugars metabolized by yeast in the fermentation process. The hydrolysis process can be carried out as the SHF process—separate hydrolysis and fermentation (enzymes operate at 50–60 °C)—or the SSF (process of simultaneous saccharification and fermentation), where enzymes must be adapted to the conditions of the fermentation process, i.e., 30–40 °C. The last stage in the process of hemp biomass conversion is the ethanol fermentation of the obtained hydrolysates. A method that combines cellulose hydrolysis with sugar fermentation in one bioreactor seems to be more effective and economical [6–11].

The aim of the presented study was to indicate which of the four Polish varieties of industrial hemp (Białobrzeskie, Tygra, Henola, and Rajan) are the most suitable raw materials for the effective production of lignocellulosic ethanol. Thus far, no literature has been published about the possibility of using these Polish varieties of hemp as a raw material in the process of obtaining bioethanol.

2. Results and Discussion

2.1. Hemp Biomass Preparation

The hemp biomass of the four varieties (Białobrzeskie, Tygra, Henola, and Rajan) was first cut into fragments up to 1 cm in size and then was comminuted by the knife mill for mesh sizes of 2 and 4 mm. In order to choose the most favorable fractions, the content of reducing sugars released during an enzymatic test was determined. It was found that the highest values of reducing sugars were obtained for fractions up to 2 mm, and two varieties, Tygra and Białobrzeskie, showed higher sugar values for the tested fractions compared to Rajan and Henola (Table 1).

Table 1. The amount of reducing sugars ($mg \cdot g^{-1}$) released after an enzymatic test depending on the fraction size.

Hemp Biomass	4 mm	2 mm
Białobrzeskie	65.1 ± 0.06	68.5 ± 0.26
Tygra	73.3 ± 0.14	76.3 ± 0.17
Rajan	51.3 ± 0.32	57.8 ± 0.04
Henola	50.4 ± 0.07	54.8 ± 0.16

2.2. Alkaline Pretreatment

The purpose of chemical pretreatment is removing lignin from materials with lignocelullose and increasing the accessibility of biomass structure. The type of a reagent used has a significant effect on the performance of the chemical pretreatment. Sodium hydroxide is one of the most popular alkaline reagents used in this process.

The optimization of the concentration of sodium hydroxide used in alkaline treatment for the four varieties of hemp biomass was carried out based on the amount of reducing sugars released after the enzymatic test. Concentrations ranging from 1.5% to 3% were tested (Table 2). It was found that, for Tygra and Białobrzeskie varieties, for 2% NaOH, the amount of released reducing sugars was about 13% higher than for 1.5%. In turn, for 3% sodium hydroxide, the content of reducing sugars was at a similar level. For Henola and Rajan varieties, a completely different correlation was observed; the lowest level of released reducing sugars was noted for the concentration of 3% NaOH, while, at the concentration of 2%, the content of reducing sugars was the highest. Moreover, it was noted that two of the varieties, i.e., Tygra and Rajan, were characterized by over 10% higher content of reducing sugars than Białobrzeskie and Henola, which proves that these are the varieties more susceptible to the alkaline pretreatment. Based on the obtained results, the concentration of sodium hydroxide at the level of 2% was selected for further research. Kumar et al. [12] conducted similar research and stated that the sodium hydroxide pretreatment of lignocellulosic biomass resulted in the highest level of delignification at 2% NaOH. In turn, Zhao et al. [10], in the research on the use of American industrial hemp for the production of bioethanol, used the alkaline treatment of 1% NaOH.

Table 2. The amount of reducing sugars ($mg \cdot g^{-1}$) released after an enzymatic test depending on the NaOH concentration.

Hemp Biomass	1.5%	2%	3%
Białobrzeskie	163 ± 0.78	178 ± 1.37	173 ± 0.23
Tygra	183 ± 1.43	206 ± 0.87	203 ± 1.70
Rajan	190 ± 2.54	180 ± 0.68	176 ± 0.55
Henola	159 ± 1.86	166 ± 1.15	147 ± 1.33

Efficient pretreatment should decrystallize cellulose, depolymerize hemicelluloses, reduce the formation of inhibitors that hinder carbohydrate hydrolysis, require low energy expenditure, and recover value-added products such as lignin.

To confirm the efficiency of the alkaline treatment, the determination of the chemical composition of hemp biomass after NaOH treatment was performed and compared to the chemical composition of the biomass before pretreatment. The results are presented in Table 3.

Table 3. The chemical composition of hemp biomass (percentage of dry matter); BP: before pretreatment; AP: after pretreatment.

Variety	Samples	Cellulose (%)	Hemicelluloses (%)	Lignin (%)
Białobrzeskie	BP	50.10 ± 0.18	32.10 ± 0.22	15.40 ± 0.03
	AP	61.46 ± 0.37	21.59 ± 0.06	15.12 ± 0.16
Tygra	BP	50.82 ± 0.12	27.79 ± 0.33	14.68 ± 0.46
	AP	62.70 ± 0.09	20.16 ± 0.16	15.12 ± 0.22
Henola	BP	46.82 ± 0.04	29.94 ± 0.45	15.48 ± 0.17
	AP	57.62 ± 0.08	20.33 ± 0.22	17.80 ± 0.06
Rajan	BP	48.69 ± 0.39	31.43 ± 0.04	16.72 ± 0.08
	AP	59.30 ± 0.33	19.91 ± 0.25	18.40 ± 0.18

The analysis of the chemical composition of the hemp biomass before and after treatment showed that, in all the four varieties, the alkaline treatment resulted in a visible increase in the cellulose content (by approximately 10%) and the partial degradation of hemicelluloses (as much as 12% for the Rajan variety). The highest content of cellulose after treatment was found in the following varieties: Tygra, Białobrzeskie, and Rajan, and its level was approximately 60%. In the case of the lignin content, only for the Białobrzeskie hemp biomass, a very slight reduction was observed after the alkaline pretreatment. In the case of the remaining varieties, the tendency was contrary. Similar observations were reported by Stevulova et al. [13], who examined the chemical composition of hemp biomass before and after the pretreatment with sodium hydroxide and proved that the content of lignin after the pretreatment was 7% higher than before.

At this stage, due to better properties, availability, and higher yield, only two types of biomass of the Tygra and the Rajan varieties were selected, for which further research on bioethanol production was continued.

The effect of the alkaline treatment on Tygra and Rajan hemp biomass was also confirmed by Fourier transform infrared spectrometer (FTIR) shown in Figure 1a,b and by scanning electron microscopy (SEM) shown in Figure 2.

An effective method for studying the structure of biomass after alkaline treatment is FT-IR [14]. Figure 1a,b show the changes in FTIR spectra after the alkali treatment of hemp biomass between 600 cm^{-1} and 4000 cm^{-1}. On the spectra of both varieties, typical vibration bands in the cellulose molecule were observed at 3300 cm^{-1}, 2900 cm^{-1}, and 1610 cm^{-1}. The broad band in the 3600–3100 cm^{-1} region, which was due to the OH stretching vibration, gave considerable information concerning the hydrogen bonds. The peaks characteristic of hydrogen bonds from the spectra of the Rajan variety AP became a little sharper and more intense compared to the Rajan variety BP. However, in the case of the Tygra variety, this band was less intense. The 2900 cm^{-1} peak corresponding to the C–H stretching vibration in the case of the Tygra variety shifted to higher wavenumber values and slightly decreased in the intensity. These changes could have resulted from both the increased amount of cellulose after the treatment (2.2, Table 3) and the reduced cellulose crystallinity [15,16]. The band at 1610 cm^{-1} from the stretching vibrations of the O–H bonds, due to the adsorbed water in the sample, decreased, especially for the Rajan variety, which could be attributed to water loss due to drying the sample [17]. The intensities of peaks at 1160−1170 cm^{-1} (asymmetric C−O−C stretching from cellulose) and 1110−1120 cm^{-1} (C−OH skeletal vibration in cellulose) increased after pretreatment in both types of biomass. Furthermore, the intensity at 1050−1060 cm^{-1} (C−O−C pyranose ring skeletal vibration ascribed to cellulose) also increased slightly after alkali pretreatment.

These changes were confirmed by the increase in the concentration of cellulose in the pretreated biomass in the chemical composition tests (2.2, Table 3) [18].

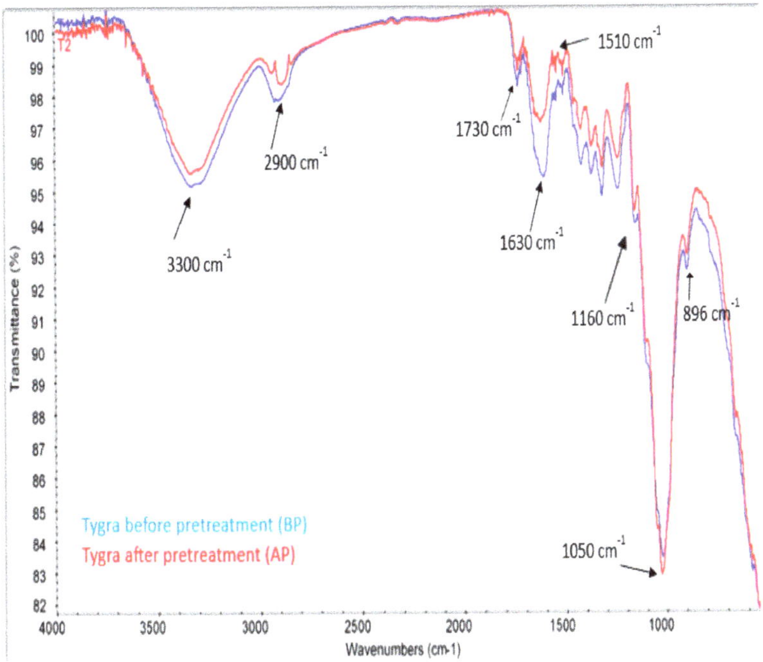

(**a**)

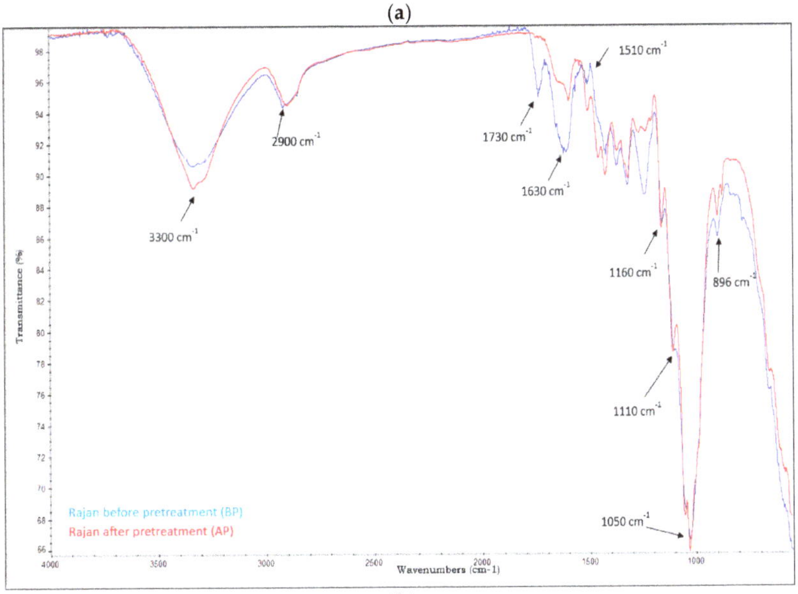

(**b**)

Figure 1. (**a**) The FTIR spectra of the Tygra biomass before and after the alkaline treatment. (**b**) The FTIR spectra of the Rajan biomass before and after the alkaline treatment.

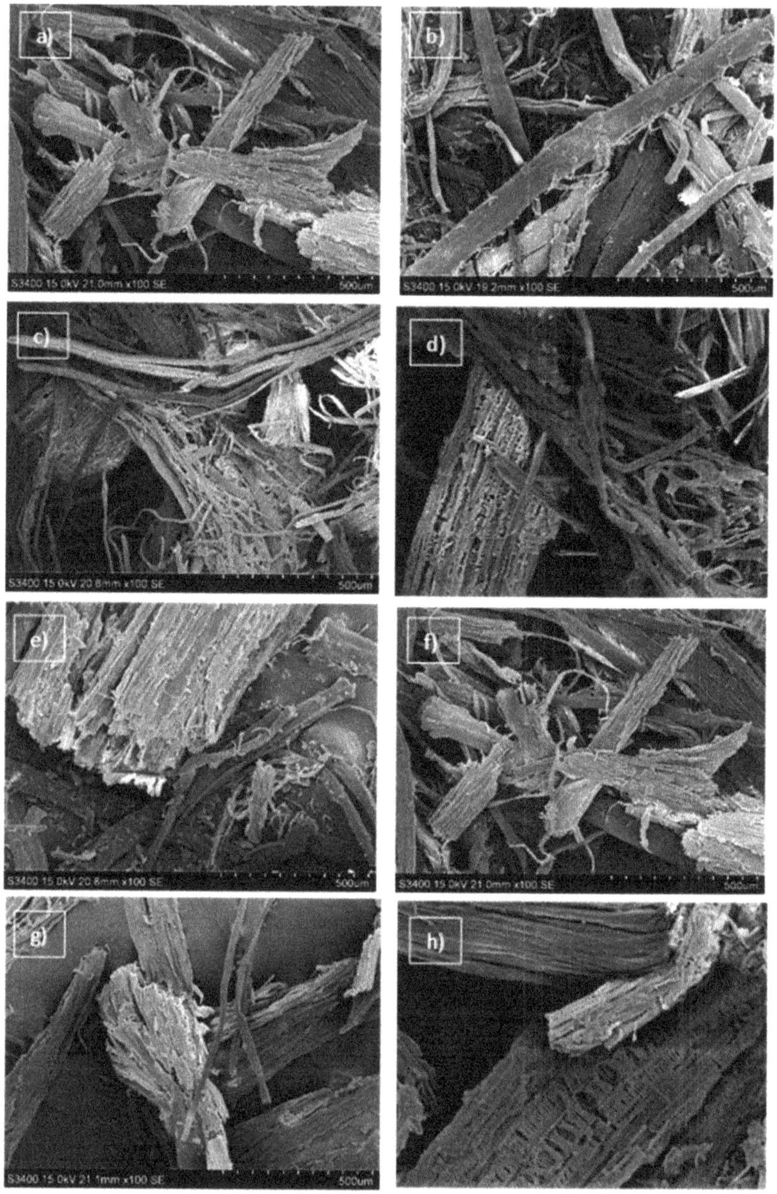

Figure 2. The SEM images of hemp biomass: (**a**) Tygra biomass before pretreatment, (**b**) Tygra biomass after pretreatment, (**c**) Tygra biomass after enzymatic hydrolysis, (**d**) Tygra biomass after enzymatic hydrolysis—selected fragment in high magnification, (**e**) Rajan biomass before pretreatment, (**f**) Rajan biomass after pretreatment, (**g**) Rajan biomass after enzymatic hydrolysis, (**h**) Rajan biomass after enzymatic hydrolysis—selected fragment in high magnification.

The vibration band visible at 1730 cm^{-1} (C=O stretching of acetyl groups in hemicelluloses and aldehydes in lignin) [19] was reduced in both hemp varieties after alkaline treatment, while the change was much more significant for the Rajan variety where the band almost disappeared. This occurred due to the decomposition of hemicelluloses and the solubilization of lignin during alkali pretreatment. This result correlates very well with the obtained results of the chemical composition of the biomass after pretreatment. According to these studies, the content of hemicelluloses after alkaline treatment in the Rajan variety decreased by as much as 12% (2.2, Table 3).

Moreover, the obtained FTIR spectra showed that the removal of lignin in the alkaline treatment process was problematic (band at 1510 cm^{-1}). This was confirmed by the results of the chemical composition of the lignin content presented in Table 2. This problem is widely described in the literature dealing with the studies of lignocellulosic biomass by infrared spectroscopy [20,21]. The process of the degradation or fragmentation of lignin is complicated due to the presence of strong C–C bonds and other functional groups, such as aromatic groups [13].

Significant changes on the surface of the biomass were observed and presented in the SEM images taken before and after the biomass pretreatment as well as after enzymatic hydrolysis (Figure 2). In the case of both varieties of hemp biomass (Tygra and Rajan), similar changes were observed in the biomass surface, which appeared as a result of subsequent processes. However, in the case of the Rajan variety, changes were more intensive, especially after the enzymatic hydrolysis process. The untreated hemp biomass was observed to have intact, rigid, and coarse structures with smooth surface and a well-ordered fiber skeleton (Figure 2a,e). This strongly blocked access to cellulose to limit enzymatic attack [22]. As a result of pretreatment, the biomass underwent various specific structural changes. The SEM images of hemp biomass after pretreatment showed that the surface area of the biomass was partially purified (Figure 2b,f). The morphological changes that indicated damage to the structure of biomass and that increased the surface area, making it more accessible to the cellulolytic enzymes [20,23,24], were observed. The enzymatic hydrolysis of the samples subjected to the previous alkaline treatment caused further significant changes in the structure of the biomass visible in the SEM pictures (Figure 2c,d,g,h). The appearance of micropores was very characteristic here.

Undoubtedly, the opening of the hemp biomass and creating the holes all over the biomass enhanced the enzyme accessibility of the structure and facilitated biomass digestibility [25]. Moreover, it was clearly visible that enzymatic hydrolysis of the Rajan hemp biomass made the fibrous structure fragile and was more successful.

2.3. Evaluation of Enzyme Preparations

The enzymatic hydrolysis process is the second main step in the process of obtaining bioethanol from plant biomass. Enzymatic hydrolysis determines the amounts of simple sugars that are metabolized by yeast in the fermentation process. The breakdown of cellulose into simple sugars requires the synergistic action of different enzymes: cellulases, endoglucanases, cellobiohydrolases, and ß-glucosidases. First, the commercial enzyme preparations of various compositions were gained (Flashzyme Plus 200, ACx8000L, Celluclast 1.5L, Cellobiase, Xylanase), and then their cellulolytic and xylanolytic activities were determined. Taking into account evaluated activity of the tested enzymes and their commercial availability, Flashzyme Plus 200 and Celluclast 1.5L preparations were selected for further research (Table 4).

In order to select the enzyme complex for the SHF and the SSF processes, enzymatic tests were performed using selected enzymes and their supplementation with glucosidase and xylanase and were partially described in our previous studies [4]. For the SHF process in the case of the Tygra biomass, an enzyme complex was selected with the composition of Flashzyme Plus 200, glucosidase, and xylanase, while composition for the Rajan was Flashzyme Plus 200:Celluclast 1.5L hemp biomass in the proportion of 70:30. For the SSF process, the enzyme complex for the Tygra biomass was selected as Flashzyme Plus 200:Celluclast 1.5L (70%:30%) and xylanase and for the Rajan was Flashzyme Plus 200:Celluclast 1.5L hemp biomass in the 50:50 ratio.

Table 4. The determination of cellulolytic and xylanolytic activities of commercial enzyme preparations (FPU·mL^{-1}).

Enzyme Preparations	Composition	Cellulolytic Activity	Xylanolytic Activity
Flashzyme Plus 200	endoglucanase, cellobiohydrolase, cellobiose, xylanase, mannase	90	2430
ACx3000L	endo-1,4-β-D-glucanase	63	487
Celluclast 1.5L	cellulase, xylanase	62	278
Accellerase 1500	exoglucanase, endoglucanase, cellobiose, β-glucosidase, hemicellulase	53	616
Alternafuel CMAX	cellulase, β-glucosidase, hemicellulase, arabinose	2.78	110
ACx8000L	cellulase	8	190
Cellobiase	glucosidase	0.18	325
Xylanase	xylanase	0.03	746

2.4. *Separate Hydrolysis and Fermentation (SHF)*

To determine the optimal conditions of the enzymatic hydrolysis method as a separate process of SHF, based on the literature data and the research experience, the following parameter ranges were selected for testing with the response surface methodology (RSM): dose of the enzyme 10–30 FPU·g^{-1} of solid, temperature 50–70 °C, pH 4.2–5.4, and time 24–72 h. The RSM method is an effective optimization tool consisting of mathematical and statistical techniques and used for the process of optimization [26–28].

Individual enzymatic tests of the hydrolysis process were performed for Tygra and Rajan biomasses, and the evaluation criterion was the amount of released glucose. Figure 3 presents various response surfaces and interaction effects of variables (temperature, time, enzymes' dose, and pH) on the glucose yield. The variable that had the most significant impact on the glucose content turned out to be the temperature. The lower the temperature was, the higher the glucose content was. The pH of the solution, the process time, and the enzymes' dose had lesser effects. However, slight differences were observed in the dependence of these variables on the glucose yield for the two biomass varieties. For the Tygra biomass, it was found that, in the SHF process, the optimal conditions for enzymatic hydrolysis were obtained for the substrate concentration of 5% using the following enzymes: Flashzyme Plus 200 30 FPU·g^{-1} of solid, glucosidase 20 CBU·g^{-1} of solid, and xylanase 500 XU·g^{-1} of solid. The process parameters were: temperature 50 °C, pH 4.2, and time 48 h. These parameters provided the opportunity to obtain a maximum glucose yield, which was 36.9 ± 0.64 (g·L^{-1}). In the SHF process for the Rajan biomass, optimal enzymatic hydrolysis conditions were obtained for a substrate concentration of 5% using the Flashzyme Plus 200:Celluclast 1.5L (70:30) enzyme complex with a dose of 10 FPU·g^{-1} of solid. The process parameters were: temperature 50 °C, pH 5.4, and time 72 h. The maximum glucose yield was 23.66 ± 0.16 (g·L^{-1}).

Similar research was conducted by Abraham [29]; during biomass hydrolysis at 50 °C and 18 FPU·g^{-1} of solid, the highest glucose yield was obtained. Salimi and others [30] optimized the enzymatic hydrolysis of lignocellulosic biomass using the RSM method. They applied the temperature range of 45–60 °C and the pH of 4.5—6.0. They obtained the highest content of monosaccharides at 45 °C and pH 6.0. Jambo et al. [31], in turn, optimized lignocellulosic biomass using similar parameters—temperature (30–60 °C), pH (3.8–5.8), and incubation time (12–72 h)—and obtained the glucose concentration at the level 24.24 g·L^{-1}.

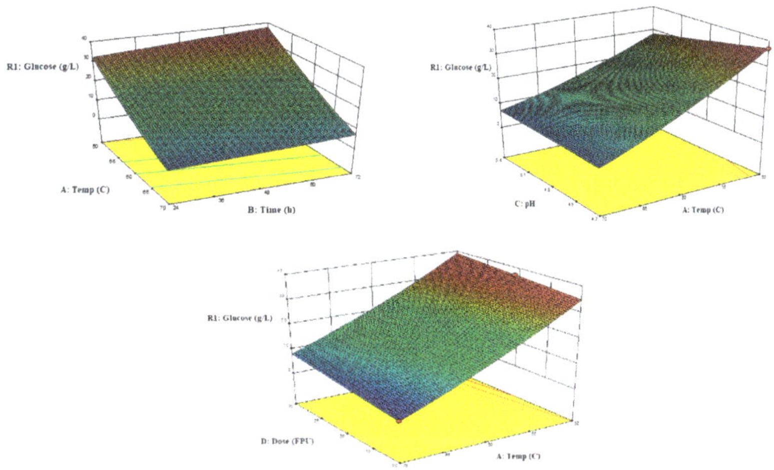

(**a**)

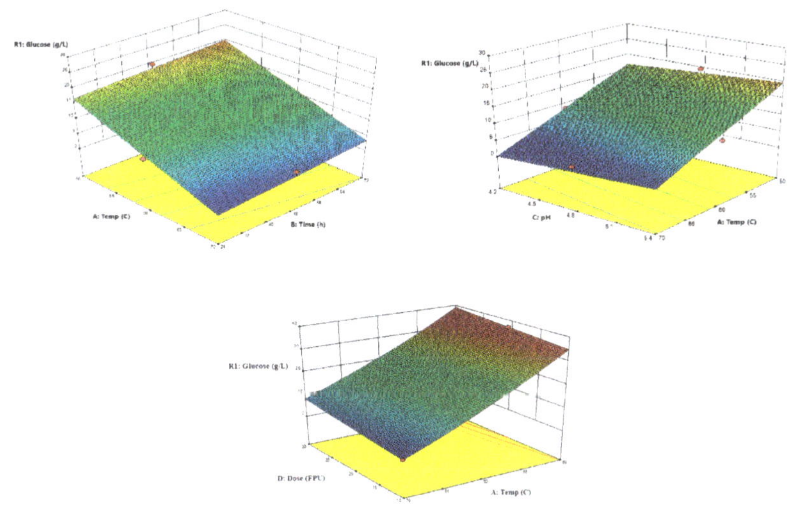

(**b**)

Figure 3. Enzymatic hydrolysis process of hemp biomass (RSM). Response surface representing the interaction effects of temperature, time, dose, and pH on glucose yield: (**a**) Tygra biomass, (**b**) Rajan biomass.

The next step in the conversion of hemp biomass to bioethanol in the SHF process was ethanol fermentation. In the SHF process for the Tygra biomass, the highest concentration of ethanol was observed at 48 hours and was 10.51 g·L^{-1}. In the following hours of the process, no significant increase in ethanol concentration was observed. In turn, for the Rajan biomass, the highest ethanol concentration was noticed at 96 h, and it was only 2.76 g·L^{-1}. Such a low concentration of ethanol in this case could be attributed to various reasons, which together may have had a significant negative impact on this parameter—for example, the chemical composition, which was characterized by a higher lignin content compared to the Tygra biomass. The reason could also have been the use of a low enzyme dose (10 FPU g^{-1} solid) in Rajan biomass as well as the low glucose concentration (23.66 g L^{-1}), which was determined immediately after the enzymatic hydrolysis step. Additionally, after optimizing this step, a pH of 5.4 was chosen based on the glucose concentration, which may have, to some extent, inhibited yeast activity in the initial phase of fermentation. Nevertheless, it was noticed that, with each passing day, the concentration of ethanol increased slightly, which ultimately indicates that the process itself was proceeding correctly. However, for Tygra biomass, no significant increase in ethanol concentration was observed with time extension of the SHF process (Figure 4). The average yield of bioethanol for the Tygra variety was equal to 253 L·Mg^{-1} (of hemp straw dry matter), i.e., 3.04 m^3·ha^{-1}. For the Rajan variety, the average yield of bioethanol was 69 L·Mg^{-1} (of hemp straw dry matter), i.e., 0.80 m^3·ha^{-1}. A similar study was presented by Kusmiyati et al. [32]. In their work, the conversion of the lignocelullosic biomass to bioethanol was carried out through pretreatment, saccharification, and fermentation processes. Their results showed that the SHF process gave a higher concentration of ethanol (8.11 g·L^{-1}). Fischer and others [33], in their research, dealt with lignocellulosic biomass and examined the SHF process, obtaining the ethanol concentration of 12.1 g·L^{-1}.

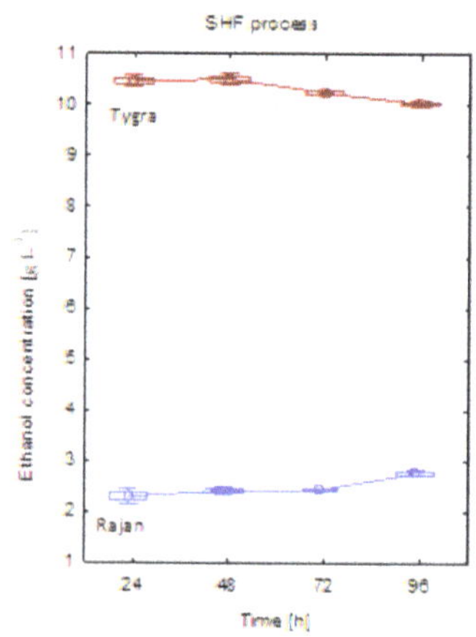

Figure 4. The ethanol concentration of hemp biomass in the SHF process. Optimized process conditions: Tygra biomass—substrate concentration 5%, enzymes: Flashzyme Plus 200 30 FPU·g^{-1} of solid, glucosidase 20 CBU·g^{-1} of solid and xylanase 500 XU·g^{-1} of solid, temperature 50 °C, pH 4.2, and time 48 h. Rajan biomass—substrate concentration 5%, Flashzyme Plus 200:Celluclast 1.5L (70:30) enzyme complex with a dose of 10 FPU·g^{-1} of solid, temperature 50 °C, pH 5.4, and time 72 h.

SHF as an alternative process in an industrial bioethanol plant manifests both potential and limitations. The main advantage of SHF is the possibility to optimize the process steps separately, especially to be able to run the enzymatic hydrolysis at an optimal temperature with respect to enzymes [34]. However, most literature reports confirm that SSF is a more promising and advantageous approach with respect to SHF because of a low production cost, less processing time, less reactor volume, higher ethanol productivity, a lower requirement of enzyme, the ability to overcome enzymatic inhibition by simultaneous end-product removal, and a lower requirement for sterile conditions, as bioethanol is produced immediately with glucose conversion [35,36].

2.5. Simultaneous Saccharification and Fermentation (SSF)

The simultaneous hydrolysis and fermentation must be carried out under conditions that ensure the optimal synergy of enzymes and distillery yeast. To optimize the SSF process according to the RSM, the following ranges of process parameters were selected: substrate content 5%–7% w/v, the dose of Flashzyme:Celluclast 1.5L enzymes (50:50)

10–30 FPU·g^{-1} of solid using *S. cerevisiae* yeast at 37 °C, pH 4.8, and 96 h. Then, the fermentation tests were carried out using the selected parameters, and the amount of ethanol (HPLC) was determined. The optimal conditions of the SSF process for the Tygra and the Rajan hemp biomasses were selected. The highest ethanol concentration for Tygra biomass was observed at a substrate content of 5% *w*/*v* and a dose of enzyme at 30 FPU·g^{-1} of solid, and it was 6.5 g·L^{-1}. In turn, for the Rajan biomass, the highest ethanol concentration equal to 7.5 g·L^{-1} was recorded for the substrate of 5% and for the enzyme 30 FPU·g^{-1} of solid. Higher substrate content above 5% *w*/*v* interfered with the effective mixing of the fermentation solution, which resulted in poorer access to biomass and thus less effective action of enzymes and yeast. It was also observed that enlarging the enzyme dose enhanced the conversion of cellulose to glucose and thus increased the concentration of ethanol. The enzyme dose of 10 FPU·g^{-1} of solid tested in the optimization process turned out to be too low to carry out efficient enzymatic hydrolysis.

For the Tygra biomass, there were no changes in the concentration of ethanol with the time lapse of the process. In turn, for the Rajan biomass, after 24 h, there was observed a decrease in ethanol concentration, and after 48 h of the SSF process, there was noticed a significant increase in ethanol concentration (Figure 5). It was observed that, for the Rajan biomass, in contrast to the Tygra biomass, a higher ethanol concentration was obtained in the SSF process than in the SHF process (Figures 4 and 5).

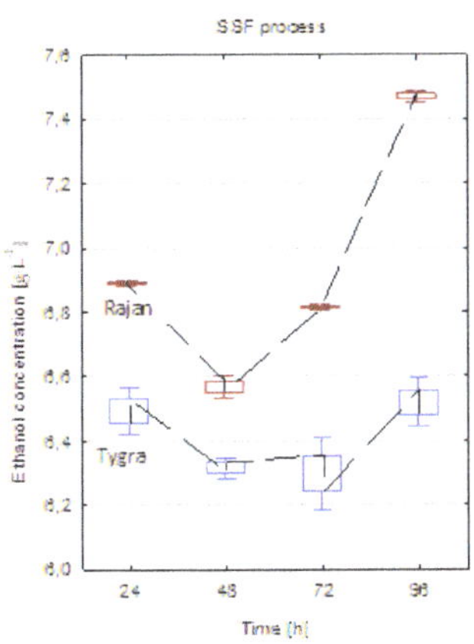

Figure 5. Ethanol concentration of hemp biomass in the SSF process. Optimized process conditions: Tygra and Rajan biomass-substrate concentration 5%, enzymes: Flashzyme:Celluclast 1.5L (50:50) 30 FPU·g^{-1} of solid, temperature 37 °C, pH 4.8, and time 96 h.

In their research, Fojas and Rosario [37] optimized the enzymatic saccharification of lignocellulosic biomass, and the SSF process was carried out with the following parameters: 3%–6% the amount of substrate, 20–25 FPU·g^{-1} of solid dose of enzyme, and temperature of 37 °C for 120 h. They achieved an ethanol content of about 9 g·L^{-1}.

Research on obtaining bioethanol from hemp biomass was also carried out by Orlygsson [38]; after the SSF process, the author obtained an ethanol concentration of approximately 1 g·L^{-1}.

On the basis of the average ethanol content, the hemp straw yield that could be obtained from 1 ha of hemp cultivation was specified. The highest average yield of bioethanol was estimated for the Rajan variety and was equal to 190 L·Mg^{-1} (of hemp straw dry matter), i.e., 2.23 m^3·ha^{-1}, while the average yield of bioethanol estimated for the Tygra variety was 165 L·Mg^{-1} (of hemp straw dry matter), i.e., 1.81 m^3·ha^{-1}.

Extensive research on industrial hemp as potential raw material for bioethanol production compared to other raw materials such as kenaf and sorghum was conducted by

Das et al. [39]. According to these studies, the ethanol yield from hemp was 250 L·Mg^{-1}, which turned out to be much higher than that of kenaf. Moreover, the cost analysis allowed the researchers to conclude that industrial hemp can generate higher gross profits per hectare than other crops. In conclusion, this scientific report emphasized that hemp has the potential to be a promising crop for the production of bioethanol.

3. Materials and Methods

3.1. Bioethanol Production Process

3.1.1. Hemp Biomass Preparation

The raw materials used in the study were Białobrzeskie, Tygra, Henola, and Rajan hemp (*Cannabis sativa* L.) biomasses from the Experimental Farm of the Institute of Natural Fibres and Medicinal Plants in Pętkowo. This material was subjected to preliminary crushing to the particles of size 20–40 mm and then dried in 50–55 °C for 24 h. Next, the material was disintegrated on a knife mill (Retsch SM-200, Germany) with the sieves of the mesh size of 2–4 mm. An enzymatic test for the crushed fractions was performed using the Celluclast 1.5L enzyme preparation, and the content of reducing sugars was determined by the Miller's method with 3,5-dinitrosalicylic acid (DNS) [40].

3.1.2. Alkaline Pretreatment

The evaluation of pretreatment conditions for hemp biomass was carried out at 5 h treatment with 1.5%–3% sodium hydroxide in 90 °C. NaOH:biomass weight ratio was 10:1. After the alkaline pretreatment was carried out, the biomass solution was filtered on a Büchner funnel, then washed with distilled water until neutralized and dried in a laboratory dryer at 50 °C for 24 h. The alkali effect on the biomass was evaluated in the enzymatic test, and content of the released reducing sugars was determined by the Miller's method. This test was performed with the use of Celluclast 1.5L (Novozymes, Bagsværd, Denmark) enzymatic preparation at the dose of 10 FPU·g^{-1} of solid. The raw material was incubated at 55 °C in 0.05 M citrate buffer of pH 4.8 for 24 h. Then, after the enzymatic test, the supernatant was diluted, a DNS reagent was added, and the mixture was incubated in a boiling water bath for 10 min. After cooling to room temperature, the absorbance of the supernatant was measured at 530 nm on UV–VIS Spectrophotometer V-630 (Jasco, Pfungstadt, Germany).

3.1.3. Enzyme Complex

The enzymes cellulolytic activity was determined according to NREL LAP Measurement of Cellulase Activities. In turn, enzymes xylanolytic activity was determined according to the Osaka University procedure (with changes) [41].

In order to select the enzyme complex for SHF and SSF processes, tests were performed using selected enzymes (Flashzyme Plus 200:Celluclast 1.5L) and their supplementation with glucosidase 20 CBU·g^{-1} of solid and xylanase 500 XU·g^{-1} of solid (Sigma-Aldrich, Darmstadt, Germany). Enzymatic tests were carried out for 5% of biomass with the enzyme in the amount of 10 FPU·g^{-1} of solid, at a pH of 4.8, and during 24 h at 55 °C for the SHF process and at 38 °C for the SSF process. The selection criterion was the content of reducing sugars determined by the Miller's method.

3.1.4. Separate Hydrolysis and Fermentation (SHF)

The optimization of the enzymatic hydrolysis of hemp biomass in the SHF process was carried out according to the response surface methodology (RSM) using the parameters: biomass content 5%–7% w/v, temperature 50–70 °C, time 24–72 h, pH 4.2–5.4, dose of enzyme 10–30 FPU·g^{-1} of solid. Then, tests of the hemp biomass hydrolysis process were performed, and the evaluation criterion was the amount of released glucose.

In the next stage, the obtained hydrolyzate was subjected to the ethanol fermentation process carried out in bioreactor Biostat B Plus (Sartorius, Göttingen, Germany) in 2 L vessel equipped with pH, temperature, stirring, and foaming controls. The temperature was maintained at 37 °C and stirring at 900 rpm, while pH was controlled at 4.2 for Tygra and 5.4 for Rajan by adding 1 M NaOH or 1 M HCl. Non-hydrated freeze-dried distillery yeast *S. cerevisiae* (Ethanol Red, Lesaffre, France) at a dose of 1 g·L^{-1} was used in the process, which corresponded to cell concentration after inoculation of about 1 × 10^7 cfu/mL. Inoculum grew for 24 h at 30 °C. After inoculation, a 96 h fermentation was carried out, and samples were taken every 24 h. All experiments were performed in triplicate.

3.1.5. Simultaneous Saccharification and Fermentation (SSF)

To optimize the SSF process according to the RSM, the ranges of process parameters were selected: substrate content 5%–7% *w*/*v*, dose of (Flashzyme:Celluclast 1.5L) enzymes 10–30 FPU·g^{-1} of solid. The SSF process was carried out in bioreactor Biostat B Plus (Sartorius, Göttingen, Germany) in 2 L vessel equipped with pH, temperature, stirring, and foaming controls. The temperature was maintained at 37 °C and stirring at 900 rpm, while pH was controlled at 4.8 by adding 1 M NaOH or 1 M HCl. In the fermentation process, non-hydrated freeze-dried distillery yeast *S. cerevisiae* (Ethanol Red, Lesaffre, France) at a dose of 1 g·L^{-1} was used, which corresponded to cell concentration after inoculation of about 1 × 10^7 cfu/mL. The duration of ethanol fermentation was 96 h. All experiments were performed in triplicate.

3.2. Analytical Methods

The chemical composition of hemp biomass before and after pretreatment was determined, i.e., cellulose according to TAPPI T17 m-55 [42], hemicelluloses as the difference of holocellulose and cellulose according to TAPPI T9 m-54 [43], and lignin according to TAPPI T13 m-54 [44].

In order to provide a more complete picture of the molecular structure of hemp biomass before and after the chemical pretreatment, the analysis of FTIR spectroscopy was performed using a Fourier Transform Infrared Spectrometer ISS 66v/S (Bruker, Bremen, Germany) at infrared wavenumbers of 400–4000 cm^{-1} [13].

The physical morphologies of hemp biomass before and after the chemical treatment and after enzymatic hydrolysis were performed using Scanning Electron Microscope S-3400N (Hitachi, Japan) in high vacuum conditions. The samples were covered with gold dust.

The contents of glucose and ethanol were determined by high performance liquid chromatography on Elite LaChrom (Hitachi, Tokio, Japan) using an RI L-2490 detector, Rezex ROA 300x7.80 mm column (Phenomenex, Torrance, CA, USA), as the mobile phase used 0.005 N H_2SO_4 at a flow rate of 0.6 mL/min at 40 °C.

3.3. Calculations

The ethanol yield from 100 g of raw material Y_s (g/100 g of raw material) was calculated according to the Equation (1) [45]:

$$Y_s = \frac{Et \times 100}{M} \tag{1}$$

where: *Et*—amount of ethanol in 1000 mL of tested sample (g); *M*—mass of material weighed in 1000 mL fermentation sample (g).

Then, based on the ethanol yield from 100 g of raw material, the amount of ethanol in L per ton of straw dry matter (L·Mg^{-1}) was calculated, and on the basis of straw yield, the ethanol yield per hectare (m^3·ha^{-1}) was determined.

3.4. Statistical Analysis

The experiments of ethanol fermentation were carried out in triplicates. Standard deviations were calculated using the analysis of variance ANOVA, Statistica 13.0 software ($p < 0.05$).

4. Conclusions

To sum up, the Tygra and the Rajan varieties of hemp were selected, which proved to be proper sources of second generation bioethanol as alternatives to petroleum-oil based fossil fuels. Pretreatment, enzymatic hydrolysis, and ethanol fermentation were optimized. Alkaline pretreatment caused an increase in cellulose content and partial degradation of hemicelluloses. Enzymatic hydrolysis allowed us to achieve glucose yield at the level up to 36.9 $g \cdot L^{-1}$. For the Tygra biomass in the SHF process, the ethanol concentration was 10.5 $g \cdot L^{-1}$ (3.04 $m^3 \cdot ha^{-1}$), and for the Rajan biomass in the SSF process, the ethanol concentration was 7.5 $g \cdot L^{-1}$ (2.23 $m^3 \cdot ha^{-1}$).

In the future, it will be important to conduct research on the mixtures of different varieties of hemp biomass in order to determine their potential for the production of lignocellulosic ethanol, which seems important in practical application, because the industrial production of biofuels occurs most often in large refineries which process the biomass of different varieties and species of plants.

Author Contributions: Conceptualization, A.W., J.B., W.G.; Methodology, A.W., J.B., W.G.; Software, A.W., W.G.; Validation, J.B.; Formal Analysis, J.B., A.W.; Investigation, A.W., W.G.; Resources, A.W., J.B. WG.; Data Curation, J.B., A.W., W.G.; Writing—Original Draft Preparation, A.W., W.G; Writing—Review & Editing, A.W., J.B., W.G.; Visualization, A.W.; Supervision, J.B.; Project Administration, J.B.; Funding Acquisition, J.B. All authors have read and agreed to the published version of the manuscript.

Funding: This research was funded by the Ministry of Agriculture and Rural Development, Poland, Multiannual Program (2017–2020).

Institutional Review Board Statement: Not applicable.

Informed Consent Statement: Not applicable.

Data Availability Statement: Data is contained within the article.

Acknowledgments: The study was conducted as a research project- Multiannual Program (2017–2020): Reconstruction and sustainable development of production and processing of natural fiber raw materials for the needs of agriculture and the economy was financed by the Ministry of Agriculture and Rural Development, Poland.

Conflicts of Interest: The authors declare no conflict of interest.

References

1. Directive 2009/28/EC of the European Parliament and of the Council of 23 April 2009 on the promotion of the use of energy from renewable sources and amending and subsequently repealing Directives 2001/77/EC and 2003/30/EC (Text with EEA relevance). Available online: https://eur-lex.europa.eu/legal-content/EN/ALL/?uri=CELEX:32009L0028 (accessed on 8 April 2021).
2. Cattaneo, C.; Givonetti, A.; Leoni, V.; Guerrieri, N.; Manfredi, M.; Giorgi, A.; Cavaletto, M. Biochemical aspects of seeds from *Cannabis sativa* L. plants grown in a mountain environment. *Sci. Rep.* **2021**, *11*, 3927. [CrossRef]
3. Zadrożniak, B.; Radwańska, K.; Baranowska, A.; Mystkowska, I. Possibility of industrial hemp cultivation in areas of high nature value. *Econ. Reg. Stud.* **2017**, *10*, 114–127. [CrossRef]
4. Wawro, A.; Batog, J.; Gieparda, W. Chemical and Enzymatic Treatment of Hemp Biomass for Bioethanol Production. *Appl. Sci.* **2019**, *9*, 5348. [CrossRef]
5. Bing, S.; Richen, L.; Chun Ho, L.; Hao, W.; To-Hung, T.; Yun, Y. Recent advances and challenges of inter-disciplinary biomass valorization by integrating hydrothermal and biological techniques. *Renew. Sustain. Energy Rev.* **2021**, *135*, 110370.
6. Bajpai, P. Pretreatment of Lignocelluloses Biomass for Bioethanol Production. In *Developments in Bioethanol, Green Energy and Technology*; Springer: Singapore, 2021. [CrossRef]
7. Vasić, K.; Knez, Ž.; Leitgeb, M. Bioethanol Production by Enzymatic Hydrolysis from Different Lignocellulosic Sources. *Molecules* **2021**, *26*, 753. [CrossRef]
8. Abo, B.O.; Gao, M.; Wang, Y.; Wu, C.; Ma, H.; Wang, Q. Lignocellulosic biomass for bioethanol: An overview on pretreatment, hydrolysis and fermentation processes. *Rev. Env. Health* **2019**, *34*, 57–68. [CrossRef] [PubMed]
9. Rahmati, S.; Doherty, W.; Dubal, D.; Atanda, L.; Moghaddam, L.; Sonar, P.; Hessel, V.; Ostrikov, K. Pretreatment and fermentation of lignocellulosic biomass: Reaction mechanisms and process engineering. *React. Chem. Eng.* **2020**, *5*, 2017–2047. [CrossRef]
10. Zhao, J.; Xu, Y.; Wang, W.; Griffin, J.; Roozeboom, K.; Wang, D. Bioconversion of industrial hemp biomass for bioethanol production: A review. *Fuel* **2020**, *281*, 118725. [CrossRef]

11. Rheay, H.T.; Omondi, E.C.; Brewer, C.E. Potential of hemp (*Cannabis sativa* L.) for paired phytoremediation and bioenergy production. *Bioenergy* **2021**, *13*, 525–536.
12. Kumar, A.K.; Sharma, S. Recent updates on different methods of pretreatment of lignocellulosic feedstocks: A review. *Bioresour. Bioprocess* **2017**, *4*, 7. [CrossRef]
13. Stevulova, N.; Cigasova, J.; Estokova, A.; Terpakova, E.; Geert, A.; Kacik, F.; Singovszka, E.; Holub, M. Properties characterization of chemically modified hemp hurds. *Materials* **2014**, *7*, 8131–8150. [CrossRef]
14. Kumar, A.; Rapoport, A.; Kunze, G.; Kumar, S.; Singh, D.; Singh, B. Multifarious pretreatment strategies for the lignocellulosic substrates for the generation of renewable and sustainable biofuels: A review. *Renew Energy* **2020**, *160*, 1228–1252.
15. Araújo, D.; Vilarinho, M.; Machado, A. Effect of combined dilute-alkaline and green pretreatments on corncob fractionation: Pretreated biomass characterization and regenerated cellulose film production. *Ind. Crop. Prod.* **2019**, *149*, 111785. [CrossRef]
16. Ciolacu, D.; Ciolacu, F.; Popa, V.I. Amorphous cellulose-structure and characterization. *Cellul. Chem. Technol.* **2011**, *45*, 13.
17. Gupta, A.D.; Pandey, S.; Kumar Jaiswal, V.; Bhadauria, V.; Singh, H. Simultaneous oxidation and esterification of cellulose for use in treatment of water containing Cu (II) ions. *Carbohydr. Polym.* **2019**, *222*, 114964. [CrossRef] [PubMed]
18. Zhao, J.; Xu, Y.; Wang, W.; Griffin, J.; Wang, D. High Ethanol Concentration (77 g/L) of Industrial Hemp Biomass Achieved Through Optimizing the Relationship between Ethanol Yield/Concentration and Solid Loading. *ACS Omega* **2020**, *5*, 21913–21921. [CrossRef]
19. Sun, X.F.; Xu, F.; Sun, R.C.; Fowler, P.; Baird, M.S. Characteristics of degraded cellulose obtained from steam-exploded wheat straw. *Carbohyd. Res.* **2005**, *340*, 97–106. [CrossRef]
20. Putnina, A.; Kukle, S.; Gravitis, J. STEX treated and untreated hemp fiber comparative structural analysis. *Mater. Science. Text. Cloth. Technol.* **2011**, *6*, 36–42.
21. Plácido, J.; Capareda, S. Analysis of alkali ultrasonication pretreatment in bioethanol production from cotton gin trash using FT-IR spectroscopy and principal component analysis. *Bioresour. Bioprocess* **2014**, *1*, 23. [CrossRef]
22. Zhao, J.; Xu, Y.; Zhang, M.; Wang, D. Integrating bran starch hydrolysates with alkaline pretreated soft wheat bran to boost sugar concentration. *Bioresour. Technol.* **2020**, *302*, 122826. [CrossRef] [PubMed]
23. Fang, G.; Chen, H.; Chen, A.; Mao, K.; Wang, Q. An efficient method of bio-chemical combined treatment for obtaining high-quality hemp fiber. *BioResources* **2017**, *12*, 1566–1578. [CrossRef]
24. Abraham, R.E.; Barrow, C.J.; Puri, M. Relationship to reducing sugar production and scanning electron microscope structure to pretreated hemp hurd biomass (*Cannabis sativa*). *Biomass Bioenerg.* **2013**, *58*, 180–187. [CrossRef]
25. Abraham, R.E.; Vongsvivut, J.; Barrow, C.J.; Puri, M. Understanding physicochemical changes in pretreated and enzyme hydrolysed hemp (Cannabis sativa) biomass for biorefinery development. *Biomass Conv. Bioref.* **2016**, *6*, 127–138. [CrossRef]
26. Lavudi, S.; Oberoi, H.S.; Mangamoori, L.N. Ethanol production from sweet sorghum bagasse through process optimization using response surface methodology. *Biotech* **2017**, *7*, 233. [CrossRef] [PubMed]
27. Campos, L.M.A.; Moura, H.O.M.A.; Cruz, A.J.G. Response surface methodology (RSM) for assessing the effects of pretreatment, feedstock, and enzyme complex association on cellulose hydrolysis. *Biomass Conv. Bioref.* **2020**, *12*, 124. [CrossRef]
28. Jaisamut, K.; Paulová, L.; Patáková, P.; Rychtera, M.; Melzoch, K. Optimization of alkali pretreatment of wheat straw to be used as substrate for biofuels production. *Plant. Soil. Environ.* **2013**, *59*, 537–542. [CrossRef]
29. Abraham, R.E. Bioprocessing of Hemp Hurd (*Cannabis sativa*) for Biofuel Production. Ph.D Thesis, Deakin University, Melbourne, Australia, 30 October 2014.
30. Salimi, M.N.; Lim, S.E.; Yusoff, A.H.M.; Jamlos, M.F. Conversion of rice husk into fermentable sugar by two stage hydrolysis. *J. Phys. Conf. Ser.* **2017**, *908*, 012056. [CrossRef]
31. Jambo, A.; Abdulla, R.; Marbawi, H.; Gansau, J.A. Response surface optimization of bioethanol production from third generation feedstock -Eucheuma cottonii. *Renew. Energy.* **2019**, *132*, 1–10. [CrossRef]
32. Maryanto, D.; Sonifa, R.; Kurniawan, S.A.; Hadiyanto, H. Pretreatment of Starch-Free Sugar Palm Trunk (*Arenga pinnata*) to Enhance Saccharification in Bioethanol Production. *MATEC Web. Conf.* **2018**, *156*, 6.
33. Fischer, J.; Lopes, V.S.; Cardoso, L.; Coutinho Filho, U.; Cardoso, V.L. Machine learning techniques applied to lignocellulosic ethanol in simultaneous hydrolysis and fermentation. *Braz. J. Chem. Eng.* **2017**, *34*, 53–63. [CrossRef]
34. Moreno, A.D.; Alvira, P.; Ibarra, D.; Tomas-Pejo, E. Production of ethanol from lignocellulosic biomass. In *Production of Platform Chemicals from Sustainable Resources*; Biofuels and Biorefineries; Feng, Z., Smith, R.L., Qi, X., Eds.; Springer: Singapore, 2017; pp. 375–410.
35. Robak, K.; Balcerek, M. Current state-of-the-art in ethanol production from lignocellulosic feedstocks. *Microbiol. Res.* **2020**, *240*, 126534. [CrossRef]
36. Dey, P.; Pal, P.; Kevin, J.D.; Das, D.B. Lignocellulosic bioethanol production: Prospects of emerging membrane technologies to improve the process—A critical review. *Rev. Chem. Eng.* **2020**, *36*, 333–336. [CrossRef]
37. Fojas, J.J.R.; Rosario, E.J.D. Optimization of pretreatment and enzymatic saccharification of cogon grass prior ethanol production. *In. Chem. Mol. Eng.* **2013**, *7*, 296–299.
38. Orlygsson, J. Ethanol production from biomass by a moderate thermophile, Clostridium AK1. *Icel. Agric. Sci.* **2012**, *25*, 25–35.
39. Das, L.; Liu, E.; Saeed, A.; Williams, D.W.; Hu, H.; Li, C.; Ray, A.E.; Shi, J. Industrial hemp as a potential bioenergy crop in comparison with kenaf, switchgrass and biomass sorghum. *Bioresour. Technol.* **2017**, *244*, 641–649. [CrossRef] [PubMed]
40. Miller, G.L. Use of dinitrosalicylic acid reagent for determination of reducing sugars. *Anal. Chem.* **1959**, *31*, 426–428. [CrossRef]

41. Measurement of Cellulase Activities-. Available online: https://www.nrel.gov/docs/gen/fy08/42628.pdf (accessed on 7 May 2021).
42. TAPPI 17 m-55. In *Cellulose in Wood*; TAPPI Press: Atlanta, GA, USA, 1955.
43. TAPPI T9 m-54. In *Holocellulose in Wood*; TAPPI Press: Atlanta, GA, USA, 1998.
44. Bagby, M.O.; Nelson, G.H.; Helman, E.G.; Clark, T.F. Determination of lignin non-wood plant fiber sources. *TAPPI J.* **1971**, *54*, 11.
45. Kawa-Rygielska, J.; Pietrzak, W. Zagospodarowanie odpadowe pieczywa do produkcji bioetanolu. *Żywność Nauka Technol. Jakość* **2011**, *79*, 105–118.

molecules

MDPI

Article

Effect of the Micronization of Pulp Fibers on the Properties of Green Composites

Bruno F. A. Valente [1], Armando J. D. Silvestre [1], Carlos Pascoal Neto [2], Carla Vilela [1] and Carmen S. R. Freire [1,*]

1 CICECO—Aveiro Institute of Materials, Department of Chemistry, University of Aveiro, Campus Universitário de Santiago, 3810-193 Aveiro, Portugal; bfav@ua.pt (B.F.A.V.); armsil@ua.pt (A.J.D.S.); cvilela@ua.pt (C.V.)

2 RAIZ—Research Institute of Forest and Paper, The Navigator Company, Rua José Estevão, 3800-783 Eixo, Portugal; Carlos.Neto@thenavigatorcompany.com

* Correspondence: cfreire@ua.pt

Abstract: Green composites, composed of bio-based matrices and natural fibers, are a sustainable alternative for composites based on conventional thermoplastics and glass fibers. In this work, micronized bleached Eucalyptus kraft pulp (BEKP) fibers were used as reinforcement in biopolymeric matrices, namely poly(lactic acid) (PLA) and poly(hydroxybutyrate) (PHB). The influence of the load and aspect ratio of the mechanically treated microfibers on the morphology, water uptake, melt flowability, and mechanical and thermal properties of the green composites were investigated. Increasing fiber loads raised the tensile and flexural moduli as well as the tensile strength of the composites, while decreasing their elongation at the break and melt flow rate. The reduced aspect ratio of the micronized fibers (in the range from 11.0 to 28.9) improved their embedment in the matrices, particularly for PHB, leading to superior mechanical performance and lower water uptake when compared with the composites with non-micronized pulp fibers. The overall results show that micronization is a simple and sustainable alternative for conventional chemical treatments in the manufacturing of entirely bio-based composites.

Keywords: poly(lactic acid); poly(hydroxybutyrate); cellulose fibers; micronization; green composites

Citation: Valente, B.F.A.; Silvestre, A.J.D.; Neto, C.P.; Vilela, C.; Freire, C.S.R. Effect of the Micronization of Pulp Fibers on the Properties of Green Composites. *Molecules* **2021**, *26*, 5594. https://doi.org/10.3390/molecules26185594

Academic Editors: Alejandro Rodriguez Pascual, Eduardo Espinosa Víctor and Carlos Martín

Received: 6 August 2021
Accepted: 9 September 2021
Published: 15 September 2021

1. Introduction

The increasing demand for eco-friendly materials associated with the implementation of legislation and policies towards a more sustainable society has triggered the replacement of synthetic and petrochemical-based materials with bio-based ones [1,2]. In the field of composite materials, as far as reinforcements are concerned, a notorious increase in the use of natural-based fibers in replacement of synthetic counterparts, such as glass or aramid, has been witnessed in the last decade. Several natural fibers such as flax, hemp, jute, kenaf, wood flour, or pulp have been thoroughly investigated [3,4]. In fact, the market of natural fiber-based composites, also commonly referred to as biocomposites, is already established (USD 22.3 billion in 2019) and some large companies, such as Stora Enso, UPM, and Sappi, have launched over the years a range of products composed of conventional fossil-based and non-biodegradable thermoplastics, such as polypropylene (PP) and polyethylene (PE), reinforced with cellulosic fibers [5–8].

PP and PE, together with other polymeric matrices, such as poly(vinyl chloride) (PVC), polystyrene (PS), and acrylonitrile butadiene styrene (ABS), are still the main thermoplastics used in the biocomposite industry [9]. However, and despite the clear environmental benefits over composites reinforced with synthetic fibers, biocomposites whose matrices are derived from fossil resources still pose some environmental threats. Specifically, the non-renewability and non-biodegradability of the matrices as well as the unfeasibility to recycle the composites are still their major drawbacks [10,11]. In this regard, the logical alternative is to manufacture fully bio-based composites by replacing the non-biodegradable petrochemical-derived matrices with bio-based polymers, the so-called bioplastics [12].

Poly(lactic acid) (PLA) and poly(hydroxybutyrates) (PHB) are among the few bioplastics currently produced at a commercial scale. They have comparable properties to some commodity plastics, can be processed with technologies applied to conventional thermoplastics, and, because of the increasing demand for bioplastics and maturing of production technologies, their prices are becoming more affordable [10,13,14]. The ever-increasing number of studies regarding the use of bioplastics in the composite field reflect the growing interest on these sustainable polymers [4]. However, these so-called green composites, for which both the matrix and reinforcement are bio-based, face some of the same challenges of their petroleum-based counterparts. Although the interfacial adhesion between for instance PLA and cellulose fibers is naturally stronger than for many other thermoplastic polymers, the lack of compatibility between the hydrophilic cellulose fibers and the hydrophobic matrices is still an issue [11,15]. The intrinsic hydrophilicity and high aspect ratio of the cellulosic materials often lead to agglomeration and poor dispersion of the fibers in the polymeric matrices [16]. Therefore, the visual aspect of the composites is inevitably impaired, as well as their mechanical performance. The strategies commonly used for composites with petrochemically based matrices have also been investigated for PLA and PHB-based counterparts to overcome these challenges and further increase the overall performance of such materials. Pre-treatments of the fibers, such as alkali treatments [17] or surface modifications including, for example, acetylation [18] or silylation [19], are among the most common. Other compatibilization strategies, such as the use of coupling agents [20–22], which can be done prior to or during melt-mixing, have also been tested. From the industrial point of view, however, the manufacturing process should be as simple, inexpensive, and efficient as possible. Thus, as an alternative to the aforementioned chemical methods, mechanical procedures may also be efficient to overcome some drawbacks. In this sense, size reduction by milling processes, such as pan milling [23], ball milling [24,25], or shear and cooling milling [26], may be used to decrease the length and width of the fibers to the micro or nano range. For instance, ball milling has been already used to reduce the size of bleached pine kraft pulp to particles sizes inferior to 120 μm [24,25]. The results showed that size reduction was an efficient method to increase the dispersion of cellulose materials in PLA or PLA/poly(3-hydroxybutyrate-*co*-3-hydroxyvalerate) (PHBV) based composites. In addition, the size reduced particles still had a reinforcing effect as proved by the increased mechanical properties. However, upon these mechanical treatments, the fibers lost their fibrillar morphology, becoming irregular particles instead. Additionally, the aspect ratio and crystallinity index (CI) of the fibers drastically decreased [24,25].

In this context, the present study aims at manufacturing fully green composites using bio-based matrices, namely PLA and PHB, reinforced with micronized bleached Eucalyptus kraft pulp (BEKP). Micronization is proposed as a mechanical treatment to reduce both the length and width of the fibers to the micrometric range, without compromising their fibrillar morphology and crystallinity but rather improving their dispersion within the polymeric matrices. This strategy has the additional advantage of being practical and free of any solvents or chemicals. The influence of the fiber load and effect of their aspect ratio were studied. The composites were evaluated regarding their interfacial morphology, mechanical performance, water uptake capacity, melt flowability, and thermal properties.

2. Results and Discussion

In the present work, the impact of the micronized fiber load and the effect of their aspect ratio on the properties of green composites with two distinct grades of both PLA and PHB matrices were evaluated. The four bio-based polymeric matrices (i.e., PLA 3D860, PLA 3100HP, PHB P209E, and PHB P226) were selected according to their mechanical properties, melt flow index, and recommended uses [27,28]. To assess the impact of the micronized fibers' load, Cel355 fibers with an intermediate aspect ratio (26.6) were melt-mixed with the thermoplastic matrices. To study the effect of the aspect ratio of the fibers, four micronized fibers with distinct aspect ratios and BEKP (Table 1) were also melt-mixed with all four thermoplastic matrices for a fixed fiber load of 40 wt.% (Figure 1).

Table 1. Average dimensions and crystallinity indexes of the cellulose fibers used in this study.

Fiber Sample	Length (μm)	Width (μm)	Aspect Ratio	Crystallinity Index (%)
Cel60	149 ± 129	13.6 ± 5.4	11.0	54.1
Cel200	257 ± 170	14.9 ± 4.6	17.2	65.4
Cel355	332 ± 211	12.5 ± 5.4	26.6	64.6
Cel500	405 ± 203	14.4 ± 4.5	28.9	68.4
BEKP	770 ± 0.006	18.2 ± 0.1	42.3	70.7

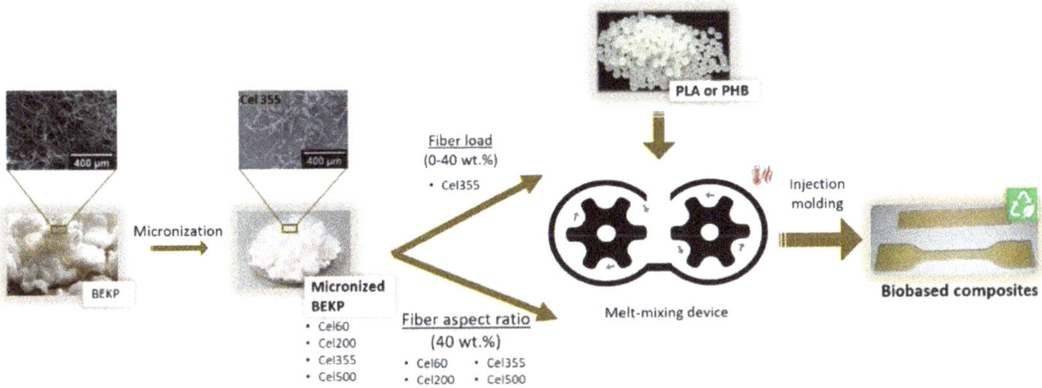

Figure 1. Schematic illustration of the experimental procedure used in the present study.

The macroscopic aspect of the composites specimens after the injection molding is shown in Figure 2. The composites with different loads of Cel355 (Figure 2A) are increasingly darker with the increment of the fiber load, particularly for the composites with PLA. This behavior can be attributed to the high temperatures used during the processing of these composites, particularly during injection molding (195 °C), which led to some thermal degradation of cellulose. A similar observation was reported by Ozyhar et al. [29] for composites of PLA reinforced with 40 wt.% of wood fibers, where the increase in the composites color intensity was also associated to some thermal degradation of the polysaccharides during processing. Besides the increased color intensity, no visible agglomerates of the fibers could be perceived for composites with different reinforcement loads or for those with 40 wt.% micronized fibers with distinct aspect ratios (Figure 2B), which might indicate that the fibers were homogeneously dispersed in the matrices. The density values of the polymeric matrices agreed with the specification of the products [27,28] (Table S1). As for the composites, as expected, an increase on the fiber load raised the density of the composite, given the superior density of the fibers [30] in comparison with the thermoplastic polymers used in the present work. Conversely, the density of the composites with fibers having different aspect ratios remained relatively unchanged (Table S1). Then, the morphology, mechanical properties, water-uptake capacity, flowability, and thermal stability of all the composites were evaluated.

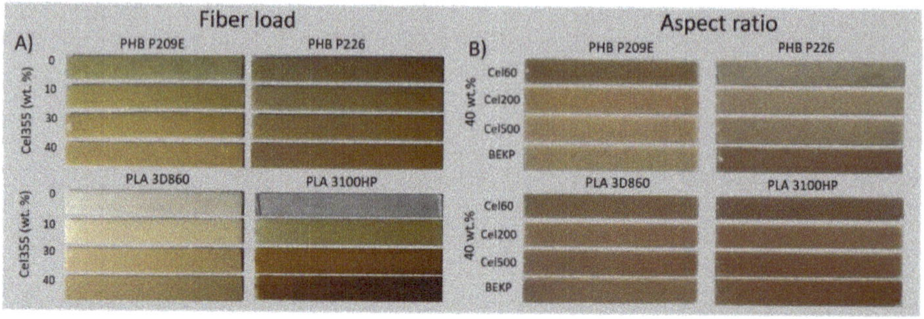

Figure 2. Digital photographs of the PHB- and PLA-based composites with: (**A**) different Cel355 loads (0, 10, 30 and 40 wt.%), and (**B**) with micronized fibers with distinct aspect ratios and non-micronized BEKP fibers.

2.1. Morphology

The mechanical properties of composites are highly dependent on the ability to efficiently transfer energy from the polymeric matrix to the fibers that in turn depends on the dispersion of the fibers and on the interfacial adhesion between them and the matrix [31]. The cross-section fracture surfaces of the neat matrices, obtained after tensile tests, and of the composites with different loads of Cel355 are displayed in Figure 3, while those of the composites with micronized fibers with different aspect ratios are shown in Figure 4. Despite the strong tendency of the hydrophilic cellulose fibers to stack together when compounded with hydrophobic thermoplastic matrices [31], no visible agglomerates or bundles of micronized fibers can be observed on the micrographs of the composites with different contents of Cel355 (Figure 3) or with fibers with different aspects ratios (Figure 4). This is a confirmation of the good dispersion of the micronized fibers in the PLA and PHB matrices, which agrees with the visual observation of the test specimens (Figure 2).

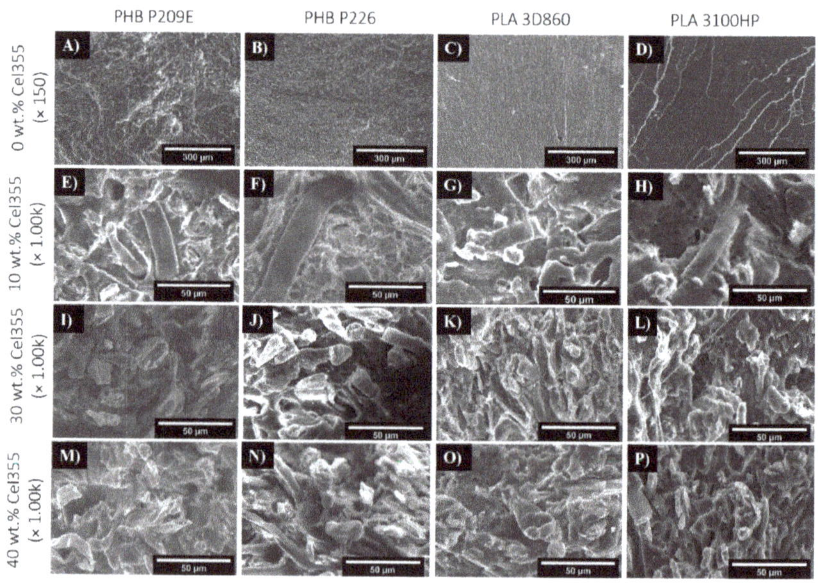

Figure 3. SEM micrographs of the fractured surfaces of (**A–D**) the neat PHB and PLA matrices and of (**E–P**) the corresponding composites reinforced with different Cel355 loads: (**E–H**) 10 wt.%, (**I–L**) 30 wt.% and (**M–P**) 40 wt.%.

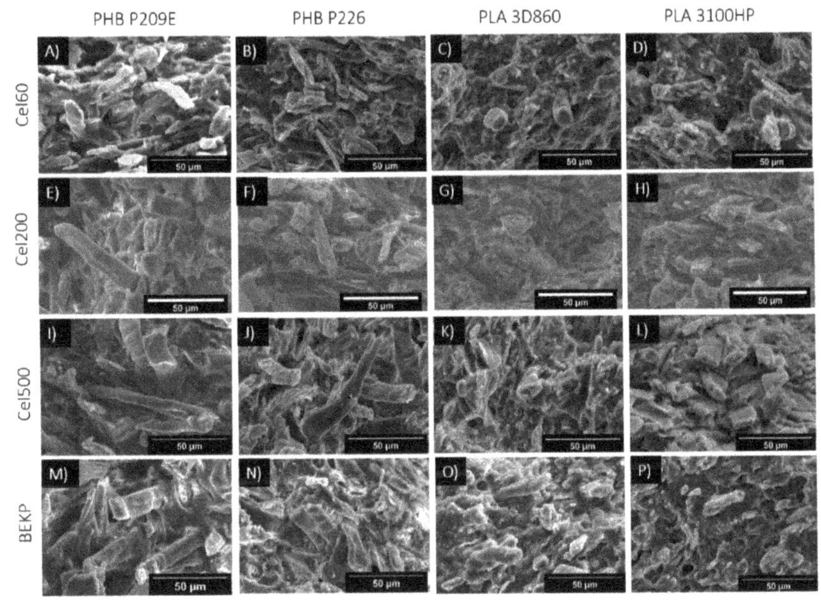

Figure 4. SEM micrographs of the fractured surfaces of composites with fibers having different aspect ratios: (**A–D**) Cel60; (**E–H**) Cel200; (**I–L**) Cel500 and (**M–P**) BEKP.

Besides this good dispersion, the fracture surfaces of the composites also revealed the existence of some fibers pull-outs and voids (Figure 3E, top). This phenomenon has been extensively described in the literature and is usually attributed to the different phobic nature of the constituents [32–34]. However, due to the carboxylic and hydroxyl end groups of PHB and PLA, there is a high degree of compatibility between these matrices and cellulose fibers when compared with common thermoplastic matrices, such as PP or PE [15,35]. The fact that fractured fibers are more prevalent on the micrographs than the pulled-out fibers is also evidence of such a compatibility. Additionally, the micrographs point to a slightly better compatibility of the fibers with PLAs than with the PHBs, which may be related to the chemical structural differences between these two polyesters, with PHBs having a longer monomeric chain.

Concerning the composites loaded with different micronized fibers (Figure 4), as the fiber aspect ratio increases, alongside with the length, the fiber pullouts and voids created during the tensile testing are more frequent and the pulled-out fibers are longer, as would be expected. In regard to the interfaces between the fibers and matrices, although only minor differences can be observed on the micrographs, shorter micronized fibers may have improved dispersion and embedment over the longer and higher aspect ratio non-treated BEKP. Such difference of the effect of the fiber dimensions on the interfacial morphology was described by Madyan et al. [36], in which larger fibers led to bigger gaps and cracks, and smaller fibers were better imbedded in the matrix.

2.2. *Mechanical Properties*

2.2.1. Tensile Properties

The Young's modulus, tensile strength, and elongation at break of the polymeric matrices and of the corresponding composites are presented in Figure 5. The results show, for both PLA- and PHB-based composites, a gradual increment in the Young's modulus with the increasing Cel355 fiber load, which is in agreement with previous findings on the effect of the fiber load, and can be easily explained by the higher stiffness of the cellulose fibers compared to the polymeric matrices [37]. Specifically, the Young's modulus of composites reinforced with 40 wt.% Cel355 increased by 1.7 GPa (2.37 ± 0.06 GPa) and 2.6 GPa (3.83 ± 0.02 GPa) in comparison with the respective PHB matrices (PHB P209E: 0.64 ± 0.02 GPa; PHB P226: 1.25 ± 0.07 GPa), and by 2.8 GPa (5.16 ± 0.05 GPa) and 3.1 GPa (5.83 ± 0.06 GPa) compared to the corresponding PLA matrices (PLA 3D860: 2.40 ± 0.04 GPa; PLA 3100HP: 2.75 ± 0.04 GPa). As the biggest increases are noted for the PLA-based composites, those findings corroborate the relatively better compatibility of the fiber with PLAs than with PHBs, as previously discussed.

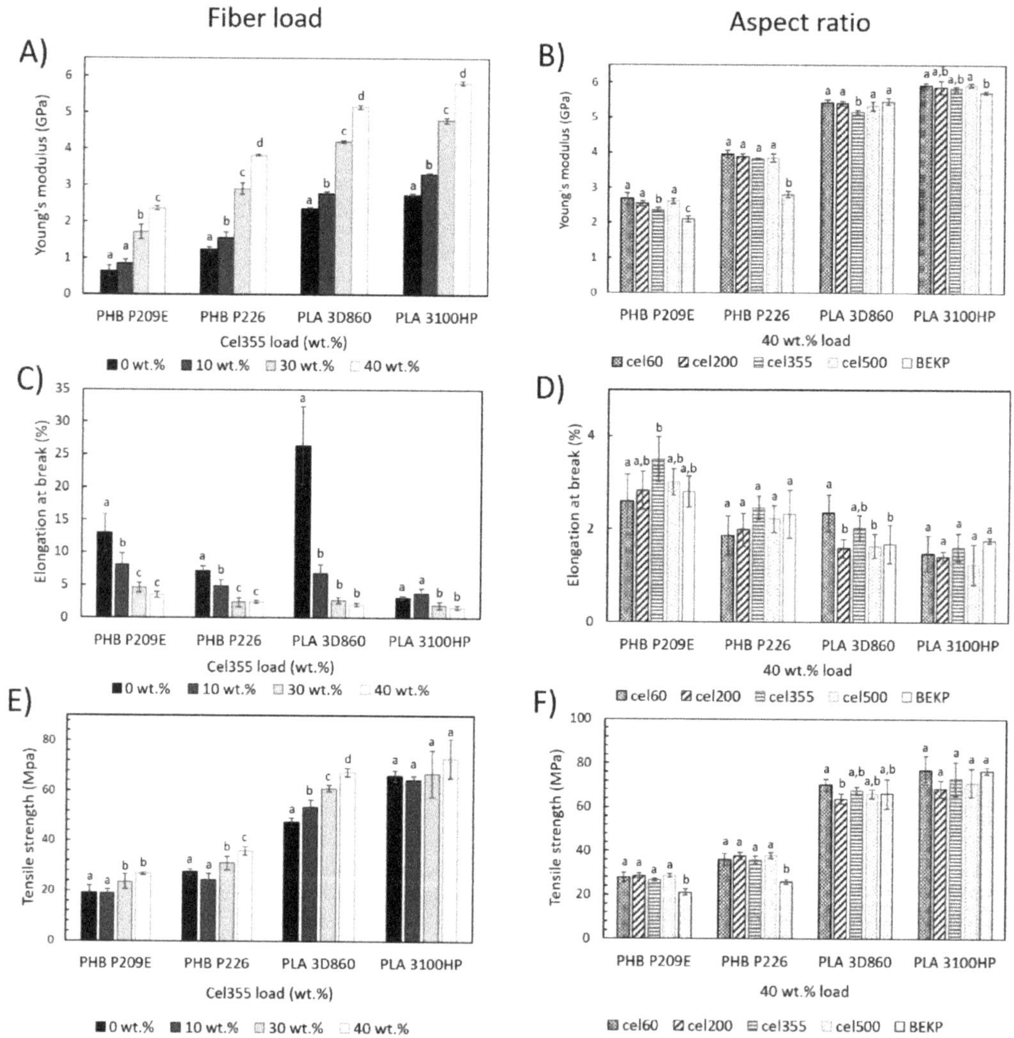

Figure 5. Tensile properties of the PHB and PLA-based composites (**A**,**C**,**E**) reinforced with different loads of Cel355 and (**B**,**D**,**F**) reinforced with fibers having different aspect ratios for a load of 40 wt.%. Different letters (a,b,c,d) indicate statistically significant differences ($p < 0.05$).

In previous studies of PLA reinforced with a 40 wt.% load of bleached softwood kraft pulp, a similar increase of 3.0 GPa was registered when compared to the matrix [38], which shows that the micronized fibers used in this work, despite their reduced sizes, still have a good reinforcing effect. The effectiveness of such a reinforcing effect is closely related to the fibrillar morphology and crystallinity (Table 1) of the micronized fibers since, unlike other mechanical treatments (e.g., ball milling), the micronization of the fibers still retained their fibrillar morphology and only a slight decrease in the crystallinity was observed [24,25]. Moreover, the Young's modulus of the composites of PHBs reinforced with 40 wt.% Cel355 were within the range of the commercial products based on PP or PE reinforced with 40 wt.% pulp fibers (1.9 to 4.6 GPa) and the Young's modulus of the PLA-based composites were clearly superior to those of the mentioned commercial products [5–7].

Concerning the effect of the aspect ratio on the Young's modulus, for the most part, changes in the micronized fiber aspect ratio (in the range between 11.0 and 28.9) had little influence on this parameter (Figure 5B). Qiang et al. studied the effect of size variations of ball-milled bleached pine kraft pulp in composites with PLA or binary mixtures of PLA and poly(3-hydroxybutyrate-*co*-3-hydroxyvalerate) (PHBV), and also concluded that the loading content contributes more to the variation of the mechanical properties than the fiber aspect ratio [24,25]. Interestingly, PHB-based composites reinforced with micronized fibers had a significantly superior Young's modulus than the composites reinforced with BEKP. In fact, the average Young's modulus of the PHB composites with the micronized fibers is superior to the corresponding composites with BEKP by 26% for PHB P209E and 29% for PHB P226.

With respect to the elongation at break, as expected, a reduction in this parameter was generally observed with the increasing fibers content [10]. All composites with 40 wt.% reinforcement had elongation at break values below 3.5% (Figure 5C), which is inferior to those of the commercial products mentioned before (4.0 to 6.5%) [5–7]. In contrast, with only a few exceptions, the aspect ratio of the fibers did not have a significant effect on the elongation at break of the composites, as portrayed in Figure 5D.

The representation of the tensile strength of the composites as a function of the fiber load (Figure 5E) shows that higher reinforcement contents of micronized fibers generally raised the tensile strength. For example, in composites with 40 wt.% Cel355, an improvement on the tensile strength of 39, 30, 42, and 10% was noted in comparison with the corresponding PHB P209E, PHB P226, PLA 3D860, and PLA 3100HP matrices. The present results contradict some published works in which often the increase in the fiber load tended to decrease the tensile strength, either for PLA or PHB matrices [37,39]. Additionally, the tensile strength values of composites of PLA 3D860 (67.5 $\pm$ 1.7 MPa) and PLA 3100HP (72.9 $\pm$ 7.7 MPa) reinforced with 40 wt.% Cel355 were superior to those of commercial products of PP reinforced with pulp fibers (47 to 64 MPa) [5–7]. These results are certainly related to the relatively good dispersion of the micronized fibers within the polymeric matrices, as well as to their interfacial adhesion, as previously observed by SEM analysis. The composites with micronized fibers having different aspect ratios had similar tensile strengths (Figure 5F), which has also been previously reported for composites of PLA and mixtures of PLA and PHBV [24,25]. Despite the almost negligible effect of the aspect ratio, the tensile strength of all the PHB composites reinforced with the micronized fibers was far superior to those with BEKP, which agrees with the Young's modulus trend previously described.

2.2.2. Flexural Properties

The effect of the reinforcement of Cel355 fibers on the flexural modulus of composites based on PLA and PHB matrices was in line with the results of the Young's modulus previously discussed, i.e., the increase in the fiber load led to a higher flexural modulus (Figure 6A). This is not surprising, though, given that the flexural modulus often follows the same pattern as the Young's modulus [39,40]. Moreover, the results obtained for PHBs reinforced with 30 wt.% Cel355 (2.5 $\pm$ 0.1 GPa for PHB P209E and 3.8 $\pm$ 0.1 GPa for PHB P226) are similar or even better to those obtained by Gunning et al., in which PHB matrices were reinforced with 30 wt.% hemp ($\approx$1.6 GPa), Lyocell ($\approx$1.9 GPa) and Jute ($\approx$3.8 GPa) fibers [31]. For PLA-based composites, the flexural modulus of the composites reinforced with 40 wt.% Cel355 (7.3 $\pm$ 0.1 GPa for PLA 3D860 and 8.1 $\pm$ 0.2 GPa for

PLA 3100HP) was even superior to the values reported for PLA reinforced with 40 wt.% of sisal (≈6.2 GPa) [41], wood (≈4.5 GPa) [37], or with pulp fibers from poplar wood (≈5.7 GPa) [37]. On the contrary to the effect observed for the fiber load, the fiber aspect ratio had little influence on the flexural modulus of the composites. However, PHB-based composites reinforced with BEKP had an inferior modulus than the composites reinforced with micronized fibers (Figure 6B), which is also in agreement with the variation in the Young's modulus. Such results are probably related to the larger dimensions of the BEKP that can lead to larger gaps and cracks, creating weak points for composites to fail [36]. On the contrary, the shorter micronized fibers may have improved distribution and embedment on the matrix, improving its reinforcing effect [42].

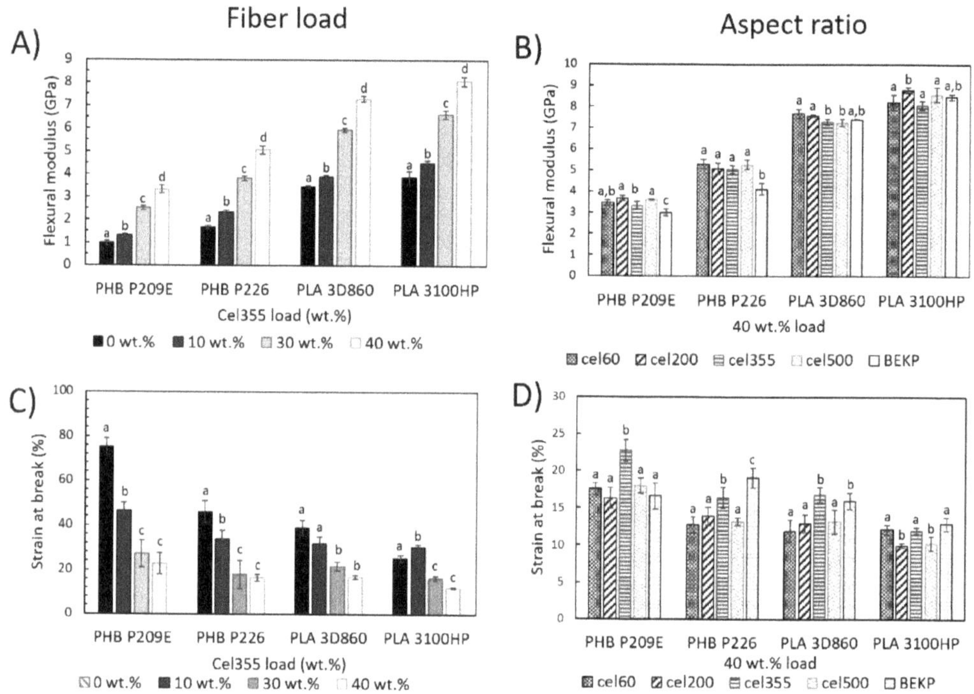

Figure 6. Flexural properties of) the PHB- and PLA-based composites reinforced (**A**,**C**) with different loads of Cel355 and (**B**,**D**) with fibers having different aspect ratios for a load of 40 wt.%. Different letters (a,b,c,d) indicate statistically significant differences ($p < 0.05$).

The strain at break, calculated as the ratio between the extension at break and the maximum deflection (20 mm), provides important information about the flexibility of the composites. As observed in Figure 6C, the polymeric matrices are more flexible than the corresponding composites and higher fiber content led to lower flexibility, which can be also credited to the high stiffness of the fibers in comparison with the matrices [10]. For example, the incorporation of only 10 wt.% of Cel355 in the most flexible matrix (PHB P209E) led to a reduction of the strain at break from 75.1 ± 4.1% to only 46.3 ± 4.9%. However, only small differences could be observed for the strain at break in the composites reinforced with fibers having different aspect ratios. The results of the strain at break are, in general, in agreement with the elongation at break determined on the tensile tests.

2.2.3. Impact Properties

The impact strength of neat polymeric matrices and composites as a function of the fiber load and aspect ratio were evaluated following the Charpy edgewise impact test (Figure 7). Regarding the polymeric matrices, the impact strength of the unnotched specimens of the PHBs, which are 38.0 ± 8.7 kJ m^{-2} for PHB P209E and 18.2 ± 5.2 kJ m^{-2} for PHB P226, were within the range of impact strengths reported in the literature, which can vary from 5 kJ m^{-2} to over 65 kJ m^{-2} [43,44]. The impact strength of PLA 3100HP (21.5 ± 1.8 kJ m^{-2}) was also near the values reported elsewhere [44,45]. However, the energy required to break an unnotched specimen of PLA 3D860 (121.4 ± 32.4 kJ m^{-2}) was more than three times higher than for any PHB matrix studied and more than five times higher than for PLA 3100HP, which can be justified by the fact that PLA 3D860 is designated by the manufacturers as a high impact polymer [28].

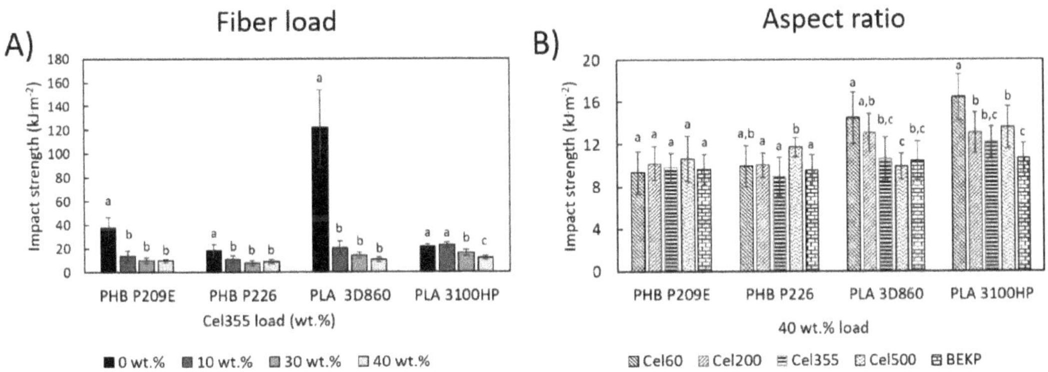

Figure 7. Impact properties of PHB- and PLA-based composites (**A**) reinforced with different loads of Cel355 and (**B**) reinforced with fibers having different aspect ratios for a load of 40 wt.%. Different letters (a,b,c) indicate statistically significant differences ($p < 0.05$).

The incorporation of micronized fibers (Cel355) significantly reduced the ability of the material to absorb the impact (Figure 7A). In fact, the most accentuated decreases were noted for composites with only 10 wt.%. of fibers. For higher cellulose contents, the decreases were not significantly different. In the literature, the effect of fiber incorporation in the impact strength of biocomposites is still unclear and contradictory. Previous studies have shown both increases and decreases in the impact strength upon incorporation of cellulosic fibers [46,47]. It's known that many factors may influence the composite's impact properties, such as the crystallinity and stiffness of the individual components. However, and despite the good homogeneity and dispersion of micronized fibers on the matrices, as observed in the SEM images (Figure 3), some defects on the interface may still be responsible for the deterioration of the impact strength [37,48]. Such defects on the interface, even at lower fiber loads, leads to crack initiation and propagation, which are responsible for the decreased amount of force the material can absorb during impact [32]. Nonetheless, the obtained values are in line with literature data. For instance, Oliver-Ortega et al. recorded an impact strength of 21.7 ± 1.2 kJ m^{-2} in composites of PLA reinforced with 10 wt.% bleached kraft softwood pulp [45], which is similar to the impact strength obtained for PLA 3D860 reinforced with 10 wt.% of Cel355 (20.4 ± 5.6 kJ m^{-2}) and PLA 3100HP (22.8 ± 2.3 kJ m^{-2}).

In Figure 7B, the results indicate that for PHB matrices, the changes in the aspect ratio of the fibers did not translate into different impact strength properties. In contrast, for composites based on PLA, smaller aspect ratio fibers seemed to favor the impact strength. The relatively better interfacial adhesion between the PLA matrices and fibers, as seen by SEM, combined with their smaller aspect ratio, reduced the formation of defects on the composite, consequently lowering sites for initiation and propagation of cracks [36]. In comparison with other materials, the obtained impact properties of composites with a fiber load of 40 wt.%, ranging from 9.0 ± 1.8 kJ m^{-2} to 16.4 ± 2.2 kJ m^{-2}, are inferior to those of the commercial products of PP or PE reinforced with 40 wt.% pulp fibers (33 to 42 kJ m^{-2}) [5–7], which still presents a challenge.

2.3. Water Uptake Capacity

The evaluation of the water absorption of the composites reinforced with natural fibers is of enormous importance because it is normally associated with dimensional stability issues and a decrease of the mechanical properties [49]. The water absorption of the neat PLA and PHB matrices used in this study were under 1.3 ± 0.1%, after 31 days of immersion (Figure 8), which is in accordance with the values reported in the literature for PLA [40] and PHB [50] matrices. Since cellulose has high affinity for water [51,52], all composites showed higher water absorptions that increased with the growing fiber contents, which is not surprising given there is unanimity in the fact that increasing the content of the hydrophilic portion of the composites leads to an increased water absorption [22,45]. Moreover, all composites followed the same uptake pattern: a rapid increase during the first few days of immersion and stabilization after reaching saturation [53]. It is also noticeable that composites with PHBs had higher water-uptakes than composites with PLAs. This observation can also be related to the apparent inferior compatibility between the fibers and PHBs, as previously discussed (Figure 3), which leads to enhanced water penetration thought the composite material [45,51]. These current findings contradict the observations made by Yatigala et al., in which wood fiber-reinforced PLA composites had higher-water uptakes (11.9%) than the PHB counterpart (10.2%) [22]. However, and more importantly, the composites prepared in this work with 30 wt.% of the micronized fiber (Cel355) had significantly inferior water-uptakes (6.4 ± 0.1% to 6.6 ± 0.1% for PHBs and 4.6 ± 0.1% to 4.9 ± 0.1% for PLAs) than the ones previously reported by the authors for the same reinforcement percentage [22].

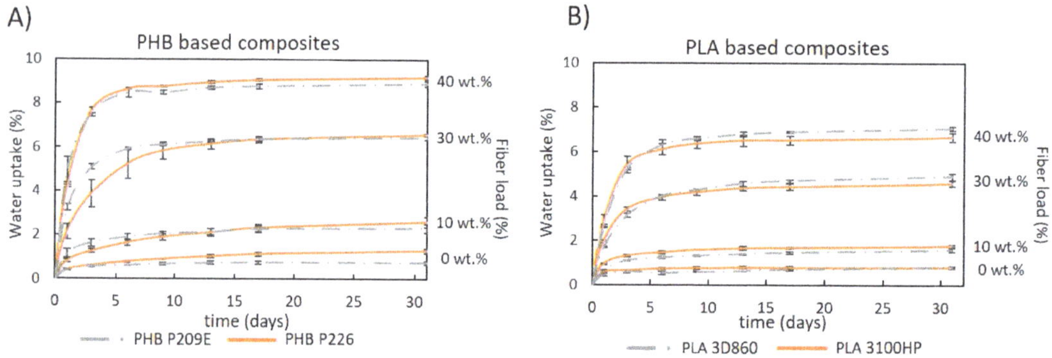

Figure 8. Water uptake as function of time for: (**A**) PHB- and (**B**) PLA-based composites reinforced with Cel355.

Interestingly, composites reinforced with micronized fibers have a clear advantage over the ones reinforced with non-micronized BEKP (Figure 9). Apart from those of PLA 3100HP, composites with BEKP not only have higher water-uptake values (up to 11.1 ± 0.2%) but also higher water absorption rates during the first days, mainly for composites with PHBs. The positive effect of micronization on the water-uptake behavior is in line with the results of the tensile and flexural moduli, which strengthens the idea that the reduced size of micronized fibers favors their embedment in the matrix, preventing water from permeating easily into the composite [51]. In reference to the aspect ratio of the micronized fibers (11.0 to 28.9), with the exception of composites based on PLA 3100HP, all composites with different micronized fibers had similar water-uptake values.

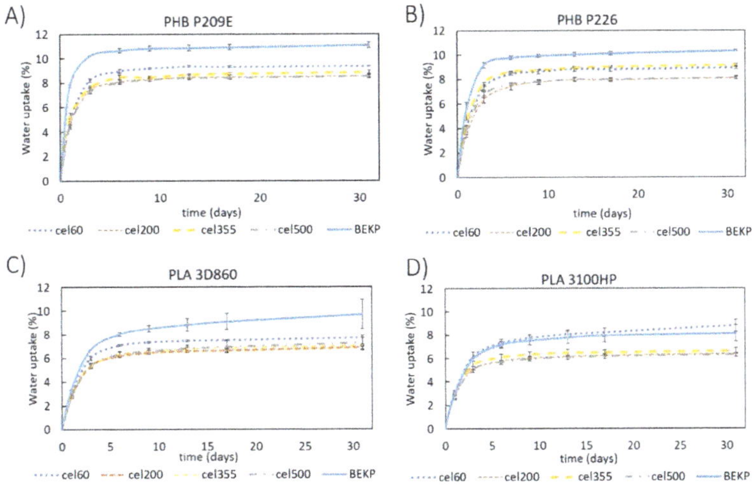

Figure 9. Water uptake as function of time for composites prepared with different micronized cellulose samples for a fiber load of 40 wt.% in matrices of: (**A**) PHB P209E; (**B**) PHB P226; (**C**) PLA 3D860 and (**D**) PLA 3100HP.

2.4. Melt Flow Rate

The melt flow rate measures the ease of a molten thermoplastic material to flow under very specific conditions, such as the diameter and length of the die and the cylinder [54]. The outcome is expressed in grams of the material that flows over the course of ten minutes when standard weights are applied at a predetermined temperature [54]. Figure 10A,C represent the melt flow rate of the PLA and PHB matrices prior to and after melt-mixing, along with the melt flow rate of the composites with different fiber loadings, while Figure 10B,D show the melt flow of the composites having fibers with different aspect ratios.

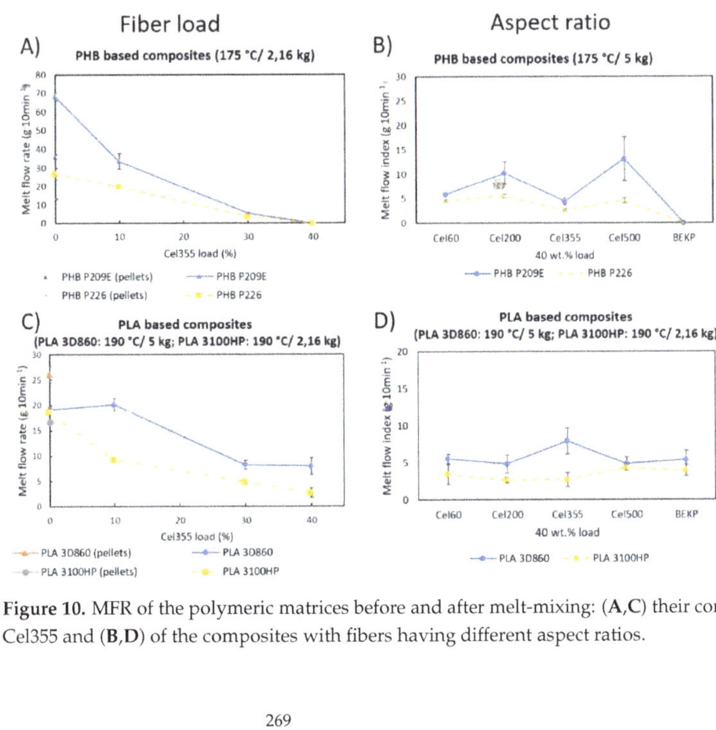

Figure 10. MFR of the polymeric matrices before and after melt-mixing: (**A**,**C**) their composites with Cel355 and (**B**,**D**) of the composites with fibers having different aspect ratios.

According to Figure 10A,C, the melt flow rate of the polymeric matrices increased after being submitted to the melt-mixing procedure, except for PLA 3D860, probably due to the presence of additives in this PLA sample. The effect is more pronounced for the PHBs, with increments of 90% for PHB P209E and a 103% increase for PHB P226. Giving that there is an inversely proportional relationship between the MFR and the molecular weight of the thermoplastic polymers, the increase in the MFR is most likely due to some thermal degradation during the melt-mixing procedure, which certainly leads to the decrease in the molecular weight of the polymeric matrices [55]. This is supported by previous studies as Carrasco F. et al., who studied the influence of melt processing on the molecular weight and melt flowability of PLA, concluding that the thermal degradation caused by melt processing decreased the molecular weight of PLA from 212.3 kDa to 162.5 kDa. As a direct consequence, the melt flow rate increased from 7.0 $g \cdot 10\ min^{-1}$ to 10.7 $g \cdot 10\ min^{-1}$ [56]. A similar outcome was verified for PHB matrices where melt processing decreased the molecular weight from $535 \times 10^3\ g \cdot mol^{-1}$ to $208 \times 10^3\ g \cdot mol^{-1}$, leading the MFR to raise from 19 $g \cdot 10\ min^{-1}$ to 26 $g \cdot 10\ min^{-1}$ [57].

When the micronized fibers (Cel355) were added to the matrices, the MFR gradually decreased. Similar observations have been reported for composites of PLA or PHB reinforced with different natural fibers (e.g., jute and hemp) [31,58] and the decrease was mainly attributed to the bad dispersion of the fibers within the matrices and to the fiber/fiber and fiber/matrix frictions [31,58]. However, since the composites in the present work show a good dispersion of the micronized fibers, as previously discussed, the decrease in the MFR probably has to do with the increase in the friction between fibers and between fibers with the matrices, which its turn decreases the ease of flow [58]. Moreover, the most pronounced decreases in the MFR were noted for PHB-based composites, which, according to the SEM micrographs, were slightly less compatible with the fibers than the PLA matrices. The poorer interfacial adhesion may increase the friction between the fibers and the matrices, leading in turn to a decreased flowability of the composite material [59].

The fiber aspect ratio had little effect on the MFR, especially for PLA-based composites. However, for the composites based on PHBs, the micronization had a positive effect on the MFR. All the composites reinforced with micronized fibers showed MFRs between $4.5 \pm 0.3\ g \cdot 10\ min^{-1}$ and $13.2 \pm 4.5\ g \cdot 10\ min^{-1}$, while those reinforced with BEKP did not flow at all. In composites with PHBs, where the compatibility with fibers is slightly inferior than with PLAs, as previously observed by SEM (Figures 3 and 4), the BEKP fibers, which are considerably longer and wider than the micronized fibers, may lead to a higher fiber entanglement, preventing the composite material from flowing [31,60]. In contrast, the shorter micronized fibers with improved dispersion and embedment may have prevented fiber entanglement. Such an observation emphasizes that the micronization of the BEKP fibers is a useful strategy to improve the flowability of the composites, which is of great importance for injection molding applications to ensure proper mold-filling and good quality of the materials [31,61].

2.5. Thermal Analysis

The thermal stability of all the thermoplastic polymers and composites were studied by thermogravimetric analysis (TGA). The thermogravimetric curves of PLA and PHB matrices, as well as of the corresponding composites with different fiber loads, are shown in Figure 11 and those of the composites with fibers having different aspect ratios are represented in Figure S5. The neat polymeric matrices showed a degradation profile with one major weight loss at 236 °C and 281 °C for PHB P209E and PHB P226, respectively, and 336 °C and 364 °C for PLA 3D860 and PLA 3100HP, respectively. These main weight losses, credited to the degradation of the polymeric backbone, are in agreement with the maximum degradation temperatures of other PHB [62,63] and PLA [34,64] grades reported in the literature. Apart from PLA 3100HP, the other matrices showed a small degradation step at 338 °C, 395 °C, and 456 °C for PHB P209E, PHB P226, and PLA 3D860, respectively, which were likely due to the degradation of the additives.

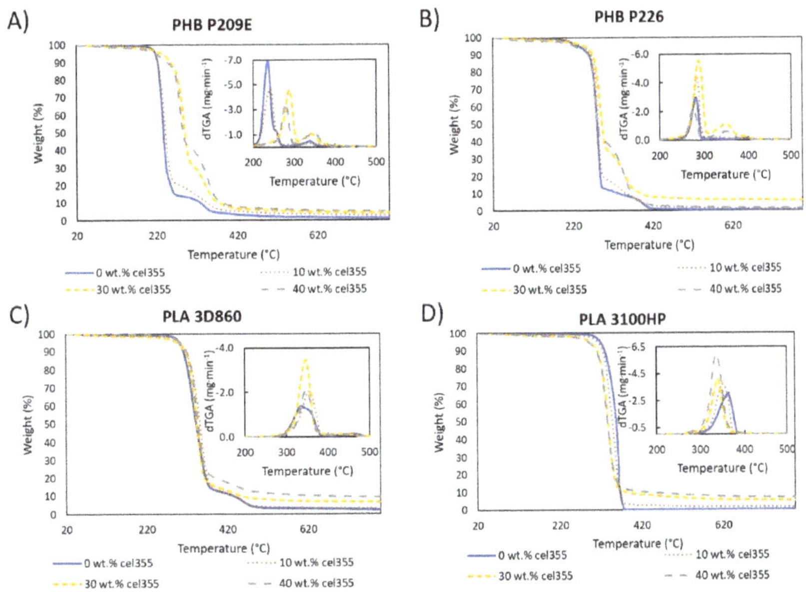

Figure 11. Thermogravimetric and derivative (dTGA) curves of the pure polymeric matrices and composites with different loads of Cel355 (10, 30, and 40 wt.%) in matrices of: (**A**) PHB P209E; (**B**) PHB P226; (**C**) PLA 3D860 and (**D**) PLA 3100HP.

PHB-based composites have three different degradation steps: (i) a small weight loss attributed to water vaporization starting at approximately 100 °C, associated with the humidity of the fibers; (ii) a main weight loss step between 236 and 287 °C, related to the degradation of PHBs' main chain; and (iii) the step related to the thermal degradation of cellulose between 335 and 350 °C (Figure S4c) [22,65]. Additionally, the weight losses associated with this last step match the different reinforcement loads of the fibers. These ternary degradation profiles have also been reported by other authors for PHB-based composites loaded with other natural fibers (e.g., wood fibers and flax) [22,66] The maximum degradation temperatures of PHB P226-based composites were similar to those of the pristine matrix (around 280 °C), but for composites with PHB P209E (Figure 11A), the incorporation of high contents of cellulose fibers, e.g., 40 wt.% load, increased the maximum thermal degradation from 236 °C to over 280 °C. This increase proves that, although the compatibility of PHB with the micronized fibers is not excellent, as seen in the SEM micrographs, there is some degree of interfacial adhesion between them [35]. Even if increases in the thermal stability have been reported in other studies, as is the example of PHB reinforced with agave fibers [39], the effects of the fiber incorporation on the thermal stability of PHB-based composites are debatable, with most studies showing decreases on the stability upon incorporation of natural fibers such as piassava [67], flax [66] almond shell, or rice husk [68].

For composites with PLA, however, given the relatively good interfacial adhesion between the micronized fibers and these thermoplastic polymers, as well as the similar thermal stabilities of PLAs and micronized cellulose fibers, only a main weight loss can be observed on the thermograms in Figure 11. Regarding the composites with PLA 3D860 as the matrix, the incorporation of fibers led to an increase in the maximum degradation temperature of the material (Figure 11C). For instance, the addition of 40 wt.% of Cel355 raised the maximum degradation temperature by 11 °C from 336 °C to 347 °C, which can be justified by the lower maximum degradation temperature of PLA 3D860 (336 °C) in comparison with Cel355 (350 °C) (Figure S4c). On the contrary, because the maximum degradation temperature of PLA 3100HP (364 °C) was superior to that of Cel355, the gradual increase on the fiber load led to a slight reduction on the maximum degradation temperature of the composite (Figure 11D). In a similar research study, Espinach et al. [64] investigated the thermal properties of composites made of PLA reinforced with bleached kraft softwood and also concluded that the decrease in the thermal stability of the composites was due to the presence of a cellulosic filler with a lower thermal stability than the PLA. The existence of only one peak also strengthens the relatively good compatibility between PLA and cellulosic fibers.

Analogous to the composites with different fiber loads, composites based on PHBs reinforced with 40 wt.% of micronized fibers with different aspect ratios displayed the same ternary degradation profile and PLA-based composites only exhibited one major single degradation step. The similar degradation patterns and degradation temperatures observed in Figure S5 allow us to withdraw the conclusion that the aspect ratio of the micronized fibers does not influence the thermal stability of the composites. This conclusion corroborates the results obtained for the tensile, flexural, and impact mechanical properties, as well as the results from MFR in the sense that different aspect ratios of the micronized fibers gave origin to composites with identical properties.

3. Materials and Methods

3.1. Materials

Two poly(lactic acid) (PLA) pellet samples: Ingeo™ Biopolymer 3D860, with a melt flow rate (MFR) of 5–7 $g \cdot 10\ min^{-1}$ (210 °C, 2.16 kg), density of 1.22 $g \cdot cm^{-3}$, and relative viscosity of 4.0, and Ingeo™ Biopolymer 3100HP with an MFR of 24 $g \cdot 10\ min^{-1}$ (210 °C, 2.16 kg), density of 1.24, relative viscosity of 3.1, and molecular weight of 148 kDa [69], were supplied by NatureWorks (Minnnetonka, Minnesota, USA). Two commercial pellet grades of poly(hydroxybutyrate) (PHB): Biomer® P209E, with an MFR of 10 $g \cdot 10\ min^{-1}$ (180 °C, 2.16 kg), density of 1.20 $g \cdot cm^{-3}$, and molecular weight of 0.6×10^6 [70], and Biomer® P226 with an MFR of 10 $g \cdot 10\ min^{-1}$ (180 °C, 5 kg), density of 1.25 $g \cdot cm^{-3}$, and number-average molecular weight of 22,200 ± 4500 [71], were purchased from Biomer Biopolyesters (Schwalbach, Germany). PHBs samples had 0–40 wt.% of a plasticizer and an unreported amount of a nucleating agent [27]. The chemical and crystalline structures of the thermoplastic matrices were confirmed through Fourier-transform infrared spectroscopy (FTIR) and X-ray diffraction, as shown in Figures S1 and S2, respectively.

Bleached Eucalyptus kraft pulp (BEKP) with a crystallinity index (CI) of 70.1% was kindly provided by The Navigator Company (Aveiro, Portugal). The micronized fibers were obtained from the same BEKP on a drum milling equipment (Pallman Fine grinding PS, Zweibrücken, Germany) armed with knives. The BEKP was fed into the cutting chamber and cut repeatedly between the rotor knifes and stator knifes against each other until the material could pass through the screen insert. Different sieve meshes were used to produce fibers with different aspect ratios, namely Cel60 (11.0), Cel200 (17.2), Cel355 (26.6), and Cel500 (28.9). The micrographs showing their fibrillar morphology as well as the length and width distributions are presented in Figure S3, and the average length, width, aspect ratio, and crystallinity indexes are shown in Table 1. The CI of BEPK and of all the microfibers (Figure S4b) were determined by X-ray diffraction based on the peak height method. A Phillips X'pert MDP diffractometer (PANalytical, The Netherlands) using CuKα radiation (λ = 1.541 Å) with a scan rate of 0.05° s^{-1} was used for this analysis [72].

3.2. Compounding and Processing of the Biocomposites

Composites with different percentage loads of Cel355 (micronized fibers with an intermediate aspect ratio) ranging from 10 to 40 wt.%, relative to the total weight of the composite, were compounded with the four distinct thermoplastic polymers: PHB P209E, PHB P226, PLA3D860, and PLA3100HP. Additionally, to evaluate the influence of the fibers aspect ratio, the four thermoplastic matrices were also compounded with the four micronized pulp fibers and the non-treated BEKP for a fixed reinforcement load of 40 wt.%.

All composites were manufactured by melt-mixing in a Brabender W 30 EHT Plastograph EC mixer (Duisburg, Germany) with a total volume capacity of 30 cm^3. The thermoplastic polymeric matrices and cellulose fibers were mixed for 15 min at a screw speed of 50 rpm at a temperature of 180 °C for PLA 3100HP and 170 °C for the remaining polymers.

Test specimens for the mechanical and water-uptake assays were prepared by injection molding in a Thermo Scientific Haake Minijet II (Waltham, MA, USA). For the PHB-based composites, the injection temperature was set at 175 °C, the mold temperature at 60–65 °C, injection pressure of 400 bar for 20 s, and post-injection pressure of 200 bar for 5 s. For the PLA-based composites, the injection temperature was 195 °C, with a mold temperature of 100–130 °C and injection pressure of 800 bar.

3.3. Characterization

The density of the polymeric matrices and composites were calculated by dividing the weight of the test specimens by their volume. At least five specimens (80 × 10 × 4 mm^3) with a volume of 3.2 cm^3 were weighted for each sample and both the mean and standard deviation calculated.

Scanning electron microscopy (SEM) analysis were performed on a FE-SEM Hitachi SU70-47 microscope (Hitachi High-Technologies Corporation, Tokyo, Japan), operated at 15.0 kV. The cross-section micrographs were obtained from the tensile test specimens after breaking. Prior to the analysis, all samples were coated with a carbon film.

Tensile and flexural assays were carried out on the universal testing machine Instron 5564 (Instron Corporation, Norwood, MA, USA). The tensile properties of at least six specimens were tested in accordance with the ISO-527-2 procedure (bar type3) at a velocity of 5 mm·min^{-1} using a 10 kN static load cell. For the flexural modulus, the three-point loading model was used according to ISO 178. Five specimens (80 × 10 × 4 mm^3) were tested at a crosshead velocity of 5 mm·min^{-1} and at a length of span between supports of 64 mm using a 500 N static load cell. To measure the unnotched Charpy (edgewise) impact strength, a Ray Ran Universal Pendulum impact system (Ray-Ran Test Equipment Ltd., Nuneaton, UK), operating a pendulum of 4 J, was used according to ISO 179/1eU. The support span was set at 62 mm. Ten specimens with dimensions of 80 × 10 × 4 mm^3 were tested for each sample and the average values were calculated.

The water-uptake capacity was assessed by immersing composite specimens (60 × 10 × 1 mm^3) in water at room temperature during a period of 31 days. The weight of the samples was periodically assessed after removing the excess water with tissue paper. The water uptake (%) at time t was calculated according to Equation (1):

$$Water\ uptake\ (\%) = \frac{(W_t - W_0)}{W_0} \times 100 \quad (1)$$

where W_0 is the specimen's initial weight and W_t is the weight of the specimens after the immersion time in grams. The mean and standard deviation were calculated for three replicates.

The melt flow rate of the different samples was evaluated using the Melt Flow Indexer Davenport (MFR-9) (Ametek, Denmark) operated at 190 °C for the PLA-based composites and at 175 °C for the PHB-based composites. The cut-off intervals were chosen according to the ASTM D1238 standard. At least five cut-offs for each sample were weighted and the melt flow rate was calculated as follows:

$$MFR\left(g \cdot 10\ min^{-1}\right) = \frac{600 \times m}{t} \quad (2)$$

where m is the average mass of the cut-offs in grams and t is the cut-off time interval in seconds.

The thermal stability of the composites was evaluated with a SETSYS Setaram TGA analyzer (SETARAM Instrumentation, Lyon, France) equipped with a platinum cell.

Approximately 10 mg of each sample was heated from room temperature to 800 °C at a constant rate of 10 $°C·m^{-1}$ under a nitrogen atmosphere.

Statistical analysis of all the mechanical properties data was performed using the analysis of variance (ANOVA) and Tukey's mean comparison test (OriginPro 9.6.5, OriginLab Corporation, Northampton, MA, USA) with the statistical significance established at $p < 0.05$.

4. Conclusions

Micronized cellulose fibers obtained from bleached Eucalyptus kraft pulp were investigated as reinforcements in composites with bio-based matrices (PLA and PHB). The influence of the fiber load and fiber aspect ratio on the performance of the composites was thoroughly studied.

The composites with micronized fibers displayed overall good homogeneity with no visible fiber agglomerates. Reinforcing the composites with increasing contents of micronized fibers significantly improved the Young's modulus, tensile strength, and flexural modulus, while decreasing the elongation at break, strain at break, and the impact strength. The increase in the water uptake and the decrease of the melt flow rate were also consequences of the increasing reinforcement content. For the most part, the maximum degradation temperatures of the composites were slightly raised with the incorporation of the fibers, especially in composites based on PHB P209E and PLA 3D860.

For composites with PHBs, the water-uptake resistance, melt flowability, and mechanical performance, namely the tensile strength and both the Young's and flexural moduli, were significantly improved when using micronized fibers rather than non-treated BEKP. However, the aspect ratio of the micronized fibers had little influence on the tensile, flexural, and impact properties, as well as on the thermal stability of the composites. In short, the overall results lead us to conclude that the micronization of the BEKP was an efficient and convenient method to further improve the performance of the composites, without the use of any hazardous solvents or chemicals. The mechanical and flow properties show the potential of the present fully sustainable composites as an alternative to the existing partially bio-based ones for injection molding applications intended for electronics, furniture, house appliances, and other applications.

Supplementary Materials: The following are available online: FTIR-ATR spectra and X-ray diffractograms of the thermoplastic matrices; SEM micrographs and size (width and length) histograms of the micronized fibers; FTIR-ATR spectra, X-ray diffractograms, and TGA thermograms of the micronized fibers; density of the matrices and composites, and TGA thermograms of the composites reinforced with 40 wt.% fibers having different aspect ratios.

Author Contributions: Conceptualization, C.P.N., C.V. and C.S.R.F.; methodology, B.F.A.V.; investigation, B.F.A.V. and C.V.; resources, A.J.D.S., C.P.N. and C.S.R.F.; data curation, B.F.A.V.; writing—original draft preparation, B.F.A.V.; writing—review and editing, B.F.A.V., A.J.D.S., C.P.N., C.V. and C.S.R.F.; supervision, C.P.N., C.V. and C.S.R.F.; project administration, A.J.D.S., C.P.N. and C.S.R.F. All authors have read and agreed to the published version of the manuscript.

Funding: This work was carried out under the project Inpactus: innovative products and technologies from eucalyptus, project number 21874, funded by Portugal 2020 through the European Regional Development Fund (ERDF) in the frame of COMPETE 2020 n°246/AXIS II/2017, and under the project CICECO-Aveiro Institute of Materials, UIDB/50011/2020 and UIDP/50011/2020, financed by national funds through the Portuguese Foundation for Science and Technology (FCT)/MCTES. FCT is also acknowledged for the research contracts under the Scientific Employment Stimulus to C.V. (CEECIND/00263/2018) and C.S.R.F. (CEECIND/00464/2017).

Institutional Review Board Statement: Not applicable. This study did not involve humans or animals.

Informed Consent Statement: Not applicable.

Data Availability Statement: Not applicable.

Conflicts of Interest: The authors declare no conflict of interest.

Sample Availability: Samples of the compounds are not available from the authors.

References

1. Ramamoorthy, S.K.; Skrifvars, M.; Persson, A. A review of natural fibers used in biocomposites: Plant, animal and regenerated cellulose fibers. *Polym. Rev.* **2015**, *55*, 107–162. [CrossRef]
2. Khan, M.Z.; Srivastava, S.K.; Gupta, M. Tensile and flexural properties of natural fiber reinforced polymer composites: A review. *J. Reinf. Plast. Compos.* **2018**, *37*, 1435–1455. [CrossRef]
3. Wang, J.; Liu, X.; Jin, T.; He, H.; Liu, L. Preparation of nanocellulose and its potential in reinforced composites: A review. *J. Biomater. Sci. Polym. Ed.* **2019**, *30*, 919–946. [CrossRef]
4. Gholampour, A.; Ozbakkaloglu, T. A review of natural fiber composites: Properties, modification and processing techniques, characterization, applications. *J. Mater. Sci.* **2020**, *55*, 829–892. [CrossRef]
5. Upmformi. Available online: www.upmformi.com (accessed on 22 April 2021).
6. Aqvacomp Composites. Available online: https://www.aqvacomp.fi/en/materials/ (accessed on 22 April 2021).
7. SAPPI Symbio. Available online: https://www.sappi.com/symbio (accessed on 22 April 2021).
8. Storaenso. Available online: https://www.storaenso.com/en/products/biocomposites (accessed on 22 April 2021).
9. Jaafar, J.; Siregar, J.P.; Mohd Salleh, S.; Mohd Hamdan, M.H.; Cionita, T.; Rihayat, T. Important Considerations in Manufacturing of Natural Fiber Composites: A Review. *Int. J. Precis. Eng. Manuf. Green Technol.* **2019**, *6*, 647–664. [CrossRef]
10. Granda, L.A.; Espinach, F.X.; Tarrés, Q.; Méndez, J.A.; Delgado-Aguilar, M.; Mutjé, P. Towards a good interphase between bleached kraft softwood fibers and poly(lactic) acid. *Compos. Part B Eng.* **2016**, *99*, 514–520. [CrossRef]
11. Peltola, H.; Immonen, K.; Johansson, L.; Virkajärvi, J.; Sandquist, D. Influence of pulp bleaching and compatibilizer selection on performance of pulp fiber reinforced PLA biocomposites. *J. Appl. Polym. Sci.* **2019**, *136*, 47955. [CrossRef]
12. Zini, E.; Scandola, M. Green composites: An overview. *Polym. Compos.* **2011**, *32*, 1905–1915. [CrossRef]
13. Plackett, D.; Södergård, A. Polylactide-based biocomposites. In *Natural Fibers, Biopolymers, and Biocomposites*; Mohanty, A.K., Mirsa, M., Drzal, L., Eds.; Taylor & Francis: Abingdon, UK, 2005; pp. 583–600.
14. Misra, M.; Pandey, J.K.; Mohanty, A.K. *Biocomposites: Design and Mechanical Performance*, 1st ed.; Woodhead Publishing: Sawston, UK, 2015; ISBN 9781782423942.
15. Faludi, G.; Dora, G.; Imre, B.; Renner, K.; Móczó, J.; Pukánszky, B. PLA/lignocellulosic fiber composites: Particle characteristics, interfacial adhesion, and failure mechanism. *J. Appl. Polym. Sci.* **2014**, *131*, 39902. [CrossRef]
16. Satyanarayana, K.G.; Arizaga, G.G.C.; Wypych, F. Biodegradable composites based on lignocellulosic fibers-An overview. *Prog. Polym. Sci.* **2009**, *34*, 982–1021. [CrossRef]
17. Mazzanti, V.; Pariante, R.; Bonanno, A.; Ruiz de Ballesteros, O.; Mollica, F.; Filippone, G. Reinforcing mechanisms of natural fibers in green composites: Role of fibers morphology in a PLA/hemp model system. *Compos. Sci. Technol.* **2019**, *180*, 51–59. [CrossRef]
18. Mukherjee, T.; Sani, M.; Kao, N.; Gupta, R.K.; Quazi, N.; Bhattacharya, S. Improved dispersion of cellulose microcrystals in polylactic acid (PLA) based composites applying surface acetylation. *Chem. Eng. Sci.* **2013**, *101*, 655–662. [CrossRef]
19. Srubar, W.V.; Pilla, S.; Wright, Z.C.; Ryan, C.A.; Greene, J.P.; Frank, C.W.; Billington, S.L. Mechanisms and impact of fiber-matrix compatibilization techniques on the material characterization of PHBV/oak wood flour engineered biobased composites. *Compos. Sci. Technol.* **2012**, *72*, 708–715. [CrossRef]
20. Yu, T.; Hu, C.; Chen, X.; Li, Y. Effect of diisocyanates as compatibilizer on the properties of ramie/poly(lactic acid) (PLA) composites. *Compos. Part A Appl. Sci. Manuf.* **2015**, *76*, 20–27. [CrossRef]
21. Solle, M.A.; Arroyo, J.; Burgess, M.H.; Warnat, S.; Ryan, C.A. Value-added composite bioproducts reinforced with regionally significant agricultural residues. *Compos. Part A Appl. Sci. Manuf.* **2019**, *124*, 105441. [CrossRef]
22. Yatigala, N.S.; Bajwa, D.S.; Bajwa, S.G. Compatibilization improves physico-mechanical properties of biodegradable biobased polymer composites. *Compos. Part A Appl. Sci. Manuf.* **2018**, *107*, 315–325. [CrossRef]
23. Niu, Z.; Chen, Y.; Feng, J. Preparation, structure, and property of wood flour incorporated polypropylene composites prepared by a solid-state mechanochemical method. *J. Appl. Polym. Sci.* **2016**, *133*, 43108. [CrossRef]
24. Qiang, T.; Wang, J.; Wolcott, M.; Qiang, T.; Wang, J.; Wolcott, M.P. Facile Fabrication of 100% Bio-based and Degradable Ternary Cellulose/PHBV/PLA Composites. *Materials* **2018**, *11*, 330. [CrossRef] [PubMed]
25. Qiang, T.; Wang, J.; Wolcott, M.P. Facile Preparation of Cellulose/Polylactide Composite Materials with Tunable Mechanical Properties. *Polym. Plast. Technol. Eng.* **2018**, *57*, 1288–1295. [CrossRef]
26. Yang, S.; Bai, S.; Wang, Q. Sustainable packaging biocomposites from polylactic acid and wheat straw: Enhanced physical performance by solid state shear milling process. *Compos. Sci. Technol.* **2018**, *158*, 34–42. [CrossRef]
27. Biomer Biopolyesters. Available online: www.biomer.de (accessed on 23 April 2021).
28. NatureWorks. Available online: https://www.natureworksllc.com/Products/3-series-for-injection-molding (accessed on 22 April 2021).
29. Ozyhar, T.; Baradel, F.; Zoppe, J. Effect of functional mineral additive on processability and material properties of wood-fiber reinforced poly(lactic acid) (PLA) composites. *Compos. Part A Appl. Sci. Manuf.* **2020**, *132*, 105827. [CrossRef]
30. Khouaja, A.; Koubaa, A.; Ben Daly, H. Dielectric properties and thermal stability of cellulose high-density polyethylene bio-based composites. *Ind. Crops Prod.* **2021**, *171*, 113928. [CrossRef]

31. Gunning, M.A.; Geever, L.M.; Killion, J.A.; Lyons, J.G.; Higginbotham, C.L. Mechanical and biodegradation performance of short natural fibre polyhydroxybutyrate composites. *Polym. Test.* **2013**, *32*, 1603–1611. [CrossRef]
32. Faruk, O.; Bledzki, A.K.; Fink, H.-P.; Sain, M. Biocomposites reinforced with natural fibers: 2000–2010. *Prog. Polym. Sci.* **2012**, *37*, 1552–1596. [CrossRef]
33. Arbelaiz, A.; Txueka, U.; Mezo, I.; Orue, A. Biocomposites Based on Poly(Lactic Acid) Matrix and Reinforced with Lignocellulosic Fibers: The Effect of Fiber Type and Matrix Modification. *J. Nat. Fibers* **2020**, *17*, 26. [CrossRef]
34. Lee, M.C.; Koay, S.C.; Chan, M.Y.; Choo, H.L.; Pang, M.M.; Chou, P.M.; Tshai, K.Y. Properties of poly(lactic acid)/durian husk fiber biocomposites: Effects of fiber content and processing aid. *J. Thermoplast. Compos. Mater.* **2020**, *33*, 1518–1532. [CrossRef]
35. Robledo-Ortíz, J.R.; González-López, M.E.; Martín del Campo, A.S.; Pérez-Fonseca, A.A. Lignocellulosic Materials as Reinforcement of Polyhydroxybutyrate and its Copolymer with Hydroxyvalerate: A Review. *J. Polym. Environ.* **2021**, *29*, 1350–1364. [CrossRef]
36. Madyan, O.A.; Wang, Y.; Corker, J.; Zhou, Y.; Du, G.; Fan, M. Classification of wood fibre geometry and its behaviour in wood poly(lactic acid) composites. *Compos. Part A Appl. Sci. Manuf.* **2020**, *133*, 105871. [CrossRef]
37. Yang, Z.; Feng, X.; Bi, Y.; Zhou, Z.; Yue, J.; Xu, M. Bleached extruder chemi-mechanical pulp fiber-PLA composites: Comparison of mechanical, thermal, and rheological properties with those of wood flour-PLA bio-composites. *J. Appl. Polym. Sci.* **2016**, *133*, 44241. [CrossRef]
38. Immonen, K.; Anttila, U.; Wikström, L. Coupling of PLA and bleached softwood kraft pulp (BSKP) for enhanced properties of biocomposites. *J. Thermoplast. Compos. Mater.* **2019**, *32*, 328–341. [CrossRef]
39. Smith, M.K.M.; Paleri, D.M.; Abdelwahab, M.; Mielewski, D.F.; Misra, M.; Mohanty, A.K. Sustainable composites from poly(3-hydroxybutyrate) (PHB) bioplastic and agave natural fibre. *Green Chem.* **2020**, *22*, 3906–3916. [CrossRef]
40. Baghaei, B.; Skrifvars, M.; Rissanen, M.; Ramamoorthy, S.K. Mechanical and thermal characterization of compression moulded polylactic acid natural fiber composites reinforced with hemp and lyocell fibers. *J. Appl. Polym. Sci.* **2014**, *131*, 40534. [CrossRef]
41. Liang, Z.; Wu, H.; Liu, R.; Wu, C. Preparation of long sisal fiber-reinforced polylactic acid biocomposites with highly improved mechanical performance. *Polymers* **2021**, *13*, 1124. [CrossRef] [PubMed]
42. Mendes, J.F.; Castro, L.S.; Corrêa, A.C.; Marconcini, J.M.; Mattoso, L.H.C.; Mendes, R.F. Effects of short fibers and processing additives on HDPE composites properties reinforced with Pinus and Eucalyptus fibers. *J. Appl. Polym. Sci.* **2021**, *138*, 50178. [CrossRef]
43. Ren, H.; Zhang, Y.; Zhai, H.; Chen, J. Production and evaluation of biodegradable composites based on polyhydroxybutyrate and polylactic acid reinforced with short and long pulp fibers. *Cellul. Chem. Technol.* **2015**, *49*, 641–652.
44. Graupner, N.; Müssig, J. A comparison of the mechanical characteristics of kenaf and lyocell fibre reinforced poly(lactic acid) (PLA) and poly(3-hydroxybutyrate) (PHB) composites. *Compos. Part A Appl. Sci. Manuf.* **2011**, *42*, 2010–2019. [CrossRef]
45. Oliver-Ortega, H.; Tarrés, Q.; Mutjé, P.; Delgado-Aguilar, M.; Méndez, J.A.; Espinach, F.X. Impact strength and water uptake behavior of bleached kraft softwood-reinforced PLA composites as alternative to PP-based materials. *Polymers* **2020**, *12*, 2144. [CrossRef]
46. Virtanen, S.; Wikström, L.; Immonen, K.; Anttila, U.; Retulainen, E. Cellulose kraft pulp reinforced polylactic acid (PLA) composites: Effect of fibre moisture content. *AIMS Mater. Sci.* **2016**, *3*, 756–769. [CrossRef]
47. Peltola, H.; Pääkkönen, E.; Jetsu, P.; Heinemann, S. Wood based PLA and PP composites: Effect of fibre type and matrix polymer on fibre morphology, dispersion and composite properties. *Compos. Part A Appl. Sci. Manuf.* **2014**, *61*, 13–22. [CrossRef]
48. Singh, S.; Mohanty, A.K. Wood fiber reinforced bacterial bioplastic composites: Fabrication and performance evaluation. *Compos. Sci. Technol.* **2007**, *67*, 1753–1763. [CrossRef]
49. Espert, A.; Vilaplana, F.; Karlsson, S. Comparison of water absorption in natural cellulosic fibres from wood and one-year crops in polypropylene composites and its influence on their mechanical properties. *Compos. Part A Appl. Sci. Manuf.* **2004**, *35*, 1267–1276. [CrossRef]
50. Gallardo-Cervantes, M.; González-García, Y.; Pérez-Fonseca, A.A.; González-López, M.E.; Manríquez-González, R.; Rodrigue, D.; Robledo-Ortíz, J.R. Biodegradability and improved mechanical performance of polyhydroxyalkanoates/agave fiber biocomposites compatibilized by different strategies. *J. Appl. Polym. Sci.* **2021**, *138*, 50182. [CrossRef]
51. Trinh, B.M.; Ogunsona, E.O.; Mekonnen, T.H. Thin-structured and compostable wood fiber-polymer biocomposites: Fabrication and performance evaluation. *Compos. Part A Appl. Sci. Manuf.* **2021**, *140*, 106150. [CrossRef]
52. Huerta-Cardoso, O.; Durazo-Cardenas, I.; Longhurst, P.; Simms, N.J.; Encinas-Oropesa, A. Fabrication of agave tequilana bagasse/PLA composite and preliminary mechanical properties assessment. *Ind. Crops Prod.* **2020**, *152*, 112523. [CrossRef]
53. Vilela, C.; Engström, J.; Valente, B.F.A.; Jawerth, M.; Carlmark, A.; Freire, C.S.R. Exploiting poly(ε-caprolactone) and cellulose nanofibrils modified with latex nanoparticles for the development of biodegradable nanocomposites. *Polym. Compos.* **2019**, *40*, 1342–1353. [CrossRef]
54. Singh, R.; Kumar, R.; Hashmi, M.S.J. Friction Welding of Dissimilar Plastic-Based Material by Metal Powder Reinforcement. *Ref. Modul. Mater. Sci. Mater. Eng.* **2017**, *101*, 77. [CrossRef]
55. Nagarajan, V.; Misra, M.; Mohanty, A.K. New engineered biocomposites from poly(3-hydroxybutyrate-co-3-hydroxyvalerate) (PHBV)/poly(butylene adipate-co-terephthalate) (PBAT) blends and switchgrass: Fabrication and performance evaluation. *Ind. Crops Prod.* **2013**, *42*, 461–468. [CrossRef]

56. Carrasco, F.; Pagès, P.; Gámez-Pérez, J.; Santana, O.O.; Maspoch, M.L. Processing of poly(lactic acid): Characterization of chemical structure, thermal stability and mechanical properties. *Polym. Degrad. Stab.* **2010**, *95*, 116–125. [CrossRef]
57. Pachekoski, W.M.; Dalmolin, C.; Agnelli, J.A.M. The influence of the industrial processing on the degradation of poly(hidroxybutyrate)-PHB. *Mater. Res.* **2013**, *16*, 327–332. [CrossRef]
58. Long, H.; Wu, Z.; Dong, Q.; Shen, Y.; Zhou, W.; Luo, Y.; Zhang, C.; Dong, X. Effect of polyethylene glycol on mechanical properties of bamboo fiber-reinforced polylactic acid composites. *J. Appl. Polym. Sci.* **2019**, *136*, 47709. [CrossRef]
59. Lee, C.H.; Khalina, A.; Lee, S.H. Importance of interfacial adhesion condition on characterization of plant-fiber-reinforced polymer composites: A review. *Polymers* **2021**, *13*, 438. [CrossRef]
60. Jaszkiewicz, A.; Meljon, A.; Bledzki, A.K.; Radwanski, M. Gaining knowledge on the processability of PLA-based short-fibre compounds-A comprehensive comparison with their PP counterparts. *Compos. Part A Appl. Sci. Manuf.* **2016**, *83*, 140–151. [CrossRef]
61. Spear, M.J.; Eder, A.; Carus, M. *Wood polymer Composites*; Woodhead Publishing: Sawston, UK, 2015; ISBN 9781782424772.
62. Reis, K.C.; Pereira, L.; Melo, I.C.N.A.; Marconcini, J.M.; Trugilho, P.F.; Tonoli, G.H.D. Particles of coffee wastes as reinforcement in polyhydroxybutyrate (PHB) based composites. *Mater. Res.* **2015**, *18*, 546–552. [CrossRef]
63. Aydemir, D.; Gardner, D.J. The effects of cellulosic fillers on the mechanical, morphological, thermal, viscoelastic, and rheological properties of polyhydroxybutyrate biopolymers. *Polym. Compos.* **2020**, *41*, 3842–3856. [CrossRef]
64. Espinach, F.X.; Boufi, S.; Delgado-Aguilar, M.; Julián, F.; Mutjé, P.; Méndez, J.A. Composites from poly(lactic acid) and bleached chemical fibres: Thermal properties. *Compos. Part B Eng.* **2018**, *134*, 169–176. [CrossRef]
65. Yeng, L.C.; Wahit, M.U.; Othman, N. Thermal and flexural properties of regenerated cellulose(RC)/poly(3-hydroxybutyrate)(PHB) biocomposites. *J. Teknol.* **2015**, *75*, 107–112. [CrossRef]
66. Ventura, H.; Claramunt, J.; Rodríguez-Pérez, M.A.; Ardanuy, M. Effects of hydrothermal aging on the water uptake and tensile properties of PHB/flax fabric biocomposites. *Polym. Degrad. Stab.* **2017**, *142*, 129–138. [CrossRef]
67. Santos, E.B.C.; Barros, J.J.P.; Moura, D.A.D.; Moreno, C.G.; Fim, F.D.C.; Silva, L.B.D. Rheological and thermal behavior of PHB/piassava fiber residue-based green composites modified with warm water. *J. Mater. Res. Technol.* **2019**, *8*, 531–540. [CrossRef]
68. Sánchez-Safont, E.L.; Aldureid, A.; Lagarón, J.M.; Gámez-Pérez, J.; Cabedo, L. Biocomposites of different lignocellulosic wastes for sustainable food packaging applications. *Compos. Part B Eng.* **2018**, *145*, 215–225. [CrossRef]
69. Mysiukiewicz, O.; Barczewski, M.; Skórczewska, K.; Matykiewicz, D. Correlation between processing parameters and degradation of different polylactide grades during twin-screw extrusion. *Polymers* **2020**, *12*, 1333. [CrossRef]
70. Taguchi, S.; Iwata, T.; Abe, H.; Doi, Y.; Aqida, S.N. Poly(hydroxyalkanoate)s. In *Reference Module in Materials Science and Materials Engineering*; Elsevier: Amsterdam, The Netherlands, 2016.
71. Srubar, W.V.; Wright, Z.C.; Tsui, A.; Michel, A.T.; Billington, S.L.; Frank, C.W. Characterizing the effects of ambient aging on the mechanical and physical properties of two commercially available bacterial thermoplastics. *Polym. Degrad. Stab.* **2012**, *97*, 1922–1929. [CrossRef]
72. Park, S.; Baker, J.O.; Himmel, M.E.; Parilla, P.A.; Johnson, D.K. Cellulose crystallinity index: Measurement techniques and their impact on interpreting cellulase performance. *Biotechnol. Biofuels* **2010**, *3*, 10. [CrossRef] [PubMed]

molecules

MDPI

Article

Effective Utilisation of Halophyte Biomass from Saline Soils for Biorefinering Processes

Jolanta Batog, Krzysztof Bujnowicz, Weronika Gieparda, Aleksandra Wawro * and Szymon Rojewski

Institute of Natural Fibres and Medicinal Plants—National Research Institute, Wojska Polskiego 71B, 60-630 Poznan, Poland; jolanta.batog@iwnirz.pl (J.B.); krzysztof.bujnowicz@iwnirz.pl (K.B.); weronika.gieparda@iwnirz.pl (W.G.); szymon.rojewski@iwnirz.pl (S.R.)
* Correspondence: aleksandra.wawro@iwnirz.pl; Tel.: +48-618455814

Abstract: The salinity of European soil is increasing every year, causing severe economic damage (estimated 1–3 million hectares in the enlarged EU). This study uses the biomass of halophytes—tall fescue (grass) and hemp of the Białobrzeskie variety from saline soils—for bioenergy, second generation biofuels and designing new materials—fillers for polymer composites. In the bioethanol obtaining process, in the first stage, the grass and hemp biomass were pretreated with 1.5% NaOH. Before and after the treatment, the chemical composition was determined and the FTIR spectra and SEM pictures were taken. Then, the process of simultaneous saccharification and fermentation (SSF) was carried out. The concentration of ethanol for both the grass and hemp biomass was approx. 7 $g \cdot L^{-1}$ (14 $g \cdot 100\ g^{-1}$ of raw material). In addition, trials of obtaining green composites with halophyte biomass using polymers (PP) and biopolymers (PLA) as a matrix were performed. The mechanical properties of the composites (tensile and flexural tests) were determined. It was found that the addition of a compatibilizer improved the adhesion at the interface of PP composites with a hemp filler. In conclusion, the grass and hemp biomass were found to be an interesting and promising source to be used for bioethanol and biocomposites production. The use of annually renewable plant biomass from saline soils for biorefinering processes opens up opportunities for the development of a new value chains and new approaches to sustainable agriculture.

Keywords: halophyte biomass; saline soils; pretreatment; SSF; biorefinering process; bioethanol; biocomposites

Citation: Batog, J.; Bujnowicz, K.; Gieparda, W.; Wawro, A.; Rojewski, S. Effective Utilisation of Halophyte Biomass from Saline Soils for Biorefinering Processes. *Molecules* **2021**, *26*, 5393. https://doi.org/10.3390/molecules26175393

Academic Editors: Alejandro Rodriguez Pascual, Eduardo Espinosa Víctor and Carlos Martín

Received: 2 August 2021
Accepted: 31 August 2021
Published: 5 September 2021

1. Introduction

The dynamic development of economic activities often causes changes in the environment and human life quality. The world population is anticipated to grow 40% within 40–50 years with unprecedented demands for energy, food, freshwater and a clean environment. At 43% of the total landmass, exploiting the Earth's arid and semi-arid lands becomes a matter of necessity [1].

Soil salinisation effects are seen as a major cause of desertification and, therefore, are a serious form of soil degradation, endangering the potential use of European soils. The contamination of the environment by the excessive salinity of soils may be induced during coastal flooding, by the intensive agriculture practices or under the pressure of climate changes (aridisation phenomena). Such soils have lower or zero suitability for traditional agriculture production, influencing the decline of regional economics [2].

The remediation of these sites or the implementation of new ways to use them are necessary to achieve the restoration of agricultural production and thus economic growth. This can be met by the use of halophytes in the development of an integrated system of soil bioremediation and biomass biorefinery [3].

Tall fescue (*Festuca arundinacea*) and hemp (*Cannabis sativa* L.) are characterised by high biomass production and wide ecological amplitude, including tolerance to soil salinity.

Tall fescue creates large, dense and deeply rooted clumps, showing considerable tolerance to unfavourable habitat conditions. It occurs on roadsides, meadows (including saline meadows) and alluvia.

Hemp is a plant with a short growing season, resistant to diseases and unfavourable environmental conditions, with a dry matter yield of 10–15 $t \cdot ha^{-1}$. They improve soil quality and are useful for the restoration of brownfield sites; 1 ha of hemp binds about 11 t of CO_2 [4–6].

This study addresses the challenge of improving the knowledge and understanding of halophyte species in order to better use the biomass from these plants for the production of new environmentally friendly materials [7].

EU Member States have been obliged to achieve a certain share of biofuels in transport and to take measures to reduce greenhouse gas emissions. According to the EU RED II Directive, the contribution of advanced biofuels and biogas produced, among others, from lignocellulosic raw materials, as a share of final energy consumption in the transport sector, it is expected to amount to at least 0.2% by 2022, 1% by 2025 and 3.5% by 2030 [8]. Therefore, it is very interesting and important to try to obtain bioethanol from the halophyte biomass from saline soils [3].

The production of lignocellulosic ethanol involves the breakdown of cell walls into individual polymers and the hydrolysis of carbohydrates into simple sugars. Plant biomass contains the lignocellulose complex (cellulose, hemicellulose, lignin), which is relatively resistant to biodegradation. The process of converting halophyte biomass to bioethanol involves several steps, from the preparation of the plant material (physical and chemical treatment), through enzymatic hydrolysis, to ethanol fermentation. The purpose of the pretreatment of biomass is to crush the solid phase and loosen the compact structure of the lignocellulose. Moreover, the method combining cellulose hydrolysis with sugar fermentation in one bioreactor (SSF process), where enzymes must be adapted to the conditions of the fermentation process (30–40 °C), seems to be effective and economical [9–11].

Another important point is that around 25 million tons of plastic waste are produced annually in the EU. The market is dominated by products made from materials such as PE, PP, PET, PS and PVC, that take several hundred years to decompose. More and more emphasis is placed on the need to significantly reduce the amount of plastic waste (Single-Use Plastics Directive) [12,13].

The plastics processing industry is looking for material innovations that meet customer expectations and legal requirements; therefore, in the next few years, an increased interest in environmentally friendly biodegradable materials is expected. An example is the production of polylactic acid (PLA), which is one of the leading bioplastics on the market [14–16].

The barrier to the wider use of biodegradable plastics is the high production costs, higher than traditionally used plastics.

A method to improve the economics may be the use of natural fillers from the biomass of annual plants (flax and hemp biomass), which are cheaper than biodegradable plastics [17,18]. Materials composed of thermoplastic polymers with natural fillers have functional properties that do not differ from pure polymers and may be used in various fields of the economy, including the automotive industry, transport, construction and the furniture industry.

Due to the expected increase in the production of biocomposite materials in the EU in the coming years, an increase in the demand for biocomponents based on natural resources, including natural fibres, should be expected.

Moreover, to fully understand the impacts of biorefineries, taking a comprehensive view of sustainability, it is necessary to consider a wider range of factors including economics and social and environmental issues. The potential benefits of strengthening the halophyte based biorefineries are certainly reduced greenhouse gas emissions, less dependence on fossil resources, better land and natural resource management and improved food security and soil quality. In addition, a significant positive effect of biobased industries is the creation of employment opportunities in rural areas, as well as the possibility of creating new markets for agriculture (biofuels, green composites) in synergy with the existing ones.

The aim of the presented study is to indicate the possibility of using tall fescue and hemp from saline soils in the process of obtaining second generation biofuels and for the design of green biocomposites. Thus far, only a few reports of the use of halophyte biomass from degraded areas in biorefining processes have been published [3,19,20]. A novelty in this manuscript is an indication of the possibility of managing degraded, saline soils through the production of biomass for industrial purposes, including bioenergy, which is in line with the idea of sustainable development, that is now a priority policy of the European Union.

2. Results and Discussion

2.1. Bioethanol Production Process

2.1.1. Halophyte Biomass Pretreatment

The biomass crushed on a knife mill with a mesh of 2 mm was treated with alkaline. One of the most popular alkaline reagents used in the treatment of raw plant material is sodium hydroxide. The main purpose of the pretreatment of lignocellulosic biomass is to crush the solid phase and loosen the compact lignocellulose structure. After a chemical pretreatment test with sodium hydroxide (1.5–3%, 90 °C), sulfuric acid (1–2%, 121 °C) and hot water (135 °C), it turned out that the most effective treatment was alkaline with 1.5% NaOH. Table 1 shows the alkaline treatment of the grass and hemp biomass from saline soils using 1.5% NaOH.

Table 1. Content of reducing sugars before (BP) and after (AP) sodium hydroxide treatment.

Halophyte	Sample	Reducing Sugars ($mg \cdot g^{-1}$)
Grass	BP	100.20 ± 0.09
	AP	354.59 ± 0.01
Hemp	BP	62.95 ± 0.10
	AP	187.95 ± 0.13

It was found that grass from the saline soil was characterised by an almost two times higher content of reducing sugars than hemp. The results indicate that grass is more susceptible to alkaline treatment than hemp.

An effective chemical treatment should ensure the destruction of the crystalline structure of cellulose, the separation of lignin from carbohydrates and, thus, an increase in the availability of the substrate for further biofuel production processes [21].

To confirm the efficiency of the alkaline treatment, a determination of the chemical composition of the halophyte biomass after the NaOH treatment was performed and compared to the chemical composition of the biomass before the pretreatment. The results are presented in Table 2.

Table 2. Chemical composition of halophyte biomass (% of dry matter); BP: before pretreatment; AP: after pretreatment.

Halophyte	Sample	Cellulose (%)	Hemicellulose (%)	Lignin (%)
Grass	BP	33.69 ± 0.40	34.74 ± 0.39	17.08 ± 0.16
	AP	50.41 ± 0.18	25.23 ± 0.37	12.35 ± 0.07
Hemp	BP	47.34 ± 0.40	33.49 ± 0.68	13.94 ± 0.05
	AP	58.46 ± 0.29	22.12 ± 0.13	17.35 ± 0.26

The analysis of the chemical composition of the halophyte biomass before and after the treatment showed that the alkaline effect caused a visible increase in the cellulose content by over 10%, especially for grass, which saw a 17% increase. In addition, a partial degradation of hemicellulose was found (as much as 11% for hemp). In the case of the lignin content for the hemp biomass, an increase was observed after the alkaline pretreatment (over 3%), and in the case of grass—an almost 5% reduction. Similar observations regarding the hemp biomass were presented by Stevulova et al. [22], who showed that the lignin content after pretreatment with sodium hydroxide was 7% higher than before.

It was found that in the case of the grass biomass, both the higher cellulose growth and the reduction in lignin content positively influenced the higher values of the reducing sugars released after the pretreatment with sodium hydroxide compared to the hemp biomass (see Table 1). It should be emphasised that one of the main goals of the chemical treatment of lignocellulosic materials is the removal of lignin, which is a strong obstacle in the process of biomass conversion.

The effect of alkaline treatment on the grass and hemp biomass from saline soils was confirmed using Fourier Transform Infrared Spectrometer (FTIR) between 600–4000 cm^{-1} shown in the Figure 1a,b and using Scanning Electron Microscopy (SEM).

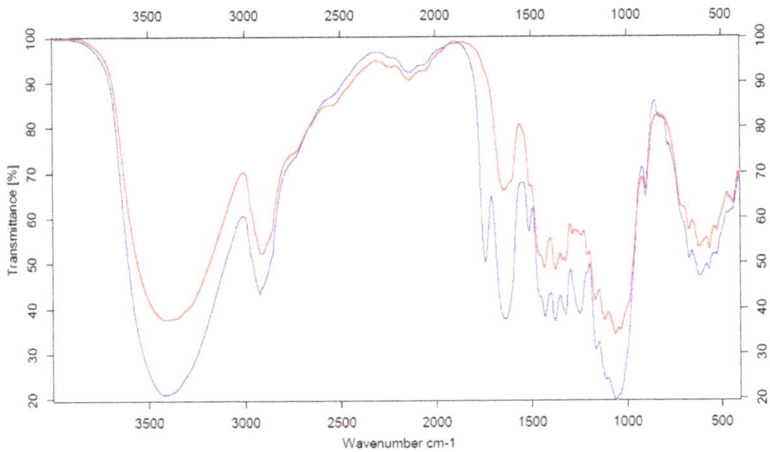

(**a**)

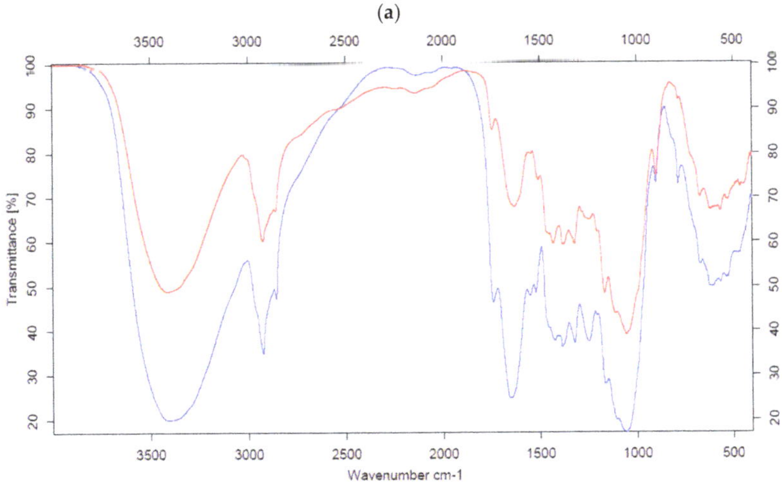

(**b**)

Figure 1. (**a**) FTIR spectra of hemp biomass from saline soil before (blue) and after (red) NaOH pretreatment. (**b**) FTIR spectra of grass biomass from saline soil before (blue) and after (red) NaOH pretreatment.

The broad band, ranging from about 3500 to 3000 cm^{-1}, comes from the O–H stretching vibrations in the cellulose molecule. In all the cases, the band after pretreatment is less intense and narrower. The band at 2900 cm^{-1}, resulting from the stretching vibrations of the C–H group (cellulose), is also less intense after pretreatment. The decrease in the intensity of these bands, despite the increase in the cellulose content after treatment (confirmed by results from the chemical composition), may be caused by the reduction in the crystalline structure of cellulose [23,24]. These bands, in the case of different materials, have a different shape (for grass, two characteristic peaks appear in this place, while for hemp, one broad band is visible). The vibration band visible at 1730 cm^{-1}, resulting from the C=O stretching vibrations of the acetyl group in hemicellulose [25], disappears in the case of hemp after treatment, while in the case of grass, it is significantly reduced. This is due to the degradation of hemicellulose during the alkaline pretreatment of biomass. In turn, the lower intensity of the band at 1600 cm^{-1} corresponding to the O–H stretching vibrations, reflecting the amount of water absorbed in the sample, is probably caused by the loss of water during the drying process of the samples after pretreatment [26]. The absorption band at 1510 cm^{-1}, resulting from the vibrations of the aromatic ring in lignin in the case of grass, decreases after alkaline treatment, which is reflected in the chemical composition of the biomass. Significant changes are visible in the band at 1230 cm^{-1}. This band is attributed to the vibration of the guaiacyl ring in lignin, as well as to the vibration of the C–O groups in pectin. After the alkaline treatment, this band disappears completely or significantly loses its intensity, which may indicate a reduction in both lignin and pectin. Going to the lower wavenumber values, the following three more characteristic bands for the cellulose molecule appear: 1160 cm^{-1} (asymmetric C–O–C stretching vibrations), 1110 cm^{-1} (C–OH skeletal vibrations) and 1050 cm^{-1} (C–O–C skeletal vibrations of the pyranose ring) [27]. These bands, despite the increase in cellulose content after chemical treatment, are reduced. This is characteristic of each biomass tested and can be attributed to the reduction in the crystalline structure of the cellulose after treatment.

Significant changes on the surface of the grass and hemp biomass were observed and presented in the SEM images taken before and after pretreatment (Figure 2).

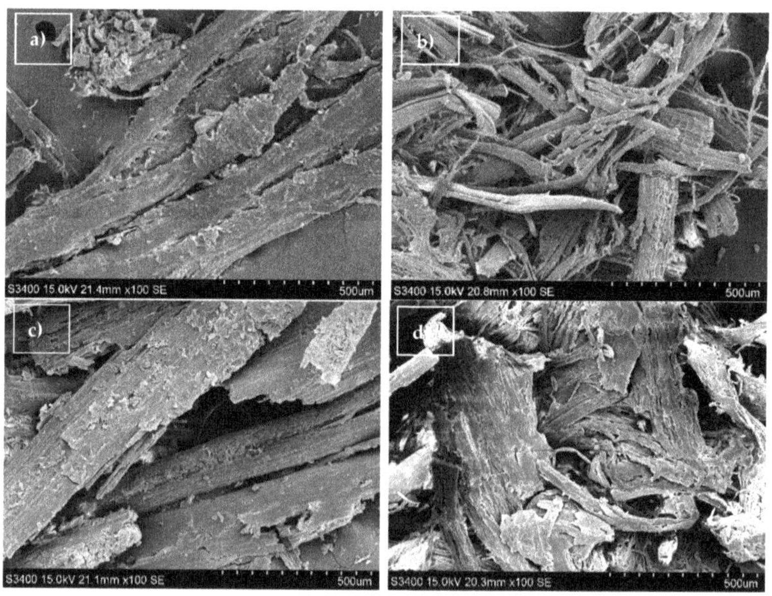

Figure 2. SEM images of halophyte biomass: (**a**) hemp biomass before NaOH pretreatment, (**b**) hemp biomass after NaOH pretreatment, (**c**) grass biomass before NaOH pretreatment, (**d**) grass biomass after NaOH pretreatment.

Figure 2a,c show that the untreated grass and hemp biomass have intact and rigid structures with well-ordered fibrous skeletons, effectively blocking access to lignocellulose [28]. After the pretreatment of the grass and hemp biomass with sodium hydroxide for, similar changes were observed on the surface of the biomass. The SEM images of the halophyte biomass after alkaline treatment (Figure 2b,d) show damage to the structure of the biomass and the appearance of hollow areas, which increases its surface and positively affects the enzymatic availability and digestibility of the biomass [29,30].

To sum up, the main goal of the chemical treatment of lignocellulosic materials for biofuel is to increase the availability of the biomass structure by the decrystallisation of cellulose and the removal of lignin.

2.1.2. Enzyme Complex

The second stage of the process of obtaining bioethanol from plant biomass is the enzymatic hydrolysis process. The breakdown of cellulose into simple sugars requires the synergistic action of cellulases—endoglucanases, cellobiohydrolases and ß-glucosidases. Two enzyme preparations were selected for the research—Flashzyme Plus 200 and Celluclast 1.5 L [31]. In order to select the enzyme complex for the SSF process, tests were performed using selected enzymes and their supplementation with glucosidase and xylanase (Table 3).

Table 3. Content of reducing sugars after the enzymatic test.

Enzyme	Reducing Sugar ($mg \cdot g^{-1}$)	
	Hemp	Grass
Flashzyme Plus 200	338 ± 0.04	846 ± 1.00
Celluclast 1.5 L	342 ± 0.05	696 ± 0.52
Flashzyme/Celluclast 1.5 L (70/30)	420 ± 0.06	892 ± 0.02
Flashzyme/Celluclast 1.5 L (50/50)	430 ± 0.05	800 ± 0.44
Flashzyme/Celluclast 1.5 L (30/70)	355 ± 0.38	810 ± 0.34
Flashzyme/Celluclast 1.5 L (50/50)/β-glucosidase	351 ± 0.14	-
Flashzyme/Celluclast 1.5 L (50/50)/xylanase	324 ± 0.65	-
Flashzyme/Celluclast 1.5 L (50/50)/β-glucosidase/xylanase	343 ± 0.16	-
Flashzyme/Celluclast 1.5 L (70/30)/β-glucosidase	-	472 ± 3.29
Flashzyme/Celluclast 1.5 L (70/30)/xylanase	-	458 ± 1.81
Flashzyme/Celluclast 1.5 L (70/30)/β-glucosidase/xylanase	-	735 ± 1.86

For the hydrolysis of the solid fraction in the SSF process, the Flashzyme Plus 200/Celluclast 1.5 L complex in the proportion of 50/50 was selected for the hemp biomass, and the Flashzyme Plus 200/Celluclast 1.5 L complex in the proportion of 70/30 for the grass biomass. On the basis of the enzymatic test, it was found that the content of the released reducing sugars for the grass biomass (892 $mg \cdot g^{-1}$) was two times higher than for the hemp biomass (430 $mg \cdot g^{-1}$).

2.1.3. Simultaneous Saccharification and Fermentation (SSF)

The SSF process, consisting of simultaneous hydrolysis and fermentation, takes place under conditions ensuring the optimal synergy of enzymes and distillery yeast. After carrying out the fermentation tests with selected parameters, the amount of ethanol was determined (HPLC). Figure 3 shows the ethanol concentration after the SSF process for the hemp and grass biomass from saline soils.

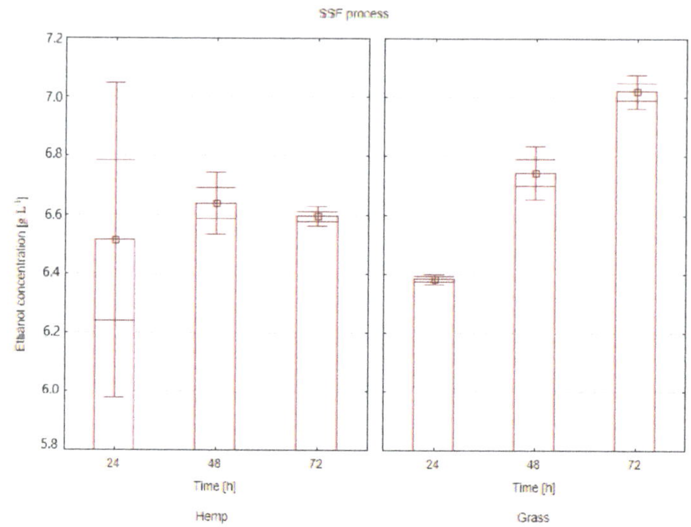

Figure 3. Ethanol concentration of hemp and grass biomass after the SSF process (Statistica 13.0).

It was noted that at 48 h, there was a slight decrease in the ethanol concentration for the hemp biomass but an increase for the grass biomass. The highest ethanol concentration for the hemp biomass was observed at 48 h and it was 6.6 $g \cdot L^{-1}$, which is 13.3 $g \cdot 100\ g^{-1}$ of raw material. In turn, for the grass biomass, the highest ethanol concentration, equal to 7.0 $g \cdot L^{-1}$ (14.0 $g \cdot 100\ g^{-1}$ of raw material), was recorded after 72 h.

According to the literature reports, SSF is a beneficial process due to its short processing time, small reactor volume and high ethanol yield, as bioethanol is produced immediately with glucose conversion [32].

Taufikurahman and Sherly [33] conducted similar research on the production of bioethanol from the biomass of Napier grass. After 96 h of enzymatic hydrolysis and fermentation, they obtained an ethanol concentration of 1.25 $g \cdot L^{-1}$. In turn, Riadi et al. [34] conducted research on lignocellulosic biomass in the form of sugarcane bagasse. They first optimised the alkaline pretreatment, followed by enzymatic hydrolysis and ethanol fermentation. After the 48-h fermentation process, they obtained an ethanol concentration of 5.84 $g \cdot L^{-1}$. Obtaining bioethanol from hemp biomass was carried out by Orlygsson [35] and after the pretreatment with 0.5% NaOH and SSF process, the ethanol concentration was approx. 1 $g \cdot L^{-1}$.

Summing up, it should be emphasised that both the hemp and grass biomass obtained from crops on saline soils have great potential in the process of obtaining bioethanol.

2.2. Biocomposite Production Process

2.2.1. Fillers from Halophyte Biomass

Natural fillers with particles less than 1 mm were obtained from the grinding of the halophyte biomass. It was assumed that the fraction below 1 mm would be a compromise, allowing for the effective processing of the obtained biocomposites in the injection moulding process, taking into account the diameters of the most commonly used injection nozzles, the simplification of the biomass grinding process and the lower costs of preparing natural fillers. The sieve analysis of the natural fillers was performed, and their humidity was determined. The detailed share of the individual fractions is presented in Table 4.

Table 4. Particle size distribution and humidity of fillers from halophyte biomass.

Plant Biomass	Humidity (%)	Particle Size Distribution (%)						
		1 mm	0.5 mm	0.4 mm	0.25 mm	0.2 mm	0.1 mm	Below 0.1 mm
Grass	8.73	1.2	48.3	8.3	30.5	2.5	2.7	6.5
Hemp	7.65	1.1	53.8	14.7	15.8	8.5	2.6	3.5

The bulk density of the hemp fillers (0.19 g·mL^{-1}) and the grass fillers (0.22 g·mL^{-1}) were determined.

2.2.2. Mechanical Properties of Biocomposites

The multipurpose test specimens—type A in accordance with ISO 3167 [36], used to determine the mechanical properties of composites and biocomposites with halophyte biomass—are shown in Figure 4.

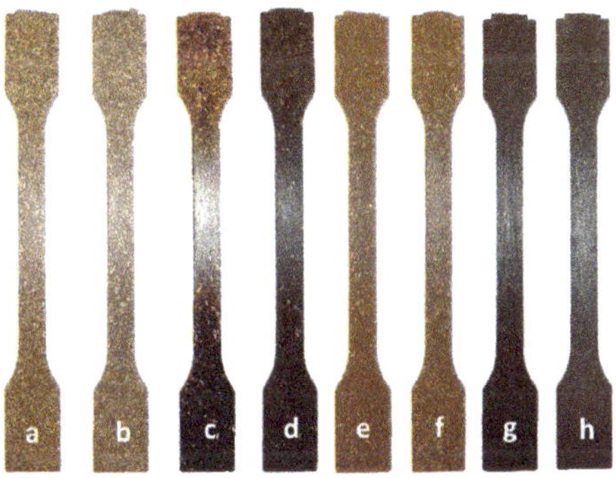

Figure 4. Test specimens of composites and biocomposites with halophyte biomass: (**a**) PP/Hemp20, (**b**) PP/Hemp30, (**c**) PP/Grass20, (**d**) PP/Grass30, (**e**) PLA/Hemp20, (**f**) PLA/Hemp30, (**g**) PLA/Grass20, (**h**) PLA/Grass30.

Tables 5 and 6 show the effect of the different fillers from halophyte biomass on both the tensile and flexural properties of the polypropylene composites.

Table 5. Tensile and flexular properties of PP/hemp composites.

Sample	Tensile Strength δ_M (MPa)	Tensile Modulus E_t (GPa)	Flexular Strength δ_{fM} (MPa)	Flexural Modulus E_f (GPa)
PP HP648T	31.0 ± 0.27	1.5 ± 0.02	41.5 ± 0.48	1.2 ± 0.13
PP-H20	23.3 ± 0.25	1.7 ± 0.04	39.5 ± 0.60	1.8 ± 0.17
PP-H20S5	27.5 ± 0.57	1.8 ± 0.05	40.7 ± 0.75	2.7 ± 0.16
PP-H30	19.3 ± 0.21	2.3 ± 0.03	37.0 ± 0.70	2.5 ± 0.22
PP-H30S5	31.4 ± 0.21	2.8 ± 0.05	51.6 ± 0.77	3.1 ± 0.14

Table 6. Tensile and flexular properties of PP/grass composites.

Sample	Tensile Strength δ_M (MPa)	Tensile Modulus E_t (GPa)	Flexular Strength δ_{fM} (MPa)	Flexural Modulus E_f (GPa)
PP HP648T	31.0 ± 0.27	1.5 ± 0.02	41.5 ± 0.48	1.2 ± 0.13
PP-G20	23.6 ± 0.19	1.7 ± 0.04	39.5 ± 0.27	1.3 ± 0.08
PP-G20S5	28.1 ± 0.25	1.8 ± 0.07	40.3 ± 0.35	1.4 ± 0.09
PP-G30	20.4 ± 0.14	2.0 ± 0.03	41.0 ± 0.40	1.9 ± 0.17
PP-G30S5	24.4 ± 0.27	2.0 ± 0.05	44.4 ± 0.27	1.9 ± 0.11

The 24–38% reduction in the tensile strength, respectively, was demonstrated for the composites based on PP with hemp fillers in the amount of 20% and 30%. The use of 5%

interfacial adhesion promoter (Scona 8112) indicates a positive effect on material adhesion, which ensures more effective mechanical properties. In the presence of a compatibilizer, the PP composites with 30% of the hemp filler showed better mechanical properties: the tensile strength is the same as that of pure PP, and the flexural strength is 24% higher.

Depending on the amount of hemp filler in the presence of the compatibilizer, the composites showed a significant increase in the tensile modulus during elongation (20–87%) and bending (17–58%).

The use of grass biomass fillers in the amount of 20 and 30% in composites based on polypropylene resulted in the reduction in tensile strength by 24 and 34%, respectively. The use of the 5% interfacial adhesion promoter (Scona 8112) improved the tensile strength of the composites by an average of 14%. However, the composites show a higher (13–33%) modulus of elasticity in relation to the pure PP. The use of grass fillers in the amount of 20 and 30% did not significantly affect the flexural strength of the composites. The modulus of elasticity determined during bending increased by almost 58% when the filler was used in the amount of 30%.

The effect of the different fillers from the halophyte biomass on both the tensile and flexural properties of the PLA composites are shown in Tables 7 and 8.

Table 7. Tensile and flexular properties of PLA/hemp composites.

Sample	Tensile Strength δ_M (MPa)	Tensile Modulus E_t (GPa)	Flexular Strength δ_{fM} (MPa)	Flexural Modulus E_f (GPa)
PLA 3260HP	64.5 ± 1.25	3.5 ± 0.07	108.6 ± 0.99	3.4 ± 0.14
PLA-H20	51.4 ± 0.99	5.5 ± 0.04	88.0 ± 2.37	4.9 ± 0.21
PLA-H30	53.0 ± 1.17	6.7 ± 0.05	94.7 ± 1.76	6.6 ± 0.18

Table 8. Tensile and flexular properties of PLA/grass composites.

Sample	Tensile Strength δ_M (MPa)	Tensile Modulus E_t (GPa)	Flexular Strength δ_{fM} (MPa)	Flexural Modulus E_f (GPa)
PLA 3260HP	64.5 ± 1.25	3.5 ± 0.07	108.6 ± 0.99	3.4 ± 0.14
PLA-G20	49.1 ± 0.30	3.9 ± 0.02	83.6 ± 0.43	3.8 ± 0.11
PLA-G30	41.2 ± 1.22	4.1 ± 0.02	78.1 ± 0.93	4.0 ± 0.20

The use of hemp fillers in the amount of 20 and 30% in composites based on PLA–Ingeo 3260HP resulted in a reduction in tensile strength by about 20 and 18%, respectively. The reduction in the flexural strength by 13–19% was demonstrated by the composites of PLA with hemp fillers. Depending on the amount of the hemp filler, the composites showed a significant increase in the modulus during elongation (57–91%) and bending (44–94%) in relation to pure PLA.

The use of grass biomass fillers in the amount of 20 and 30% in composites based on PLA–Ingeo 3260HP resulted in a reduction in tensile strength by about 24 and 36%, respectively. The composites show, respectively, an 11 and 17% increase in the modulus of elasticity in relation to pure PLA. The flexural strength of the composites decreased by about 28% for the 30% filler content. The modulus of elasticity determined during bending increased by 12 and 18% in relation to pure PLA.

It should be noted that the interaction between the matrix and the filler is an important factor influencing the mechanical properties. Despite the hydrophilic nature of the polylactide polymer matrix surface, similar to natural fillers, PLA composites with hemp and grass fillers do not show effective interface adhesion. The interfacial bonding strength of the natural fillers with the polymer matrix were lower, and at a lower value of the force acting on the samples, the fillers were detached from the matrix, and the entire load acted on the polymer matrix [37,38].

Moreover, it was found that the value of the reduction in tensile strength can be correlated with the content of natural fillers. For composites with 20% by weight of hemp filler, the reduction in tensile strength was 24.8% for PP composites and 20% for

PLA composites, respectively. For PP composites with 30% by weight of hemp filler, the reduction in tensile strength was 38%. For composites containing 20 and 30% by weight of grass filler, the reduction in tensile strength was 24% and 34–36% for PP and PLA composites, respectively.

Summing up, it should be emphasised that the addition of a compatibilizer improved the adhesion at the interface of the PP composites with the hemp filler, which is confirmed by the better mechanical properties of composites in terms of elongation and bending.

3. Materials and Methods

3.1. Halophyte Biomass

The raw material used in the study was the biomass of grass—tall fescue and hemp of the Białobrzeskie variety from a saline field in the Kuyavian-Pomeranian Voivodeship (Poland), the average salinity of which, calculated as sodium chloride, was 3.5 g·L^{-1}.

3.2. Bioethanol Production Process

3.2.1. Halophyte Biomass Pretreatment

Grass and hemp biomass was subjected to preliminary crushing to particles of size 20–40 mm and dried at 50–55 °C for 24 h. Then, the material was disintegrated on knife mill SM-200 (Retsch, Hann, Germany) with a sieve of the mesh size of 2 mm.

The next step was the alkaline treatment of the halophyte biomass for 5 h with 1.5% sodium hydroxide at 90 °C [39]. NaOH/biomass weight ratio was 10:1. After the alkaline pretreatment was carried out, the biomass solution was filtered on a Büchner funnel, then washed with distilled water until neutralised, and dried in a laboratory dryer at 50 °C for 24 h. The alkali effect on the content of the released reducing sugar was determined using Miller's method with 3,5-dinitrosalicylic acid (DNS) [40]. The raw material was incubated at 40 °C in 0.05 M citrate buffer pH 4.8 for 2 h using the enzyme preparation Flashzyme Plus 200 (AB Enzyme) at the dose of 20 FPU·g^{-1}. The absorbance of the supernatant was measured at 530 nm on UV–VIS Spectrophotometer V-630, (Jasco, Pfungstadt, Germany).

3.2.2. Enzyme Complex

In order to select the enzyme complex for the SSF process, tests were performed using selected enzymes—Flashzyme Plus 200 and Celluclast 1.5 L (Novozymes, Bagsværd, Denmark) and their supplementation with glucosidase 20 CBU·g^{-1} and xylanase 500 XU·g^{-1} (Sigma-Aldrich, Darmstadt, Germany).

The composition of Flashzyme Plus 200 (90 FPU·mL^{-1}, 2430 XU·mL^{-1}) is endoglucanase, cellobiohydrolase, cellobiase, xylanase and mannanase, and Celluclast 1.5 L (62 FPU·mL^{-1}, 278 XU·mL^{-1}) consists of cellulase from *Trichoderma reesei*.

Enzymatic tests were carried out for 5% of biomass with the enzyme in the amount of 10 FPU·g^{-1}, at pH 4.8 and for 24 h at 38 °C. The selection criterion was the content of reducing sugars determined using the Miller's method.

3.2.3. Simultaneous Saccharification and Fermentation (SSF)

The SSF process was carried out in bioreactor Biostat B Plus (Sartorius, Goettingen, Germany) in 2-litre vessel equipped with pH, temperature, stirring and foaming controls. The temperature was maintained at 37 °C and stirred at 900 rpm; pH was controlled at 4.8 by adding 1 M NaOH or 1 M HCl. The mixture of Flashzyme Plus 200 and Celluclast 1.5 L enzymes in the amount of 10 FPU g^{-1} was used for the hydrolysis process of biomass. The fermentation process was performed with the use of not hydrated, freeze-dried yeast *S. cerevisiae* at a dose of 1 g·L^{-1}, which corresponded to cell concentration after inoculation of about 1×10^7 cfu·mL^{-1}. Duration of the SSF process was 24, 48 and 72 h.

Ethanol yield from 100 g of raw material Y_s (g·100 g^{-1} of raw material) was calculated according to the following Equation (1) [41]:

$$Y_s = \frac{Et \times 100}{M} \tag{1}$$

where *Et*—amount of ethanol in 1000 mL of tested sample (g); and *M*—mass of material weighed in 1000 mL fermentation sample (g).

3.3. Biocomposite Production Process

3.3.1. Natural Fillers from Halophyte Biomass

Samples of dried halophyte biomass were grounded using a Rekord A (Jehmlich, Nossen, Germany) mill with a sieve separator with a mesh diameter of 1 mm. The sieve analysis of the natural fillers obtained in the milling process was performed using Analysette 3 Spartan (Fritsch, Idar-Oberstein, Germany) and their humidity were determined with moisture analyser MA.X2.A (Radwag, Radom, Poland).

3.3.2. Polymer Matrix

The polymer matrix consisted of traditional thermoplastic polymer—polypropylene Moplen HP648T (Basell Orlen Polyolefins, Płock, Poland): density 0.9 g·mL^{-1} and mass flow rate (MFR) 53 g·10 min^{-1} (190 °C, 2.16 kg) and biodegradable polymer—poly(lactic acid) Ingeo 3260HP (NatureWorks, Blair, NE, USA): density 1.24 g·mL^{-1} and MFR 65 g·10 min^{-1} (210 °C, 2.16 kg).

To improve adhesion between the natural filler and the polypropylene matrix, the interfacial adhesion promoter Scona 8112 (S) (BYK-Chemie GmbH, Germany)—PP grafted with maleic anhydride—was used in an amount of 5% by weight.

3.3.3. Preparation of Composites

Composites and biocomposites contained 20 and 30 wt% hemp (H) and grass (G) fillers were compounded with polypropylene (PP) and polylactide (PLA) in co-rotating twin screw extruder Leistritz MICRO 27 GL/GG-44D (Leistritz Extrusionstechnik, Nürnberg, Germany) with Brabender gravimetric feeding system (Brabender Technologie, Duisburg, Germany). Compounding parameters: barrel temperature profile 170–200 °C, extruder rotation speed of 150 rpm, throughput 16 kg·h^{-1}.

Composites pellets were dried to achieve the appropriate process humidity. Polypropylene composites were dried in flow dryer EHD-25BT (Enmair Automation Machinery, Guangdong, China) at temperatures of 105 °C to a humidity level below 0.15%. Biocomposites based on biodegradable polymer were dried in molecular dehumidifier Drywell DW25/40 (Digicolor, Herford, Germany) at 70 °C (dew point −40 °C) to a humidity level below 0.02%.

Multipurpose test specimens—type A in accordance with ISO 3167 [36]—were moulded by hydraulic injection moulding machine Haitian Mars II Eco 600 kN (Haitian Plastics Machinery, Zhejiang, China). Barrel temperature profile: 180 °C (hopper), 185 °C, 190 °C and 190 °C (nozzle). Mould temperature was set at 40 °C.

3.4. Analytical and Testing Methods

The chemical composition of halophyte biomass before and after pretreatment was determined, i.e., cellulose acc. to TAPPI T17 m-55 [42], hemicellulose as the difference holocellulose acc. to TAPPI T9 m-54 [43] and cellulose, and lignin acc. to TAPPI T13 m-54 [44].

In order to provide a more complete picture of the molecular structure of grass and hemp biomass before and after the alkaline pretreatment, the analysis of FTIR spectroscopy was performed using a Fourier Transform Infrared Spectrometer ISS 66v/S (Bruker, Bremen, Germany) at wavenumbers of 400–4000 cm^{-1} [22].

The physical morphologies of halophyte biomass before and after the chemical treatment were performed by using Scanning Electron Microscope S-3400N (Hitachi, Japan) in high vacuum conditions. The samples were covered with gold dust.

The content of ethanol was determined using High Performance Liquid Chromatography on Elite LaChrom (Hitachi, Tokio, Japan) using an RI L-2490 detector, Rezex ROA

300 × 7.80 mm column (Phenomenex, Torrance, CA, USA), as the mobile phase used 0.005N H_2SO_4 at a flow rate of 0.6 mL·min^{-1}, at 40 °C.

Composites' tensile and flexural tests were carried out at room temperature with a universal testing machine Inspekt Table 50 (Hegewald & Peschke MPT, Nossen, Germany) as recommended by ISO 527 [45,46] and ISO 178 [47], respectively. A crosshead speed was set to 5 mm·min^{-1} in both tests. Tensile tests were performed using an MFA clip-on extensometer (Mess- & Feinwerktechnik, Velbert, Germany) with a nominal length of 50 mm.

3.5. Statistical Analysis

The experiments of ethanol fermentation were carried out in triplicates. Standard deviations were calculated using the analysis of variance ANOVA, Statistica 13.0 software ($p < 0.05$).

4. Conclusions

This study took on the challenge of broadening the knowledge and understanding of the halophyte species to achieve better use of the biomass from these plants. To sum, the biomass of both the tall fescue and the hemp of the Białobrzeskie variety from saline areas turned out to be a suitable source for the production of second-generation bioethanol and a natural filler for polymer composites based on traditional polymers (polypropylene) and biodegradable polymers (polylactide).

The SSF process made it possible to obtain an ethanol concentration for grass and hemp biomass at the level of approx. 7 g·L^{-1} (14 g·100 g^{-1} of raw material). In the case of composites, studies have shown that despite lowering the mechanical strength in terms of stretching and bending, it is possible to improve interfacial adhesion by modifying the compatibility between the phases of the natural filler and the polymer matrix.

This study will certainly have a positive impact on raising public opinion about the social and economic benefits of the bioproducts obtained from biomass sources using land unsuitable for agriculture.

Author Contributions: Conceptualisation, J.B. and K.B.; methodology, J.B., K.B., W.G., A.W. and S.R.; software, A.W. and W.G.; validation, J.B. and K.B.; formal analysis, J.B. and K.B.; investigation, K.B., W.G., A.W. and S.R.; resources, J.B., K.B., W.G., A.W. and S.R.; data curation, J.B., K.B., W.G., A.W. and S.R.; writing—original draft preparation, J.B. and K.B.; writing—review and editing, J.B., K.B., W.G. and A.W.; visualisation, J.B.; supervision, J.B.; project administration, J.B.; funding acquisition, J.B. All authors have read and agreed to the published version of the manuscript.

Funding: This research was funded by National Centre for Research and Development, Poland, grant ERA-NET CO-FOUND FACCE SURPLUS 2 (2018-2021).

Institutional Review Board Statement: Not applicable.

Informed Consent Statement: Not applicable.

Data Availability Statement: Not applicable.

Acknowledgments: The study was conducted as research project ERA-NET CO-FOUND FACCE SURPLUS 2 (2018–2021): Integrated system of bioremediation—biorefinering using halophyte species, and financed by National Centre for Research and Development, Poland.

Conflicts of Interest: The authors declare no conflict of interest.

Sample Availability: Samples of the compounds are not available from the authors.

References

1. Hendricks, R.C.; Bushnell, D.M. Atmospheric and soil carbon and halophytes. In Proceedings of the 13th International Symposium on Transport Phenomena and Dynamics of Rotating Machinery (ISROMAC-13), Honolulu, HI, USA, 4–7 April 2010; p. 113.
2. Shrivastava, P.; Kumar, R. Soil salinity: A serious environmental issue and plant growth promoting bacteria as one of the tools for its alleviation. *Saudi J. Biol. Sci.* **2015**, *22*, 123–131. [CrossRef]
3. Hameed, A.; Khan, M.A. Halophytes: Biology and economic potentials. *Karachi Univ. J. Sci.* **2011**, *39*, 40–44.

4. Zadrożniak, B.; Radwańska, K.; Baranowska, A.; Mystkowska, I. Possibility of industrial hemp cultivation in areas of high nature value. *Econ. Reg. Stud.* **2017**, *10*, 114–127. [CrossRef]
5. Wawro, A.; Batog, J.; Gieparda, W. Chemical and enzymatic treatment of hemp biomass for bioethanol production. *Appl. Sci.* **2019**, *9*, 5348. [CrossRef]
6. Hu, H.; Liu, H.; Liu, F. Seed germination of hemp (*Cannabis sativa* L.) cultivars responds differently to the stress of salt type and concentration. *Ind. Crop. Prod.* **2018**, *123*, 254–261. [CrossRef]
7. Mohapatra, S.; Mishra, C.; Behera, S.S.; Thatoi, H. Application of pretreatment, fermentation and molecular techniques for enhancing bioethanol production from grass biomass—A review. *Renew. Sustain. Energy Rev.* **2017**, *78*, 1007–1032. [CrossRef]
8. Available online: https://eur-lex.europa.eu/legal-content/EN/ALL/?uri=CELEX:32009L0028 (accessed on 8 April 2021).
9. Abo, B.O.; Gao, M.; Wang, Y.; Wu, C.; Ma, H.; Wang, Q. Lignocellulosic biomass for bioethanol: An overview on pretreatment, hydrolysis and fermentation processes. *Rev. Environ. Health* **2019**, *34*, 57–68. [CrossRef]
10. Rahmati, S.; Doherty, W.; Dubal, D.; Atanda, L.; Moghaddam, L.; Sonar, P.; Hessel, V.; Ostrikov, K. Pretreatment and fermentation of lignocellulosic biomass: Reaction mechanisms and process engineering. *React. Chem. Eng.* **2020**, *5*, 2017–2047. [CrossRef]
11. Vasić, K.; Knez, Ž.; Leitgeb, M. Bioethanol production by enzymatic hydrolysis from different lignocellulosic sources. *Molecules* **2021**, *26*, 753. [CrossRef] [PubMed]
12. European Council. Directive (EU) 2019/904 of the European Parliament and of the Council of 5 June 2019 on the Reduction of the Impact of Certain Plastic Products on the Environment. Available online: https://eur-lex.europa.eu/legal-content/EN/TXT/HTML/?uri=CELEX:32019L0904&from=HU (accessed on 1 February 2021).
13. Moore, C. Plastic Pollution. Encyclopedia Britannica. Available online: https://www.britannica.com/science/plastic-pollution (accessed on 18 June 2021).
14. Getme, A.S.; Patel, B. A review: Bio-fiber's as reinforcement in composites of polylactic acid (PLA). *Mater. Today Proc.* **2019**, *26*, 2116–2122. [CrossRef]
15. Das, P.P.; Chaudhary, V. Moving towards the era of bio fibre based polymer composites. *Clean. Eng. Technol.* **2021**, *4*, 100182. [CrossRef]
16. Manral, A.; Ahmad, F.; Chaudhary, V. Static and dynamic mechanical properties of PLA bio-composite with hybrid reinforcement of flax and jute. *Mater. Today Proc.* **2020**, *25*, 577–580. [CrossRef]
17. Papadopoulou, E.; Bikiaris, D.; Chrysafis, K.; Władyka-Przybylak, M.; Wesołek, D.; Mankowski, J.; Kołodziej, J.; Baraniecki, P.; Bujnowicz, K.; Gronberg, V. Value-added industrial products from bast fiber crops. *Ind. Crop. Prod.* **2015**, *68*, 116–125. [CrossRef]
18. Gąsiorowski, R.; Rojewski, S.; Wesołek, D.; Maciejewski, H.; Bujnowicz, K. The influence of lignocellulosic filler modification with silicon compounds on flammability and adhesion with a composite polymer matrix. *Polym. Process.* **2013**, *19*, 336–339.
19. Akinshina, N.; Toderich, K.; Azizov, A.; Saito, L.; Ismail, S. Halophyte biomass—A promising source of renewable energy. *J. Arid Land Stud.* **2014**, *24*, 231–235.
20. Turcios, A.E.; Cayenne, A.; Uellendahl, H.; Papenbrock, J. Halophyte plants and their residues as feedstock for biogas production-chances and challenges. *Appl. Sci.* **2021**, *11*, 2746. [CrossRef]
21. Kumar, D.; Murthy, G.S. Impact of pretreatment and downstream processing technologies on economics and energy in cellulosic ethanol production. *Biotechnol. Biofuels* **2011**, *4*, 27. [CrossRef]
22. Stevulova, N.; Cigasova, J.; Estokova, A.; Terpakova, E.; Geert, A.; Kacik, F.; Singovszka, E.; Holub, M. Properties characterization of chemically modified hemp hurds. *Materials* **2014**, *7*, 8131–8150. [CrossRef] [PubMed]
23. Araújo, D.; Vilarinho, M.; Machado, A. Effect of combined dilute-alkaline and green pretreatments on corncob fractionation: Pretreated biomass characterization and regenerated cellulose film production. *Ind. Crop. Prod.* **2019**, *141*, 111785. [CrossRef]
24. Ciolacu, D.; Ciolacu, F.; Popa, V.I. Amorphous cellulose—Structure and characterization. *Cellul. Chem. Technol.* **2011**, *45*, 13–21.
25. Sun, X.F.; Xu, F.; Sun, R.C.; Fowler, P.; Baird, M.S. Characteristics of degraded cellulose obtained from steam-exploded wheat straw. *Carbohydr. Res.* **2005**, *340*, 97–106. [CrossRef] [PubMed]
26. Gupta, A.D.; Pandey, S.; Kumar Jaiswal, V.; Bhadauria, V.; Singh, H. Simultaneous oxidation and esterification of cellulose for use in treatment of water containing Cu (II) ions. *Carbohydr. Polym.* **2019**, *222*, 114964. [CrossRef] [PubMed]
27. Zhao, J.; Xu, Y.; Wang, W.; Griffin, J.; Wang, D. High Ethanol concentration (77 g/L) of industrial hemp biomass achieved through optimizing the relationship between ethanol yield/concentration and solid loading. *ACS Omega* **2020**, *5*, 21913–21921. [CrossRef] [PubMed]
28. Zhao, J.; Xu, Y.; Zhang, M.; Wang, D. Integrating bran starch hydrolysates with alkaline pretreated soft wheat bran to boost sugar concentration. *Bioresour. Technol.* **2020**, *302*, 122826. [CrossRef] [PubMed]
29. Abraham, R.E.; Barrow, C.J.; Puri, M. Relationship to reducing sugar production and scanning electron microscope structure to pretreated hemp hurd biomass (*Cannabis sativa*). *Biomass Bioenerg.* **2013**, *58*, 180–187. [CrossRef]
30. Kumar, D.; Murthy, G.S. Pretreatments and enzymatic hydrolysis of grass straws for ethanol production in the Pacific Northwest U.S. *Biol. Eng.* **2011**, *3*, 97–110. [CrossRef]
31. Wawro, A.; Batog, J.; Gieparda, W. Efektywność enzymatycznej konwersji biomasy sorgo i konopi do glukozy. *Przemysł Chem.* **2020**, *99*, 1731–1734.
32. Robak, K.; Balcerek, M. Current state-of-the-art in ethanol production from lignocellulosic feedstocks. *Microbiol. Res.* **2020**, *240*, 126534. [CrossRef]

33. Taufikurahman, T.; Xie, S. Production of Bioethanol and Crude Cellulase Enzyme Extract from Napier Grass (*Pennisetum purpureum* S.) through Simultaneous Saccharification and Fermentation. *Bio J. Biol. Sci. Technol. Man.* **2020**, *2*, 18–29. [CrossRef]
34. Riadi, L.; Hansen, Y.; Pratiwi, J.; Goretti, M.; Purwanto, M. Mild Alkaline Pretreatment on Sugarcane Bagasse: Effects of Pretreatment Time and Lime to Dry Bagasse Ratio. *B Life Environ. Sci.* **2020**, *57*, 43–50.
35. Orlygsson, J. Ethanol production from biomass by a moderate thermophile, Clostridium AK1. *Icel. Agric. Sci.* **2012**, *25*, 25–35.
36. International Standards Organization. *PN-EN ISO 3167:2014, Plastics—Multipurpose Test Specimens*; Polish Committee for Standardization: Geneva, Switzerland, 2014.
37. Yan, Z.L.; Wang, H.; Lau, K.T.; Pather, S.; Zhang, J.C.; Lin, G.; Ding, Y. Reinforcement of polypropylene with hemp fibres. *Compos. Part B Eng.* **2013**, *46*, 221–226. [CrossRef]
38. Yu, T.; Jiang, N.; Li, Y. Study on short ramie fiber/poly(lactic acid) composites compatibilized by maleic anhydride. *Compos. Part A Appl. Sci. Manuf.* **2014**, *64*, 139–146. [CrossRef]
39. Batog, J.; Wawro, A. Chemical and biological deconstruction in the conversion process of sorghum biomass for bioethanol. *J. Nat. Fibers* **2021**, 1–12. [CrossRef]
40. Miller, G.L. Use of dinitrosalicylic acid reagent for determination of reducing sugars. *Anal. Chem.* **1959**, *31*, 426–428. [CrossRef]
41. Kawa-Rygielska, J.; Pietrzak, W. Zagospodarowanie odpadowe pieczywa do produkcji bioetanolu. *Żywność Nauka Technol. Jakość* **2011**, *79*, 105–118.
42. TAPPI 17 m-55. In *Cellulose in Wood*; TAPPI Press: Atlanta, GA, USA, 1955.
43. TAPPI T9 m-54. In *Holocellulose in Wood*; TAPPI Press: Atlanta, GA, USA, 1998.
44. Bagby, M.O.; Nelson, G.H.; Helman, E.G.; Clark, T.F. Determination of lignin non-wood plant fiber sources. *TAPPI J.* **1971**, *54*, 11.
45. International Standards Organization. *PN-EN ISO 527-1:2020, Plastics—Determination of Tensile Properties—Part 1: General Principles*; Polish Committee for Standardization: Geneva, Switzerland, 2020.
46. International Standards Organization. *PN-EN ISO 527-2:2012, Plastics—Determination of Tensile Properties—Part 2: Test Conditions for Moulding and Extrusion Plastics*; Polish Committee for Standardization: Geneva, Switzerland, 2012.
47. International Standards Organization. *PN-EN ISO 178:2019, Plastics—Determination of Flexural Properties*; Polish Committee for Standardization: Geneva, Switzerland, 2019.

MDPI AG
Grosspeteranlage 5
4052 Basel
Switzerland
Tel.: +41 61 683 77 34

Molecules Editorial Office
E-mail: molecules@mdpi.com
www.mdpi.com/journal/molecules

www.ingramcontent.com/pod-product-compliance
Lightning Source LLC
Chambersburg PA
CBHW061809210926
51850CB00006B/651
9783725822560